Cutting Edge Earth Sciences: Three Decades of Cosmogenic Nuclides

Cutting Edge Earth Sciences: Three Decades of Cosmogenic Nuclides

Editors

Naki Akçar
Susan Ivy-Ochs
Fritz Schlunegger

MDPI • Basel • Beijing • Wuhan • Barcelona • Belgrade • Manchester • Tokyo • Cluj • Tianjin

Editors
Naki Akçar
University of Bern
Bern, Switzerland

Susan Ivy-Ochs
ETH Zurich
Zurich, Switzerland

Fritz Schlunegger
University of Bern
Bern, Switzerland

Editorial Office
MDPI
St. Alban-Anlage 66
4052 Basel, Switzerland

This is a reprint of articles from the Special Issue published online in the open access journal *Geosciences* (ISSN 2076-3263) (available at: https://www.mdpi.com/journal/geosciences/special_issues/Cosmogenic_Nuclides).

For citation purposes, cite each article independently as indicated on the article page online and as indicated below:

LastName, A.A.; LastName, B.B.; LastName, C.C. Article Title. *Journal Name* **Year**, *Volume Number*, Page Range.

ISBN 978-3-0365-7148-5 (Hbk)
ISBN 978-3-0365-7149-2 (PDF)

Cover image courtesy of Naki Akçar

Contents

geosciences

Editorial

A Special Issue of *Geosciences*: Cutting Edge Earth Sciences—Three Decades of Cosmogenic Nuclides

Naki Akçar [1,*], Susan Ivy-Ochs [2] and Fritz Schlunegger [1]

1 Institute of Geological Sciences, University of Bern, CH-3012 Bern, Switzerland
2 Laboratory of Ion Beam Physics, Department Earth Sciences, ETH Zurich, CH-8093 Zurich, Switzerland
* Correspondence: akcar@geo.unibe.ch

Citation: Akçar, N.; Ivy-Ochs, S.; Schlunegger, F. A Special Issue of *Geosciences*: Cutting Edge Earth Sciences—Three Decades of Cosmogenic Nuclides. *Geosciences* **2022**, *12*, 409. https://doi.org/10.3390/geosciences12110409

Received: 30 October 2022
Accepted: 1 November 2022
Published: 5 November 2022

"What we know is a drop, what we don't know is an ocean."

Isaac Newton

A key breakthrough in Earth Sciences was triggered in 1955 when Davis and Schaeffer [1] measured cosmogenic ^{36}Cl *for the first time in history* in two igneous rock surface samples: one specimen of phonolite from a high-altitude (ca. 3300 m a.s.l.) unglaciated cliff in Colorado, and one specimen of syenite from a quarry located close to sea level (ca. 300 m a.s.l.) and within the extent of the Wisconsin glaciation in New Hampshire. This spark gave rise to the unique "cosmogenic nuclide tool" that enables Earth scientists to disentangle the unsolved pieces of Earth's history. The application of cosmogenic nuclides to geomorphological problems emerged gradually right after the publication by Davis and Schaeffer in 1955 [1]; expanded almost exponentially in the early 1990s; and has reached a prominent position during the last decade, with about more than 100 publications coming out per year. The success of the unique "cosmogenic nuclide tool" is reflected in the literature, with more than 2000 publications since conception (available online: scopus.com, accessed on 20 September 2022, searched for "cosmogenic nuclide"). *Why during the past three decades?* During this period, the physical processes responsible for the production of cosmogenic nuclides became better understood. In addition, sampling strategies, analytical sample preparation, and the accelerator and noble gas mass spectrometric analyses were astoundingly improved. As a consequence, the wide applicability of cosmogenic nuclides in solving geological problems in Earth Sciences rapidly increased. Today, cosmogenic nuclides are an amazingly versatile tool for dating landforms and deposits and for deciphering landscape evolution processes during the Quaternary. Cosmogenic nuclides have been widely applied in dating Quaternary ice volume fluctuations, and volcanic and palaeoseismic events; in quantifying surface and/or rock uplift and denudation rates; and in locating sediment sources in highly dynamic landscapes. Moreover, due to the sensitivity of cosmogenic nuclide accumulation to surface erosion and depths below the surface, the application of the technique has led to significant breakthroughs in establishing terrace chronologies, the rates and styles of local and large-scale erosion, and soil development.

Over the past three decades, the "cosmogenic nuclide tool" did not only become a universal and standard method, but it also has kept its momentum gained during this time and is at the forefront of cutting-edge research in Earth Sciences. Scientists from all over the world are exploring new challenges that require improvement and additional knowledge and are still ambitiously diving into the unknown ocean of methodologies to tackle these challenges. It was under these circumstances that this Special Issue of *Geosciences* arose. We launched this Special Issue with a call for contributions illustrating the novel applications of cosmogenic nuclides (^{3}He, ^{10}Be, ^{14}C, ^{21}Ne, ^{26}Al, and ^{36}Cl, including new and less frequently used nuclides such as ^{38}Ar and ^{53}Mn) in diverse disciplines in the field of Geosciences, as well as contributions from purely methodological and measurement arenas (AMS and noble gas mass spectrometry). This Special Issue contains 12 papers that well portray the realm of cosmogenic nuclide applications in Geosciences. What follow are the contributions grouped into the following topics: dating Quaternary ice volume fluctuations, gauging erosion rates (glacial, catchment-wide, and soil erosion), dating the landforms created by natural hazards (landslides and earthquakes), and specific themes of the cosmogenic nuclide methodology and measurement infrastructure.

We introduce this Special Issue with three papers about the application of cosmogenic nuclides in establishing the timing of events in Quaternary glacial settings. Dieleman et al. [2] focused on one of the longest standing questions in Swiss Quaternary geology: When did glaciers from the Alps reach their farthest extent? To solve this mystery, the authors explored the Bünten Till layer exposed in a gravel pit in Möhlin (Canton of Aargau, Switzerland). Prior studies there had suggested the presence of moraine ridges, which have recently been shown to be loess swales [3]. In this Special Issue, Dielmann et al. [2] combined field sedimentology and cosmogenic $^{26}Al/^{10}Be$ isochron-burial dating on clasts in the till. The authors concluded that an alpine glacier reached its farthest position at 500 $\pm$ 100 ka during the Middle Pleistocene, which is contemporaneous with the maximum expansion of glaciers in the northern hemisphere. The study by Reber et al. [4] illustrated the use of cosmogenic nuclides in reconstructing the chronology of the Quaternary glacier fluctuations in northeastern Anatolia. In particular, they map the extent of paleoglaciers in the Barhal Valley in the southern ranges of the Kaçkar Mountains by using field and photogrammetric data. They established the glacial chronology by exposure dating 32 glacially transported boulders of volcanic origin with cosmogenic ^{36}Cl. The data point to three independent periods when the glaciers were stable, which occurred at 34.0 $\pm$ 2.3 ka, 22.2 $\pm$ 2.6 ka, and 18.3 $\pm$ 1.7 ka during the global Last Glacial Maximum (LGM; after [5]). In addition, they noted the occurrence of an early advance phase of the LGM glacier. They also showed the rapid downwasting of the glaciers at the end of the LGM. Evidence for Lateglacial advances is absent, and the Holocene is dominated by rock glacier activity that developed in the cirques. An equilibrium line altitude (ELA) of 2900 m a.s.l. for the LGM, which is a 600 m drop with respect to modern ELA, is documented in this study. Reber et al. [4] closed their work by proposing that, during the LGM, lakes in central–western Siberia, and the Aral and Caspian Seas served as moisture sources for the build-up of ice and that a southward migration of the Polar Jet Front and the Siberian High Pressure System controlled moisture transport during this time. In their paper, Anjar et al. [6] called into question ^{36}Cl exposure age calculations of volcanic rocks with high native Cl. They instead presented a noteworthy example of how much young exposure ages can be impacted when modeling the amount of cosmogenic ^{36}Cl produced through non-cosmogenic neutron capture reactions. Anjar et al. [6] determined 89 cosmogenic ^{36}Cl exposure ages from rock surfaces primarily exposed to glacial and volcanic events on Jan Mayen Island, located 550 km north of Iceland. For accurate age calculations, they first updated the CRONUScalc code; then, they recalculated the ages with an assumption of non-equilibrated background ^{36}Cl production [7] using the independently determined eruption ages for the volcanic rocks. They showed that almost 30% of the exposure ages underestimate the real age by up to four times. The only briefly exposed rock surfaces suffered most from the equilibrated background assumption (correction for background non-cosmogenic production was too high). Anjar et al. [6] closed their contribution by recommending the exclusion of assumed equilibrium conditions for background production when calculating exposure ages for young volcanic rock surfaces containing high native Cl, or high U and/or Th.

In the second group of contributions, Steinemann et al. [8], da Silva Guimarães et al. [9], and Musso et al. [10] delved into the realm of estimating erosion rates using cosmogenic nuclides. Upon quantifying erosion, Steinemann et al. [8] showed that a significant role of glaciers in sculpting mountain landscapes is not only their potential to deeply carve the landscape but also, surprisingly, their ability to not erode the bedrock. In their striking study, they investigated the abrasion of limestone bedrock in the forefield of the Vorab glacier in the eastern Swiss Alps. Based on measured cosmogenic ^{36}Cl concentrations, a numerical model was used to quantify subglacial erosion rates over the last 15 ka. The modeled abrasion rates were at the submillimeter scale, varying from 0.01 to 0.16 mm per year, which were astonishingly low in comparison to the rates measured on crystalline bedrock (typically more than 1 mm per year) [11]. They explained the low erosion rates with the immediate drainage of subglacial meltwater into the karst passages at the base of the glacier. They concluded that the sudden escape of water hinders basal sliding, allowing only limited subglacial erosion. As a consequence, broad and flat limestone plateaus arose. da Silva Guimarães et al. [9] exemplified the unique application of cosmogenic ^{10}Be to riverbed sediments to gauge basin-wide denudation rates and sediment fluxes, a methodology that has been successfully applied for more than two decades. They explored one of the tributary streams to the Alpine Rhine, namely the Plessur basin in the eastern Swiss Alps. The combination of the cosmogenic ^{10}Be-derived denudation rates with the geomorphological and sedimentological analysis of the drainage basin enabled them to reveal the adjustment of the Plessur basin to the landscape perturbation created by the thick Alpine Ice Sheet during the Last Glacial Maximum. The results from the highly erodible North Penninic flysch and Bündnerschist in the downstream portion of the basin indicate the most efficient adjustment there, where glacial landforms from the Last Glacial Maximum are completely absent. In contrast, hardly erodible South Penninic and Austroalpine bedrock in the upstream part promote good preservation of the glacial landforms. They concluded that the bedrock geology, geodynamics, and glacial molding are substantial factors in the processes of local uplift and denudation. Musso et al. [10] tracked the fingerprint of chemical weathering recorded by the evolution of calcareous and siliceous soils in two proglacial areas in the Swiss Alps by analyzing the meteoric ^{10}Be. They showed that the chemical weathering rate in siliceous soils is high in the early stage of formation and that it rapidly decelerates after a few thousand years. Erosion rates in calcareous soils are compensated by the soil production rates, resulting in a delay in the development of soil and vegetation cover. They concluded that vegetation is an important factor in the evolution of soil because it augments the rate of chemical weathering and surface stabilization, and it modifies the hydrogeological properties.

Mozafari et al. [12], Aksay et al. [13], and Ruggia et al. [14] validated the novel use of cosmogenic nuclides in disentangling the timing of events in natural hazard research and highlighted the importance of such an analysis for risk assessment and hazard mitigation. Mozafari et al. [12] showed the potential of cosmogenic ^{36}Cl analysis to gather crucial information required for a precise evaluation of seismic risk by exploring normal faults for unknown major prehistorical earthquakes in Western Anatolia, one of the seismically most active extensional regimes of our planet. On scarps of the Manastır and Mugırtepe faults in the Manisa Fault Zone, they modeled the occurrence of three major earthquakes at 6.5 ± 1.6 ka, 3.5 ± 0.9 ka, and 2.0 ± 0.5 ka with vertical displacements of 2.7 ± 0.4 m, 3.3 ± 0.5 m, and 3.6 ± 0.5 m, respectively. Combining their results with the existing geological and paleoseismological data [15], they demonstrated that the reconstructed seismic activity resulted in a syn-depositional rotation in the Manastır fault during the Late Pleistocene–Early Holocene, which was followed by the formation of secondary faults during the Early–Late Holocene. Aksay et al. [13] investigated the Sennwald landslide, located in the Rhine Valley. They combined detailed field mapping of the rock avalanche deposits with dynamic run-out modelling and cosmogenic ^{36}Cl dating of limestone boulders. The data point to a single catastrophic failure at 4.3 ± 0.5 ka. This coincides with a past earthquake identified in lake sediments within the region by [16], implicating a seismic origin for the Sennwald event. This provides further support for a major earthquake at the Mid–Late Holocene transition. Ruggia et al. [14] combined field mapping, runout modeling, and surface exposure dating with cosmogenic ^{36}Cl to reconstruct the evolution of the giant Gorte rock avalanche at the northeastern end of Lake Garda in northern Italy. They successfully simulated the rock avalanche and reproduced the size and the thickness of the landslide deposits. They documented a single collapse in a rock mass of about 70–75 Mm^3 at 6.1 ± 0.8 ka, which resulted in a deposit volume of about 85–95 Mm^3. The initial collapse of bedrock was followed by rapid disintegration and spreading. They argued that heavy precipitation might have triggered the rock avalanche because the timing of the rock avalanche falls into a relatively warm and wet period of the Middle Holocene when a period of frequent flooding 6900–6200 years ago was already identified for the region. However, they did not exclude a seismic trigger because of the occurrence of three contemporaneous landslides in the nearby region (within 15 km distance).

Finally, we close this Special Issue with three contributions on the brass tacks of cosmogenic nuclide methodologies. Halsted et al. [17] investigated one of the recent critical topics of the cosmogenic nuclide community: the cosmogenic $^{26}Al/^{10}Be$ surface production ratio, which is vital for two-isotope applications. To test the overlap of the theoretical production ratio with that from analyzed data, they scrutinized the Informal Cosmogenic-nuclide Exposure-age Database (ICE-D) and selected 313 samples from ice-molded bedrock and glacial boulders located between 53° S and 70° N and at altitudes up to 5000 m above sea level, which are assumed to have experienced the same exposure history. They determined insignificant interlaboratory systematic differences and a negative correlation between the $^{26}Al/^{10}Be$ production ratio and elevation, which agrees well with the assumptions based on the measured energy dependence of nuclear reaction cross sections and the spatial variability of the cosmic ray cascade. They also identified a positive correlation between the production ratio and latitude, but they noted the occurrence of a bias in the dataset, which they related to the altitude of the samples. They concluded that a global value of 6.75 for the $^{26}Al/^{10}Be$ production ratio is not appropriate for high-altitude samples and maybe even for high latitude ones. They suggested employing a nuclide-specific production rate spatial scaling scheme, such as the LSDn [18], to avoid potential biases. Ángel Rodés [19] addressed the challenge of applying the cosmogenic nuclide methodology in glacial landscapes, which particularly emerges when the exposure ages of the landforms deviate from the timing of deglaciation. This occurs as a consequence of post-depositional and post-exposure geological processes. Researchers applying cosmogenic nuclides to nunataks to reconstruct the history of ice thinning face such challenges especially in continental polar regions, where extreme aridity and cold conditions prevail. Ángel [19] provided a new user-friendly tool to model cosmogenic nuclide accumulation along the elevation profiles: The NUNAtak Ice Thinning model (NUNAIT). In brief, NUNAIT calculates cosmogenic nuclide concentrations by tailoring the exposure time, pre-exposure, subaerial weathering, subglacial erosion, uplift, and subglacial muon production to an array of apparent exposure ages gathered from nunataks along an elevation profile. Györe et al. [20] introduced a new infrastructure for high-precision analysis of cosmogenic Ne in terrestrial and extra-terrestrial rocks. They connected a Thermo Fisher ARGUS VI mass spectrometer with an automated laser gas extraction and purification system and adapted this line for high-throughput and high-precision analysis. The stunning outcome of this new system is its ability to measure Ne isotopic ratios in very small samples (~20 mg) with very low uncertainties. For example, while measuring the extra-terrestrial material, they were able to reduce the overall uncertainty of the Ne isotope ratio down to 0.5% (four times less with respect to earlier systems) by analyzing two to six times less material. In addition, they can successfully analyze samples that are two to five times smaller than what is commonly used for noble gas spectrometry, and they are able to measure Ne isotopes at high precision even if the samples have low noble gas concentrations, which is generally the case for the terrestrial material. They discussed how their system can further be improved particularly for the analysis of terrestrial samples. This new system expands our capabilities and paves the way for novel research, as it enables us to analyze sample material that is extremely valuable and limited, such as the rock samples from space missions (perhaps even rock samples from Mars).

In brief, the studies compiled in this Special Issue show how cosmogenic nuclides are, at present, in full bloom as an amazingly versatile tool after the past three decades and nearly 70 years after the first spark by Davis and Schaeffer [1]. As quoted by Sir Isaac Newton: "What we know is a drop, what we don't know is an ocean"; there are certainly more drops in the ocean of cosmogenic nuclides to be discovered in the future that will keep cosmogenic nuclides at the cutting edge of Earth Sciences research.

Funding: This research received no external funding.

Conflicts of Interest: The authors declare no conflict of interest.

References

1. Davis, R.; Schaeffer, O.A. Chlorine-36 in Nature. *Ann. N. Y. Acad. Sci.* **1955**, *62*, 107–121. [CrossRef]
2. Dieleman, C.; Christl, M.; Vockenhuber, C.; Gautschi, P.; Graf, H.R.; Akçar, N. Age of the Most Extensive Glaciation in the Alps. *Geosciences* **2022**, *12*, 39. [CrossRef]
3. Gaar, D.; Preusser, F. Age of the Most Extensive Glaciation of Northern Switzerland: Evidence from the scientific drilling at Möhliner Feld. *E&G Quat. Sci. J.* **2017**, *66*, 1–5.
4. Reber, R.; Akçar, N.; Tikhomirov, D.; Yesilyurt, S.; Vockenhuber, C.; Yavuz, V.; Ivy-Ochs, S.; Schlüchter, C. LGM Glaciations in the Northeastern Anatolian Mountains: New Insights. *Geosciences* **2022**, *12*, 257. [CrossRef]
5. Shakun, J.D.; Carlson, A.E. A global perspective on Last Glacial Maximum to Holocene climate change. *Quat. Sci. Rev.* **2010**, *29*, 1801–1816. [CrossRef]
6. Anjar, J.; Akçar, N.; Larsen, E.A.; Lyså, A.; Marrero, S.; Mozafari, N.; Vockenhuber, C. Cosmogenic Exposure Dating (^{36}Cl) of Landforms on Jan Mayen, North Atlantic, and the Effects of Bedrock Formation Age Assumptions on ^{36}Cl Ages. *Geosciences* **2021**, *11*, 390. [CrossRef]
7. Marrero, S.M.; Phillips, F.M.; Borchers, B.; Lifton, N.; Aumer, R.; Balco, G. Cosmogenic nuclide systematics and the CRONUScalc program. *Quat. Geochronol.* **2016**, *31*, 160–187. [CrossRef]
8. Steinemann, O.; Martinez, A.; Picotti, V.; Vockenhuber, C.; Ivy-Ochs, S. Glacial Erosion Rates Determined at Vorab Glacier: Implications for the Evolution of Limestone Plateaus. *Geosciences* **2021**, *11*, 356. [CrossRef]
9. da Silva Guimarães, E.; Delunel, R.; Schlunegger, F.; Akçar, N.; Stutenbecker, L.; Christl, M. Cosmogenic and Geological Evidence for the Occurrence of a Ma-Long Feedback between Uplift and Denudation, Chur Region, Swiss Alps. *Geosciences* **2021**, *11*, 339. [CrossRef]
10. Musso, A.; Tikhomirov, D.; Plötze, M.L.; Greinwald, K.; Hartmann, A.; Geitner, C.; Maier, F.; Petibon, F.; Egli, M. Soil Formation and Mass Redistribution during the Holocene Using Meteoric ^{10}Be, Soil Chemistry and Mineralogy. *Geosciences* **2022**, *12*, 99. [CrossRef]
11. Steinemann, O.; Ivy-Ochs, S.; Hippe, K.; Christl, M.; Haghipour, N.; Synal, H.A. Glacial erosion by the Trift glacier (Switzerland): Deciphering the development of riegels, rock basins and gorges. *Geomorphology* **2021**, *375*, 107533. [CrossRef]
12. Mozafari, N.; Özkaymak, Ç.; Tikhomirov, D.; Ivy-Ochs, S.; Alfimov, V.; Sözbilir, H.; Schlüchter, C.; Akçar, N. Seismic Activity of the Manisa Fault Zone in Western Turkey Constrained by Cosmogenic ^{36}Cl Dating. *Geosciences* **2021**, *11*, 451. [CrossRef]
13. Aksay, S.; Ivy-Ochs, S.; Hippe, K.; Grämiger, L.; Vockenhuber, C. Slope Failure in a Period of Increased Landslide Activity: Sennwald Rock Avalanche, Switzerland. *Geosciences* **2021**, *11*, 331. [CrossRef]
14. Ruggia, G.; Ivy-Ochs, S.; Aaron, J.; Steinemann, O.; Martin, S.; Rigo, M.; Rossato, S.; Vockenhuber, C.; Monegato, G.; Viganò, A. Reconstructing the Gorte and Spiaz de Navesele Landslides, NE of Lake Garda, Trentino Dolomites (Italy). *Geosciences* **2021**, *11*, 404. [CrossRef]
15. Özkaymak, Ç.; Sözbilir, H.; Uzel, B. Neogene–Quaternary evolution of the Manisa Basin: Evidence for variation in the stress pattern of the İzmir-Balıkesir Transfer Zone, western Anatolia. *J. Geodyn.* **2013**, *65*, 117–135. [CrossRef]
16. Oswald, P.; Strasser, M.; Hammerl, C.; Moernaut, J. Seismic control of large prehistoric rockslides in the Eastern Alps. *Nat. Commun.* **2021**, *12*, 1059. [CrossRef] [PubMed]
17. Halsted, C.T.; Bierman, P.R.; Balco, G. Empirical Evidence for Latitude and Altitude Variation of the In Situ Cosmogenic 26Al/10Be Production Ratio. *Geosciences* **2021**, *11*, 402. [CrossRef]
18. Lifton, N.; Sato, T.; Dunai, T.J. Scaling in situ cosmogenic nuclide production rates using analytical approximations to atmospheric cosmic-ray fluxes. *Earth Planet. Sci. Lett.* **2014**, *386*, 149–160. [CrossRef]
19. Rodés, Á. The NUNAtak Ice Thinning (NUNAIT) Calculator for Cosmonuclide Elevation Profiles. *Geosciences* **2021**, *11*, 362. [CrossRef]
20. Györe, D.; Di Nicola, L.; Currie, D.; Stuart, F.M. New System for Measuring Cosmogenic Ne in Terrestrial and Extra-Terrestrial Rocks. *Geosciences* **2021**, *11*, 353. [CrossRef]

geosciences

Article

Age of the Most Extensive Glaciation in the Alps

Catharina Dieleman [1,*], Marcus Christl [2], Christof Vockenhuber [2], Philip Gautschi [2], Hans Rudolf Graf [3] and Naki Akçar [1]

1 Institute of Geological Sciences, University of Bern, 3012 Bern, Switzerland; akcar@geo.unibe.ch
2 Laboratory of Ion Beam Physics, ETH Zurich, 8093 Zurich, Switzerland; mchristl@phys.ethz.ch (M.C.); vockenhuber@phys.ethz.ch (C.V.); philipg@phys.ethz.ch (P.G.)
3 Dr. von Moos AG, 8214 Gächlingen, Switzerland; graf@geomv.ch
* Correspondence: catharina.dieleman@geo.unibe.ch

Abstract: Previous research suggested that the Alpine glaciers of the Northern Swiss Foreland reached their maximum extensive position during the Middle Pleistocene. Relict tills and glaciofluvial deposits, attributed to the Most Extensive Glaciation (MEG), have been found only beyond the extents of the Last Glacial Maximum (LGM). Traditionally, these sediments have been correlated to the Riss glaciation sensu Penck and Brückner and have been morphostratigraphically classified as the Higher Terrace (HT) deposits. The age of the MEG glaciation was originally proposed to be intermediate to the Brunhes/Matuyama transition (780 ka) and the Marine Isotope Stage 6 (191 ka). In this study, we focused on the glacial deposits in Möhlin (Canton of Aargau, Switzerland), in order to constrain the age of the MEG. The sediments from these deposits were analyzed to determine the provenance and depositional environments. We applied isochron-burial dating, with cosmogenic ^{10}Be and ^{26}Al, to the till layer in the Bünten gravel pit near Möhlin. Our results indicate that a glacier of Alpine origin reached its most extensive position during the Middle Pleistocene (500 ± 100 ka). The age of the MEG thus appears to be synchronous with the most extensive glaciations in the northern hemisphere.

Keywords: isochron-burial dating; cosmogenic nuclides; Swiss northern Alpine Foreland; Middle Pleistocene; Möhlin glaciation; Bünten Till

Citation: Dieleman, C.; Christl, M.; Vockenhuber, C.; Gautschi, P.; Graf, H.R.; Akçar, N. Age of the Most Extensive Glaciation in the Alps. *Geosciences* **2022**, *12*, 39. https://doi.org/10.3390/geosciences12010039

Academic Editors: Philippe Claeys and Jesus Martinez-Frias

Received: 17 November 2021
Accepted: 10 January 2022
Published: 13 January 2022

1. Introduction

The Most Extensive Glaciation (MEG), locally known as Möhlin glaciation, Hosskirch, Mindel, or Most Extensive Helvetic Glaciation (*Grösste Helvetische Vergletscherung* in German; GHV), is proposed to have occurred during the Middle Pleistocene (774–129 ka; [1]) [2–6]. Previous studies reconstructed its extent by mapping erratic boulders detected beyond the extents of the LGM, along with few relict glacial deposits [5,7,8]. Further, it was suggested that this glaciation reached the interiors of the northern Alpine Foreland and advanced at least until Möhlin (Canton of Aargau), close to Basel in northern Switzerland (Figure 1), where the Rhone, Reuss, Linth, and Rhaetian paleoglaciers coalesced [9–11]. In Möhlin, the Bünten Till, one of the few preserved glacial relicts attributed to the MEG outcrops in a gravel pit. This is perceived to be an important locality for reconstructing the MEG in the Swiss northern Alpine Foreland [3,6,8]. The MEG is also esteemed as the first glacier advance that formed overdeepened valleys in the northern Alpine Foreland [4,12]. Moreover, the MEG was tentatively correlated with the complex of Upper Terraces (HT; *Hochterrasse* in German) and, therefore, with the Riss glaciation (ca. 130–185 ka) [4,8,9,13], using the four-fold glaciation schemes by Penck and Brückner [9]. Despite several attempts, the exact age of the Alpine glacial expansion still remains a question of debate.

Earlier studies demonstrate that the MEG took place between the Brunhes/Matuyama transition (780 ka) [14] and the Marine Isotope Stage (MIS) 6 (130 ka–191 ka [15]). Based on the morphostratigraphy of the Swiss northern Alpine Foreland, Schlüchter [2,16] proposed that the MEG followed the Deckenschotter glaciations and therefore should have occurred just after the Brunhes/Matuyama transition. Analyses of pollen assemblages from the glacial and interglacial sediments of the Lower Aare Valley reveal the MEG to be older than the Holsteinian, an interglacial period generally attributed to MIS 11 (360–420 ka) [4]. In Southern Germany, the Hosskirch glaciation gravel deposits underlie the Holsteinian sediments [5]. According to Keller [6], the MEG is older than the Riss glaciation but younger than the Deckenschotter deposits, thus implying an approximate age of 350 ka. The age of the loess layer (19 ka–ca. 60 ka) overlying the glacial and glaciofluvial sediments in Möhlin, derived from the optically stimulated luminescence (OSL) technique, indicates that the MEG occurred prior to the MIS 6 [17]. U/Th and OSL dating of the Landiswil gravels (Canton of Bern), attributed to the MEG, assigns them an age comparable to MIS 6 [18].

The major objectives of this study were to: (1) put constraints on the age of the MEG in the Alps in context of the northern hemispheric glaciation by focusing on the Bünten Till, which is attributed to the MEG [3,8,18,19] and outcropping in a gravel pit near Möhlin (Figure 1); and (2) reconstruct whether the glacier that deposited the Bünten Till originated from the Alps or the Black Forest. We established the age of the Bünten Till by using isochron-burial dating with cosmogenic ^{10}Be and ^{26}Al. The source of the till was determined by analyzing the sedimentology of clasts within the Bünten Till.

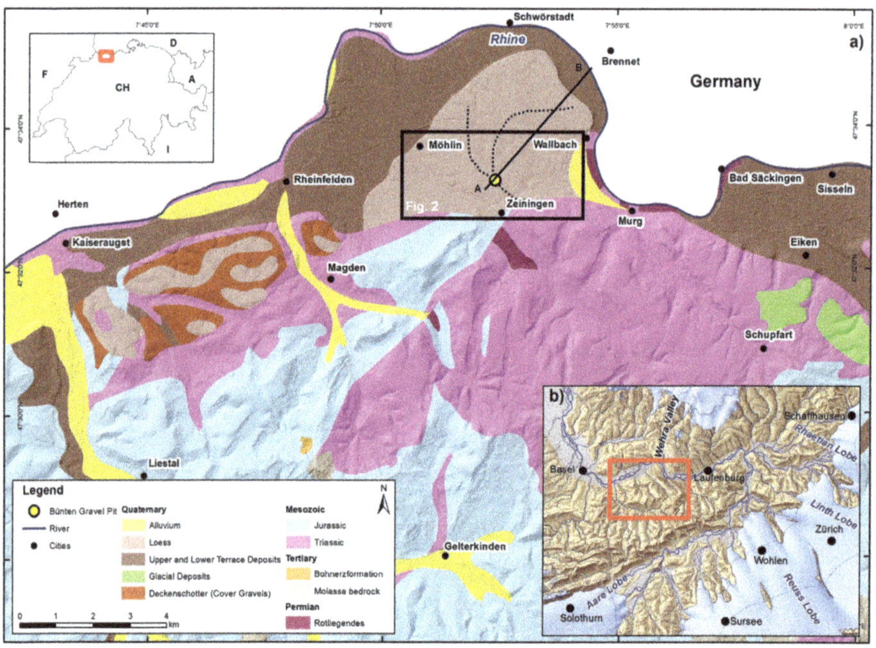

Figure 1. (**a**) Simplified geological map of the Möhlin area. The black dotted lines indicate the ridges that have been interpreted as moraines by Penck and Brückner [9]. The black rectangle indicates the extent of Figure 2. The cross-section A–B is given in Figure 3. (**b**) Extent of the Aare, Reuss, Linth, and the Rhaetian Lobes in the Swiss northern Alpine Foreland during the Last Glacial Maximum (LGM) [20]. The background map and the geological 1:500,000 map are reproduced with the authorization of the Swiss Federal Office of Topography (swisstopo).

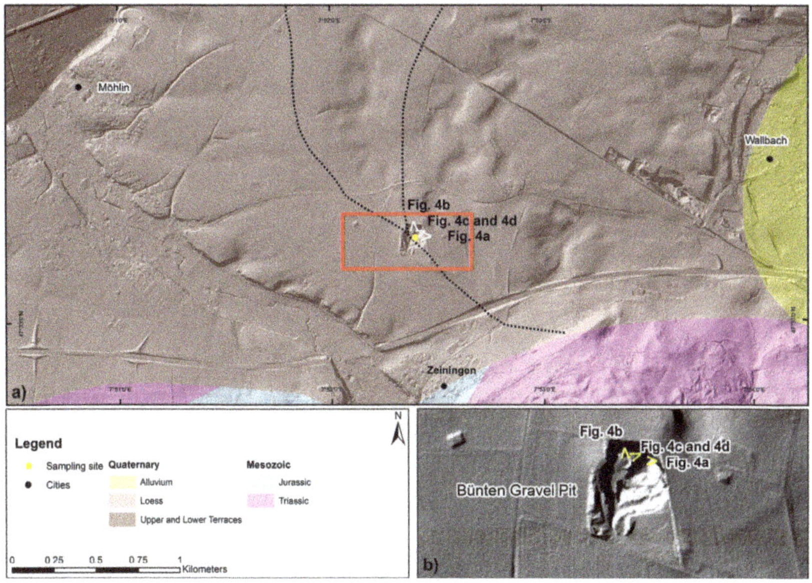

Figure 2. (**a**) Simplified geological map of the area around the Bünten gravel pit. For its extent, please refer to Figure 1. The yellow dot indicates the sampling site. The white angles in (**a**) and the yellow ones in (**b**) display the locations where the pictures in Figure 4 were captured. (**b**) DEM (Digital Elevation Model) of the Bünten gravel pit. The background map and the geological 1:500,000 map are reproduced with the authorization of the Swiss Federal Office of Topography (swisstopo).

2. Study Site

The Bünten gravel pit on the Möhlinerfeld, an approximately 4 km long and 3.5 km wide plateau, is located close to Möhlin, approximately 20 km east of Basel, and to the south of the River Rhine (Figures 1 and 2), with a maximum altitude of 379 m above mean sea level (m a.s.l.). Permian, Jurassic, and Triassic bedrocks can be observed to have outcropped towards the south of the Möhlinerfeld. HT gravels and Lower Terraces (NT; *Niederterrasse* in German) have been reported towards the north of the plateau. In addition, the Möhlinerfeld lies far beyond the extents of the Last Glacial Maximum (LGM) (Figure 1). Penck and Brückner [9] have described two shallow ridges of lengths 3 km and 4 km, respectively, on the plateau, which were previously interpreted as terminal moraines deposited by an Alpine glacier during the Riss glaciation [9,19] (Figures 1 and 2); recent studies have, however, interpreted these ridges as loess rather than terminal moraines [3,16].

The stratigraphy of the Bünten gravel pit has been illustrated in Figure 3a,b [21,22]. Although no bedrock is exposed in the gravel pit or has been reached by drilling [22], Müller-Dick [22] has predicted at least a 30 m thick gravel layer (Figure 3a) overlying the Triassic bedrocks [9,23]. The Bünten Till is presently outcropping as a 50 cm thick reddish till layer containing clasts derived from the Alps and the Black Forest. This layer is not in its original position but is part of a push moraine [23]. The Bünten Till, attributed to the MEG [4], is located approximately 30 m below the surface, at an elevation of 345 m a.s.l. [21,22] (Figure 3b). Earlier, the Bünten Till was observed to be located at a lower level of the gravel pit in an autochthonous position, where it is overlain by the glaciofluvial Bünten Gravel. This has later partially been glaciotectonically deformed owing to broken clasts and folding events [21–23] (Figure 3a,b). These gravels predominantly consist of Alpine clasts. However, clasts derived from the Black Forest are also present [4]. The Bünten Gravel is overlain by a paleosol, which is considered to likely represent interglacial climatic conditions [4,21] (Figure 3b). The Wallbach Gravel, located approximately 15 m below the surface, is composed of glaciofluvial gravels of Alpine origin [4,21] (Figure 3a,b). The Bünten and the Wallbach Gravel were together glaciotectonically deformed resulting in a folding of the paleosol. A till layer with sediments derived from the Black Forest envelops the Wallbach Gravel [21]. This entire sequence has been interpreted as the Zeiningen Till [3,4]. Petrographic analyses of this till indicated its deposition by a glacier that originated from the Black Forest [3,4] (Figure 3a,b). The glacier advance depositing the Zeiningen Till has also been suggested to have deformed the underlying sedimentary units [4]. The uppermost unit of the Bünten gravel pit is a 10 m thick loess layer [3,4,21]. The youngest deposit of the Möhlinerfeld is the Möhlinerfeld Gravels, which partly stem from the Black Forest [3,4] (Figure 3a). These gravels overlie the Wallbach Gravels and the Bünten Gravels in the Möhlinerfeld, but are absent in the Bünten gravel pit [3,4,21].

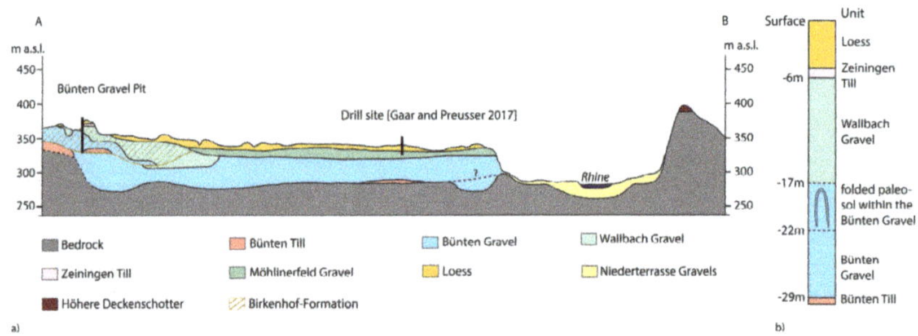

Figure 3. (**a**) Cross-section A–B showing the different units in the Möhlinerfeld (modified after [23]). (**b**) Stratigraphic column of the Bünten Gravel Pit (modified after [21]).

3. Methodology

3.1. Sedimentary Analyses

Petrographic analyses of the clasts provide information regarding the provenance of the sedimentary sequences [24–30]. A sample set of at least 250 clasts was collected from the field utilizing a bucket, to avoid any visual bias [26,30]. The freshly sampled material from the till was sieved into the pebble fraction (2–6 cm). The clasts were petrographically classified into the following lithology classes: (1) light colored limestone, (2) dark colored limestone, (3) gray colored limestone, (4) sparitic limestone, (5) ocher colored limestone, (6) oosparite, (7) siliceous limestone, (8) vein quartz, (9) quartzite, (10) chert/hornstone, (11) radiolarite, (12) sandstone, and (13) crystallines [8,25,26,30]. The crystalline components are utilized to distinguish the Black Forest and the Alpine origin clasts. Red colored granite clasts characterize the Black Forest origin [31], whereas the greenish and white colored crystallines characterize the Alpine origin.

The clast morphometric analysis is a commonly utilized method to distinguish between glacial and fluvial transport mechanisms [32]. Among the several methods developed for these analyses [32–34], we applied the Cailleux method [33], appropriate for glacial and melt water environments [26,33]. As the transport resistance varies with the lithologies, it is essential to analyze clasts from the same lithology [26,32]. The dark colored limestone clasts account for the most abundant lithology and hence 100 such clasts were segregated from the pebble fraction for analytical purposes. Moreover, flattening (A_i) and roundness (Z_i) indices for these clasts were determined in the following manner:

$$A_i = \frac{a+b}{2c} \times 100 \quad (1)$$

$$Z_i = \frac{2r_1}{a} \times 1000 \quad (2)$$

where a is the length of the clast, b the width, c the thickness, and r_1 the radius of the smallest curvature [33,35].

The depositional environment and the transport mechanism were determined by analyzing the clast fabric [32,34,36–38]. The clast fabric provides information on the orientation of a single clast, which forms the basis for reconstruction of paleoflow directions [29,38–40]. Elongated clasts were examined for this purpose [40]; that is, only the clasts with an a-axis > 6.3 cm and a:b ratio of >1.5. Thus, it was ensured that measurements of elongated clasts were utilized for the paleoflow reconstruction. The orientation and the inclination of 25 clasts fulfilling these criteria were measured in the field.

3.2. Isochron-Burial Dating

The isochron-burial dating technique is generally utilized to establish a chronology for 0.1 Ma to 5 Ma old terrestrial deposits such as fluvial terraces, paleosols, and glaciofluvial gravels [30,41–54]. It can be further applied to deposits composed of clasts, with a more complex pre-burial history or which experienced post-burial production [41,48,51,55]. The technique is based on the radioactive decay of cosmogenic ^{10}Be and ^{26}Al and uses the difference between the production ratio at the surface at the point of deposition and the measured ^{26}Al/^{10}Be ratio of the deposit to calculate the depositional age [41,45,47,48,51]. The samples should have different pre-burial histories in order to calculate their isochron-burial age; that is, they should differ in the inherited nuclide concentrations. To fulfill this criterion, samples of different lithologies, sizes, and shapes were collected [45–49,51,54].

Sampling, Sample Preparation, and Measurements

A total of 12 samples, 11 clasts (quartzite, sandstone, and granite) and a sediment sample (of 50 small quartz pebbles) were collected from the Bünten Till. The samples for cosmogenic ^{10}Be and ^{26}Al analysis were prepared at the Surface Exposure Laboratory of the University of Bern. The sediment and nine clast samples were leached and purified ([51], and references therein) to obtain 50 g pure quartz. Ideally, Al concentrations <30 ppm are required for high quality ^{26}Al/^{27}Al Accelerator Mass Spectrometry (AMS) analysis [51,54]. However, the new AMS facility (MILEA), recently developed at ETH Zurich, measures ^{10}Be and ^{26}Al with higher efficiency [56], allowing us to use samples with higher total Al concentrations (>100 ppm). Before dissolving, the total Al concentrations of the samples were checked utilizing inductively coupled plasma optical emission spectrometry (ICP-OES) at the Institute of Geological Sciences, University of Bern. In addition, samples with extremely high total Al concentrations (>100 ppm) were subjected to additional leaching steps or were abandoned. Six samples with lowest total Al concentrations and a sufficient amount of pure quartz were dissolved (Table 1). A full process ^{10}Be blank was processed in a batch of nine samples. Samples were spiked with 200 μL and the full ^{10}Be process blank with 400 μL of 1 g/L ^{9}Be carrier. The cosmogenic ^{10}Be and ^{26}Al was extracted following the protocol by Akçar et al. [51]. The ^{10}Be/^{9}Be and the ^{26}Al/^{27}Al ratios were measured in the MILEA AMS facility at ETH Zurich [57,58]. An error weighted average full process blank ratio of $(2.76 \pm 0.18) \times 10^{-15}$ was utilized to correct the measured ^{10}Be/^{9}Be ratios. ICP-OES, at the Institute of Geological Sciences, University of Bern, was utilized to determine the total Al concentrations. The CRONUS-Earth exposure age calculator was utilized to calculate the ^{26}Al/^{10}Be ratios, which were referenced to 07KNSTD ([59] and the updates from v. 2.2 to v. 2.3 published by Balco in June 2016; http://hess.ess.washington.edu/math/al_be_v23/al_be_multiple_v23.html; accessed on 13 January 2021). The isochron-burial ages were calculated with the MatLab® code provided by ([45]; personal communication with Darryl Granger) considering 1σ measurement uncertainties. To calculate an isochron-burial age, we applied a production rate of 4.00 ± 0.32 atoms/gSiO_2/a cosmogenic ^{10}Be at the surface, due to spallation at sea level high latitude (SLHL) [60]. In addition, a surface production ratio of 6.75 was applied for the ^{26}Al/^{10}Be ratio [41]. The time dependent Lm scheme [61,62] was utilized to calculate the altitude/latitude scaling of the surface production rate. Half-lives of 1.387 Ma for ^{10}Be [63,64] and 0.705 Ma for ^{26}Al [65,66] were used. For the determination of an isochron-burial age, initially measured ^{26}Al concentrations were plotted against the measured ^{10}Be concentrations. Furthermore, a regression was calculated through the measured ^{10}Be and ^{26}Al concentrations. Subsequently, the slope of the regressed line was utilized to estimate an initial isochron-burial age based on the offset from the initial surface production ratio [41,45]. Based on this initial age estimate, the post-burial production component is determined, subtracted from the measured concentrations and the resulting inherited concentrations are corrected for isotope decay [41,45]. After determining the pre-burial erosion rates based on the corrected inherited ^{10}Be concentrations, an inherited ^{26}Al/^{10}Be ratio was calculated [45]. The inherited isotope ratios were applied to estimate a linearization factor, which was applied to correct for post-burial production [45]. The corrected ^{10}Be and ^{26}Al concentrations were again plotted against each other. Finally, these steps have been iterated until age convergence [41,45]. For fluvial depositional environments, the initial ratio equals the surface production ratio, as fluvial clasts are assumed to stem from the surface. Therefore, the surface production ratio of 6.75 [41] is commonly utilized as the initial ratio in calculating isochron-burial ages. This technique has often been applied for determining the age of fluvial terraces (e.g., [43], among others). In landscapes sculptured repeatedly by deep erosion, such as glacial landscapes, the production ratio at depth becomes equal to the initial ratio [51]. As muogenic production becomes dominant with depth, the ^{26}Al/^{10}Be ratio increases. For example, Braucher et al. [67,68] and Margreth et al. [69] reported a value up to 8.3. In the following, we briefly outline how a glacially created, transported, and deposited clast can account for an initial ratio higher than the surface ratio (6.75). During glacial erosion, the clasts are first excavated by the glacier from the bedrock at depth (deep erosion) (cf. Figure 7 in [51]). Subsequently, these clasts are transported either subglacially or englacially, i.e., completely shielded from cosmic rays. Later, they are embedded in a glacial deposit, such as the Bünten till in this study, or in a glaciofluvial deposit. In both cases, such clasts will never be exposed at the surface prior to burial and their initial ratio will be >6.75. As one cannot determine the original depth at the source, from which a clast originates, Akçar et al. [51] suggested to use the production ratio at depth (between 6.75 and 8.4, average value is 7.6) for the isochron-burial age calculations for landscapes dominated by deep erosion. In addition, Knudsen et al. [42] applied the P-PINI (Particle Pathway Inversion of Nuclide Inventories) method to model burial ages, utilizing a source to sink approach. Their modelling demonstrated that the majority of initial ^{26}Al/^{10}Be ratios at the source were larger than 7.2 (cf. Figure 8 in [42]). They concluded that these samples were derived from environments which experienced fast and deep erosion due to glacier activity. Based on these lines of evidence, the isochron-burial age of the Bünten Till was calculated with initial ratios of 6.75, 7.6, and 8.4, respectively. The isochron-burial age calculated with an initial ratio of 7.6 has been considered in the ensuing discussion.

Table 1. Sample information for the Möhlin site.

Sample Name	Lithology	Weight (g)	a-Axis (cm)	Amount of Quartz after Leaching (g)	Al Concentration after Leaching (ppm)
MÖHL-1	Quartzite	1150	12.7	34	76
MÖHL-4	Quartzite	660	12.3	44	161
MÖHL-5	Quartzite	800	9.3	54	145
MÖHL-7	Quartzite	520	8.2	36	94
MÖHL-10	Sandstone	1630	15.3	46	138
[1] MÖHL-12	Quartz pebbles	1610	-	67	49

[1] Sediment sample. Coordinates of the sampling site: 47.5533° N, 7.8716° E, 345 m a.s.l.

4. Results

4.1. Sedimentary Analyses

4.1.1. Sediments of the Bünten Gravel Pit

The till in the Bünten gravel pit is located approximately 30 m below the surface (Figure 4a). The 50 cm thick till layer is observed to be tilted (Figure 4b). The till is poorly sorted, with clast sizes ranging up to 20 cm (Figure 4c,d). The clasts are matrix supported, consisting of reddish colored clay and silt (Figure 4c,d). Two groups of paleoflow directions were measured, one towards N and another towards W, resulting in a NW mean paleoflow direction. The Bünten gravels overlying the till are poorly sorted with a maximum clast size of approximately 25 cm. A sandy to silty matrix can be observed. The horizontally bedded glaciofluvial gravels, located to the west of the Bünten Till, are nearly 10 m thick (Figure 4b). They contain a matrix predominantly composed of sand and minor amounts of silt. A maximum clast size of 25 cm has been reported from these gravels. At a depth of 10 m below the surface, the glaciofluvial gravel sequence exhibits cross bedding (Figure 4a).

Figure 4. Field photographs of the Möhlin site. (**a**) Overview of the Bünten gravel pit and spatial positioning of Figure 4b,d; (**b**) the Bünten Till (indicated by the red line) and spatial positioning of Figure 4c; (**c**) samples MÖHL-4 and MÖHL-5; and (**d**) sample MÖHL-2.

4.1.2. Clast Petrography

The results of the clast petrography for 275 samples are shown in Figure 5. Most of the clasts are dark colored limestones (21%), successively followed by gray colored limestones (14%), quartzites (13%), and crystalline clasts (9%). The crystalline clasts can be further categorized as those of the Black Forest origin (4%) and those of the Alpine origin (5%). The ocher-colored limestones and oosparites account for 8% and 6% of the total clasts, respectively. The sparitic and siliceous limestones contain relative abundances of 7% each. Sandstones, vein quartz clasts, and cherts exhibit a relative abundance of 4% each. The light colored limestone clasts and the weathered components represent 2% and 1% of the total clasts, respectively. Since only one radiolarite clast was found, its abundance was too small to be represented on the pie diagram.

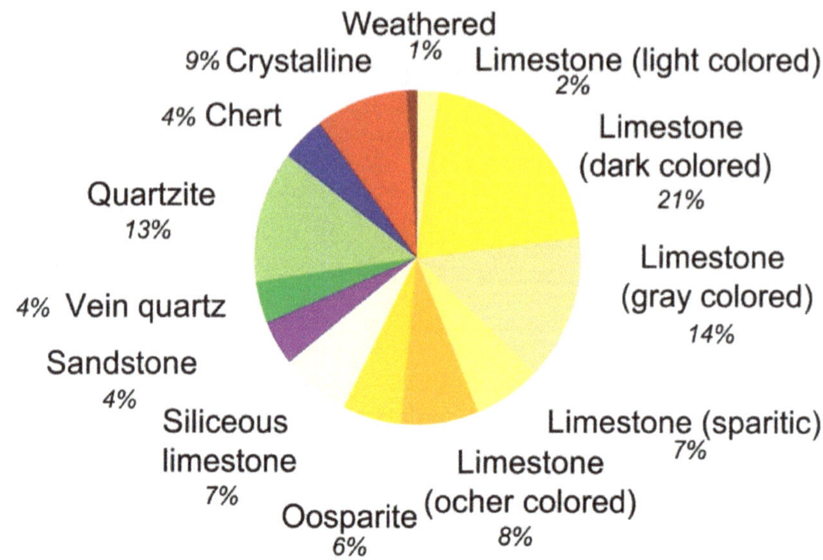

Figure 5. Clast petrography of the Bünten Till analyzed at the Möhlin site.

4.1.3. Clast Morphometry

A total of 110 clasts were measured to calculate the roundness index (Z_i) and the flattening index (A_i) (Figures 6 and 7). The Z_i values range from 50 to 550, with a few clasts displaying values between 600 and 700 (Figure 6). The median of the roundness index ($Md(Z_i)$) is 244. The clasts exhibit bimodal distribution, the first highest mode being represented between 100 and 150 and the second between 250 and 300. The calculated flattening indices vary between 100 and 450 with a median ($Md(A_i)$) of 180 (Figure 7).

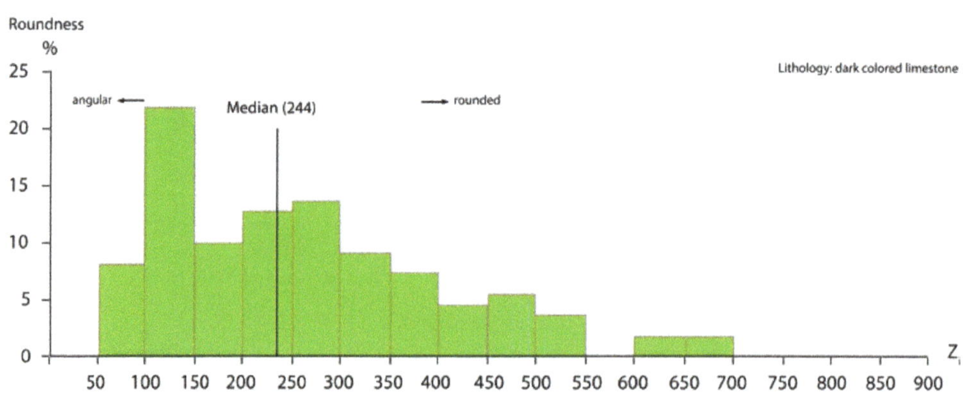

Figure 6. Histogram illustrating roundness index of clasts at the Möhlin site.

4.2. Isochron-Burial Dating

At the Möhlin site twelve samples were collected, of which six samples were processed to extract cosmogenic ^{10}Be and ^{26}Al. After leaching, these samples exhibited a total Al concentration between 49 ppm and 161 ppm (Table 1). The processed samples contain lithologies of quartzite (MÖHL-1, MÖHL-4, MÖHL-5, and MÖHL-7) and sandstone (MÖHL-10). MÖHL-12 is a sediment sample.

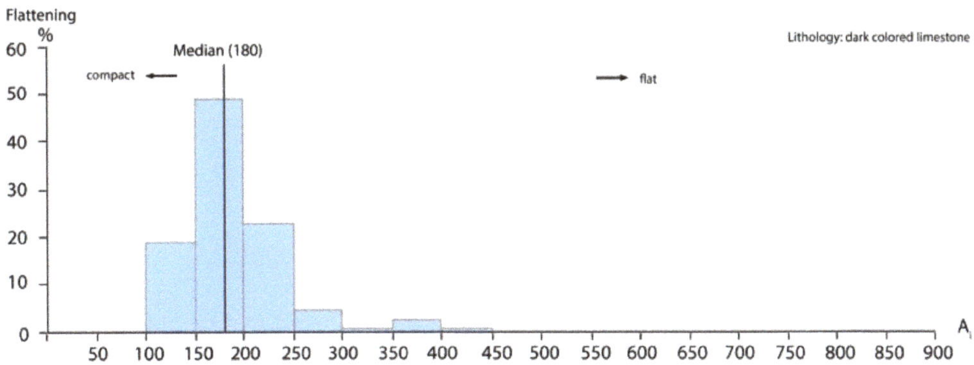

Figure 7. Histogram illustrating flattening index of clasts at the Möhlin site.

The results of the cosmogenic ^{10}Be and ^{26}Al measurements are displayed in Table 2. The ^{10}Be/^{9}Be ratios range from 1.24×10^{-14} to 8.60×10^{-14}. The relative measurement uncertainty of the ^{10}Be/^{9}Be ratios lies between 4% and 11%. The full process blank accounts for 3% to 23% of the measured ^{10}Be/^{9}Be ratios. The calculated, blank corrected ^{10}Be concentrations vary between $(3.8 \pm 0.5) \times 10^3$ atoms/g and $(25.3 \pm 1.3) \times 10^3$ atoms/g. The total Al amount varies between 3 and 9 mg, and the total Al concentrations between 60 and 190 ppm, respectively. The measured ^{26}Al/^{27}Al ratios range from 1.25×10^{-14} to 9.56×10^{-14} with relative uncertainties of 3% to 11%. The calculated concentrations of the ^{26}Al are between $(44.3 \pm 3.2) \times 10^3$ atoms/g and $(172.0 \pm 13.4) \times 10^3$ atoms/g. The ^{26}Al/^{10}Be ratio ranges from 6.8 ± 0.4 to 13.3 ± 2.3. The sample MÖHL-10 was excluded from the modeling of the isochron-burial age, as it lies beyond the two-sigma solution space [49].

For the Bünten Till, a lower isochron-burial age limit of 260 ± 110 ka, with an initial ratio of 6.75, was calculated using the code provided by [45] and personal communication with Darryl Granger. The mean initial ratio of 7.6 yielded an age of 500 ± 100 ka. An upper age boundary of 700 ± 100 ka was calculated using an initial ratio of 8.4 (Figure 8). In order to explore the contribution of post-burial nuclide production in the measured concentrations, we re-calculated the isochron-burial age by using St [62] and LSDn scaling scheme [70] and based on the Bender approach, which do not include the post burial component (cf. [48]). Use of both calculations with different scaling schemes and Bender code did not alter the isochron-burial age, which indicates a minimum contribution of post-burial production. The 500 ± 100 ka age will be henceforth utilized with regards to the MEG.

Table 2. Cosmogenic ^{10}Be and ^{26}Al results of the samples from the Möhlin site.

Sample Name	Quartz Dissolved (g)	^{9}Be Spike (mg)	^{10}Be/^{9}Be ($\times 10^{-14}$)	Relative Uncertainty (%)	Blank Correction (%)	^{10}Be Concentration ($\times 10^{3}$ Atoms/g)	Total Al (ppm)	Total Al (mg)	^{26}Al/^{27}Al ($\times 10^{-14}$)	Relative Uncertainty (%)	^{26}Al Concentration ($\times 10^{3}$ Atoms/g)	^{26}Al/^{10}Be
MÖHL-1	34.0800	0.1990	1.24	10.2	22.2	3.8 ± 0.5	158	5.38	1.42	10.6	50.1 ± 5.3	13.3 ± 2.3
MÖHL-4	43.5431	0.1980	8.60	4.8	3.2	25.3 ± 1.3	185	8.04	4.17	7.8	172.0 ± 13.4	6.8 ± 0.4
MÖHL-5	49.9954	0.1991	1.75	7.4	15.8	3.9 ± 0.4	159	7.93	1.25	7.2	44.3 ± 3.2	11.3 ± 1.3
MÖHL-7	35.5300	0.1990	3.55	6.3	7.8	12.2 ± 0.8	106	3.78	3.85	5.5	91.5 ± 5.0	7.5 ± 0.7
MÖHL-10	45.6933	0.1993	5.71	4.6	4.8	15.8 ± 0.8	157	7.16	4.12	10.4	144.0 ± 15.0	9.1 ± 0.6
[1] MÖHL-12	50.0300	0.2000	7.14	4.2	3.9	18.3 ± 0.8	63	3.16	9.56	3.9	134.5 ± 5.3	7.3 ± 0.4

[1] Sediment sample. Accelerator mass spectrometry (AMS) measurement errors at 1σ level, including the statistical (counting) error, the uncertainty of standard normalization, and the propagated error of blank correction. The error weighted average ^{10}Be/^{9}Be full-process blank ratio was (2.76 ± 0.18) $\times 10^{-15}$. ^{26}Al/^{10}Be ratios were calculated with the CRONUS-Earth exposure age calculator and were referenced to 07KNST (http://hess.ess.washington.edu/math/al_be_v23/al_be_multiple_v23.html; accessed on 13 January 2021); see [59] and update from v.2.2 to v.2.3 published by Balco in June 2016).

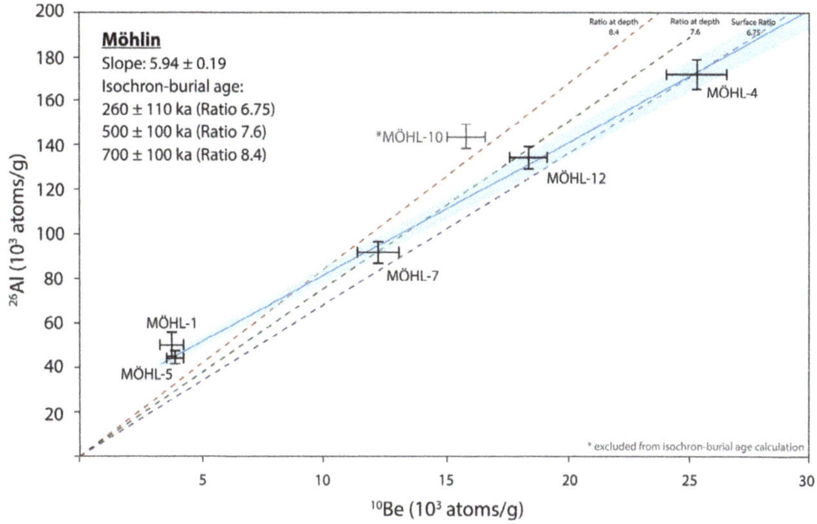

Figure 8. Isochron plot of the samples from the Möhlin site (the samples are plotted with 1σ uncertainties and the isochron-burial ages calculated with the initial ratios 6.75, 7.6, and 8.4). The best fit isochron-line is indicated in blue and the light blue envelope shows the 1σ solution space. The sample indicated in gray and labeled with an asterisk (*) is defined as an outlier and therefore excluded from the isochron-burial age calculation.

5. Discussion

5.1. Provenance of the Sediments

Given the proximity of Möhlin to the Black Forest, it is important to establish whether the till in the Bünten gravel pit represents the most extensive position of Alpine glaciers or has been deposited by a paleoglacier from the Black Forest. The petrography of the clasts is essential to understand the origin of the paleoglacier. Several varieties of limestone (60%) reported from Möhlin, including the dark colored, gray colored, light colored, and siliceous limestone clasts, possibly have their origins in the Helvetic and Penninic Nappes of the Alps [71–73]. These nappes cover extensive areas and therefore the precise provenance cannot be determined. The ocher colored limestone, oosparite, and sparitic limestone clasts probably originated from the Jura Mountains in the northern and western parts of the northern Alpine Foreland and/or south of the Black Forest [71–73]. The quartzite clasts are currently exposed in the Valaisian Alps and eastern Central Alps, and have also been observed in the Molasse Conglomerates [74–76]. The composition of the crystalline components from the till indicates a Black Forest as well as an Alpine origin.

The presence of Alpine and Black Forest origin clasts in the Bünten Till can be explained by the reworking of previously deposited Alpine clasts by a glacier initiating from the Black Forest or vice versa. Previous studies reported the presence of Alpine and Black Forest lithologies at various locations between Brennet and Laufenburg (Figure 1) and thus, numerous theories were proposed for the origin of the paleoglacier. In 1895, Gutzwiller [19] observed the coexistence of the Alpine and Black Forest material in a glacial deposit, but did not make a clear statement regarding the origin of this paleoglacier. Reichelt [77] concluded that the two glaciers merged to the east, close to the city of Laufenburg (Figure 1). However, Pfannenstiel [78] suggested that the glacier initiated from the Black Forest and coalesced with the Alpine glacier approximately 5 km east of Möhlin (Figure 1). A few studies proposed that the glacier from the Wehra Valley advanced close to Möhlin, but did not reach the Möhlinerfeld [79,80] (Figure 1). Müller-Dick [22] suggested that the glacier depositing the Bünten Till was of Alpine origin, while a second glacier advance depositing the Zeiningen Till originated from the Black Forest.

The lithology of a paleoglacier from the Black Forest should ideally contain contributions of: red colored granites, minor amounts of Mesozoic carbonates, and a few Tertiary rocks [71–73,81]. A southbound advancement of the Black Forest paleoglacier potentially enabled its encounter with some Alpine clasts, reworked from the deposits along the course of the River Rhine. In contrast, theoretically, a paleoglacier from the Alps would have predominantly transported carbonate clasts from the Helvetic and Penninic Nappes, along with quartzite clasts, crystalline clasts (such as Julier granite, Aare granite, and Serpentinites, among others), and a few Mesozoic rocks from the Jura Mountains, allowing limited contribution of the clasts from the Black Forest. Clast petrographic analysis during this study revealed coexistence of sediments from both provenances. The components from the Black Forest demonstrate a rather small relative abundance.

The morphometry of the clasts embedded in the Bünten Till points towards a glacially influenced sediment [35,82]. According to Cailleux and Tricart [35], and Schlüchter [82], clasts with a roundness index between 150 and 300 were deposited in the proximity of a glaciofluvial environment. Two third of the clasts from the Bünten Till show Z_i values below 300, of which a third contain values between 50 and 150, which indicate a glacial deposition. The clasts with Z_i values above 300, therefore, probably represent better rounded clasts and are interpreted as indicators for reworked sediment [82]. We accordingly conclude that one third of the quartz vein clasts in the Bünten Till are fresh material delivered from the Alps, whereas, two thirds appear to bear evidence of reworking by the glacier. The A_i values point towards a compact shape and, therefore, glacial and glaciofluvial transport in contrast to the flat and disc shaped clasts, which are interpreted as evidence of a fluvial transport [34]. In brief, we propose that the Bünten Till was deposited by a glacier descending from the Alps based on the petrographical composition of the sediment, the morphometry, and the measured paleoflow direction towards the northwest. The Black Forest lithologies encountered in the sediment were most probably eroded from outcrops located further east of Möhlin and close to the River Rhine (Figure 1).

5.2. Age of the MEG in the Northern Hemisphere

Previous studies have tentatively reconstructed the age of the MEG based on the morphostratigraphy of the northern Alpine Foreland, whereas the obtained 500 $\pm$ 100 ka corresponds to the age of the most extensive position of Alpine glaciers. The chronology of the MEG lies within the time range suggested by previous studies and implied by the morphostratigraphy of the northern Alpine Foreland [2,4,16,17]. Schlüchter [2,16] suggested that this advance occurred after the Deckenschotter glaciations and the Brunhes/Matuyama transition. Based on the OSL ages from the loess cover, Gaar et al. [17] tentatively attributed the deposition of the Zeiningen Till to MIS 6, thus implying that the MEG predates the Zeiningen Till (Figure 3a).

Based on the existing data and results obtained from this study, we suggest the following chronostratigraphy for the Möhlinerfeld area. At approximately 500 ka, the Alpine glaciers reached their most extensive position. The Bünten Till indicates that a glacier lobe covered the Möhlinerfeld; however, evidence for the thickness of the ice and the position of the ice margins at that time is lacking. The measured paleoflow directions suggest that the ice margin was located NW of the Bünten gravel pit. According to Frei [10], the Rhone Lobe covered the Möhlinerfeld during the MEG. Such an assumption implies that the Rhone Lobe reached an approximate minimum length of 315 km. This further implies that the glacier during the MEG was nearly 35 km longer than that during the LGM. The deposition of the gravels and the Zeiningen Till located on top of the Bünten Till occurred between 500 ka and 60 ka, as per the age of the loess coverage [17] (Figure 3a). An age of 160 ka, corresponding to the MIS 6, was suggested for the Zeiningen Till [17]. Assuming that the Zeiningen Till is of the MIS 6 age, the Bünten Gravel would have been either deposited during the MEG or the Habsburg glaciation, with the Wallbach Gravel overlying the Bünten Gravel during the Habsburg or Hagenholz/first advance of the Beringen glaciation, respectively (Figure 3a,b). The deposition of the Möhlinerfeld Gravel can be tentatively attributed to the Beringen glaciation (Figure 3a).

Glaciers played an important role in shaping the Quaternary landscapes of the northern Swiss Foreland. Glaciers that advanced onto the northern Alpine Foreland sculpted the overdeepened valleys (up to 300 m in depth) (see [83] and references therein). The MEG is considered responsible for the commencement of the overdeepened valley formation [4,12]. Therefore, we suggest that the first overdeepened valley formed not later than approximately 500 ka. Recently, several drill cores were obtained from overdeepened valleys in the Swiss northern Alpine Foreland to comprehensively analyze the infill and to reconstruct the glaciation history [83–87]. Sediments from the base of the investigated overdeepened valley fills were dated to approximately 180 ka [83–87]. Some of these also represent an older sedimentary infill [74]. According to these findings, the beginning of the overdeepening has been assigned to a glacial advance at 260 ka or older [83]. In the Lower Aare Valley, the presence of different sediment units implies that during 160 ka to 180 ka, the area was dominated by a periglacial setting; the lowermost sands covering a glacial diamicton are older than 180 ka [87]. The presence of glaciolacustrine sediments, dated by applying OSL, indicate that glacial lakes dominated the Wehn Valley and the Lower Glatt Valley between ca. 130 ka and ca. 180 ka [83,86] as well as between ca. 180 ka and >260 ka [84,86]. Assuming the challenges involved in OSL dating of proglacial sediments, the ages of roughly 260 ka might also be related to the upper limit of the OSL dating technique and can be reliable up to 200 ka [88,89]. The upper dating limit with OSL is given by the saturation of the dose, usually resulting in an age of 150 ka, but few deposits can be dated up to ca. 400 ka [89–92]. Based on these results, the MEG still possibly remains responsible for the first overdeepened valleys, albeit inconclusively.

Owing to the limitations of the OSL technique, with a few exceptions a chronology of only up to ca. 400 ka can be dated; that is, the OSL helps reconstruct chronology of deposits older than the LGM [93–104]. It is possible, however, that evidence of glaciation at 500 ka exists somewhere in the Alps. In the Upper Rhine Graben (URG) about 300 km north of Möhlin, recently deposited sediments in a fluvial environment partially influenced by gravitational processes were dated to 454 ± 29 ka and attributed to the MEG [96]. Two phases for the deposition of HT-complex sediments were revealed by OSL ages in the northern Alpine Foreland: one at approximately 160 ka and another at 260 ka (Figure 9). HT deposits, 20 km to the west of Möhlin, were dated at approximately 236 ka by OSL, suggesting that the underlying gravel units were deposited by a glaciation older than ca. 240 ka [93]. The ^{10}Be depth-profile age indicates that at 270 ka this area was characterized by a distal glaciofluvial environment [94]. These two phases were also identified based on glaciofluvial sediments in Southern Germany [98]. In Austria, the glaciofluvial sediments from the penultimate glaciation, attributed to the HT, were dated to 140 ka [99,100]. In the Southern Alps, a cold phase was determined at ca. 250 ka [101], but there is some evidence that these sediments might be of earlier glaciations [102,103]. No deposits older than the LGM have been dated in the French Alps; however, there is evidence that there were glaciers present during the Middle Pleistocene [104]. These ages of the HT exhibit that the MEG is clearly older and should therefore be classified separately. In addition, very few MEG deposits have been dated so far. Therefore, the ca. 500 ka of the Bünten Till represents the only time constraint for the MEG in the Alps.

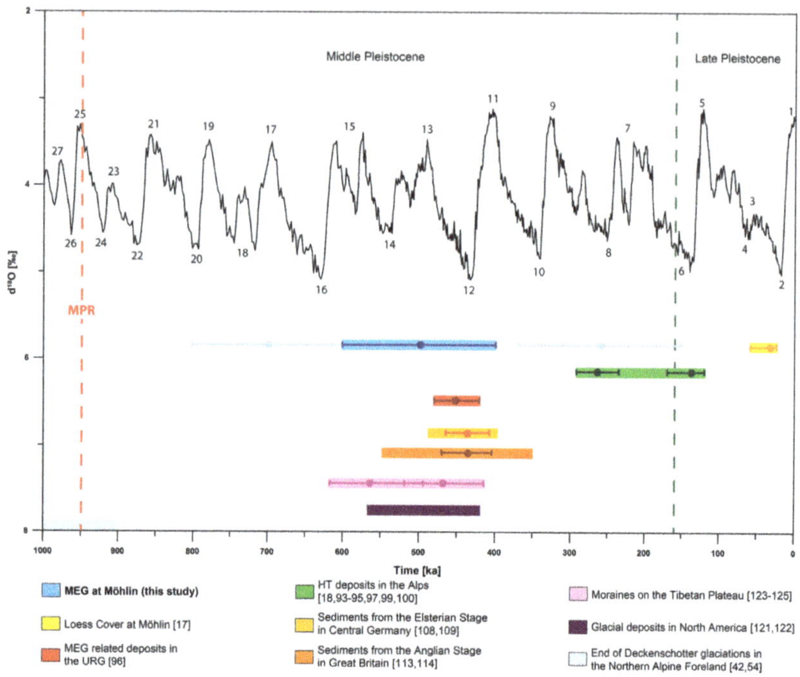

Figure 9. $\delta^{18}O$ variation in the last 1 Ma (modified after [15]). The upper error ranges of the MEG, with an age of 500 ± 100 ka, overlap with those of the Elsterian stage in Northern Europe, the Anglian stage in Great Britain, moraines on the Tibetan Plateau, while its lower error ranges overlap with glaciofluvial sediments of North America. The Alpine glaciations of 250 ka and 160 ka represent separate glaciation events that do not correlate with the MEG. The light blue bar indicates the end of Deckenschotter glaciations, which took place between ca. 2.6 and 0.9 Ma [42,54]. The red dashed line indicates the Mid Pleistocene Revolution (MPR) occurring at around 0.95 Ma [105] and the green one the boundary between the Middle and Late Pleistocene [1].

At approximately 500 ± 100 ka, the glaciers apparently reached their most extensive position, not only in the Alps but also in other parts of Europe and of the northern hemisphere [106] (Figure 9). The Fennoscandian ice sheet reached its most extensive position during the Elsterian glaciation, covering Northern Europe and advancing up to Central Germany [107–109]. Fluvial sediments overlying Elsterian till were analyzed with the luminescence technique and dated between 447 ± 52 ka and 387 ± 48 ka, indicating that the Elsterian stage occurred during MIS 12 [108]. Dated glaciofluvial sediments indicate a glacier advance during the Elsterian glaciation between 461 ± 34 ka and 421 ± 25 ka [109] (Figure 9). In the Netherlands and the western part of Germany, archives of the Elsterian stage exist, which are not considered to represent the most extensive glaciation [110,111]. The MEG is considered as a glacier advance that initiated the overdeepening of valleys not only in the Swiss northern Alpine Foreland but also in Northern Europe (Elsterian glaciation) [112]. Glaciofluvial deposits indicate that the British ice sheet had the maximum extent (the Anglian stage) during 440 ka [113,114] (Figure 9). The presence of the Eurasian ice sheet (500 ka) can be observed in Russia. A till layer overlying interglacial sediments has been dated to 510 ka using the thermoluminescence technique [115]. This till layer corresponds to the Oka glaciation (tentatively correlating with the Elsterian glaciation) at 500–460 ka [116]. Although the Oka glaciation occurred comparably with the Möhlin glaciation, it did not reach its most extensive position in that area. The Don glaciation, considered to have had the largest extent, occurred prior to the Oka glaciation and therefore predates 500 ka [111,115]. The U/Th analysis of the secondary carbonates precipitated in the pores of the glaciofluvial deposits in the Balkan Mountains suggests that the MEG occurred between 470 and 420 ka [117–120]. Glaciations dating 500 ka have also been reported from North America [121,122]. There are, for instance, glacial deposits overlain by ca. 470 ka old basalts and underlain by marine deposits of 570 ka [121] or a till deposit with a suggested minimum age of 424–478 ka [122] (Figure 9). Evidence of a 500 ka old glaciation have been retrieved from the Tibetan Plateau, where glacial deposits were dated to 460–571 ka by the electron spin resonance (ESR) technique [123–125] (Figure 9). This indicates that the age of the MEG is consistent with other glacier advances in the northern hemisphere.

6. Conclusions

The Bünten gravel pit close to Möhlin, deposited by the MEG, was comprehensively examined during this study. Based on the petrographic analyses and the results of paleoflow direction, we conclude that the glacier originated from the Alps and that the Black Forest clasts were incorporated into the till due to reworking of the nearby sediments. Moreover, the isochron-burial dating of the Bünten Till to 500 $\pm$ 100 ka provides the first direct chronology for the MEG, thus addressing the complex chronostratigraphy of the Swiss northern Alpine Foreland. However, for improved understanding regarding the age and the extent of the most extensive glaciation in the Alps, further studies on stratigraphy and chronostratigraphy of MEG deposits are essential. We thus infer that the MEG is apparently synchronous with other glacier advances in the northern hemisphere.

Author Contributions: Conceptualization, N.A. and C.D.; methodology, C.D. and N.A.; fieldwork, C.D., N.A. and H.R.G.; sample preparation, C.D. and N.A.; AMS-measurements, M.C., C.V. and P.G.; investigation, C.D. and N.A.; writing, original draft preparation, C.D. with contributions from all authors; supervision, N.A.; project administration, N.A.; and funding acquisition, N.A. All authors have read and agreed to the published version of the manuscript.

Funding: This research was funded by the Swiss Federal Nuclear Safety Inspectorate (ENSI), grant number CTR 00314.

Data Availability Statement: All generated and analyzed data during this study has been included in this article.

Acknowledgments: We would like to thank Susan Ivy Ochs at ETH Zurich and Marius Büchi at the University of Bern for their help and support during the sampling campaign. We are grateful to Julia Gajic at the University of Bern for her help in preparing the samples. We are thankful to Niklaus Waber, Priska Bähler, and Christopher Pichler at the University of Bern for the ICP measurements. We would also like to thank the AMS facility team at ETH Zurich for the AMS measurements.

Conflicts of Interest: The authors declare no conflict of interest. The funders had no role in the design of the study, collection, analyses, or interpretation of data, writing of the manuscript, or in the decision to publish the results.

References

1. Cohen, K.M.; Finney, S.C.; Gibbard, P.L.; Fan, J.-X. The ICS International Chronostratigraphic Chart. *Episodes* **2013**, *36*, 199–204. [CrossRef] [PubMed]
2. Schlüchter, C. The Swiss glacial record—A schematic summary. In *Quaternary Glaciations-Extent and Chronology*; Elsevier B.V.: Amsterdam, The Netherlands, 2004; Volume 2, pp. 413–418. ISBN 9780444514622.
3. Graf, H.R. *Stratigraphie von Mittel- und Spätpleistozän in der Nordschweiz. Beiträge zur Geol. Karte der Schweiz*; Bundesamt für Landestopographie, Swisstopo: Wabern, Switzerland, 2009; Volume 168.
4. Preusser, F.; Graf, H.R.; Keller, O.; Krayss, E.; Schlüchter, C. Quaternary glaciation history of northern Switzerland. *E&G Quat. Sci. J.* **2011**, *60*, 282–305. [CrossRef]
5. Ellwanger, D.; Wielandt-Schuster, U.; Franz, M.; Simon, T. The Quaternary of the southwest German Alpine Foreland (Bodensee-Oberschwaben, Baden-Württemberg, Southwest Germany). *E&G Quat. Sci. J.* **2011**, *60*, 306–328. [CrossRef]
6. Keller, O. Erwägungen zur Korrelation mittelpleistozäner Relikte des Rheingletschers mit der nordschweizer Stratigraphie. *E&G Quat. Sci. J.* **2014**, *63*, 19–43. [CrossRef]
7. Schlüchter, C. *Das Eiszeitalter in der Schweiz*; Stiftung Landschaft und Kies: Rubigen, Switzerland, 2010; p. 4.

8. Keller, O.; Krayss, E. Mittel-und spätpleistozäne Stratigraphie und Morphogenese in Schlüsselregionen der Nordschweiz. *E&G Quat. Sci. J.* **2010**, *59*, 88–119. [CrossRef]
9. Penck, A.; Brückner, E. *Die Alpen im Eiszeitalter*; H. Tauchnitz: Leipzig, Germany, 1909; p. 1199.
10. Frei, R. *Über die Ausbreitung der Diluvialgletscher in der Schweiz. Beiträge zur Geol. Karte der Schweiz*; Bundesamt für Landestopographie, Swisstopo: Wabern, Switzerland, 1912; Volume 41.
11. Hantke, R. *Eiszeitalter Kalt-/Warmzeit-Zyklen und Eistransport im alpinen und voralpinen Raum*; Hep Verlag Ag: Bern, Switzerland, 2011; ISBN 978-3-7225-0121-5.
12. Ellwanger, D. Eine landschaftsübergreifende Lockergesteinsgliederung vom Alpenrand zum Oberrhein. In *Landschaftsgeschichte im europäischen Rheinland*; Schirmer, W., Ed.; Lit Verlag: Münster, Germany, 2003; pp. 81–124.
13. Fiebig, M.; Preusser, F. Pleistocene glaciations of the northern Alpine Foreland. *Geogr. Helv.* **2008**, *63*, 145–150. [CrossRef]
14. Spell, T.L.; Mcdougall, I. Revisions to the Age of the Brunhes-Matuyama boundary and the Pleistocene geomagnetic polarity timescale. *Geophys. Res. Lett.* **1992**, *19*, 1181–1184. [CrossRef]
15. Lisiecki, L.E.; Raymo, M.E. A Pliocene-Pleistocene stack of 57 globally distributed benthic $\delta^{18}O$ records. *Paleoceanography* **2005**, *20*, 1–17. [CrossRef]
16. Schlüchter, C. The Deglaciation of the Swiss-Alps: A paleoclimatic event with chronological problems. *Bull. L'association Française Pour L'étude Quat.* **1988**, *2*, 141–145. [CrossRef]
17. Gaar, D.; Preusser, F. Age of the Most Extensive Glaciation of Northern Switzerland: Evidence from the scientific drilling at Möhliner Feld. *E&G Quat. Sci. J.* **2017**, *66*, 1–5. [CrossRef]
18. Dehnert, A.; Preusser, F.; Kramers, J.D.; Akçar, N.; Kubik, P.W.; Reber, R.; Schlüchter, C. A multi-dating approach applied to proglacial sediments attributed to the Most Extensive Glaciation of the Swiss Alps. *Boreas* **2010**, *39*, 620–632. [CrossRef]
19. Gutzwiller, A. Die Diluvialbildungen der Umgebung von Basel. *Verh. Nat. Ges. Basel* **1895**, *10*, 512–690. [CrossRef]
20. Bini, A.; Buoncristiani, J.F.; Couterrand, S.; Ellwanger, D.; Felber, M.; Florineth, D.; Graf, H.R.; Keller, O.; Kelly, M.; Schlüchter, C.; et al. *Die Schweiz während dem letzteiszeitlichen Maximums (LGM)*; Bundesamt für Landestopographie, Swisstopo: Wabern, Switzerland, 2009.
21. Dick, K.A.; Graf, H.-R.; Müller, B.U.; Hartmann, P.; Schlüchter, C. Das nordalpine Wasserschloss und seine eiszeitgeologische Umgebung. *Eclogae Geol. Helv.* **1996**, *89*, 635–645.
22. Müller-Dick, K. Das Möhlinerfeld im Hochrheintal. In *DEUQUA 2000, Bern, Exkursionsführer, Exk. A1*; Kelly, M., Linden, U., Schlüchter, C., Eds.; DEUQUA: Hannover, Germany, 2000; pp. 10–14.
23. Pfirter, U.; Jordan, P.; Graf, H.R.; Burger, H.; Pietsch, J.; Huber, M.; Kiefer, S.; Grezet, C.; Maise, C.; Burkhalter, R. *Geologischer Atlas der Schweiz 1:25'000. Blatt Sissach (1068) mit Südteil von Blatt Rheinfelden (1048)–Erläuterungen*; Bundesamt für Landestopographie, Swisstopo: Wabern, Switzerland, 2019; ISBN 9783302401003.
24. Frei, R. *Monographie des Schweizerischen Deckenschotters. Beiträge zur Geol. Karte der Schweiz*; Bundesamt für Landestopographie, Swisstopo: Wabern, Switzerland, 1912; Volume 37.
25. Graf, H.R. Die Deckenschotter der Zentralen Nordschweiz. Ph.D. Thesis, Swiss Federal Institute of Technology Zurich, Zurich, Switzerland, 1993.
26. Schlüchter, C. *Eiszeitliche Lockergesteine-Geologie, Genese und Eigenschaften. Ein Beitrag zu den Beziehungen zwischen fundamentaler und angewandter Eiszeitgeologie, Habilitationsschrift*; Swiss Federal Institute of Technology Zurich: Zurich, Switzerland, 1989.
27. Weltje, G.J.; von Eynatten, H. Quantitative provenance analysis of sediments: Review and outlook. *Sediment. Geol.* **2004**, *171*, 1–11. [CrossRef]
28. Lindsey, D.A.; Langer, W.H.; Van Gosen, B.S. Using pebble lithology and roundness to interpret gravel provenance in piedmont fluvial systems of the Rocky Mountains, USA. *Sediment. Geol.* **2007**, *199*, 223–232. [CrossRef]
29. Adelsberger, K.A. Sedimentology. In *Encyclopedia of Geoarchaeology*; Gilbert, A.S., Ed.; Springer: Dordrecht, The Netherlands, 2017; pp. 764–772. ISBN 978-1-4020-4409-0.
30. Claude, A.; Akçar, N.; Ivy-Ochs, S.D.; Schlunegger, F.; Kubik, P.W.; Dehnert, A.; Kuhlemann, J.; Rahn, M.; Schlüchter, C. Timing of early Quaternary gravel accumulation in the Swiss Alpine Foreland. *Geomorphology* **2017**, *276*, 71–85. [CrossRef]
31. Günther, D. *Der Schwarzwald und seine Umgebung. Geologie-Mineralogie-Bergbau-Umwelt und Geotourismus*; Rothe, P., Ed.; Gebr Borntraeger: Stuttgart, Germany, 2010; 302p.
32. Benn, D.I. *Clast Morphology. A Practical Guide to the Study of Glacial Sediments*; Evans, D.J.A., Benn, D.I., Eds.; Routledge, Taylor & Francis Group: London, UK, 2004; pp. 78–92.
33. Cailleux, A. L'indice d'émoussé: Définition et première application. *Société Géologique Fr.* **1947**, *13*, 250–252.
34. Sneed, E.D.; Folk, R.L. Pebbles in the Lower Colorado River, Texas a study in particle morphogenesis. *J. Geol.* **1958**, *66*, 114–150. [CrossRef]
35. Cailleux, A.; Tricart, J. *Initiation a l'étude des Sables et des Galets (3 Volumes)*; Centre de Documentation Universitaire: Paris, France, 1959; 772p.
36. Boulton, G.S. Boulder shapes and grain-size distributions of debris as indicators of transport paths through a glacier and till genesis. *Sedimentology* **1978**, *25*, 773–799. [CrossRef]
37. Benn, D.I.; Ballantyne, C.K. Reconstructing the transport history of glacigenic sediments: A new approach based on the co-variance of clast form indices. *Sediment. Geol.* **1994**, *91*, 215–227. [CrossRef]

38. Lukas, S.; Graf, A.; Coray, S.; Schlüchter, C. Genesis, stability and preservation potential of large lateral moraines of Alpine valley glaciers-towards a unifying theory based on Findelengletscher, Switzerland. *Quat. Sci. Rev.* **2012**, *38*, 27–48. [CrossRef]
39. Chandler, D.M.; Hubbard, B. Quantifying sample bias in clast fabric measurements. *Sedimentology* **2008**, *55*, 925–938. [CrossRef]
40. Boggs, S. *Petrology of Sedimentary Rocks*, 2nd ed.; Cambridge University Press: Cambridge, UK, 2009; ISBN 9780521897167.
41. Balco, G.; Rovey, C.W. An isochron method for cosmogenic-nuclide dating of buried soils and sediments. *Am. J. Sci.* **2008**, *308*, 1083–1114. [CrossRef]
42. Knudsen, M.F.; Nørgaard, J.; Grischott, R.; Kober, F.; Egholm, D.L.; Hansen, T.M.; Jansen, J.D. New cosmogenic nuclide burial-dating model indicates onset of major glaciations in the Alps during Middle Pleistocene Transition. *Earth Planet. Sci. Lett.* **2020**, *549*, 116491. [CrossRef]
43. van Buuren, U.; Prins, M.A.; Wang, X.; Stange, M.; Yang, X.; van Balen, R.T. Fluvial or aeolian? Unravelling the origin of the silty clayey sediment cover of terraces in the Hanzhong Basin (Qinling Mountains, central China). *Geomorphology* **2020**, *367*, 107294. [CrossRef]
44. Darling, A.L.; Karlstrom, K.E.; Granger, D.E.; Aslan, A.; Kirby, E.; Ouimet, W.B.; Lazear, G.D.; Coblentz, D.D.; Cole, R.D. New incision rates along the Colorado River system based on cosmogenic burial dating of terraces: Implications for regional controls on Quaternary incision. *Geosphere* **2012**, *8*, 1020–1041. [CrossRef]
45. Erlanger, E.D.; Granger, D.E.; Gibbon, R.J. Rock uplift rates in South Africa from isochron burial dating of fluvial and marine terraces. *Geology* **2012**, *40*, 1019–1022. [CrossRef]
46. Balco, G.; Soreghan, G.S.; Sweet, D.E.; Marra, K.R.; Bierman, P.R. Cosmogenic-nuclide burial ages for Pleistocene sedimentary fill in Unaweep Canyon, Colorado, USA. *Quat. Geochronol.* **2013**, *18*, 149–157. [CrossRef]
47. Çiner, A.; Doğan, U.; Yildirim, C.; Akçar, N.; Ivy-Ochs, S.; Alfimov, V.; Kubik, P.W.; Schlüchter, C. Quaternary uplift rates of the Central Anatolian Plateau, Turkey: Insights from cosmogenic isochron-burial nuclide dating of the Kizilirmak River terraces. *Quat. Sci. Rev.* **2015**, *107*, 81–97. [CrossRef]
48. Bender, A.M.; Amos, C.B.; Bierman, P.; Rood, D.H.; Staisch, L.; Kelsey, H.; Sherrod, B. Differential uplift and incision of the Yakima River terraces, central Washington State. *J. Geophys. Res. Solid Earth* **2016**. [CrossRef]
49. Schaller, M.; Ehlers, T.A.; Stor, T.; Torrent, J.; Lobato, L.; Christl, M.; Vockenhuber, C. Timing of European fluvial terrace formation and incision rates constrained by cosmogenic nuclide dating. *Earth Planet. Sci. Lett.* **2016**, *451*, 221–231. [CrossRef]
50. Zhao, Z.; Granger, D.; Zhang, M.; Kong, X.; Yang, S.; Chen, Y.; Hu, E. A test of the isochron burial dating method on fluvial gravels within the Pulu volcanic sequence, West Kunlun Mountains, China. *Quat. Geochronol.* **2016**, *34*, 75–80. [CrossRef]
51. Akçar, N.; Ivy-Ochs, S.; Alfimov, V.; Schlunegger, F.; Claude, A.; Reber, R.; Christl, M.; Vockenhuber, C.; Dehnert, A.; Rahn, M.; et al. Isochron-burial dating of glaciofluvial deposits: First results from the Swiss Alps. *Earth Surf. Process. Landf.* **2017**, *42*, 2414–2425. [CrossRef]
52. Tu, H.; Shen, G.; Granger, D.; Yang, X.; Lai, Z. Isochron 26Al/10Be burial dating of the Lantian hominin site at Gongwangling in Northwestern China. *Quat. Geochronol.* **2017**, *41*, 174–179. [CrossRef]
53. Litty, C.; Schlunegger, F.; Akçar, N.; Lanari, P.; Christl, M.; Vockenhuber, C. Possible climatic controls on the accumulation of Peru's most prominent alluvial fan: The Lima Conglomerate. *Earth Surf. Process. Landf.* **2018**, *44*, 991–1003. [CrossRef]
54. Claude, A.; Akçar, N.; Ivy-Ochs, S.; Schlunegger, F.; Kubik, P.W.; Christl, M.; Vockenhuber, C.; Kuhlemann, J.; Rahn, M.; Schlüchter, C. Changes in landscape evolution patterns in the northern Swiss Alpine Foreland during the mid-Pleistocene revolution. *Geol. Soc. Am. Bull.* **2019**, *131*, 2056–2078. [CrossRef]
55. Granger, D.E.; Muzikar, P.F. Dating sediment burial with in situ-produced cosmogenic nuclides: Theory, techniques, and limitations. *Earth Planet. Sci. Lett.* **2001**, *188*, 269–281. [CrossRef]
56. Maxeiner, S.; Synal, H.A.; Christl, M.; Suter, M.; Müller, A.; Vockenhuber, C. Proof-of-principle of a compact 300 kV multi-isotope AMS facility. *Nucl. Instrum. Methods Phys. Res. Sect. B Beam Interact. Mater. At.* **2019**, *439*, 84–89. [CrossRef]
57. Christl, M.; Gautschi, P.; Synal, H.A. 10Be on the 0.3 MV MILEA System. In *Annual Report 2020–Laboratory of Ion Beam Physics*; Librum Publishers and Editors: Basel, Switzerland, 2021; p. 12.
58. Gautschi, P.; Vockenhuber, C.; Synal, H.A. 26Al on the 0.3 MV MILEA System. In *Annual Report 2020–Laboratory of Ion Beam Physics*; Librum Publishers and Editors: Basel, Switzerland, 2021; p. 13.
59. Balco, G.; Stone, J.O.; Lifton, N.A.; Dunai, T.J. A complete and easily accessible means of calculating surface exposure ages or erosion rates from 10Be and 26Al measurements. *Quat. Geochronol.* **2008**, *3*, 174–195. [CrossRef]
60. Borchers, B.; Marrero, S.; Balco, G.; Caffee, M.; Goehring, B.; Lifton, N.; Nishiizumi, K.; Phillips, F.; Schaefer, J.; Stone, J. Geological calibration of spallation production rates in the CRONUS-Earth project. *Quat. Geochronol.* **2016**, *31*, 188–198. [CrossRef]
61. Lal, D. Cosmic ray labeling of erosion surfaces: In situ nuclide production rates and erosion models. *Earth Planet. Sci. Lett.* **1991**, *104*, 424–439. [CrossRef]
62. Stone, J.O. Air pressure and cosmogenic isotope production. *J. Geophys. Res. Solid Earth* **2000**, *105*, 753–759. [CrossRef]
63. Chmeleff, J.; von Blanckenburg, F.; Kossert, K.; Jakob, D. Determination of the10Be half-life by multicollector ICP-MS and liquid scintillation counting. *Nucl. Instrum. Methods Phys. Res. Sect. B: Beam Interact. Mater. At.* **2010**, *268*, 192–199. [CrossRef]
64. Korschinek, G.; Bergmaier, A.; Faestermann, T.; Gerstmann, U.C.; Knie, K.; Rugel, G.; Wallner, A.; Dillmann, I.; Dollinger, G.; von Gostomski, C.L.; et al. A new value for the half-life of 10Be by Heavy-Ion Elastic Recoil Detection and liquid scintillation counting. *Nucl. Instrum. Methods Phys. Res. Sect. B Beam Interact. Mater. At.* **2010**, *268*, 187–191. [CrossRef]
65. Norris, T.L.; Gancarz, A.J.; Rokop, D.J.; Thomas, K.W. Half-Life of 26Al. *J. Geophys. Res.* **1983**, *88*, B331–B333. [CrossRef]

66. Nishiizumi, K. Preparation of 26Al AMS standards. *Nucl. Instrum. Methods Phys. Res. Sect. B Beam Interact. Mater. At.* **2004**, *223–224*, 388–392. [CrossRef]
67. Braucher, R.; Merchel, S.; Borgomano, J.; Bourlès, D.L. Production of cosmogenic radionuclides at great depth: A multi element approach. *Earth Planet. Sci. Lett.* **2011**, *309*, 1–9. [CrossRef]
68. Braucher, R.; Bourlès, D.; Merchel, S.; Vidal Romani, J.; Fernadez-Mosquera, D.; Marti, K.; Léanni, L.; Chauvet, F.; Arnold, M.; Aumaître, G.; et al. Determination of muon attenuation lengths in depth profiles from in situ produced cosmogenic nuclides. Nucl. Instruments Methods Phys. Res. Sect. B Beam Interact. with Mater. *Atoms* **2013**, *294*, 484–490. [CrossRef]
69. Margreth, A.; Gosse, J.C.; Dyke, A.S. Quantification of subaerial and episodic subglacial erosion rates on high latitude upland plateaus: Cumberland Peninsula, Baffin Island, Arctic Canada. *Quat. Sci. Rev.* **2016**, *133*, 108–129. [CrossRef]
70. Lifton, N.; Sato, T.; Dunai, T.J. Scaling in situ cosmogenic nuclide production rates using analytical approximations to atmospheric cosmic-ray fluxes. *Earth Planet. Sci. Lett.* **2014**, *386*, 149–160. [CrossRef]
71. swisstopo. *Tektonische Karte der Schweiz 1:500,000*; Bundesamt für Landestopographie, Swisstopo: Wabern, Switzerland, 2005.
72. Pfiffner, A.O. *Geologie der Alpen*; 2. Auflage; Haupt Verlag: Bern, Switzerland, 2010; ISBN 978-3-8252-8416-9.
73. Weissert, H.; Stössel, I. *Der Ozean im Gebirge: Eine geologische Zeitreise durch die Schweiz*; 2. Auflage; VDF Hochschulverlag AG an der ETH Zürich: Zurich, Switzerland, 2010.
74. Pavoni, N. Geologie der Zürcher Molasse zwischen Albiskamm und Pfannenstiel. *Vierteljahresschr. Nat. Ges. Zürich* **1957**, *102*, 119–315.
75. Matter, A. Sedimentologische Untersuchungen im östlichen Napfgebiet. *Ecol. Geol. Helv.* **1964**. [CrossRef]
76. Sartori, M.; Gouffon, Y.; Marthaler, M. Harmonisation et définition des unités lithostratigraphiques briançonnaises dans les nappes penniques du Valais. *Eclogae Geol. Helv.* **2006**, *99*, 363–407. [CrossRef]
77. Reichelt, G. Zur Frage einer Rissvereisung des Südschwarzwaldes. *Erdkunde* **1960**, *14*, 53–58. [CrossRef]
78. Pfannenstiel, M.; Rahm, G. Die Vergletscherung des Wehratales und der Wiesetäler während der Rißeiszeit. *Ber. Naturf. Ges. Freibg.* **1964**, *54*, 209–278.
79. Leser, H. Zur Glazialproblematik auf Blatt Freiburg-Süd der Geomorphologischen Karte 1: 100 000 der Bundesrepublik Deutschland (GMK 100, Blatt 2). *Eiszeitalt. Ggw.* **1987**, *37*, 139–144. [CrossRef]
80. Schreiner, A. Zur Quartärgeologie des unteren Wehratals und zur Frage der Vergletscherung des Dinkelberges in der Rißeiszeit (SW Deutschland). *E&G Quat. Sci. J.* **1995**, *45*, 62–74. [CrossRef]
81. Suter, H.H. Zur Petrographie des Grundgebirges von Laufenburg und Umgebung (Südschwarzwald). Ph.D. Thesis, University of Zurich, Zurich, Switzerland, 1924.
82. Schlüchter, C. *Geologische Untersuchungen im Quartär des Aaretals südlich von Bern. Beiträge zur Geol. Karte der Schweiz*; Bundesamt für Landestopographie, Swisstopo: Wabern, Switzerland, 1976; Volume 148.
83. Anselmetti, F.S.; Drescher-Schneider, R.; Furrer, H.; Graf, H.R.; Lowick, S.E.; Preusser, F.; Riedi, M.A. A ~180,000 years sedimentation history of a perialpine overdeepened glacial trough (Wehntal, N-Switzerland). *Swiss J. Geosci.* **2010**, *103*, 345–361. [CrossRef]
84. Dehnert, A.; Lowick, S.E.; Preusser, F.; Anselmetti, F.S.; Drescher-Schneider, R.; Graf, H.R.; Heller, F.; Horstmeyer, H.; Kemna, H.A.; Nowaczyk, N.R.; et al. Evolution of an overdeepened trough in the northern Alpine Foreland at Niederweningen, Switzerland. *Quat. Sci. Rev.* **2012**, *34*, 127–145. [CrossRef]
85. Büchi, M.W.; Graf, H.R.; Haldimann, P.; Lowick, S.E.; Anselmetti, F.S. Multiple Quaternary erosion and infill cycles in overdeepened basins of the northern Alpine foreland. *Swiss J. Geosci.* **2018**, *111*, 133–167. [CrossRef]
86. Büchi, M.W.; Lowick, S.E.; Anselmetti, F.S. Luminescence dating of glaciolacustrine silt in overdeepened basin fills beyond the last interglacial. *Quat. Geochronol.* **2017**, *37*, 55–67. [CrossRef]
87. Müller, D.; Preusser, F.; Büchi, M.W.; Gegg, L.; Deplazes, G. Luminescence properties and dating of glacial to periglacial sediments from northern Switzerland. *Geochronology* **2020**, *2*, 305–323. [CrossRef]
88. Lowick, S.E.; Büchi, M.W.; Gaar, D.; Graf, H.R.; Preusser, F. Luminescence dating of Middle Pleistocene proglacial deposits from northern Switzerland: Methodological aspects and stratigraphical conclusions. *Boreas* **2015**, *44*, 459–482. [CrossRef]
89. Klasen, N.; Fiebig, M.; Preusser, F. Applying luminescence methodology to key sites of Alpine glaciations in Southern Germany. *Quat. Int.* **2016**, *420*, 249–258. [CrossRef]
90. Rhodes, E.J. Optically stimulated luminescence dating of sediments over the past 200,000 years. *Annu. Rev. Earth Planet. Sci.* **2011**, *39*, 461–488. [CrossRef]
91. Chapot, M.S.; Roberts, H.M.; Duller, G.A.T.; Lai, Z.P. A comparison of natural- and laboratory-generated dose response curves for quartz optically stimulated luminescence signals from Chinese Loess. *Radiat. Meas.* **2012**, *47*, 1045–1052. [CrossRef]
92. Wallinga, J.; Cunningham, A.C. Encyclopedia of Scientific Dating Methods. *Encycl. Sci. Dating Methods* **2015**, 1–9. [CrossRef]
93. Rentzel, P.; Preusser, F.; Pümpin, C.; Wolf, J.J. Loess and palaeosols on the High Terrace at Sierentz (France), and implications for the chronology of terrace formation in the Upper Rhine Graben. *Swiss J. Geosci.* **2009**, *102*, 387–401. [CrossRef]
94. Claude, A.; Akçar, N.; Ivy-Ochs, S.; Schlunegger, F.; Rentzel, P.; Pümpin, C.; Tikhomirov, D.; Kubik, P.W.; Vockenhuber, C.; Dehnert, A.; et al. Chronology of Quaternary terrace deposits at the locality Hohle Gasse (Pratteln, NW Switzerland). *Swiss J. Geosci.* **2017**, *110*, 793–809. [CrossRef]
95. Klasen, N. Lumineszenzdatierung Glazifluvialer Sedimente im Nördlichen Alpenvorland ("Luminescence Dating of Glaciofluvial Sediments in the Northern Alpine Foreland"). Ph.D. Thesis, Universität zu Köln, Köln, Germany, 2008.

96. Preusser, F.; Büschelberger, M.; Kemna, H.A.; Miocic, J.; Mueller, D.; May, J.-H. Exploring possible links between Quaternary aggradation in the Upper Rhine Graben and the glaciation history of northern Switzerland. *Int. J. Earth Sci.* **2021**. [CrossRef]
97. Graf, A.; Akçar, N.; Ivy-Ochs, S.; Strasky, S.; Kubik, P.W.; Christl, M.; Burkhard, M.; Wieler, R.; Schlüchter, C. Multiple advances of Alpine glaciers into the Jura Mountains in the Northwestern Switzerland. *Swiss J. Geosci.* **2015**, *108*, 225–238. [CrossRef]
98. Schielein, P.; Schellmann, G.; Lomax, J.; Preusser, F.; Fiebig, M. Chronostratigraphy of the Hochterrassen in the lower Lech Valley (Northern Alpine Foreland). *E&G Quat. Sci. J.* **2015**, *64*, 15–28. [CrossRef]
99. Bickel, L.; Lüthgens, C.; Lomax, J.; Fiebig, M. Luminescence dating of glaciofluvial deposits linked to the penultimate glaciation in the Eastern Alps. *Quat. Int.* **2015**, *357*, 110–124. [CrossRef]
100. Bickel, L.; Lüthgens, C.; Lomax, J.; Fiebig, M. The timing of the penultimate glaciation in the northern Alpine Foreland: New insights from luminescence dating. *Proc. Geol. Assoc.* **2015**, *126*, 536–550. [CrossRef]
101. Monegato, G.; Lowick, S.E.; Ravazzi, C.; Banino, R.; Donegana, M.; Preusser, F. Middle to Late Pleistocene palaeoenvironmental evolution of the southeastern Alpine Valeriano Creek succession (northeastern Italy). *J. Quat. Sci.* **2010**, *25*, 617–632. [CrossRef]
102. Monegato, G.; Poli, M.E. The Pleistocene activity of the eastern Southalpine chain unravelled by the terraced staircase of the Meduna valley (NE Italy). *Rend. Online Soc. Geol. Ital.* **2013**, *29*, 108–111.
103. Gianotti, F.; Forno, M.G.; Ivy-Ochs, S.; Monegato, G.; Pini, R.; Ravazzi, C. Stratigraphy of the Ivrea Morainic Amphitheatre (NW ITALY): An updated synthesis. *Alp. Mediterr. Quat.* **2015**, *28*, 29–58.
104. Buoncristiani, J.F.; Campy, M. Quaternary glaciations in the French Alps and Jura. In *Developments in Quaternary Science*; Elsevier Inc.: Amsterdam, The Netherlands, 2011; Volume 15, pp. 117–126. ISBN 9780444534477.
105. Maslin, M.A.; Ridgwell, A.J. Mid-Pleistocene revolution and the 'eccentricity myth'. *Geol. Soc. Lond. Spec. Publ.* **2005**, *247*, 19–34. [CrossRef]
106. Ehlers, J.; Gibbard, P.L. The extent and chronology of Cenozoic Global Glaciation. *Quat. Int.* **2007**, *164–165*, 6–20. [CrossRef]
107. Eissmann, L. Quaternary geology of eastern Germany (Saxony, Saxon-Anhalt, South Brandenburg, Thüringia), type area of the Elsterian and Saalian Stages in Europe. *Quat. Sci. Rev.* **2002**, *21*, 1275–1346. [CrossRef]
108. Lauer, T.; Weiss, M. Timing of the Saalian- and Elsterian glacial cycles and the implications for Middle-Pleistocene hominin presence in central Europe. *Sci. Rep.* **2018**, *8*, 1–13. [CrossRef]
109. Roskosch, J.; Winsemann, J.; Polom, U.; Brandes, C.; Tsukamoto, S.; Weitkamp, A.; Bartholomäus, W.A.; Henningsen, D.; Frechen, M. Luminescence dating of ice-marginal deposits in northern Germany: Evidence for repeated glaciations during the Middle Pleistocene (MIS 12 to MIS 6). *Boreas* **2015**, *44*, 103–126. [CrossRef]
110. Laban, C.; van der Meer, J.J.M. Pleistocene glaciation in the Netherlands. In *Developments in Quaternary Science*; Elsevier Inc.: Amsterdam, The Netherlands, 2011; pp. 247–260. Volume 15, ISBN 9780444534477.
111. Hughes, P.D.; Gibbard, P.L. Global glacier dynamics during 100 ka Pleistocene glacial cycles. *Quat. Res.* **2018**, *90*, 222–243. [CrossRef]
112. Preusser, F.; Reitner, J.M.; Schlüchter, C. Distribution, geometry, age and origin of overdeepened valleys and basins in the Alps and their foreland. *Swiss J. Geosci.* **2010**, *103*, 407–426. [CrossRef]
113. Pawley, S.M.; Bailey, R.M.; Rose, J.; Moorlock, B.S.P.; Hamblin, R.J.O.; Booth, S.J.; Lee, J.R. Age limits on Middle Pleistocene glacial sediments from OSL dating, north Norfolk, UK. *Quat. Sci. Rev.* **2008**, *27*, 1363–1377. [CrossRef]
114. Pawley, S.M.; Toms, P.; Armitage, S.J.; Rose, J. Quartz luminescence dating of Anglian Stage (MIS 12) fluvial sediments: Comparison of SAR age estimates to the terrace chronology of the Middle Thames valley, UK. *Quat. Geochronol.* **2010**, *5*, 569–582. [CrossRef]
115. Astakhov, V. Ice Margins of Northern Russia revisited. In *Developments in Quaternary Science*; Elsevier Inc.: Amsterdam, The Netherlands, 2011; Volume 15, pp. 323–336. ISBN 9780444534477.
116. Velichko, A.A.; Faustova, M.A.; Pisareva, V.V.; Gribchenko, Y.U.N.; Sudakova, N.G.; Lavrentiev, N.V. Glaciations of the East European Plain: Distribution and Chronology. *Dev. Quat. Sci.* **2011**, *15*, 337–359. [CrossRef]
117. Woodward, J.C.; Hughes, P.D. Glaciation in Greece. A new record of cold stage environments in the Mediterranean. In *Developments in Quaternary Sciences*; Elsevier Inc.: Amsterdam, The Netherlands, 2011; Volume 15, pp. 175–198. ISBN 9780444534477.
118. Hughes, P.D.; Woodward, J.C. Quaternary glaciation in the Mediterranean mountains: A new synthesis. *Geol. Soc. Spec. Publ.* **2017**, *433*, 1–23. [CrossRef]
119. Hughes, P.D.; Woodward, J.C.; Gibbard, P.L.; Macklin, M.G.; Gilmour, M.A.; Smith, G.R. The glacial history of the Pindus Mountains, Greece. *J. Geol.* **2006**, *114*, 413–434. [CrossRef]
120. Hughes, P.D.; Woodward, J.C.; van Calsteren, P.C.; Thomas, L.E. The glacial history of the Dinaric Alps, Montenegro. *Quat. Sci. Rev.* **2011**, *30*, 3393–3412. [CrossRef]
121. Duk-Rodkin, A.; Barendregt, R.W.; Froese, D.G.; Weber, F.; Enkin, R.; Rod Smith, I.; Zazula, G.D.; Waters, P.; Klassen, R. Timing and extent of Plio-Pleistocene glaciations in north-western Canada and east-central Alaska. In *Quaternary Glaciations-Extent and Chronology, Part II.*; Ehlers, J., Gibbard, P.L., Eds.; Elsevier B.V.: Amsterdam, The Netherlands, 2004; Volume 2, pp. 313–345. ISBN 9780444515926.
122. Rovey, C.W.; Siemens, M. Age constraints of the Middle Pleistocene till and loess sequence in northeast Missouri, USA, based on pedostratigraphy within a polygenetic paleosol. *Catena* **2021**, *203*, 105294. [CrossRef]
123. Zhou, S.; Li, J.; Zhao, J.; Wang, J.; Zheng, J. Quaternary Glaciations: Extent and Chronology in China. In *Developments in Quaternary Science*; Elsevier Inc.: Amsterdam, The Netherlands, 2011; Volume 15, pp. 981–1002. ISBN 9780444534477.

124. Zhou, S.; Wang, X.; Wang, J.; Xu, L. A preliminary study on timing of the oldest Pleistocene glaciation in Qinghai-Tibetan Plateau. *Quat. Int.* **2006**, *154–155*, 44–51. [CrossRef]
125. Xu, L.; Zhou, S. Quaternary glaciations recorded by glacial and fluvial landforms in the Shaluli Mountains, Southeastern Tibetan Plateau. *Geomorphology* **2009**, *103*, 268–275. [CrossRef]

geosciences

MDPI

Article

LGM Glaciations in the Northeastern Anatolian Mountains: New Insights

Regina Reber [1,*], Naki Akçar [1], Dmitry Tikhomirov [1,2], Serdar Yesilyurt [1,3], Christof Vockenhuber [4], Vural Yavuz [5], Susan Ivy-Ochs [4] and Christian Schlüchter [1]

1 Institute of Geological Sciences, University of Bern, Baltzerstrasse 1+3, 3012 Bern, Switzerland; naki.akcar@geo.unibe.ch (N.A.); dmitry.tikhomirov@geo.uzh.ch (D.T.); yesilyurt@ankara.edu.tr (S.Y.); christian.schluechter@geo.unibe.ch (C.S.)
2 Department of Geography, University of Zurich, Winterthurerstrasse 190, 8057 Zürich, Switzerland
3 Department of Geography, Ankara University, 06230 Ankara, Turkey
4 Institute for Particle Physics, ETH Hönggerberg, Schafmattstrasse 20, 8093 Zürich, Switzerland; vockenhuber@phys.ethz.ch (C.V.); ivy@phys.ethz.ch (S.I.-O.)
5 Civil Engineering Department, Turkish-German University, 34820 Istanbul, Turkey; vural.yavuz@tau.edu.tr
* Correspondence: regina.reber@geo.unibe.ch

Abstract: Barhal Valley belongs to the Çoruh Valley System in the Kaçkar Mountains of northeastern Anatolia. This 13 km long valley is located to the south of the main weather divide and to the east of Mt. Kaçkar, with the highest peak of the mountain range being 3932 m. Today, source of an average yearly precipitation of 2000 mm of moisture is the Black Sea, situated approximately 40 km to the north of the study site. Glaciers of the Last Glacial Maximum (LGM) descended directly from Mt. Kaçkar and reached an altitude of ca. 1850 m a.s.l. (above sea level). In this study, we are exploring whether the position of Barhal Valley to the south of the main weather divide and its east–west orientation have an influence on the existence and expansion of paleoglaciers. Here, we present 32 new cosmogenic ^{36}Cl dates on erratic boulders from the Çoruh Valley System. We reconstructed three geomorphologically well-contained glacier advances in the Barhal Valley, namely at 34.0 ± 2.3 ka, 22.2 ± 2.6 ka, and 18.3 ± 1.7 ka within the time window of the global LGM. Field evidence shows that the glacier of the 18.3 ± 1.7 ka advance disappeared rapidly and that by the latest time, at 15.6 ± 1.8 ka, the upper cirques were ice-free. No evidence for Lateglacial glacier fluctuations was found, and the Neoglacial activity is restricted to the cirques with rock glaciers. A range of 2700 to 3000 m for the Equilibrium Line Altitude (ELA) at the LGM was reported based on modeling of the glacial morphology. We determined that the most likely position of the LGM ELA in the Çoruh Valley System was at 2900 m a.s.l. We suggest an alternative moisture source to the direct transport from the Black Sea for the ice accumulation in the Eastern Black See Mountains. The shift of the Polar Front and of the Siberian High Pressure System to the south during the LGM resulted in the domination of easterly airflow to the Caucasus and Kaçkar Mountains with moisture from expanded lakes in central–western Siberia and from the enlarged Aral- and Caspian Seas.

Keywords: erratic boulders; cosmogenic ^{36}Cl; LGM Glaciations; eastern Anatolia; glacier retreat

Citation: Reber, R.; Akçar, N.; Tikhomirov, D.; Yesilyurt, S.; Vockenhuber, C.; Yavuz, V.; Ivy-Ochs, S.; Schlüchter, C. LGM Glaciations in the Northeastern Anatolian Mountains: New Insights. *Geosciences* **2022**, *12*, 257. https://doi.org/10.3390/geosciences12070257

Academic Editors: James A. Smith and Jesus Martinez-Frias

Received: 16 May 2022
Accepted: 20 June 2022
Published: 22 June 2022

1. Introduction

Glaciers are the key element for the reconstruction of paleocirculation patterns and, therefore, of transport of moisture to an area, as described e.g., in [1]. Today, climate indicators such as temperature, barometric pressure, and precipitation are recorded instrumentally. However, for records of the past, these need to be extracted from geoarchives. In mountain areas with sufficient elevation and precipitation, the cryosphere, which mainly comprises glaciers, provides the most sensitive archives for climate change [2]. To read these archives, fieldwork is essential, and state-of-the-art dating techniques need to be applied. Glaciers react dynamically and rapidly to physical changes to the environment, namely to temperature, and they provide long-term records of thousand, ten thousand, and hundred thousand years, as described e.g., in [3]. The disadvantage of glacial records is their complex sedimentary structure, especially in mountainous areas. Fieldwork, therefore, can be challenging as glacial landforms and sediments in mountains are subject to gravity-induced surface reorganization and instabilities [4]. Well-preserved glacial landforms (moraines) or sediments (tills) may be rare, and careful evaluation of erratic boulders is needed with respect to their morphological stability since deposition [5].

The mountains of Turkey and their paleoclimate records constitute a critical link between the Alps, the Balkans, and the southwest Asian mountain ranges. For this reason, Anatolia is ideally located within a zone of frontal weather dynamics and seasonal oscillations in the eastern Mediterranean (Figure 3 in [6] and [7]). The Eastern Black Sea Mountain Range in northeastern Anatolia is at close proximity to the direct moisture source of the Black Sea. For example, the Kavron Valley in the Kaçkar Mountains (elevation of 3932 m and 40 km from the coast only) receives an average annual precipitation of 1784 mm, and the coastal city of Rize receives 1989 mm [8]. In contrast, Erzurum, the city located ca. 140 km from the coast and to the south of the main mountain range, receives only 676 mm. The precipitation differences between Rize at the coast, the high mountains of Kaçkar, and inland Erzurum is a typical orography-controlled situation. With the present circulation, moisture arrives directly from the Black Sea and shows a steep precipitation gradient on the southern side of the mountain range; see, e.g., [7].

Few and small relict glaciers only exist in the Kaçkar Mountains of NE Anatolia today ([9] and references therein). This is in marked contrast to the landscape during the Last Glacial Maximum when extensive valley glaciers descended at least twice from the cirque areas. We have investigated key valleys from a glacial geological point of view over the past two decades in the Kaçkar Mountains to reconstruct the past glaciers' advances (Figure 1a).

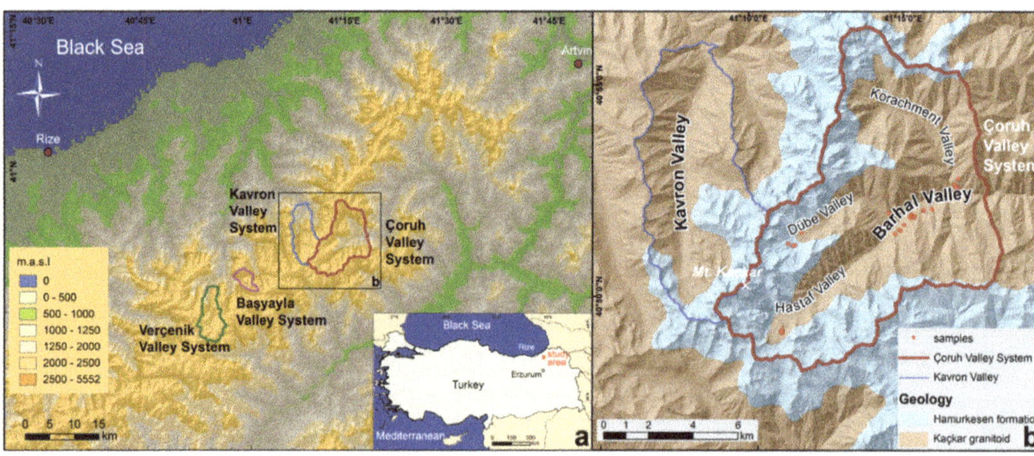

Figure 1. (**a**) Index map for the Çoruh Valley System in the Kaçkar Mountains of NE Anatolia: Kavron Valley System [10], Verçenik Valley System [11], Başyayla Valley System [12], Çoruh Valley system (this paper). (**b**) Neighboring Kavron and Barhal Valley descending from the highest peak of Mt. Kaçkar 3932 m. Çoruh Valley System with Hastaf, Dübe, and Körahmet tributaries, and Barhal Valley, referred to as the main valley (sample clusters red dots).

Based on literature surveys ([6] and references therein) and contacts with geologists working in the area, we started in Kavron Valley [10], followed by Verçenik Valley [11], and most recently Başyayla Valley [12]. All these valleys bear clear and extensive evidence of former glaciers with comparable chronologies from valley to valley. Kavron, Verçenik, and Başyayla Valleys, where our investigations were focused earlier, are open to the north and are, therefore, direct collectors of moisture moving from the Black Sea to the high Kaçkar Mountains.

The target of this study is to investigate the existence, size, and age of potential paleoglaciers to the south of the main mountain divide and, as a consequence, subject to potentially changing atmospheric circulation during glaciations (Figure 1). Additionally, it is a follow-up study of earlier investigations conducted in neighboring valleys in the north [10,11]. Therefore, the east-trending Barhal Valley, located in the southern side of the Kaçkar Mountains, was selected with a special focus on the geometry of the LGM glacier extension. Did an LGM-glaciation really occur in Barhal Valley? Additionally, then, what is the paleoclimatic context of an important glacier to the south of the main divide, but still directly connected to Mt. Kaçkar, the highest peak of the mountain range? In this study, we report field investigations, sampling campaigns, and terrestrial cosmogenic ^{36}Cl analysis of the resting time of erratic boulders in Barhal Valley.

2. Field Area and Field Work

2.1. Study Area

The Barhal Valley is one of the main tributaries of the Çoruh Valley System. The uppermost part of the southwest–northeast-trending Barhal Valley is composed of three main tributaries. Yaylalar Village is at the junction of the Körahmet Valley from the north and of Hastaf Valley from the west. Olgunlar, the uppermost village, is at the junction of Dübe Valley from the north with Hastaf Valley from the west (Figures 1b and 2). Both Körahmet and Dübe Valleys are southwest–northeast trending in the upper reaches and join the main Barhal Valley after they turn to the east at right angle. Dübe and Hastaf Valleys originate in the eastern flanks of Mt. Kaçkar, the highest peak of the mountain range (Figures 1 and 2).

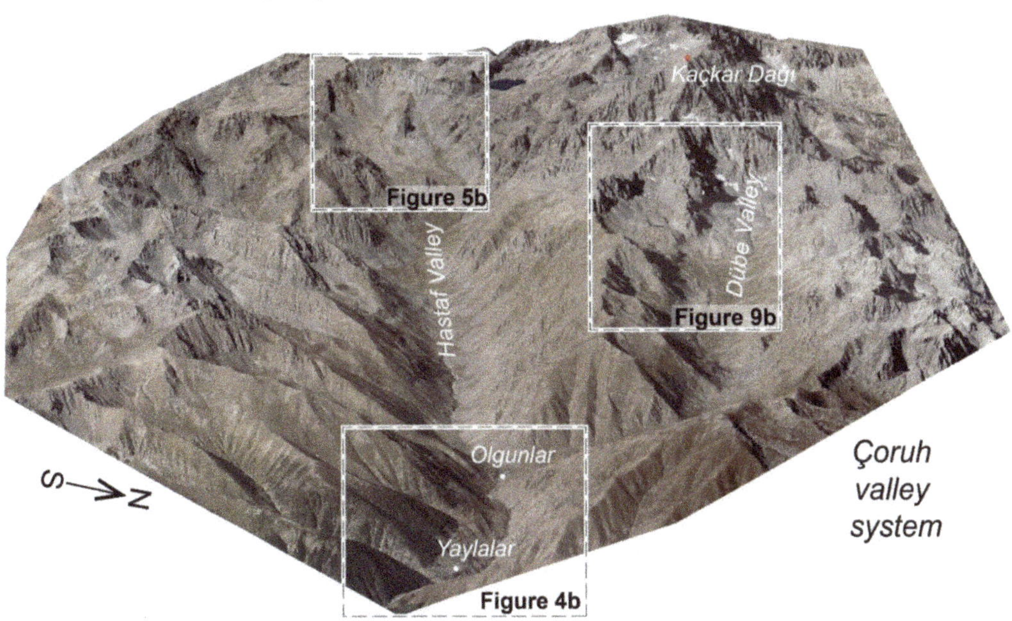

Figure 2. The Çoruh Valley System from an oblique aerial view based on orthophoto, to the east, and southeast of Mt. Kaçkar.

The Barhal Valley is a typical and broad U-shape valley in the sector of former glacier extensions (Figure 2). The open valley morphology ends about half a kilometer down valley from Yaylalar Village in a gorge. The distance from the highest peak to the gorge entrance is about 13 km. The bedrock geology is complex (Figure 1b): Mt. Kaçkar is at the contact of Upper Cretaceous volcanic and volcanoclastic rocks (to the east and northeast; Hamurkesen formation, Figure 1b) and of granitic rocks of the Kaçkar composite batholith of the Late Cretaceous-Eocene age. The Kaçkar batholith complex is part of the eastern Pontide igneous terrain [13,14]. The rocks of this batholith are a wide range of quartz-rich granitoids in a complex tectonic interrelationship.

2.2. Sampling

Field campaigns took place in August 2010 and in August 2013. We collected a total of 31 surface samples from erratic boulder tops and one sample from a decomposing bedrock ridge beyond the extent of glaciations (Table 1) for cosmogenic surface exposure dating. After careful inspection of the boulder surface and lithology, boulder stability and its relation to an ice-contact sediment or morphology, sampling was performed using a hammer and chisel following Akçar et al. [15].

Table 1. Sample location and description.

Sample Name	Altitude (m)	Latitude, °N (DD.DD WGS84)	Longitude, °E (DD.DD WGS84)	Boulder Height (cm)	Sample Thickness (cm)	Shielding Correction Factor [a]
TRYAY-1	2310	40.85623	41.24587	100	3	0.9745
TRYAY-2	2330	40.85495	41.24291	120	3	0.9815
TRYAY-3	2990	40.81733	41.1812	140	4	0.9836
TRYAY-4	2990	40.81791	41.18103	80	3	0.9887
TRYAY-5	2980	40.81851	41.18088	80 *dh 60	3	0.9872
TRYAY-6	2990	40.81759	41.17999	100	5	0.9879
TRYAY-7	2880	40.85152	41.18879	115	3	0.9727
TRYAY-8	2905	40.85141	41.1876	100	2.5	0.9624
TRYAY-9	2960	40.85221	41.18504	160	4	0.9724
TRYAY-10	2805	40.85643	41.19236	120	3.5	0.9834
TRYAY-11	2380	40.86354	41.23956	180 *dh 50	3	0.9927
TRYAY-12	1945	40.87334	41.2763	250 *dh 40	2.5	0.9595
TRYAY-13	1940	40.8733	41.27599	400 *dh 50	3	0.9804
TRYAY-14	1910	40.87257	41.27751	200	5	0.9552
TRYAY-15	2150	40.87624	41.27764	Tor	5	0.9943
TRYAY-16	2295	40.85649	41.24531	340 *dh 160	5	0.9816
TRYAY-17	2285	40.85663	41.24546	340 *dh 80	2	0.9816
TRYAY-18	2205	40.8587	41.24792	320 *dh 100	3	0.9450
TRYAY-19	2190	40.86144	41.25145	240	4	0.9768
TRYAY-20	2155	40.86191	41.25106	280	3	0.9727
TRYAY-21	2195	40.8615	41.25346	380	2	0.9685
TRYAY-22	2180	40.86205	41.25377	260	3	0.9723
TRYAY-23	2115	40.86266	41.25207	300 *dh 200	3	0.9747
TRYAY-24	2090	40.86411	41.25843	480	3	0.9768
TRYAY-25	2090	40.86429	41.25865	270 *dh 130	3	0.9768
TRYAY-26	2115	40.86452	41.26329	290 *dh 100	5	0.9729
TRYAY-27	2005	40.86704	41.26197	640	4	0.9730
TRYAY-28	1950	40.87103	41.27026	200	3	0.9591
TRYAY-29	1935	40.87314	41.27612	200	3	0.9646
TRYAY-30	1930	40.87307	41.27623	120 *dh 40	2	0.9646
TRYAY-31	1960	40.87354	41.27586	180	2	0.9630
TRYAY-32	1990	40.87418	41.27545	340	3	0.9745

*dh = difference in height from top of the boulder to the nearest sediment cover. [a] Calculated for topographic shielding and dip of the surface, following Dunne et al. [16].

2.3. Methodology and Lab Analytical Work

In this study, we analyzed the cosmogenic isotope ^{36}Cl because most of the suitable boulders for surface exposure dating are of volcanic lithologies. The interaction of Ca, K, Ti, Fe, and Cl in the mineral lattice of the rock surface with cosmic rays results in the production of ^{36}Cl [17,18]. We apply this physical principle to determine the exposure age of the sampled erratic boulders. As cosmogenic ^{36}Cl is produced through several production channels [19–25], we have determined the elemental composition of the whole rock by analyzing major and trace elements on a sample aliquot at SGS Mineral Services in Toronto, Canada (Table S1). The aliquot was taken after the samples were crushed, sieved to the fraction of 0.250–0.400 mm, and leached to avoid meteoric contamination. The extraction of cosmogenic ^{36}Cl was performed following a modified laboratory protocol by Akçar et al. [26]. In this protocol, only the carbonate fraction of the whole rock was dissolved. We combined nitric and hydrofluoric acids to fully dissolve the samples. Prior to dissolution, the samples were spiked with 2.5 mg of ^{35}Cl to apply chlorine isotope dilution method for AMS measurement at ETH [27]. Assuming a natural ratio of ^{35}Cl/^{37}Cl in the sample, the method allows determining chlorine-35, -36, and -37 concentrations, and as a result, more precise and accurate dating with ^{36}Cl [28].

The concentrations of natural Cl and ^{36}Cl were determined from one target at the ETH TANDEM AMS facility using the gas-filled magnet method to remove the isobar ^{36}S [29,30]. The ratio of ^{36}Cl/Cl was normalized to the ETH internal standard K382/4N with a value of $17.36 \pm 0.35 \cdot 10^{-12}$ [30]. The concentration of stable Cl was calculated using a ^{37}Cl/^{35}Cl ratio of 31.98% of K382/4N standard and background ratio of a machine blank. The resulting ^{36}Cl/Cl ratio varies among the samples from $0.053 \cdot 10^{-12}$ to $0.595 \cdot 10^{-12}$, while the ratio of the three preparation blanks has a range from $0.002 \cdot 10^{-12}$ to $0.005 \cdot 10^{-12}$. The final concentrations of ^{36}Cl in the rock are corrected to preparation blanks. The concentration error includes the uncertainty of the AMS standard and the blanks (Table 2).

Table 2. Cosmogenic nuclide data and calculated ^{36}Cl exposure ages.

Sample Name	Weight of Sample	Cl Conc. in Rock (ppm)	^{36}Cl Conc. (10^6 ^{36}Cl g(rock)$^{-1}$)	Erosion Corrected (ε = 1.0 mm/ka) Exposure Age (ka)
TRYAY-1	34.2269	11.45 ± 0.10	0.25 ± 0.03	16.9 ± 1.9
TRYAY-2	29.4083	26.30 ± 0.39	0.37 ± 0.02	16.8 ± 1.4
TRYAY-3	29.136	18.81 ± 0.25	0.21 ± 0.02	9.5 ± 1.0
TRYAY-4	29.0564	19.92 ± 0.13	0.44 ± 0.02	11.3 ± 0.7
TRYAY-5	27.7074	21.41 ± 0.14	0.34 ± 0.02	11.4 ± 0.9
TRYAY-6	28.6207	29.71 ± 0.15	0.92 ± 0.04	16.1 ± 0.9
TRYAY-7	28.7826	66.22 ± 0.68	0.79 ± 0.04	15.3 ± 1.3
TRYAY-8	28.6532	95.01 ± 6.46	0.92 ± 0.08	15.6 ± 1.8
TRYAY-9	30.0012	135.30 ± 0.84	1.01 ± 0.05	12.2 ± 1.2
TRYAY-10	29.1489	55.16 ± 2.59	0.78 ± 0.05	14.2 ± 1.3
TRYAY-11	29.1831	237.70 ± 9.32	1.64 ± 0.10	16.4 ± 1.8
TRYAY-12	28.4173	17.61 ± 0.26	0.41 ± 0.02	34.0 ± 2.6
TRYAY-13	34.0625	15.51 ± 0.12	0.28 ± 0.02	20.6 ± 1.7
TRYAY-14	35.4375	14.94 ± 0.09	0.15 ± 0.01	19.3 ± 1.9
TRYAY-15	28.9831	76.64 ± 1.11	0.68 ± 0.03	25.7 ± 4.0
TRYAY-16	30.1203	15.46 ± 0.17	0.16 ± 0.01	14.5 ± 1.3
TRYAY-17	30.8286	19.84 ± 0.70	0.38 ± 0.03	22.7 ± 2.0
TRYAY-18	30.3030	18.20 ± 0.29	0.48 ± 0.02	22.8 ± 1.5
TRYAY-19	30.1839	14.63 ± 0.30	0.39 ± 0.03	25.2 ± 2.2
TRYAY-20	30.1838	15.55 ± 0.10	0.40 ± 0.02	18.7 ± 1.1
TRYAY-21	30.0951	8.37 ± 0.20	0.19 ± 0.01	19.4 ± 1.7
TRYAY-22	28.0466	10.52 ± 0.32	0.19 ± 0.01	13.4 ± 1.1
TRYAY-23	27.9336	16.45 ± 0.15	0.36 ± 0.03	17.9 ± 1.5
TRYAY-24	30.3931	16.08 ± 0.11	0.37 ± 0.02	21.1 ± 1.4
TRYAY-25	30.2444	15.42 ± 0.11	0.15 ± 0.01	19.0 ± 1.9
TRYAY-26	30.2529	18.68 ± 0.24	1.31 ± 0.05	105.6 ± 8.1
TRYAY-27	30.3185	17.44 ± 0.26	0.27 ± 0.02	17.3 ± 1.6

Table 2. *Cont.*

Sample Name	Weight of Sample	Cl Conc. in Rock (ppm)	^{36}Cl Conc. (10^6 ^{36}Cl g(rock)$^{-1}$)	Erosion Corrected (ε = 1.0 mm/ka) Exposure Age (ka)
TRYAY-28	30.4064	12.55 ± 0.16	0.10 ± 0.01	9.5 ± 0.9
TRYAY-29	30.3650	16.47 ± 0.24	0.19 ± 0.02	18.8 ± 2.1
TRYAY-30	30.1825	92.66 ± 0.58	0.45 ± 0.02	14.6 ± 1.3
TRYAY-31	30.2605	15.34 ± 0.22	0.45 ± 0.02	25.6 ± 1.6
TRYAY-32	30.7316	15.86 ± 0.11	0.36 ± 0.02	33.9 ± 2.6

Analytical errors are at the 1σ level, including the statistical (counting) error and the combined counting uncertainty and uncertainty due to the normalization of standards and blanks. To calculate exposure ages, we used 48.8 ± 1.7 atoms ^{36}Cl g(Ca)$^{-1}$ a^{-1} SLHL production rate from Ca spallation, 5.3 ± 0.5 ^{36}Cl g(Ca)$^{-1}$ a^{-1} SLHL production due to muon capture ([22,23], one sigma errors), and scaled after Stone [31] to 2.47 (spallation) and 1.61 (muonic) of the SLHL values. Production rate on K, 162 at g^{-1} yr^{-1} [32]; on Ti, 13 at g^{-1} yr^{-1} [33]; and on Fe, 1.9 at g^{-1} yr^{-1} [34]. Low-energy capture of thermal and epithermal neutrons was computed following Liu et al. [19] and Phillips et al. [21] using the production rate of epithermal neutrons above the surface 760 ± 150 neutrons g^{-1} a^{-1} (see Alfimov and Ivy Ochs, [24]). Exposure ages are corrected for shielding of surrounding topography, and sample thickness.

For the calculation of the exposure ages, we used an in-house Matlab code based on Alfimov and Ivy-Ochs [24]. Sample-specific parameters are listed in Table 1. We used the following production rates of cosmogenic ^{36}Cl by spallation: on Ca 48.8 ± 1.7 atoms ^{36}Cl g(Ca)$^{-1}$ a^{-1} [22], on K 162 g^{-1} yr^{-1} [32], on Ti 13 g^{-1} yr^{-1} [33] and on Fe 1.9 g^{-1} yr^{-1} [34]. An attenuation length of high-energy neutrons of 160 g cm^{-2} [16] was used together with a rock density of 2.7 g cm^{-3}. For the production rate of epithermal and thermal neutrons in the atmosphere at the land/atmosphere interface, we used 757 n g^{-1} yr^{-1} [24]. Muonic production of ^{36}Cl was calculated following Heisinger et al. [35,36]. The local production rate was calculated with scaling scheme of Stone [31]. An erosion rate of 1 mm per thousand years was applied. The non-cosmogenic production of ^{36}Cl by neutrons from spontaneous fission of U and Th was calculated by Alfimov and Ivy-Ochs [24].

3. Results and Interpretation

Barhal Valley (Figures 1 and 2), called the main valley in the following, displays a distinct morphological break at 1850 m a.s.l. Below the break point, it has a V-shape and a narrow valley (Figure 3a), whereas it has a U-form above this point (Figure 2), and more open and broader towards the uppermost part, which is called the Hastaf Valley.

Figure 3. (**a**) View from boulder TRYAY-14 down-valley towards the gorge entrance and to the most likely ice-contact terrasses. The gorge entrance is interpreted based on the morphology and on the reconstruction of lateral ice margins (Figure 4) as the approximate LGM glacier terminus. (**b**) The perfectly perched erratic boulder TRYAY-14 (map position see Figure 4).

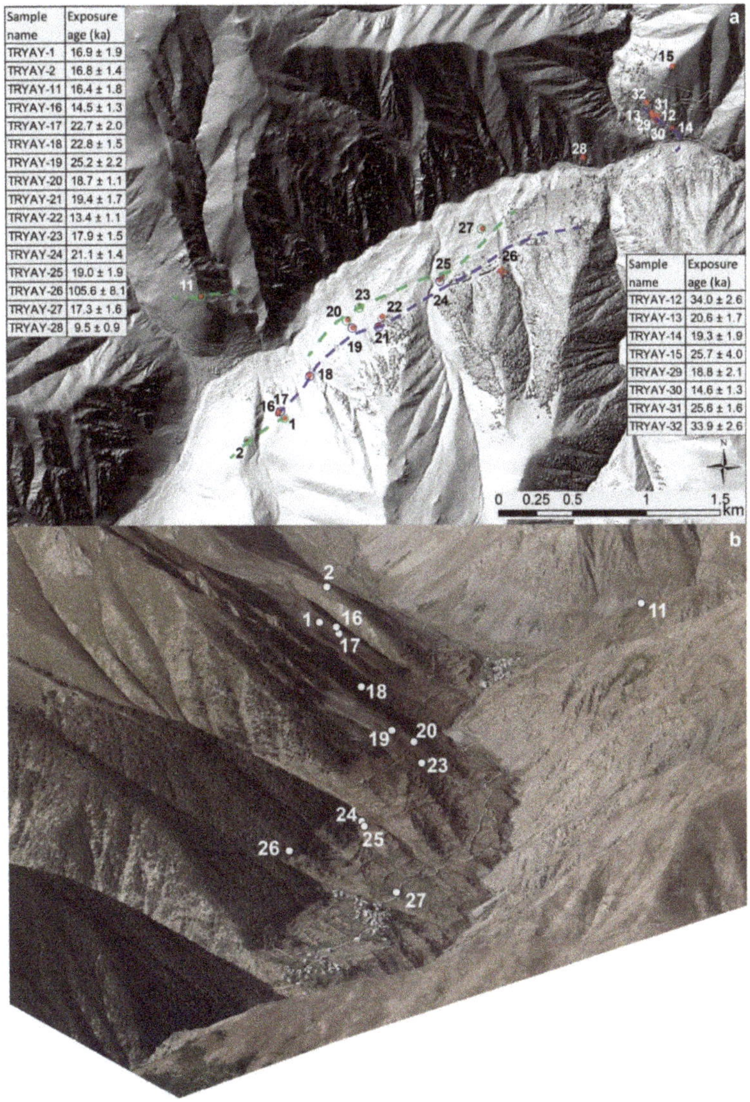

Figure 4. (**a**) Barhal Valley sample locations with summary table of cosmogenic ^{36}Cl ages. Reconstructed right lateral ice margins: lower green line = 18.3 ± 1.7 ka advance, blue line = 22.2 ± 2.6 ka advance. (**b**) Barhal Valley sample locations between Yaylalar (bottom of the map) and Olgunlar (at the valley confluence), as seen in digital relief reconstruction.

This break in valley morphology, from V- to U-shaped, is interpreted as the terminus of the LGM glacier (Figures 3a and 4) and, tentatively, of former glaciations as well.

The field evidence of this formerly large valley glacier with lateral relict moraine ridges, abraded bedrock surfaces, and abundant erratic boulders is omnipresent up-valley of the Yaylalar Gorge. There is a broad high alpine scenery surrounded by steep rock walls and peaks (e.g., Figure 5) in the upper Hastaf and the Dübe Valley. Glacial morphological features such as moraines or ice contact slopes are present in the main valley and rare in the tributaries of the Körahmet and Dübe valleys. There, only Lateglacial features are mapped (Figure 5).

Figure 5. Reconstructions based on digital elevation data of uppermost Hastaf Valley with a spectacular view into the cirque to the SE of Mt. Kaçkar. (**a**) Detailed map with sample locations. (**b**) View of detailed reconstructions of land surface expressions. Samples are from areas with little periglacial modifications.

It is difficult to morphologically constrain the confluence of a paleoglacier from Körahmet valley with the main paleoglacier one kilometer upstream of the gorge entrance.

An important paleoglacial feature is at the confluence of Körahmet and the main valley on the left-lateral slope of Körahmet: it is a boulder alignment with a nicely defined upper and a more diffuse lower limit, as can be seen from Yaylalar Village (Figure 6).

Figure 6. Left-lateral frontal position of Barhal Valley glacier, view to the N/NE from Yaylalar. The white line is the 19.7 ± 2.1 ka ice margin. TRYAY-12 and -31 represent the 34.0 ± 2.3 ka advance. For sample locations on a map, see Figure 4a. The 18.3 ± 1.7 ka advance did not reach the level of the white line at that site. Details of TRYAY-14: see Figure 3.

The position of the boulders on the slope and the overall geometry at the confluence make it a depositional feature of a glacier in the main valley because the adjoining up-valley sectors of Körahmet Valley are almost free of true glacial vestiges, and the morphology is more V-shaped (Figure 1b). We conclude, therefore, that the faint boulder terrace marks the left-lateral depositional ice margin close or corresponding to the maximum glacier extension of the LGM limit of the Barhal Paleoglacier, and that the Körahmet Paleoglacier did not reach down to the main valley.

In Figures 4 and 6, the sample locations and data for the boulder alignment are given. The results of our measurements are listed in Tables 1 and 2 and Figure 6. The seven boulder samples on this slope (TRYAY-12, -13, -14, -29, -30, -31, and -32) range from 14.6 ± 1.3 to 34.0 ± 2.6 ka. The youngest exposure age of 14.6 ± 1.3 ka (TRYAY-30) we consider as too young for a depositional age of the boulder and contemplate this surface exposure age as a post-depositional exhumation of this boulder (Table 1). Sample TRYAY-14 needs to be mentioned in particular for its methodological uniqueness and beauty (Figure 3). This boulder is delicately perched on three bedrock knobs. The boulder itself is of a subrounded form and of foreign erratic lithology. Its exposure age is 19.3 ± 1.9 ka, and its position is in the lower part of the slope. Together, TRAY-13, -14, and -29 produce a mean exposure age of 19.7 ± 2.1 ka. This exposure age allows for the decision that these samples belong to the glacier advance during the global LGM taking place at 22.1 ± 4.3 ka after Shakun and Carlson, [37]. All three boulders (TRYAY-13, -14, and -29) are situated in the lowermost part of the sampled slope (Figure 6).

TRYAY-12 and -32 are of a clearly older age (34.0 ± 2.6 and 33.9 ± 2.6 ka). The boulder surface of sample TRYAY-12 is slightly pitted, and the removal of the sample was easy due to embryonic spalling; this is in an agreement with the longer exposure of the boulder. A break in slope morphology below the sample TRYAY-12 is obvious. The boulders' location of TRYAY-12 is about 10 m above the upper limit of the "fresher-looking" boulder alignment, and TRYAY-32 is slightly up valley and upslope as well (Figure 3 and to the left and beyond in Figure 6). Samples TRYAY-12 and TRYAY-32 do not represent the boulder alignment in a strict sense and are considered to represent a different and older depositional age than the other samples at that site. The mean exposure age of 34.0 ± 2.3 ka is older than the LGM time span for the Northern Hemisphere [37]. This is a challenging situation as it relates to the question of an Early LGM or of an older independent advance. Sample TRYAY-31 is in a similar position above the boulder line as TRYAY-12 and TRYAY-32. Its statistically significant younger exposure age of 25.6 ± 1.6 ka years suggests either a later advance at around 25 ka or that this boulder was deposed with the boulder TRYAY-12 and -32, and post-depositional slope processes, not visible in the field today, are responsible for a later exhumation of this boulder. Both scenarios are possible. However, the case of deposition by a later advance at around 25 ka could explain the similar exposure ages that we find on a right lateral position farther up-valley (e.g., TRYAY-19, 25.2 ± 2.2 ka, Figure 4). This would suggest a composite deposition of the till by two distinct advances at a comparable elevation in this frontal left lateral position above the LGM advance; cf. Schneebeli [38].

Sample TRYAY-15 was collected as a reference beyond and about 200 m above the boulder alignment, representing ice-free areas at the time of maximum expansion to the boulder alignment. It did produce an age of 25.7 ± 4.0 ka. This age provides evidence that the higher slopes, especially this highly exposed sampled bedrock knob, may have experienced considerable erosion and "slope-cleaning"—cf. Mair et al. [39]—at the time of the glacier presence in the lower parts of the valley.

Sample TRYAY-28 was sampled because it is the only large boulder in this part of the slope (Figure 4) and was considered in the field as an up-valley extension of the boulder alignment. One should note that this boulder is resting downslope against a low bedrock ridge. The exposure age of 9.5 ± 0.9 ka points to a likely post-depositional movement such as sliding and turning of the boulder since deposition; cf. Akçar et al. [15]. Therefore, we excluded this boulder from further discussion.

The intermediate sector of the southern slope of the main valley between the tributaries Körahmet Valley in the east and Dübe Valley in the west is characterized by a series of ridge segments with down-valley sloping extensions (Figures 4 and 7). These segments can be interpreted as morainic complex and ice-contact terraces. However, it is difficult to track single ridges for more than a few hundred meters down-valley. The whole slope is modified by human activity, and the clearly developed ridges have an agricultural overprint. Large boulders are elements of stability. Some are at least partially covered by gravel and small boulder fractions from land cleaning (Figure 7b,c).

Therefore, it is difficult to draw a clear limit for the extent of the paleoglacier occupying the main valley.

The sample TRYAY-26 yielded the oldest date in the valley so far with 105.6 ± 8.1 ka (Figure 7d). We sampled this boulder in perspective to collect an older advance than the morphologically better constrained LGM advance. The generic higher lateral position and the decomposing-looking stage of the rounded boulder (Figure 7d) speak for an older erratic boulder sample in the field. However, the exposure age is so old that we cannot rely on this single boulder's age to propose a pre-LGM advance earlier than the 34 ka phase detected in the frontal left lateral position. To construct a much earlier advance with this single boulder age would be highly speculative, and therefore, we conclude that TRYAY-26 contains likely inherited nuclide concentrations because of an earlier exposure in a nunatak high up in the accumulation area, which was above the LGM trimline. Therefore, this exposure age is excluded from further discussions and speculations until more evidence for such an early glacier advance is found.

Figure 7. (**a**) Down-valley view from sample station TRYAY-1 at the upper glacier margin. Note the difference in slope morphology with ice contact to the right and superficially decomposing bedrock to the left (see also Figure 7b,c). (**b**,**c**) Middle part of Barhal Valley with sampled boulders TRYAY-24 (22.2 ± 2.6 ka advance) and TRYAY-25 (18.3 ± 1.7 ka advance). (**d**) Sample TRYAY-26, a decomposing boulder in a steep slope affected by slope processes.

Most of the samples from the intermediate sector of the main valley can be subdivided into two groups: (1) an older age group and higher in the slope with samples TRYAY-17, TRYAY-18, TRYAY-19, TRYAY-21, and TRYAY-24, and (2) a younger age group and in lower parts of the slope with samples TRYAY-20, TRYAY-23, TRYAY-25, and TRYAY-27 (Figure 4). Sample TRYAY-17 is exposed for 22.7 ± 2.0 ka, TRYAY-18 for 22.8 ± 1.5 ka, TRYAY-19 for 25.2 ± 2.2 ka, TRYAY-21 for 19.4 ± 1.7 ka, and TRYAY-24 for 21.1 ± 1.4 ka. A mean age for the older group (blue dash line in Figure 4) is 22.2 ± 2.6 ka. The younger group comprises the samples TRYAY-20 with 18.1 ± 1.1 ka, TRYAY-23 with 17.9 ± 1.5 ka, TRYAY-25 with 19.0 ± 1.9 ka, and TRYAY-27 with 17.3 ± 1.6 ka, respectively. This makes an average for this group of 18.3 ± 1.7 ka. Sample TRYAY-22 (Figure 4) is in a morphological position that makes it part of the younger group. Its exposure age is, however, only 13.4 ± 1.1 ka; we identify this sample as an outlier, probably due to spalling as detected in the filed on the lower part of the boulder but not obvious on the sampling spot.

TRYAY-1 and -2 are from boulder tops on a prominent terrace to the south of Olgunlar Village at the junction of the Dübe Valley with the main valley. Both boulders are embedded in till with characteristically striated clasts. The boulders are part of the till cover (Figures 7a and 8).

Figure 8. Ice contact morphologies at confluence of Dübe and Hastaf Valleys with sample and glacial sediment locations.

TRYAY-1 is on a flat terrain, and TRAYAY-2 is on a ridge towards the little gully incision of the outlet from one of the smaller southern tributaries (Figures 4 and 8). Exposure ages of these boulders are 16.9 ± 1.9 and 16.8 ± 1.4 ka. In the northern slope above the Olgunlar Village at the entrance to Dübe Valley, abundant boulders are present. One boulder was sampled there at the comparable altitude to TRYAY-1 and TRYAY-2 at a moderately defined morphological break-in-slope, assuming that this corresponds to the ice limit of the maximum glacial extent. The exposure time of the sample TRYAY-11 is 16.4 ± 1.8 ka (Figure 4). The average age of TRYAY-1, -2, and -11 is 16.7 ± 1.7 ka. Based on our observations in the field, we consider the highest stand of the innermost lateral position at 2300 m a.s.l. and the continuation of this extent, mapped as a green dash-line in Figure 4, with the boulders TRYAY-20, TRYAY-23 and TRYAY-27,as a last stand still of the main valley glacier with a higher gradient of the glacier surface in this confluence area than during the earlier advance (blue dashed line with mean age of 22.2 ± 2.6 ka). The average age from TRYAY-1, -2, -11, -20, -23, and -27 is 17.7 ± 1.9 ka and suggests a main valley occupation by glacier ice at that time before the ice retreat toward the upper Dübe and Hastaf cirques.

However, we favor the interpretation that puts the ages of TRYAY-1, -2, and -11 with an average of 16.7 ± 1.7 ka in an already ongoing down wasting phase of a glacier from the tributary system of the Barhal Valley, and the exposure ages hint to exhumation or late deposition of the boulders in context with the stand-still or melting phase of the Dübe and smaller southern Tributary arm of the Barhal valley (Figures 4 and 8, respectively). In brief, we argue that these boulders were deposited by the tributary glacier during the down-wasting phase. With this hypothesis, the boulders TRYAY-20, TRYAY-23, and TRYAY-27 would reveal a slightly older average age from this lower green dashed line of 18.3 ± 1.7 ka as reported above, and the age difference between TRYAY-1 (16.9 ± 1.9 ka) on a higher position than the older TRYAY-17 (22.7 ± 2.0 ka) could then be explained accordingly. We consider TRYAY-16 in the vicinity of these two boulders with an exposure age of 14.5 ± 1.3 ka as an outlier because of the too young exposure age for this position in the valley system. In addition, a tilting of this boulder to a later stage is likely, as there are hints for this process in the field.

Well-preserved glacial landforms are rare in the upper reaches of Hastaf and Dübe Valleys (Figure 2). Between the confluence of Dübe and the main valley at about 2000 m a.s.l. and 2800 m at the entrance to the cirques, no unquestionable paleoglacier vestiges were observed. This fact is obviously the result of rapid ice down-melt and glacier retreat to the cirque area. In addition, above about 2800 m, the broad cirque floors are characterized by complex rock glaciers (Figures 5, 9 and 10).

The uppermost part of the Hastaf Valley is a broad open landscape closed by steep cirque headwalls (Figures 2 and 10a). A complex system of rock glaciers occupies the extensive cirque floors (Figures 5 and 10a,b). Morainic ridges can only be mapped with acceptable certainty in their frontal part before the valley drops off and where glacial features were not yet completely reworked by periglacial processes. A set of four samples were taken from there (Figure 5). Three samples were collected from boulders on the same ridge which marks a moderate readvance or at last a phase of ice margin stabilization. Samples TRYAY-3 with 9.5 ± 1.0 ka, TRYAY-4 with 11.3 ± 0.7 ka, and TRYAY-5 with 11.4 ± 0.9 ka exposure time are, within errors, exposed for roughly the same period of time with a bandwidth estimation of 11.0 ± 1.9 ka. It is not easy to interpret the sample TRYAY-6 from a boulder on the next ridge, which is more up-valley 90 m long with an exposure age of 16 ± 0.9 ka. When considering the broader morphological context of the sampled ridges, it must be explained as most likely being the frontal part of the still-active rock glaciers, and the location was therefore subject to former rock glacier activity and boulder mixing. The interpretation of the age of sample TRYAY-6 as evidence for glacier-free cirques not later than 16.1 ± 0.9 ka is a hypothesis based on this one date and therefore has to be taken with caution. TRYAY-6 could also contain inherited nuclide concentration from previous exposures at its source on the high peaks of the surrounding scenery prior to the erosion, transportation, and deposition by the glacier.

The glacial morphological configurations in the uppermost Dübe Valley are identical to the Hastaf Valley. Bedrock scenery is even more spectacular, with Mt. Kaçkar forming the high headwalls of the cirques (Figures 9 and 10c,d).

There, three boulders were sampled (TRYAY-7, -8, and -9) just outside the pronounced and most likely still active rock glaciers at the comparable altitude to Hastaf of 2900 m a.s.l., and one more sample TRYAY-10 was collected about 100 m lower in elevation. In the uppermost part, moraines are rare, and many of the rock glaciers appear to originate from remobilized glacial sediments. Samples TRYAY-7 and -8 were collected from boulders on defined ridges of several meters in height, resulting in identical exposure ages of 15.3 ± 1.3 and 15.6 ± 1.8 ka. Sample TRYAY-9 is from an angular rock slab on top of a ridge of a blocky moraine with thermokarst features and yielded a surface exposure age of 12.2 ± 1.2 ka. This clearly younger age can be explained by the presence of thermokarst features at that location and likely represents the re-arrangement of this boulder after deposition. Therefore, we identify this boulder as an outlier and exclude it from further discussion. Several hundred meters down valley from the sample site of TRYAY-7 and on the continuation of the same ridge, the surface of a huge boulder was sampled (TRYAY-10), with the resulting age of 14.2 ± 1.3 ka. We calculated a mean exposure age of 15.1 ± 1.6 ka for the three boulders from this upper region of the Dübe Valley. We conclude that this date marks the timing of glacier reorganization during the retreat in this high cirque. Based on the surface exposure dates that we gathered in the upper cirques of the Hastaf and Dübe Valley, we cannot suggest Lateglacial glacier advances. It is more likely that the exposure ages represent ice-free cirques and random morphological arrangements in the final phase of decaying glaciers at the transition to dominance by rock glaciers.

Sample name	Altitude (m)	Exposure age (ka)
TRYAY-7	2880	15.3 ± 1.3
TRYAY-8	2905	15.6 ± 1.8
TRYAY-9	2960	12.2 ± 1.2
TRYAY-10	2805	14.2 ± 1.3

Figure 9. Reconstructions based on digital elevation data of uppermost Dübe Valley with a spectacular view into the cirque area of the eastern ridge of Mt. Kaçkar. (**a**) Detailed map view with sample locations. (**b**) Detailed reconstruction of the land surface expressions in the zone of contact between glacial (down-valley) and periglacial (up-valley) morphologies with delicate sample positions. Sample TRYAY-9 is from a boulder on a pronounced ridge with periglacial modifications.

Figure 10. Landscape and sampled boulders in the upper cirques. (**a**,**b**) Uppermost Hastaf Valley as seen from sample position TRYAY-3 and TRYAY-6, respectively. (**a**) View to the west with a pronounced horn morphology and with undifferentiated glacial morphologies throughout the broad cirque. (**c**,**d**) Sampled boulders and landscape in uppermost Dübe Valley.

4. Discussion

In this study, we present follow-up results to our earlier investigations in northeastern Anatolian mountains: in Kavron [10], Verçenik [11], and Başyayla Valleys ([12] and Figure 1) and discuss them in a Mediterranean context (Hughes and Woodward [40], among others). Earlier investigations focused on the main valleys descending directly to the north from the high mountain ranges in the Eastern Black Sea Mountains [10–12]. The landscape above around 1800 m a.s.l. in the Kavron, Verçenik, and Başyayla Valleys are characterized by glacial morphology [9]. As stated in the introduction, the Kaçkar Mountain Range is characterized by a pronounced precipitation gradient. It has been hypothesized that this gradient also operated during the ice ages, resulting in more extensive glaciers to the north of the mountain divide than to the south in the precipitation shadow. However, paleoglaciation has been reported as well from the drier interior of the Anatolian Plateau (for instance from Mount Erçiyes; [41]), and changes in the circulation patterns during the LGM have been reconstructed for western Anatolia (Mount Uludağ; [42–44], the Balkans, e.g., [45,46]; and the Alps, e.g., [47] and references therein).

The Barhal Valley is directly descending from the eastern ridges of Mt. Kaçkar (Figures 5, 9 and 11) to the east. It is connected to the highest peak and "disappears" in the precipitation shadow as it extends to the east. Our study revealed evidence for extensive yet complex glacial features. They are morphologically poorly preserved, for example, compared to the Başyayla Valley [12].

Figure 11. Glacier reconstruction in the Central Çoruh Valley System for the 22.2 ± 2.6 ka advance. The 34 ka advance did not reach the same ice volume in the middle valley sector (Figure 3). Reconstructed maximum ELA is at approx. 2700 m a.s.l. Dübe and Hastaf glaciers merged at Olgunlar (Figure 7, sample TRYAY-11). The Körachmet glacier did not reach the Barhal Valley. The reconstructed ice volume and glacier extension in the uppermost Hastaf Cirque are a first minimum approximation.Due to the Barhal Valley orientation from west to east, there are also several north-facing tributary valleys in the Hastaf Valley beside the main valley and the Dübe branch. The exposition of the tributary valleys and their hypsometry probably played a considerable role in the build-up of the paleoglacier volume. Furthermore, these tributary valleys have the potential to host independent resting cirque glaciers (i.e., not connected to the glacier in the main valley) by an intermediate ELA (Equilibrium Line Altitude) depression, and therefore, they might have still been present when the main valley was ice-free after 18.3 ± 1.7 ka and before 15.6 ± 1.8 ka. The direct comparison of ice-covered area during the LGM of the Barhal Valley (34.4 km^2) to the ice-covered area of the opposing Kavron Valley (22 km^2) has therefore been considered with caution but still can be counted as a considerable volume.

The paleoglacier extension in the Barhal Valley is delineated by glacial morphological arguments at the gorge entrance down-valley from Yayalar (Figures 3 and 4). Additionally, less than a kilometer up-valley from the estimated glaciation limit and just at the confluence with the Köhrahmet Valley, there is a conspicuous boulder limit in the left lateral slope. We found evidence for two glacial advances in this part of the valley, including one Early LGM with an average age of 34.0 ± 2.3 ka. This glaciation is not morphologically well-constrained, but a similar advance is found in the Başyayla Valley at 32.6 ± 1.3 ka [12], as well as in the Western Taurus Mountains in the lower Kuruova Valley at 35.1 ± 2.5 ka [48], which agrees with an Early LGM as proposed, for example, by Starnberger et al. [49] for the Eastern Alps. The younger glaciation detected in this frontal part of the Barhal valley is geomorphologically well-constrained with an array of boulders, and this glaciation is synchronous with the global LGM. The average age of these boulders in the frontal left lateral part is 19.7 ± 2.1 ka, excluding the TRYAY-31 (25.6 ± 1.6 ka), which lies beyond two sigma uncertainty. The boulder line (Figure 6) marks a clear geomorphological break with different soil development, a stable slope surface, and an advanced vegetation cover. Accordingly, it is difficult to consider phase 34.0 ± 2.3 ka just as an earlier advance of a major event also producing the 19.7 ± 2.1 ka phase.

The average exposure ages of the two clusters in the middle part of the valley (blue dashed line Figure 4), with a mean age of 22.2 ± 2.6 ka, and the lower ice margin (lower green dashed line Figure 4), with a mean of 18.3 ± 1.7 ka, constrain the LGM advance as the main morphologically evident advances of the main valley. We propose that the Barhal Paleoglacier occupied the main valley during the period from 22.2 ± 2.6 ka to 18.3 ± 1.7 ka based on the ice margin evidence from the right lateral position and with a prominent phase around 19.7 ± 2.1 ka when the boulder line in the left frontal position was formed.

We calculated an average age of 21.9 ± 2.7 ka for the LGM in the Barhal Valley in order to compare with the existing LGM chronologies. To do so, we took the average of all 12 boulders geomorphologically attributed to the LGM (TRYAY-13, -14, -17, -18, -19, -20, -21, -23, -24, -25, -27, and -29) and excluded the Early LGM boulders (TRYAY-12, -31, and -32) and the young uppermost lateral boulders (TRYAY-1, -2, and -11 in Figure 4) from these calculations.

The amount of ice in Barhal Valley was important; the Hastaf and Dübe cirques were filled, and an ice plateau was formed at an elevation of about 3500 m, which allowed glacial landforms, such as the horns shown in Figure 10, to be sculptured by flowing ice.

A connected glacier system around Mt. Kaçkar can be postulated for the maximum LGM ice. However, it is difficult to reconstruct the Dübe Paleoglacier, as clear glacial morphologies have not yet been recognized, throughout the valley. Enormous amounts of snow must have accumulated during the LGM, as even today, snow avalanche ridges are actively formed in the valley, cf. Akçar et al. [50]. Such processes clean the high slopes from all glacigenic "horizontal" sedimentary landforms such as moraine ridges or ice contact slopes. The total length of the reconstructed LGM paleoglacier (Figure 11) is approximately 13 km from the farthest peak of the Hastaf Valley to the terminus. With the upper tributary of Dübe and Hastaf Valley, the reconstructed Barhal Paleoglacier has a surface area of 34.4 km^2, whereas the Dübe Paleoglacier accounts for about a quarter of this area.

We argue that the Barhal Paleoglacier reached a similar ice thickness in the middle part of upper Barhal Valley and close to identical frontal positions at the gorge entrance at around 1850 m a.s.l. during the Early LGM and LGM, as reconstructed by the stabilization phases. This position is roughly 450 m lower than the Başyayla Paleoglacier [12]. Is this because the catchments of the cirques are broader and more effective in catching precipitation in Hastaf-Barhal, or did the eastern slope of Mt. Kaçkar receive more precipitation during the Early LGM and LGM? The broad open and flat cirque morphologies in the Barhal Valley catch large amounts of snow and ice—contrary to Başyayla—which then descend into the funnel shape middle part of Barhal Valley, possibly explaining the advance of the glaciers to lower altitudes there.

Another important factor to compare the glacier extents of the Eastern Black Sea Mountains is the estimated modeled ELA. Based on the reconstructed extent of the Barhal paleoglacier (Figure 11), the LGM ELA was located at about 2900 m a.s.l., with an AAR (accumulation area ratio) value of 0.67. Considering Messerli's [51] estimation of the modern

ELA of about 3500 m a.s.l. in the Eastern Black Sea Mountains, we estimate an LGM ELA depression of about 600 m. Akçar [52] provided a compilation of LGM ELA depressions in the Anatolian Mountains. Accordingly, the ELA depression in the Barhal Valley is comparable with LGM ELA in the Başyayla valley, while for the directly northwards opposing Kavron Valley, the LGM ELA is estimated at 2700 m a.s.l. and the Verçenik Valley, with the estimated LGM ELA at 2800 m a.s.l. (Akçar et al. [44] and references therein). Hence, this local LGM ELA comparison reveals that the ice accumulation and glacier advance in the Barhal Valley, south of the main weather divide, are in line with its neighbors to the north of the weather divide.

The timing of the LGM advance in the broader context of Anatolia is dated at 22.0 ± 0.4 (Mount Uludağ; [43,44]), 21.5 ± 0.4 ka (Kavron Valley; [10]), 20.1 ± 1.4 ka (Verçenik Valley; [11]), 24.2 ± 0.4 (Başyayla Valley; [12]), > 19.2 ± 1.2 ka (Karçal Valley; [53]), 20.7 ± 2.2 ka (Aksu Valley; [41]), 20.4 ± 1.5 ka (Üçker Valley; [41]), 19.8 ± 0.8 (Muslu Valley; [54]), two advances in Kartal Valley at 22.9 ± 3.3 ka and 20.6 ± 3.1 ka [55], >19.1 ± 3.5 ka (Namaras Valley, Geyikağ Mountains; [56]), 20.6 ± 0.6 ka (Çimi Valley, Geyikağ Mountains; [57]), and 18.9 ± 3.3 ka (Karagöl Valley, Bolka Mountains [58]). More details of the reported ages from Anatolia are given in Akçar [52] with a general overview. Our mean age of 21.9 ± 2.7 ka for the Barhal Paleoglacier fits well in this overall picture and agrees with the global LGM that occurred at 22.1 ± 4.3 ka in the northern Hemisphere [37].

The Barhal Valley is almost at the far eastern end of the Eastern Black Sea Mountains. At the western edge of the Anatolian Peninsula, there is another climatically sensitive mountain. It is Mount Uludağ, with the highest peak at 2542 m a.s.l., a mountain surrounded by lowlands. With comprehensive glacial geological mapping and cosmogenic surface exposure dating, it was possible to reconstruct timing and glacier dimensions during the LGM: maximum ice extent is dated at 20.3 ± 1.3 ka, with a readvance at 19.3 ± 1.2 ka [44]. These dates agree within two sigma uncertainties with the 22.2 ± 2.6 ka and 18.3 ± 1.7 ka phases and within one sigma uncertainty with the latero-terminal position at 19.7 ± 2.1 ka in Barhal Valley.

The down-wasting of the Barhal glacier likely started shortly after 18.3 ± 1.7 ka, which is marked by the lowermost set of ridges in the middle part of the main valley (Figure 4). Based on the exposure ages around 16.7 ± 1.7 ka (Figure 4), one can argue that some slope stabilization and probably a reorganization of the paleoglacier system might have taken place at this time, which resulted in the disconnection of the main valley glacier from the ice tongues in the tributary valleys (Figure 11).

In the Barhal Valley, no glacial morphologies are preserved further up-valley below the periglacial landscape in the cirques. Additionally, therefore, the oldest surface exposure ages measured so far in the Hastaf-Cirque of 16.1 ± 0.9 ka and Dübe-Cirque of 15.6 ± 1.8 ka (Figures 5 and 10b) are interpreted to have been an ice-free cirque the latest at 15.6 ± 1.8 ka. An equivalent exposure age is found in the upper cirque of Başyayla at 17.0 ± 1.0 ka; cf. Reber et al. [12]. This means that the Barhal Paleoglacier disappeared within 2000 years (as in the Başyayla valley), based on the data available.

Based on the existing chronology of the glaciations in the Anatolian Mountains, we conclude that the glacier expansion during the Last Global Glacial Maximum was recorded in this peninsula in the Eastern Mediterranean, and its collapse was as rapid as in the Alps (e.g., Kamleitner et al. [47] and references therein). However, undisputable Lateglacial records are rare in the Anatolian Mountains. The rapid collapse of the glaciers during the LGM points to temperature sensitivity of the system and to an obvious rapid rise in the equilibrium line altitude.

Today, moisture transport mainly occurs directly from the Black Sea to the Kaçkar Mountains, where it rains throughout the orographic control; e.g., [6]. In a scenario with moisture transport directly from only the Black Sea, the glaciers in Kavron and Başyayla Valleys should have been gigantic with respect to the Barhal Paleoglacier. However, for equally extensive glaciers in the Çoruh valley system, as argued above, an alternative moisture source should be available. During globally cold phases, such as the LGM, the Polar Front migrated to the south. In Western Europe, it was as far south as 40° N (Kamleitner et al. [47] and references therein), which would place it in the southern part of the Caspian Sea for the Caucasus area. Anticyclonic circulation controlled by the southerly shifted Siberian High-Pressure System causes airflow from the Aral and Caspian Sea areas across the Caucasus and westward to northeastern Anatolia, dumping precipitation on the high mountains there. The moisture feeding system in addition to the expanded Caspian and Aral seas were huge flooded areas in central Asia, as huge lakes in front of the southward expanding polar glaciers were damming the northward-flowing Siberian rivers ([59–61]; Figure 12).

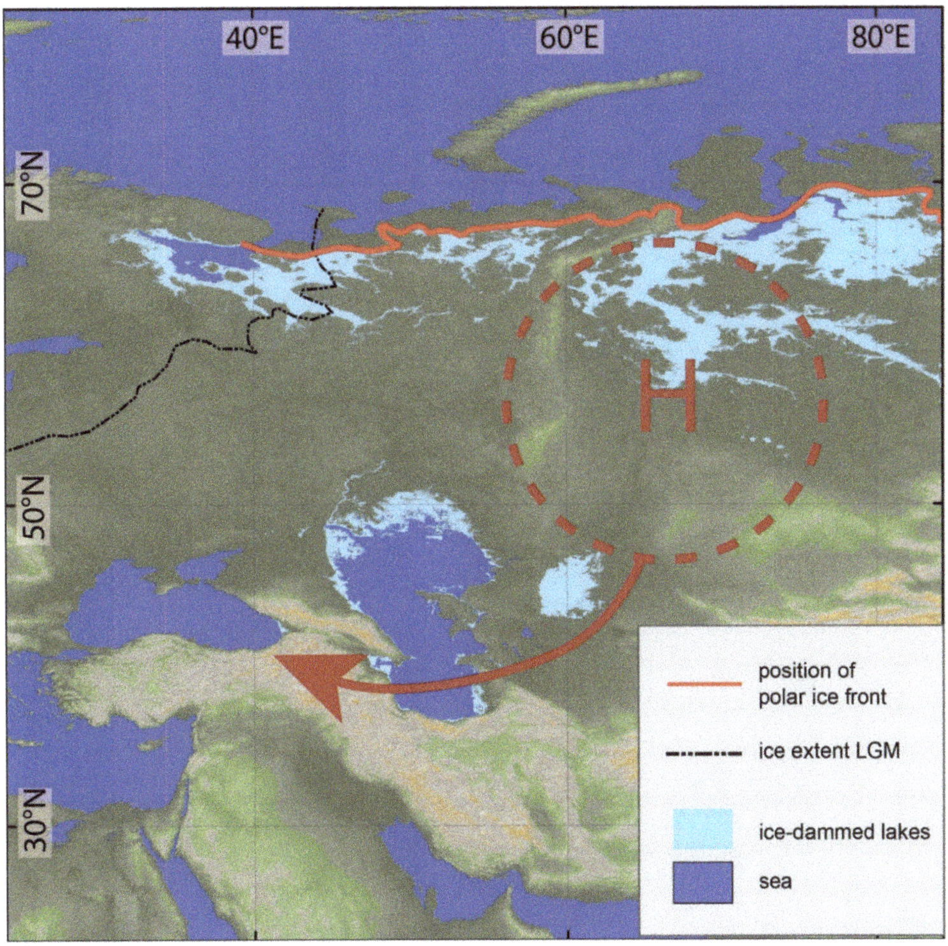

Figure 12. Reconstructed map of expanded water bodies between the Black Sea in the south and the White Sea in the north, modified from Mangerud et al. [60]. The age of 90 ka for this reconstruction by Mangerud et al. [60] is based on OSL dates on sediments of raised beaches, defining the large ice-dammed lakes, in this Figure calculated after Mangerud et al. [60]. This dating implies the early last glacial advance of Kara Sea ice onto the land. This chronology is under debate. Here, this reconstruction is combined with the fact that the Polar Front and, as a consequence, the Siberian High were pushed much to the south which activated northeasterly airflow to bring moisture to the Caucasian and Kaçkar Mountains from the northeast.

Available absolute age determinations on the huge Siberian Lakes, however, point to earlier damming events (and advances of polar ice onto to the continent) at 80–90 and 50–60 ka. However, these dates are not independently confirmed.

5. Conclusions

The field evidence in Barhal Valley favors the distinction of independent stabilization phases at 22.2 ± 2.6 ka, 19.7 ± 2.1 ka, and 18.3 ± 1.7 ka within the global LGM and an Early LGM phase at 34.0 ± 2.3 ka. The timing and expansion of these Barhal Paleoglacier extents to the south of the main weather divide are comparable to the paleoglacier extents in the neighboring valleys on the orographic Luv-side, considering the Black Sea as the main precipitation source. These findings require a more systematic comparison of glacier records from the todays Luv and Lee sides of the main weather divide in the Eastern Black Sea Mountains and the Caucasus to achieve a clearer picture of the local circulation patterns and moisture sources during glacial periods.

The deglaciation in Barhal Valley took place without stagnations producing geomorphic landmarks. The geomorphology and the exposure dates from the broad open cirque areas allow the interpretation of a complete ice retreat to the cirques at the latest at 15.6 ± 1.8 ka, when chaotic ice marginal sediment aggradation with multiphase transitions to rock glacier formations and minimal morphological reorganization Figures 5, 9 and 10 interacted.

Supplementary Materials: The following supporting information can be downloaded at: https://www.mdpi.com/article/10.3390/geosciences12070257/s1, Table S1.

Author Contributions: Conceptualization, R.R., N.A., D.T., S.Y., C.V., V.Y., S.I.-O. and C.S.; methodology, D.T., S.Y., C.V. and N.A.; formal analysis, R.R.; investigation, R.R., D.T., S.Y. and N.A.; resources, C.S., V.Y. and N.A.; writing—original draft preparation, R.R., C.S., N.A. and D.T.; visualization, R.R. and S.Y.; supervision, C.S. and N.A.; project administration, N.A.; funding acquisition, C.S. and N.A. All authors have read and agreed to the published version of the manuscript.

Funding: This research was funded by the Swiss National Science Foundation Project No. 200020-125203, by the Surface Exposure Dating Laboratory, University of Bern, Switzerland, and by the Bern University Research Foundation.

Acknowledgments: We would like to express special thanks to the Laboratory of Ion Beam Physics operated by the Swiss Federal Institute of Technology, Zurich, Switzerland, to the Swiss National Science Foundation (Project No 200020-125203), to the Cosmolab Berne.

Conflicts of Interest: The authors declare no conflict of interest.

References

1. Benn, D.; Evans, D.J. *Glaciers and Glaciation*, 2nd ed.; Taylor & Francis: London, UK, 2010; ISBN 978-0-340-90579-1.
2. Oerlemans, J. Extracting a Climate Signal from 169 Glacier Records. *Science* **2005**, *308*, 675–677. [CrossRef] [PubMed]
3. Ohmura, A.; Kasser, P.; Funk, M. Climate at the Equilibrium Line of Glaciers. *J. Glaciol.* **1992**, *38*, 397–411. [CrossRef]
4. Lukas, S.; Graf, A.; Coray, S.; Schlüchter, C. Genesis, Stability and Preservation Potential of Large Lateral Moraines of Alpine Valley Glaciers—Towards a Unifying Theory Based on Findelengletscher, Switzerland. *Quat. Sci. Rev.* **2012**, *38*, 27–48. [CrossRef]
5. Ivy-Ochs, S.; Schaller, M. Examining Processes and Rates of Landscape Change with Cosmogenic Radionuclides. *Radioact. Environ.* **2009**, *16*, 231–294.
6. Akçar, N.; Schlüchter, C. Paleoglaciations in Anatolia: A Schematic Review and First Results. *Eiszeitalt. Ggw. EG* **2005**, *55*, 102–121.
7. Türkeş, M.; Erlat, E. Winter Mean Temperature Variability in Turkey Associated with the North Atlantic Oscillation. *Meteorol. Atmos. Phys.* **2009**, *105*, 211–225. [CrossRef]
8. Climate-Data.Org. Available online: https://en.climate-data.org/asia/turkey-67/ (accessed on 15 May 2022).
9. Akçar, N. The Anatolian Peninsula. In *European Glacial Landscapes*; Elsevier: Amsterdam, The Netherlands, 2022; pp. 149–157.
10. Akçar, N.; Yavuz, V.; Ivy-Ochs, S.; Kubik, P.W.; Vardar, M.; Schlüchter, C. Paleoglacial Records from Kavron Valley, NE Turkey: Field and Cosmogenic Exposure Dating Evidence. *Quat. Int.* **2007**, *164*, 170–183. [CrossRef]
11. Akçar, N.; Yavuz, V.; Ivy-Ochs, S.; Kubik, P.W.; Vardar, M.; Schlüchter, C. A Case for a Downwasting Mountain Glacier during Termination I, Verçenik Valley, Northeastern Turkey. *J. Quat. Sci. Publ. Quat. Res. Assoc.* **2008**, *23*, 273–285. [CrossRef]
12. Reber, R.; Akçar, N.; Yesilyurt, S.; Yavuz, V.; Tikhomirov, D.; Kubik, P.W.; Schlüchter, C. Glacier Advances in Northeastern Turkey before and during the Global Last Glacial Maximum. *Quat. Sci. Rev.* **2014**, *101*, 177–192. [CrossRef]

13. Yilmaz-Şahin, S. Transition from Arc-to Post-collision Extensional Setting Revealed by K–Ar Dating and Petrology: An Example from the Granitoids of the Eastern Pontide Igneous Terrane, Araklı-Trabzon, NE Turkey. *Geol. J.* **2005**, *40*, 425–440. [CrossRef]
14. Ketin, İ. Tectonic Units of Anatolia (Asia Minor). *Bull. Miner. Res. Explor.* **1966**, *66*, 20–37.
15. Akçar, N.; Ivy-Ochs, S.; Kubik, P.W.; Schlüchter, C. Post-Depositional Impacts on 'Findlinge'(Erratic Boulders) and Their Implications for Surface-Exposure Dating. *Swiss J. Geosci.* **2011**, *104*, 445–453. [CrossRef]
16. Dunne, J.; Elmore, D.; Muzikar, P. Scaling Factors for the Rates of Production of Cosmogenic Nuclides for Geometric Shielding and Attenuation at Depth on Sloped Surfaces. *Geomorphology* **1999**, *27*, 3–11. [CrossRef]
17. Reedy, R.C. Nuclide Production by Primary Cosmic-ray Protons. *J. Geophys. Res. Solid Earth* **1987**, *92*, E697–E702. [CrossRef]
18. Lal, D. In Situ-Produced Cosmogenic Isotopes in Terrestrial Rocks. *Annu. Rev. Earth Planet. Sci.* **1988**, *16*, 355–388. [CrossRef]
19. Liu, B.; Phillips, F.M.; Fabryka-Martin, J.T.; Fowler, M.M.; Stone, W.D. Cosmogenic 36Cl Accumulation in Unstable Landforms: 1. Effects of the Thermal Neutron Distribution. *Water Resour. Res.* **1994**, *30*, 3115–3125. [CrossRef]
20. Phillips, F.M.; Zreda, M.G.; Flinsch, M.R.; Elmore, D.; Sharma, P. A Reevaluation of Cosmogenic 36Cl Production Rates in Terrestrial Rocks. *Geophys. Res. Lett.* **1996**, *23*, 949–952. [CrossRef]
21. Phillips, F.M.; Stone, W.D.; Fabryka-Martin, J.T. An Improved Approach to Calculating Low-Energy Cosmic-Ray Neutron Fluxes near the Land/Atmosphere Interface. *Chem. Geol.* **2001**, *175*, 689–701. [CrossRef]
22. Stone, J.O.; Allan, G.L.; Fifield, L.K.; Cresswell, R.G. Cosmogenic Chlorine-36 from Calcium Spallation. *Geochim. Cosmochim. Acta* **1996**, *60*, 679–692. [CrossRef]
23. Stone, J.O.H.; Evans, J.M.; Fifield, L.K.; Allan, G.L.; Cresswell, R.G. Cosmogenic Chlorine-36 Production in Calcite by Muons. *Geochim. Cosmochim. Acta* **1998**, *62*, 433–454. [CrossRef]
24. Alfimov, V.; Ivy-Ochs, S. How Well Do We Understand Production of 36Cl in Limestone and Dolomite? *Quat. Geochronol.* **2009**, *4*, 462–474. [CrossRef]
25. Schimmelpfennig, I.; Benedetti, L.; Finkel, R.; Pik, R.; Blard, P.-H.; Bourles, D.; Burnard, P.; Williams, A. Sources of In-Situ 36Cl in Basaltic Rocks. Implications for Calibration of Production Rates. *Quat. Geochronol.* **2009**, *4*, 441–461. [CrossRef]
26. Akçar, N.; Tikhomirov, D.; Özkaymak, Ç.; Ivy-Ochs, S.; Alfimov, V.; Sözbilir, H.; Uzel, B.; Schlüchter, C. 36Cl Exposure Dating of Paleoearthquakes in the Eastern Mediterranean: First Results from the Western Anatolian Extensional Province, Manisa Fault Zone, Turkey. *Bulletin* **2012**, *124*, 1724–1735. [CrossRef]
27. Ivy-Ochs, S.; Synal, H.-A.; Roth, C.; Schaller, M. Initial Results from Isotope Dilution for Cl and 36Cl Measurements at the PSI/ETH Zurich AMS Facility. *Nucl. Instrum. Methods Phys. Res. Sect. B Beam Interact. Mater. At.* **2004**, *223*, 623–627. [CrossRef]
28. Desilets, D.; Zreda, M.; Almasi, P.F.; Elmore, D. Determination of Cosmogenic 36Cl in Rocks by Isotope Dilution: Innovations, Validation and Error Propagation. *Chem. Geol.* **2006**, *233*, 185–195. [CrossRef]
29. Synal, H.-A.; Bonani, G.; Döbeli, M.; Ender, R.M.; Gartenmann, P.; Kubik, P.W.; Schnabel, C.; Suter, M. Status Report of the PSI/ETH AMS Facility. *Nucl. Instrum. Methods Phys. Res. Sect. B Beam Interact. Mater. At.* **1997**, *123*, 62–68. [CrossRef]
30. Vockenhuber, C.; Miltenberger, K.-U.; Synal, H.-A. 36Cl Measurements with a Gas-Filled Magnet at 6 MV. *Nucl. Instrum. Methods Phys. Res. Sect. B Beam Interact. Mater. At.* **2019**, *455*, 190–194. [CrossRef]
31. Stone, J.O. Air Pressure and Cosmogenic Isotope Production. *J. Geophys. Res. Solid Earth* **2000**, *105*, 23753–23759. [CrossRef]
32. Evans, J.M.; Stone, J.O.H.; Fifield, L.K.; Cresswell, R.G. Cosmogenic Chlorine-36 Production in K-Feldspar. *Nucl. Instrum. Methods Phys. Res. Sect. B Beam Interact. Mater. At.* **1997**, *123*, 334–340. [CrossRef]
33. Fink, D.; Vogt, S.; Hotchkis, M. Cross-Sections for 36Cl from Ti at Ep=35–150 MeV: Applications to in-Situ Exposure Dating. *Nucl. Instrum. Methods Phys. Res. Sect. B Beam Interact. Mater. At.* **2000**, *172*, 861–866. [CrossRef]
34. Stone, J. Cosmogenic Chlorine-36 Ages for the 1975 and 1984 Eruptions of Mauna Loa, Hawaii. In Proceedings of the 2005 Salt Lake City Annual Meeting, Salt Lake City, UT, USA, 16–19 October 2005.
35. Heisinger, B.; Lal, D.; Jull, A.J.T.; Kubik, P.; Ivy-Ochs, S.; Neumaier, S.; Knie, K.; Lazarev, V.; Nolte, E. Production of Selected Cosmogenic Radionuclides by Muons: 1. Fast Muons. *Earth Planet. Sci. Lett.* **2002**, *200*, 345–355. [CrossRef]
36. Heisinger, B.; Lal, D.; Jull, A.J.T.; Kubik, P.; Ivy-Ochs, S.; Knie, K.; Nolte, E. Production of Selected Cosmogenic Radionuclides by Muons: 2. Capture of Negative Muons. *Earth Planet. Sci. Lett.* **2002**, *200*, 357–369. [CrossRef]
37. Shakun, J.D.; Carlson, A.E. A Global Perspective on Last Glacial Maximum to Holocene Climate Change. *Quat. Sci. Rev.* **2010**, *29*, 1801–1816. [CrossRef]
38. Schneebeli, W. Untersuchungen von Gletscherschwankungen Im Val de Bagnes. In *Die Alpen*; Schweizer Alpen-Club: Bern, Switzerland, 1976; pp. 5–57.
39. Mair, D.; Lechmann, A.; Delunel, R.; Yeşilyurt, S.; Tikhomirov, D.; Vockenhuber, C.; Christl, M.; Akçar, N.; Schlunegger, F. The Role of Frost Cracking in Local Denudation of Steep Alpine Rockwalls over Millennia (Eiger, Switzerland). *Earth Surf. Dyn.* **2020**, *8*, 637–659. [CrossRef]
40. Hughes, P.D.; Woodward, J.C. Timing of Glaciation in the Mediterranean Mountains during the Last Cold Stage. *J. Quat. Sci. Publ. Quat. Res. Assoc.* **2008**, *23*, 575–588. [CrossRef]
41. Sarıkaya, M.A.; Zreda, M.; Çiner, A. Glaciations and Paleoclimate of Mount Erciyes, Central Turkey, since the Last Glacial Maximum, Inferred from 36Cl Cosmogenic Dating and Glacier Modeling. *Quat. Sci. Rev.* **2009**, *28*, 2326–2341. [CrossRef]
42. Zahno, C.; Akçar, N.; Yavuz, V.; Kubik, P.W.; Schlüchter, C. Chronology of Late Pleistocene Glacier Variations at the Uludağ Mountain, NW Turkey. *Quat. Sci. Rev.* **2010**, *29*, 1173–1187. [CrossRef]

43. Akçar, N.; Yavuz, V.; Ivy-Ochs, S.; Reber, R.; Kubik, P.W.; Zahno, C.; Schlüchter, C. Glacier Response to the Change in Atmospheric Circulation in the Eastern Mediterranean during the Last Glacial Maximum. *Quat. Geochronol.* **2014**, *19*, 27–41. [CrossRef]
44. Akçar, N.; Yavuz, V.; Yeşilyurt, S.; Ivy-Ochs, S.; Reber, R.; Bayrakdar, C.; Kubik, P.W.; Zahno, C.; Schlunegger, F.; Schlüchter, C. Synchronous Last Glacial Maximum across the Anatolian Peninsula. *Geol. Soc. Lond. Spec. Publ.* **2017**, *433*, 251–269. [CrossRef]
45. Kuhlemann, J.; Milivojević, M.; Krumrei, I.; Kubik, P.W. Last Glaciation of the Šara Range (Balkan Peninsula): Increasing Dryness from the LGM to the Holocene. *Austrian J. Earth Sci.* **2009**, *102*, 146–158.
46. Hughes, P.D.; Woodward, J.C.; Van Calsteren, P.C.; Thomas, L.E.; Adamson, K.R. Pleistocene Ice Caps on the Coastal Mountains of the Adriatic Sea. *Quat. Sci. Rev.* **2010**, *29*, 3690–3708. [CrossRef]
47. Kamleitner, S.; Ivy-Ochs, S.; Monegato, G.; Gianotti, F.; Akçar, N.; Vockenhuber, C.; Christl, M.; Synal, H.-A. The Ticino-Toce Glacier System (Swiss-Italian Alps) in the Framework of the Alpine Last Glacial Maximum. *Quat. Sci. Rev.* **2022**, *279*, 107400. [CrossRef]
48. Sarıkaya, M.A.; Çiner, A.; Haybat, H.; Zreda, M. An Early Advance of Glaciers on Mount Akdağ, SW Turkey, before the Global Last Glacial Maximum; Insights from Cosmogenic Nuclides and Glacier Modeling. *Quat. Sci. Rev.* **2014**, *88*, 96–109. [CrossRef]
49. Starnberger, R.; Rodnight, H.; Spötl, C. Chronology of the Last Glacial Maximum in the Salzach Palaeoglacier Area (Eastern Alps). *J. Quat. Sci.* **2011**, *26*, 502–510. [CrossRef]
50. Akçar, N.; Yavuz, V.; Ivy-Ochs, S.; Kubik, P.W.; Vardar, M.; Schlüchter, C. Cosmogenic Exposure Dating of Snow-Avalanche Ridges Eastern Black Sea Mountains, NE Turkey. *Quat. Int.* **2007**, *167*, 4–11. [CrossRef]
51. Messerli, B. Die Eiszeitliche Und Die Gegenwärtige Vergletscherung Im Mittelmeerraum. *Geogr. Helv.* **1967**, *22*, 105–228. [CrossRef]
52. Akçar, N. The Anatolian Mountains: Glacial Landforms from the Last Glacial Maximum. In *European Glacial Landscapes*; Elsevier: Amsterdam, The Netherlands, 2022; pp. 497–504.
53. Dede, V.; Çiçek, İ.; Sarıkaya, M.A.; Çiner, A.; Uncu, L. First Cosmogenic Geochronology from the Lesser Caucasus: Late Pleistocene Glaciation and Rock Glacier Development in the Karçal Valley, NE Turkey. *Quat. Sci. Rev.* **2017**, *164*, 54–67. [CrossRef]
54. Zahno, C.; Akçar, N.; Yavuz, V.; Kubik, P.W.; Schlüchter, C. Surface Exposure Dating of Late Pleistocene Glaciations at the Dedegöl Mountains (Lake Beyşehir, SW Turkey). *J. Quat. Sci. Publ. Quat. Res. Assoc.* **2009**, *24*, 1016–1028. [CrossRef]
55. Sarıkaya, M.A.; Zreda, M.; Çiner, A.; Zweck, C. Cold and Wet Last Glacial Maximum on Mount Sandıras, SW Turkey, Inferred from Cosmogenic Dating and Glacier Modeling. *Quat. Sci. Rev.* **2008**, *27*, 769–780. [CrossRef]
56. Çiner, A.; Sarıkaya, M.A.; Yıldırım, C. Late Pleistocene Piedmont Glaciations in the Eastern Mediterranean; Insights from Cosmogenic 36Cl Dating of Hummocky Moraines in Southern Turkey. *Quat. Sci. Rev.* **2015**, *116*, 44–56. [CrossRef]
57. Sarıkaya, M.A.; Çiner, A.; Yıldırım, C. Cosmogenic 36Cl Glacial Chronologies of the Late Quaternary Glaciers on Mount Geyikdağ in the Eastern Mediterranean. *Quat. Geochronol.* **2017**, *39*, 189–204. [CrossRef]
58. Çiner, A.; Sarıkaya, M.A. Cosmogenic 36Cl Geochronology of Late Quaternary Glaciers in the Bolkar Mountains, South Central Turkey. *Geol. Soc. Lond. Spec. Publ.* **2017**, *433*, 271–287. [CrossRef]
59. Thiede, J. Quaternary Environments of the Eurasian North. *Quat. Sci. Rev.* **2004**, *23*, 1225–1511. [CrossRef]
60. Mangerud, J.; Astakhov, V.; Jakobsson, M.; Svendsen, J.I. Huge Ice-age Lakes in Russia. *J. Quat. Sci. Publ. Quat. Res. Assoc.* **2001**, *16*, 773–777. [CrossRef]
61. Mangerud, J.; Jakobsson, M.; Alexanderson, H.; Astakhov, V.; Clarke, G.K.; Henriksen, M.; Hjort, C.; Krinner, G.; Lunkka, J.-P.; Möller, P. Ice-Dammed Lakes and Rerouting of the Drainage of Northern Eurasia during the Last Glaciation. *Quat. Sci. Rev.* **2004**, *23*, 1313–1332. [CrossRef]

geosciences

MDPI

Article

Cosmogenic Exposure Dating (^{36}Cl) of Landforms on Jan Mayen, North Atlantic, and the Effects of Bedrock Formation Age Assumptions on ^{36}Cl Ages

Johanna Anjar [1,*], Naki Akçar [2], Eiliv A. Larsen [3], Astrid Lyså [3], Shasta Marrero [4], Nasim Mozafari [2] and Christof Vockenhuber [5]

1 Department of Natural Sciences and Environmental Health, University of South-Eastern Norway, Gullbringvegen 36, 3800 Bø, Norway
2 Institute of Geological Sciences, University of Bern, Baltzerstrasse 1-3, 3012 Bern, Switzerland; akcar@geo.unibe.ch (N.A.); nasim.mozafari@geo.unibe.ch (N.M.)
3 Geological Survey of Norway, P.O. Box 6315, Torgarden, 7491 Trondheim, Norway; eiliv.larsen@ngu.no (E.A.L.); astrid.lysa@ngu.no (A.L.)
4 School of Earth and Environmental Sciences, Cardiff University, Main Building, Park Place, Cardiff CF10 3AT, UK; marreros@cardiff.ac.uk
5 Laboratory of Ion Beam Physics, ETH Zürich, 8093 Zürich, Switzerland; vockenhuber@phys.ethz.ch
* Correspondence: johanna.anjar@usn.no

Citation: Anjar, J.; Akçar, N.; Larsen, E.A.; Lyså, A.; Marrero, S.; Mozafari, N.; Vockenhuber, C. Cosmogenic Exposure Dating (^{36}Cl) of Landforms on Jan Mayen, North Atlantic, and the Effects of Bedrock Formation Age Assumptions on ^{36}Cl Ages. *Geosciences* **2021**, *11*, 390. https://doi.org/10.3390/geosciences11090390

Academic Editors: Ángel Rodés and Jesus Martinez-Frias

Received: 14 July 2021
Accepted: 10 September 2021
Published: 15 September 2021

Publisher's Note: MDPI stays neutral with regard to jurisdictional claims in published maps and institutional affiliations.

Abstract: Jan Mayen is a small volcanic island situated 550 km north of Iceland. Glacial sediments and landforms are relatively common on the island but, so far, only a few of them have been dated. In this study, we present and discuss 89 ^{36}Cl dates of primarily glacial and volcanic events on Jan Mayen. Calculations of sample exposure ages were complicated by young exposure ages, young rock formation age, and high native Cl contents, leading to updates in CRONUScalc to enable accurate exposure age calculations. The samples provide good evidence against an equilibrium assumption when subtracting background production (e.g., ^{36}Cl produced by neutron capture from fission of U or Th) for samples on young bedrock, with younger exposure ages most significantly affected. Exposure ages were calculated with a range of assumptions of bedrock formation ages appropriate for Jan Mayen, including the assumption that the rock formation age equaled the exposure age (i.e., the youngest age it could possibly have), and we found that although the effect on most of the ages was small, the calculated ages of 25 of the samples increased by more than 1 standard deviation from the age calculated assuming equilibrium background production, with a maximum deviation of 6.1 ka. Due to the very young bedrock on Jan Mayen, we consider the nonequilibrium ages to be the most reliable ages from the island and conclude that large-scale deglaciation on the south and central, lower-lying, parts of the island, started around 20 ka and lasted until ~7 ka. On northern Jan Mayen, the slopes of the 2277 m high stratovolcano Beerenberg are currently partly glaciated; however, outside of the Little Ice Age moraines, all but two samples give ages between 14 and 5.7 ka.

Keywords: cosmogenic surface exposure dating; ^{36}Cl; Jan Mayen; background production

1. Introduction

Jan Mayen is a small volcanic island. Its isolated position 550 km north of Iceland and 450 km east of Greenland in the Norwegian–Greenland Sea (Figure 1) makes it an interesting location for investigations of the climate history in the North Atlantic. A research campaign to reconstruct the glaciation and climate history of the island was therefore started in 2014 [1,2]. As part of that campaign, 89 samples for cosmogenic nuclide exposure age dating (^{36}Cl) were collected with the aim of dating glacial and volcanic events on the island.

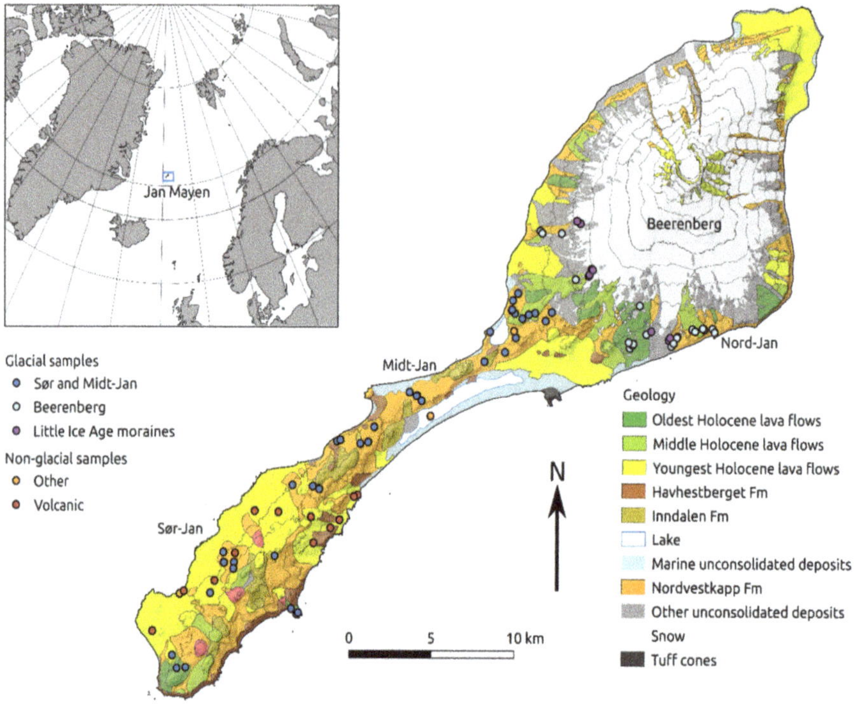

Figure 1. Geological map of Jan Mayen [3]. The different map unit colors indicate the different ages of the lava flows. Sample locations are marked by circles and the colors indicate the sample setting: dark blue indicates samples from glacial settings on south and central Jan Mayen; light blue indicates glacial samples from the slopes of Beerenberg; purple indicates samples from the young Little Ice Age (LIA) moraines; red indicates samples that are related to volcanism; samples that are unrelated to deglaciation or volcanism are classified as "other" and indicated in orange.

Exposure dating in an active volcanic landscape comes with its own challenges. In addition to being created through cosmogenic processes, ^{36}Cl is also created when ^{35}Cl absorbs neutrons produced from fission and (alpha, n) reactions, including uranium and thorium decay (for simplicity, all these reactions are referred to as "background production" herein). These neutron capture reactions occur in both the thermal and epithermal energy ranges [4], and the general formula is provided in Equation (1). The total ^{36}Cl atoms attributed to background production ($N_{36,\text{ background}}$) for any given sample is dependent on the formation age of the rock (t_{form}) and the composition of the sample (through the calculation of the total background production rate, $P_{36,\text{ background}}$, in ^{36}Cl atoms g^{-1} y^{-1}). The elemental composition of the sample determines both the creation rate of low-energy neutrons and the total absorption of those neutrons by ^{35}Cl and other elements present in the sample. Although the timescale for background production is dependent only on the half-life of ^{36}Cl ($t_{\frac{1}{2}}$, related to the decay constant, λ_{36}, through $\lambda_{36} = \ln(2)/t_{\frac{1}{2}}$), elemental composition determines the magnitude of the background production. Full details of the reactions considered in these calculations and the effects of sample composition can be found in Gosse and Phillips (2001) [4] and Marrero et al. (2016a) [5].

$$N_{36,background} = P_{36,\ background}\left[\frac{1 - exp\left(-t_{form}\lambda_{36}\right)}{\lambda_{36}}\right] \tag{1}$$

To estimate the cosmogenically produced ^{36}Cl in the sample, the background production is subtracted from the measured ^{36}Cl concentration. For many samples, the rock formation age is sufficiently old that production of ^{36}Cl from background processes balances radioactive decay (Figure 2, see also Equation (1) as t_{form} becomes large), i.e., equilibrium conditions have been achieved. Standard methods for age calculations assume that this equilibrium condition has been reached for all samples [5,6]. However, the assumption that the background production has had time to reach equilibrium is unlikely to be true in areas with young volcanic rocks [7–9], such as Jan Mayen [1,2,10]. If equilibrium conditions are assumed incorrectly, the background production subtraction will be too large, resulting in an exposure age that is too young. This incorrect equilibrium assumption is most likely to affect samples that meet three criteria: short exposure duration, young rock formation age, and susceptible composition (high in both native Cl and neutron-producing elements such as uranium and thorium). Although there are other models that do not make this assumption (e.g., Schimmelpfennig et al. (2009) [7]), they require manual iteration and were not available for a range of production rate scaling models.

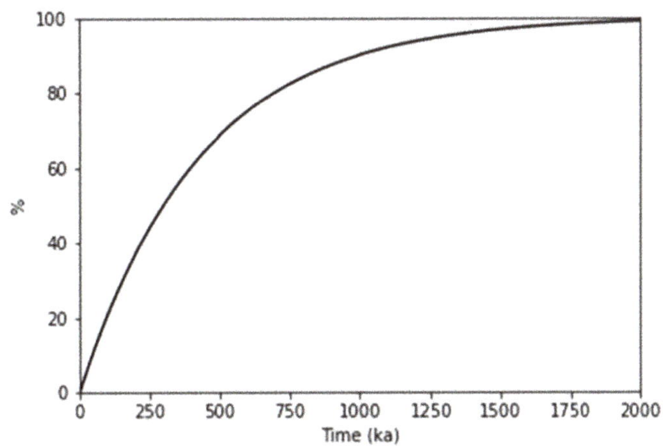

Figure 2. Modeled background production over time, shown as percent of the full equilibrium amount, based on Equation (1). Time to reach equilibrium is dependent on the half-life of ^{36}Cl.

In this article, we present 89 cosmogenic ^{36}Cl exposure ages, sampled from both glacial and volcanic landforms, all adjusted for the young rock formation ages. In addition, we present the results of the geochemical analyses for each sample, thereby adding to the previously available geochemical data from Jan Mayen [11].

A subset of the Jan Mayen samples provides evidence that the equilibrium assumption for background production should not be universally applied, with two samples even yielding measured ^{36}Cl concentrations smaller than the calculated equilibrium background production. This provides an opportunity to investigate the influence of young bedrock formation ages on young exposure ages in the application of ^{36}Cl.

Study Area

Jan Mayen is situated on the Jan Mayen fracture zone in the North Atlantic (70°5′–72°N; 7°5′–8°5′W; Figure 1). Nord-Jan, the northeast part of the small island, is dominated by the 2277 m high mountain Beerenberg, an active stratovolcano. The slopes of the volcano are covered by glaciers, a few of which reach down to sea level. In front of these glaciers are moraine ridges and other glacial deposits, with the most prominent of these assumed to be from the Little Ice Age, LIA, based on morphology, (lack of) vegetation cover, and historical observations [12]. In contrast, the southwestern part of the island, Sør-Jan, has a rougher and more variable topography, with its highest peak, Rudolftoppen, extending to

769 m. Nord-Jan and Sør-Jan are connected by Midt-Jan, a narrow (2–3 km wide) strip of land extending southwest from Beerenberg (Figure 1).

The oldest K/Ar age from the bedrock on Jan Mayen gives an age of only 564 ± 6 ka [13], whereas the youngest bedrock on the island was formed during an eruption in 1985 [10,14]. Three other eruptions were observed in 1732, 1818, and 1970 [15,16], and it is likely that there were also at least two other eruptions in the 1650–1882 period [17].

The bedrock is dominated by trachybasaltic and ankarmitic lava flows [11,16,18], which have been grouped into three separate formations. The oldest of these, Havhestberget Formation, has limited surface exposure as it is mostly covered by the younger Nordvestkapp and Inndalen Formations. Of the two latter formations, the Nordvestkapp Formation predates the Last Glacial Maximum glaciation on Jan Mayen, and thus frequently shows signs of glacial reshaping, whereas the Inndalen Formation is largely of Holocene age and postdates the deglaciation of south and central Jan Mayen.

During the Last Glacial Maximum, all of Jan Mayen was covered by an ice cap, which most likely also reached out onto the surrounding continental shelf (Lyså et al. (2021) [1], partially based on a subset of the ages presented here).

2. Materials and Methods

2.1. Sampling Strategy

In total, 89 samples were collected for ^{36}Cl dating on Jan Mayen between 2014 and 2018. The sampling focus of the early campaigns was on glacial landforms (e.g., boulders on moraine ridges, erratic boulders on till surfaces, and glacially abraded bedrock surfaces), but in the later campaigns, the young lava flows were also targeted (Figure 3).

The sample locations were recorded using a handheld GPS. All samples were preferentially collected from large, flat surfaces (dip <20° for all samples). The samples were retrieved using hammer and a chisel (2014) and an electrical rock saw (2015–2018; Figure 3D). In addition to being much more efficient, we found that the use of an electrical rock saw resulted in thinner and more consistent sample thicknesses than what was achieved with only a hammer and chisel.

The topographic shielding at each sampling location was measured with a clinometer and compass. Topographic shielding was calculated from these measurements using the CRONUS Topographic Shielding Calculator v2.1. To enable topographic shielding measurements, sampling for cosmogenic surface exposure dating was preferentially performed on fair weather days, but the difficult logistics and maritime climate, which caused plenty of fog, meant that full visibility was not always possible (Figure 3C). In particular, the peak of Beerenberg was often hidden. Missing observations were replaced by the nearest measurements of the horizon angle or, when in the direction of Beerenberg, by our best estimate of the angle towards the peak. The sample elevations were measured by GPS and/or retrieved from the available small-scale map by the Norwegian Polar Institute (40 m contour interval).

Figure 3. Photographs showing selected sample locations. (**A**) JM2015-11, erratic boulder on till, Midt-Jan. (**B**) JM2018-74, lava, tindar on Sør-Jan. (**C**) JM2015-08, striated erratic boulder on till, Nord-Jan. (**D**) JM2015-01, erratic boulder on Nord-Jan. A battery-powered rock saw was used for sampling. (**E**) JM2018-113, lava flow. (**F**) Thin layer with plant material overlying a lava flow.

2.2. Sample Treatment and Measurements

Due to the fine-grained nature of the sampled volcanic rocks, they were processed as whole rock samples. The samples were prepared at the University of Bern Surface Exposure Dating Laboratory, based on the isotope dilution technique [19–21] and following the protocols by Stone et al. [22] and Ivy-Ochs et al. [20,23]. Before chemical treatment, the samples were crushed and sieved to a grain size of 250–400 μm at the Geological Survey of Norway. Afterwards, the crushed material was leached twice to remove any possible non-in situ Cl contamination (e.g., Zreda et al. (1991) [24]). During leaching, 75 mL of 2M HNO_3 and 500 mL ultrapure water were added to the samples, which were then left overnight before being successively rinsed four times with ultrapure water and dried on a hotplate at 60 °C. A ~10 g aliquot was split from every sample for geochemical analysis, which was performed by Actlabs Analytical Services, Ontario, Canada. The major elements and relevant trace elements were measured using ICP and ICP-MS to enable the calculation of sample production rate (Table S1). A further 30 g of leached material was processed in preparation for accelerator mass spectrometry (AMS) analysis. One chemical blank was processed along with batches of maximum 15 samples in order to determine the chemical background to be subtracted from the samples. Samples were spiked with roughly 2.5–3.5 mg of pure ^{35}Cl carrier (99.63 atom %) in order to calculate the total Cl concentration (^{35}Cl, ^{37}Cl) [20,23], and were then gradually dissolved with 30 mL of 14 M HNO_3 and 120 mL of 40% HF. To remove the impurities and recover supernatant, the samples were centrifuged. Afterwards, 10 mL of 0.4 M $AgNO_3$ solution was added in the dark to precipitate AgCl. The precipitated AgCl was collected and dissolved with 2 mL of NH_4OH (16% solution). In order to suppress the unwanted isobar of ^{36}S from ^{36}Cl through AMS measurements, $BaSO_4$ precipitation was performed by adding $Ba(NO_3)_2$. At the final stage, AgCl was recovered in the form of a solid pill, rinsed with ultrapure water, and then dried. The AgCl pills were finally pressed into tantalum-lined copper targets for subsequent AMS measurements.

Concentrations of total Cl and ^{36}Cl were measured by a single target at the ETH 6 MV Tandem AMS facility using a gas-filled magnet in combination with a gas-ionization chamber for separation of ^{36}Cl from the isobar ^{36}S, in accordance with the isotope dilution technique [20,25–27]. The stable ratio of $^{37}Cl/^{35}Cl$ was normalized to the neutral ratio $^{37}Cl/^{35}Cl = 31.98\%$ of the K382/4N standard and the machine blank. ETH internal standard K382/4N with a value of $^{36}Cl/Cl = (17.36 \pm 0.35) \times 10^{-12}$ [26] was applied to normalize yielded ratios of $^{36}Cl/^{35}Cl$. The sulfur correction of the measured $^{36}Cl/^{35}Cl$ ratio was not substantial. Moreover, measured $^{36}Cl/^{35}Cl$ ratios of the sample were corrected for a procedural blank of $(1 \pm 0.02) \times 10^{-15}$. AMS results for blanks and carrier are included in Supplementary Table S3.

2.3. Calculations

Although initial exposure ages were calculated using the CRONUScalc web interface v. 2.0 [5], the model was not accurate for young samples on young volcanic material due to the built-in equilibrium assumption about background production. The CRONUScalc code was therefore modified directly in MATLAB to enable a flexible entry style allowing for calculation of three scenarios: (1) the rock formation age is sufficiently old that the equilibrium assumption for background production is reasonable (equivalent to CRONUScalc 2.1); (2) the rock formation age is known from independent constraints so the model implements Equation 1; and (3) the formation age of the rock is equal to the exposure age of the sample, so the exposure/formation age is solved iteratively. The inclusion of rock formation age as a required input is similar to the method used in the calculation spreadsheet in Schimmelpfennig et al. (2009) [7], but the calculations in CRONUScalc are fully automated rather than manual iteration on a sample-by-sample basis, and CRONUScalc allows easy access to time-dependent production rate scaling models. All CRONUScalc code, including these modifications, is publicly available on Bitbucket at https://bitbucket.org/cronusearth/cronus-calc/ [accessed on 14 September 2021; v2.2, tag: Anjar_et_al_2021_JanMayen]. The CRONUScalc code was also updated and recalibrated to use the recently published rates for cosmogenic ^{36}Cl production from iron published in Moore et al. (2019) [28]. These new rates are based on modern, robust analyses of multiple high-iron samples and can be incorporated in a similar format to the other spallation production rates already present in the code, unlike the previous Stone et al. (2005) rates [29]. This recalibration builds on the updated rates presented in Marrero et al. (2021) [30] and is discussed in more detail in Leontaritis (2021) [31]. Production rates of ^{36}Cl from iron were derived for each scaling model by A. Moore [32], and these were fixed values in the recalibrations. The other production parameters were then recalibrated following the full procedure outlined in Marrero et al. (2016b) [9]. The new production rates are very similar to earlier rates and are given in Table 1. Iron oxide concentrations are >13 wt % in some samples, meaning that approximately 1–4% of production in the Jan Mayen samples is from this pathway. This has the potential to make a small but measurable difference in sample ages compared to results from CRONUScalc v2.1.

Table 1. ^{36}Cl production rates for LM scaling model used to calculate these exposure ages. Percent change from Marrero et al. (2021) [30] (CRONUScalc v2.1) is shown for comparison. See Marrero et al. (2016b) [9] for definitions of pathways. Muon parameter changes were all <0.07%.

Pathway	Production Rate (at ^{36}Cl g^{-1} y^{-1})	% Change From v2.1
Ca—Spallation	51.686 ± 3.3	−0.0003
K—Spallation	150.996 ± 10	0.66
$P_f(0)$	647.705 ± 231	0.37

Final results for exposure ages presented in this paper use the geomagnetically corrected Lal (1991) [33]/Stone (2000) [34] scaling (LM) [5,9] and ^{36}Cl production rates from Marrero et al. (2021) [30] and Moore et al. (2019) [28], as described above. A range of rock formation ages ranging from infinite (yielding youngest exposure age) to "equal to exposure age" (yielding oldest possible exposure age) was used to evaluate a range of scenarios. Where nothing else is stated, the ages were calculated assuming a density of 2.1 ± 0.5 g/cm^3, based on average density of Icelandic lava measured by Licciardi et al. (2008) [35], and assuming an erosion rate of 1 ± 0.5 mm/ky. The input data used is presented in Supplementary Table S2.

2.4. Radiocarbon Dating

One of the exposure-dated lava flows had been partially buried by sediments (in a different part than where the exposure samples were taken), and a pit was dug through the sediments and down into the lava flow. Unidentified plant material found directly overlying the lava flow (Figure 3F) was sampled for radiocarbon dating and dated at the National Laboratory for Age Determination, Trondheim, Norway. It was calibrated using OxCal v. 4.4 and the IntCal20 atmospheric curve [36].

3. Results and Interpretations

3.1. Geochemistry

As part of the dating process, major elements and relevant trace elements were measured for all 89 samples (Supplementary Table S1) and the rock types were identified based on the geochemistry, following the classification in Le Maitre et al. (2002) [37]. A total of 18% of the samples were identified as Mg-rich rock types, mostly picrites, whereas the remaining samples were dominated by basalt, trachybasalt, and basaltic trachyandesite (Figure 4).

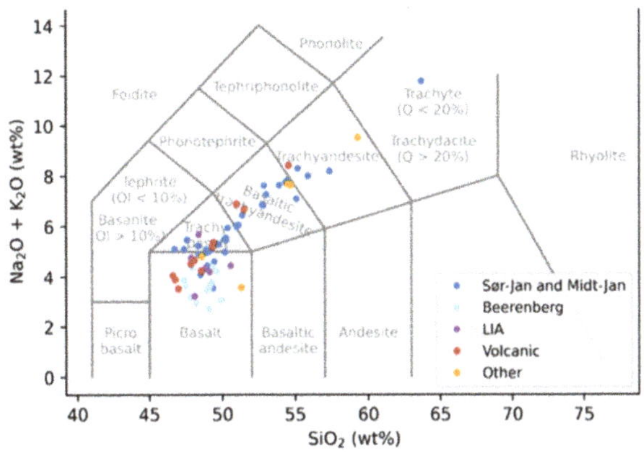

Figure 4. Rock-type classification based on the geochemistry of the samples. High-Mg (n = 15) samples have been excluded as those are not classified with a TAS-plot. Additionally, samples JM2014-01 and JM2014-02 have been excluded due to methodological problems with their geochemistry measurements (see Supplementary Table S1). The colors are the same as in Figure 1.

3.2. Radiocarbon Date

The single radiocarbon sample was dated to 2.761–2.747 cal. ka BP (sample id: JM2019-122; Lab. id: Tra-14525; ^{14}C age: 2.643 ± 0.017 ^{14}C ka BP; calibrated age given with 1 sigma uncertainty). As the organic material was found directly overlying the lava flow, the radiocarbon age represents a minimum limiting age for both the lava flow and the two samples collected from the lava flow, JM2017-69 and JM2017-70.

3.3. Cosmogenic Nuclide Surface Exposure Age Dating

Samples were divided into three main groups: "glacial" samples, taken in settings where the exposure age is expected to reflect deglaciation; "volcanic" samples, e.g., from lava flows; and "other" samples, taken in settings where the exposure age is not directly related to the deglaciation or to volcanic activity (Figure 5). The glacial samples include samples taken from erratic boulders and striated bedrock. They were further subdivided into regions, with the lower lying areas in southern and central Jan Mayen (Sør-Jan and Midt-Jan, including some low-lying areas just north of Lake Nordlaguna) in one group (n = 41), the slopes of Beerenberg in the second subgroup (n = 24), and the samples taken from the fresh moraine ridges interpreted to be from the Little Ice Age in the third group (n = 8). The volcanic samples (n = 11) include samples from lava flows and from a tindar. The "other" samples (n = 5) include a volcanic boulder deposited by a rock fall and two samples from a fluvially eroded setting.

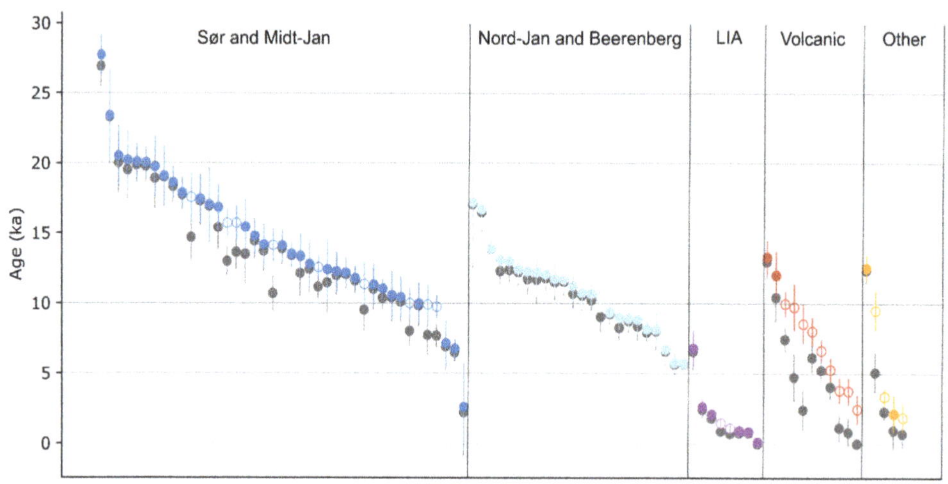

Figure 5. Exposure ages calculated assuming that the rock formation age equals the exposure age (i.e., young rock ages) in color, and below (in grey) the ages of the same sample calculated with the assumption that the background ^{36}Cl had reached equilibrium. The samples have been separated into sample setting and region and sorted from the oldest to the youngest within each region. The uncertainties are the internal uncertainties (1 standard deviation). Samples where the equil. age differs by less than 1 standard deviation from the young rock age are indicated by filled circles, and if they differ by more than 1 standard deviation, indicated by unfilled circles.

Due to the young bedrock on Jan Mayen, we suspected that the background production might not have reached equilibrium, and all ages were therefore calculated with five different background production scenarios: (1) equilibrium assumption (equivalent to CRONUScalc 2.1), (2) rock formation at 564 ka (oldest dated rock on Jan Mayen [13]), (3) rock formation at 250 ka, (4) rock formation at 50 ka (250–50 ka represents the range of rock formation ages for the likely source formation for the glacial material), and (5) rock formation age equal to the exposure age of the sample (youngest possible rock formation age) (Table 2).

Table 2. Exposure ages calculated using five different scenarios for the background production. Our best estimate ages are indicated in bold. Calculated using LM scaling and assuming a density of 2.1 ± 0.5 g/cm^3 and an erosion rate of 1 ± 0.5 mm/ky. See Supplementary Table S2 for the full set of sample information used in the calculations.

Name	Latitude WGS84	Longitude WGS84	Elevation	Equilibrium Assumption * (*Equil.*)			Formation Age = 564 ka			Formation Age = 250 ka			Formation Age = 50 ka			Formation Age = Exposure Age (*Young Rock*)		
	[dd]	[dd]	[m]	Age [ka]	Unc (ext)	Unc (int)	Age [ka]	Unc (ext)	Unc (int)	Age [ka]	Unc (ext)	Unc (int)	Age [ka]	Unc (ext)	Unc (int)	Age [ka]	Unc (ext)	Unc (int)
Glacial samples–Sør Jan and Midt Jan																		
JM2014-01	70.9751	−8.6210	114	19.8	1.6	1.2	19.9	1.6	1.2	19.9	1.6	1.2	**20.0**	1.6	1.2	20.0	1.6	1.2
JM2014-18	71.0144	−8.4514	52	7.7	1.6	0.8	8.3	1.6	0.8	8.9	1.7	0.8	**9.6**	1.8	0.8	9.8	1.9	0.8
JM2014-19	71.0104	−8.4334	119	13.4	1.1	0.9	13.4	1.1	0.9	13.4	1.1	0.9	**13.5**	1.1	0.9	13.5	1.1	0.9
JM2014-20	71.0123	−8.4226	144	11.6	1.2	0.9	11.7	1.2	0.9	11.7	1.2	0.9	**11.8**	1.2	0.9	11.8	1.2	0.9
JM2014-21	71.0129	−8.4121	162	19.0	2.4	2.2	19.0	2.4	2.2	19.1	2.4	2.2	**19.1**	2.4	2.2	19.1	2.4	2.2
JM2014-22	71.0227	−8.4404	55	11.2	1.2	0.8	11.5	1.3	0.8	12.0	1.3	0.9	**12.4**	1.3	0.9	12.6	1.3	0.8
JM2014-23	70.9896	−8.4967	71	19.9	1.6	1.3	19.9	1.6	1.3	20.0	1.6	1.3	**20.1**	1.6	1.3	20.1	1.6	1.3
JM2015-11	71.0193	−8.4489	41	10.4	1.8	1.7	10.5	1.8	1.7	10.5	1.8	1.7	**10.6**	1.8	1.7	10.6	1.8	1.7
JM2015-73	70.8673	−8.8091	45	7.7	1.6	1.5	8.4	1.6	1.5	9.0	1.6	1.5	**9.7**	1.7	1.5	9.9	1.7	1.4
JM2015-74	70.8680	−8.8063	32	10.7	1.7	1.2	11.7	1.8	1.2	12.7	1.9	1.2	**13.9**	2.0	1.2	14.1	2.0	1.2
JM2015-99	70.9707	−8.6012	115	10.1	1.5	1.4	10.2	1.5	1.4	10.3	1.5	1.4	**10.4**	1.5	1.4	10.5	1.5	1.4
JM2015-100	70.9734	−8.6085	109	2.2	3.2	3.2	2.3	3.2	3.2	2.5	3.2	3.2	**2.6**	3.2	3.2	2.6	3.2	3.2
JM2015-101	70.9732	−8.6095	100	12.4	1.8	1.7	12.5	1.8	1.7	12.6	1.8	1.7	**12.8**	1.8	1.7	12.8	1.8	1.7
JM2016-22	70.8974	−8.9312	310	20.0	2.6	2.2	20.2	2.6	2.2	20.3	2.6	2.2	**20.5**	2.6	2.2	20.5	2.6	2.2
JM2016-23	70.8924	−8.9137	412	17.3	1.9	1.7	17.3	1.9	1.7	17.4	1.9	1.7	**17.5**	1.9	1.7	17.5	1.9	1.7
JM2016-24	70.8894	−8.9146	437	13.5	3.0	2.1	14.0	3.1	2.1	14.6	3.1	2.1	**15.3**	3.2	2.1	15.5	3.2	2.1
JM2016-25	70.8924	−8.9304	258	19.5	2.6	2.1	19.7	2.7	2.1	19.9	2.7	2.1	**20.2**	2.7	2.1	20.2	2.7	2.1
JM2016-26	70.8923	−8.9300	258	18.9	2.8	2.1	19.1	2.8	2.1	19.4	2.9	2.1	**19.7**	2.9	2.1	19.8	2.9	2.1
JM2016-29	70.9281	−8.7716	393	12.1	2.1	1.7	12.5	2.2	1.7	12.8	2.2	1.7	**13.2**	2.2	1.7	13.3	2.2	1.7
JM2016-30	70.9294	−8.7822	366	12.0	1.5	1.4	12.1	1.5	1.4	12.1	1.5	1.4	**12.2**	1.5	1.4	12.3	1.5	1.4
JM2016-31	70.9581	−8.6804	265	14.4	1.5	1.3	14.5	1.5	1.3	14.6	1.5	1.3	**14.8**	1.5	1.3	14.8	1.5	1.3
JM2016-32	70.9501	−8.7037	337	14.7	3.1	1.7	15.5	3.2	1.7	16.4	3.4	1.7	**17.4**	3.5	1.7	17.6	3.5	1.7
JM2016-33	70.9513	−8.7423	202	13.7	1.7	1.5	13.8	1.7	1.5	14.0	1.7	1.5	**14.1**	1.7	1.5	14.2	1.7	1.5
JM2016-34	70.9519	−8.7368	217	11.5	2.4	2.2	11.7	2.4	2.2	12.0	2.5	2.2	**12.3**	2.5	2.2	12.4	2.5	2.2
JM2016-35	70.9508	−8.6906	318	13.9	1.4	1.2	13.9	1.4	1.2	14.0	1.4	1.2	**14.1**	1.4	1.2	14.1	1.4	1.2
JM2016-39	70.8699	−8.8181	126	9.5	1.7	1.6	10.0	1.8	1.6	10.6	1.8	1.6	**11.2**	1.8	1.6	11.4	1.8	1.6
JM2016-40	70.9300	−8.8159	205	11.0	1.6	1.5	11.1	1.6	1.5	11.2	1.6	1.5	**11.3**	1.6	1.5	11.3	1.6	1.5
JM2016-42	70.8776	−8.9529	321	23.3	3.5	3.3	23.3	3.5	3.3	23.4	3.5	3.3	**23.4**	3.5	3.3	23.4	3.5	3.3
JM2017-34	70.8474	−9.0162	180	18.3	1.5	1.1	18.4	1.5	1.1	18.5	1.5	1.1	**18.6**	1.5	1.1	18.6	1.5	1.1
JM2017-35	70.8410	−9.0084	211	10.4	1.0	0.8	10.6	1.0	0.8	10.8	1.0	0.8	**11.0**	1.0	0.8	11.0	1.0	0.8
JM2017-36	70.8413	−8.9938	231	16.9	2.8	2.7	17.0	2.8	2.7	17.0	2.8	2.7	**17.0**	2.8	2.7	17.0	2.8	2.7
JM2017-42	71.0011	−8.4445	320	17.8	1.4	1.1	17.8	1.4	1.1	17.8	1.4	1.1	**17.9**	1.4	1.1	17.9	1.4	1.1
JM2017-45	70.8955	−8.8458	605	6.5	0.8	0.7	6.6	0.8	0.7	6.7	0.8	0.7	**6.8**	0.8	0.7	6.8	0.8	0.7
JM2017-56	71.0041	−8.4871	51	12.1	1.2	1.1	12.1	1.2	1.1	12.1	1.2	1.1	**12.2**	1.2	1.1	12.2	1.2	1.1
JM2017-57	71.0041	−8.4871	51	9.8	1.6	1.5	9.9	1.6	1.5	9.9	1.6	1.5	**10.0**	1.6	1.5	10.0	1.6	1.5
JM2017-58	71.0091	−8.3948	230	15.4	2.0	1.6	15.8	2.1	1.6	16.3	2.1	1.6	**16.8**	2.1	1.6	16.9	2.1	1.6

Table 2. *Cont.*

Name	Latitude WGS84	Longitude WGS84	Elevation	Equilibrium Assumption * (*Equil.*)			Formation Age = 564 ka			Formation Age = 250 ka			Formation Age = 50 ka			Formation Age = Exposure Age (*Young Rock*)		
	[dd]	[dd]	[m]	Age [ka]	Unc (ext)	Unc (int)	Age [ka]	Unc (ext)	Unc (int)	Age [ka]	Unc (ext)	Unc (int)	Age [ka]	Unc (ext)	Unc (int)	Age [ka]	Unc (ext)	Unc (int)
JM2017-59	71.0133	−8.3844	334	26.9	2.3	1.5	27.2	2.4	1.5	27.4	2.4	1.5	**27.7**	2.4	1.5	27.8	2.4	1.5
JM2017-64	70.9941	−8.4626	29	13.0	1.5	1.0	13.8	1.6	1.0	14.6	1.6	1.0	**15.5**	1.7	1.0	15.7	1.7	1.0
JM2017-65	71.0129	−8.4459	71	6.9	1.7	1.7	7.0	1.7	1.7	7.1	1.7	1.7	**7.1**	1.7	1.7	7.2	1.7	1.7
JM2017-66	71.0146	−8.4501	41	13.6	2.7	1.2	14.2	2.8	1.2	14.9	2.9	1.2	**15.6**	3.1	1.2	15.7	3.1	1.2
JM2017-67	71.0148	−8.4506	41	8.0	1.7	1.1	8.6	1.8	1.1	9.2	1.9	1.1	**9.8**	2.0	1.1	10.0	2.0	1.1
Glacial samples–Beerenberg																		
JM2014-02	70.9955	−8.2571	116	13.9	1.3	0.9	13.9	1.3	0.9	13.9	1.3	0.9	**13.9**	1.3	0.9	13.9	1.3	0.9
JM2014-03	70.9971	−8.2528	140	11.8	1.4	0.8	11.9	1.4	0.8	11.9	1.4	0.8	**12.0**	1.4	0.8	12.1	1.4	0.8
JM2014-04	70.9976	−8.2515	144	10.2	1.8	0.8	10.4	1.8	0.8	10.5	1.8	0.8	**10.7**	1.8	0.8	10.7	1.8	0.8
JM2014-05	71.0021	−8.2298	170	8.8	0.9	0.8	8.8	0.9	0.8	8.8	0.9	0.8	**8.9**	0.9	0.8	8.9	0.9	0.8
JM2014-10	71.0008	−8.1782	156	5.7	0.8	0.8	5.7	0.8	0.8	5.7	0.8	0.8	**5.8**	0.8	0.8	5.8	0.8	0.8
JM2014-11	71.0003	−8.1790	153	6.6	1.1	1.1	6.6	1.1	1.1	6.7	1.1	1.1	**6.7**	1.1	1.1	6.7	1.1	1.1
JM2014-12	70.9962	−8.1883	81	9.3	0.9	0.7	9.3	0.9	0.7	9.3	0.9	0.7	**9.4**	0.9	0.7	9.4	0.9	0.7
JM2014-13	71.0040	−8.1531	223	8.3	1.7	0.9	8.5	1.7	0.9	8.7	1.7	0.9	**8.9**	1.8	0.9	9.0	1.8	0.9
JM2014-14	71.0043	−8.1532	233	12.4	2.1	0.9	12.6	2.1	0.9	12.8	2.1	0.9	**13.0**	2.1	0.9	13.0	2.2	0.9
JM2014-15	71.0040	−8.1331	179	16.5	1.6	1.4	16.5	1.6	1.4	16.6	1.6	1.4	**16.7**	1.6	1.4	16.7	1.6	1.4
JM2014-16	71.0041	−8.1333	181	7.9	0.8	0.7	8.0	0.8	0.7	8.1	0.8	0.7	**8.2**	0.8	0.7	8.2	0.8	0.7
JM2014-17	70.9976	−8.1844	47	10.6	1.2	1.1	10.6	1.2	1.1	10.7	1.2	1.1	**10.7**	1.2	1.1	10.8	1.2	1.1
JM2015-01	71.0291	−8.3451	460	12.3	1.9	0.9	12.5	1.9	0.9	12.8	1.9	0.9	**13.0**	2.0	0.9	13.1	2.0	0.9
JM2015-02	71.0291	−8.3451	460	8.0	1.4	1.3	8.1	1.4	1.3	8.1	1.4	1.3	**8.1**	1.4	1.3	8.2	1.4	1.3
JM2015-03	71.0307	−8.3239	552	5.7	0.7	0.6	5.7	0.7	0.6	5.7	0.7	0.6	**5.7**	0.7	0.6	5.7	0.7	0.6
JM2015-07	71.0530	−8.4027	122	8.4	1.2	1.0	8.5	1.2	1.0	8.6	1.2	1.0	**8.8**	1.2	1.0	8.8	1.2	1.0
JM2015-08	71.0519	−8.3994	122	11.7	2.0	1.7	11.8	2.0	1.7	12.0	2.0	1.7	**12.2**	2.1	1.7	12.2	2.1	1.7
JM2015-33	71.0032	−8.1487	213	9.0	2.0	1.1	9.4	2.1	1.1	9.8	2.1	1.1	**10.3**	2.2	1.1	10.4	2.2	1.1
JM2015-34	71.0046	−8.1360	195	11.5	1.5	1.3	11.6	1.5	1.3	11.6	1.5	1.3	**11.7**	1.5	1.3	11.7	1.5	1.3
JM2015-42	71.0517	−8.3676	306	10.7	1.5	1.1	10.8	1.5	1.1	11.0	1.5	1.1	**11.2**	1.6	1.1	11.2	1.6	1.1
JM2017-14	70.9983	−8.2588	145	11.5	1.3	1.0	11.6	1.3	1.0	11.6	1.3	1.0	**11.7**	1.3	1.0	11.7	1.3	1.0
JM2017-15	71.0161	−8.2402	367	12.2	1.2	0.9	12.3	1.2	0.9	12.3	1.2	0.9	**12.4**	1.2	0.9	12.4	1.2	0.9
JM2017-18	71.0043	−8.1198	84	17.1	4.6	4.6	17.1	4.6	4.6	17.2	4.6	4.6	**17.2**	4.6	4.6	17.2	4.6	4.6
JM2017-19	71.0026	−8.1165	67	11.7	1.4	1.1	11.9	1.5	1.1	12.0	1.5	1.1	**12.2**	1.5	1.1	12.3	1.5	1.1
Glacial samples–Little Ice Age moraines																		
JM2014-06	71.0035	−8.2217	191	6.6	1.4	1.3	6.6	1.4	1.3	6.7	1.4	1.3	**6.8**	1.4	1.3	6.8	1.4	1.3
JM2014-08	70.9992	−8.1929	162	0.8	0.4	0.4	0.8	0.4	0.4	0.9	0.4	0.4	**0.9**	0.4	0.4	0.9	0.4	0.4
JM2014-09	71.0003	−8.1905	191	0.0	0.0	0.0	0.0	0.0	0.0	0.0	0.4	0.4	**0.1**	0.4	0.4	0.1	0.4	0.4
JM2015-04	71.0314	−8.3225	565	0.9	0.3	0.3	1.0	0.3	0.3	1.2	0.3	0.3	**1.4**	0.3	0.3	1.5	0.3	0.3
JM2015-05	71.0341	−8.3190	602	2.4	0.6	0.6	2.5	0.6	0.6	2.5	0.6	0.6	**2.6**	0.6	0.6	2.6	0.6	0.6
JM2015-06	71.0336	−8.3171	604	0.8	0.3	0.3	0.8	0.3	0.3	0.8	0.3	0.3	**0.9**	0.3	0.3	0.9	0.3	0.3
JM2015-43	71.0564	−8.3367	491	0.7	0.2	0.2	0.8	0.2	0.2	0.9	0.2	0.2	**1.1**	0.2	0.2	1.1	0.2	0.2
JM2015-44	71.0576	−8.3424	462	1.8	0.4	0.4	1.9	0.4	0.4	2.0	0.4	0.4	**2.1**	0.4	0.4	2.1	0.4	0.4

Table 2. *Cont.*

Name	Latitude WGS84	Longitude WGS84	Elevation	Equilibrium Assumption * (*Equil.*)			Formation Age = 564 ka			Formation Age = 250 ka			Formation Age = 50 ka			Formation Age = Exposure Age (*Young Rock*)		
	[dd]	[dd]	[m]	Age [ka]	Unc (ext)	Unc (int)	Age [ka]	Unc (ext)	Unc (int)	Age [ka]	Unc (ext)	Unc (int)	Age [ka]	Unc (ext)	Unc (int)	Age [ka]	Unc (ext)	Unc (int)
Other samples																		
JM2014-24	70.9145	−8.7853	428	2.3	0.7	0.6	2.6	0.8	0.6	2.9	0.8	0.6	**3.2**	0.9	0.6	3.3	0.9	0.6
JM2015-62	71.0043	−8.4476	63	5.1	1.5	1.4	6.3	1.6	1.4	7.6	1.6	1.4	**9.1**	1.7	1.4	9.5	1.8	1.4
JM2016-27	70.8772	−9.0040	26	0.7	1.0	1.0	1.0	1.0	1.0	1.4	1.0	1.0	**1.7**	1.0	1.0	1.9	1.0	1.0
JM2016-28	70.8772	−9.0040	42	0.9	1.3	1.3	1.3	1.3	1.3	1.6	1.3	1.3	**2.0**	1.3	1.3	2.1	1.3	1.3
JM2018-74	70.9633	−8.5863	165	12.4	1.1	1.0	12.4	1.1	1.0	12.5	1.1	1.0	**12.5**	1.1	1.0	12.5	1.1	1.0
Volcanic samples																		
JM2017-37	70.8591	−9.0499	2	4.8	2.1	1.7	6.2	2.3	1.7	7.6	2.5	1.7	9.3	2.9	1.7	**9.7**	2.9	1.7
JM2017-39	70.8785	−8.9971	21	2.4	1.5	1.4	4.1	1.7	1.4	6.0	2.0	1.4	8.0	2.3	1.4	**8.6**	2.4	1.4
JM2017-43	70.9018	−8.7814	181	7.5	1.6	0.9	8.2	1.7	0.9	8.9	1.9	0.9	9.8	2.0	0.9	**10.0**	2.0	0.9
JM2017-69	70.9251	−8.7082	18	1.1	1.0	0.9	1.9	1.0	0.9	2.7	1.1	0.9	3.5	1.2	0.9	**3.8**	1.2	0.9
JM2017-70	70.9244	−8.7145	26	0.8	1.0	1.0	1.6	1.0	1.0	2.5	1.1	1.0	3.5	1.2	1.0	**3.7**	1.3	1.0
JM2018-101	70.9170	−8.8396	290	6.2	1.8	1.0	6.7	1.9	1.0	7.2	2.0	1.0	7.9	2.1	1.0	**8.0**	2.2	1.0
JM2018-107	70.8969	−8.9121	285	13.0	1.4	1.2	13.1	1.4	1.2	13.2	1.4	1.2	13.3	1.4	1.2	**13.3**	1.4	1.2
JM2018-109	70.8835	−8.9459	250	5.3	1.4	0.8	5.6	1.4	0.8	6.1	1.5	0.8	6.5	1.6	0.8	**6.7**	1.6	0.8
JM2018-111	70.9175	−8.8796	70	10.4	2.8	1.7	10.9	2.8	1.7	11.4	2.9	1.7	11.9	3.0	1.7	**12.0**	3.0	1.7
JM2018-113	70.9131	−8.7384	60	4.0	1.1	0.8	4.4	1.2	0.8	4.8	1.2	0.8	5.2	1.3	0.8	**5.3**	1.3	0.8
JM2018-115	70.9092	−8.7533	150	0.0	0.0	0.0	0.0	0.0	0.0	0.8	1.1	1.1	2.1	1.2	1.1	**2.5**	1.2	1.0

* Original using updated Fe

Proportions of cosmogenic ^{36}Cl attributed to the three major production pathways (K, Ca, and Cl) varies substantially between the samples (Figure 6). Samples with a higher proportion of production from the Cl pathway are more likely to be affected by the formation age assumption. The range of possible scenarios surrounding the background production is discussed in detail later.

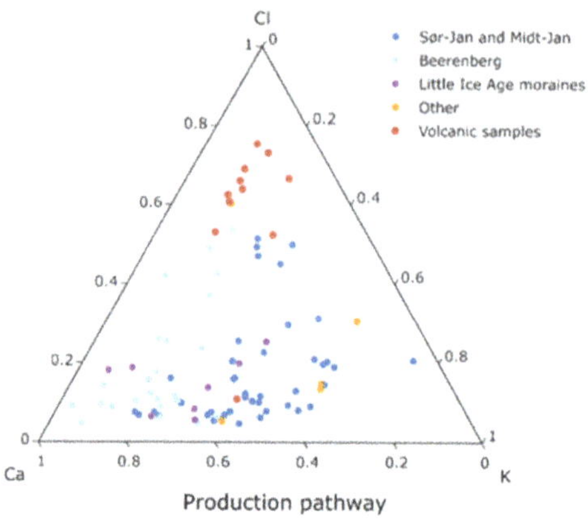

Figure 6. Ternary plot indicating the contributions from each major production pathway. The samples have been normalized so that the Ca production + K production + Cl production = 100%. The colors indicate sample region (the same colors as in Figure 1).

4. Discussion

4.1. Evaluation of General Uncertainties

The dominant reactions that produce cosmogenic ^{36}Cl target K, Ca, and ^{35}Cl in the rock (Figure 6). Whereas ^{36}Cl production through spallation of K and Ca is reasonably well constrained, the production of ^{36}Cl by ^{35}Cl neutron capture is more complicated, and sensitive to, e.g., composition, water content, and snow cover [7–9,38]. This means that higher Cl concentrations lead to larger age uncertainties for those samples. Marrero et al. (2016) [9] defined a high Cl sample as a sample with more than 80 ppm Cl, which applies to 27 of our samples, with a maximum of 835 ppm in JM2018-115. Although all samples are taken from volcanic rocks, the samples collected to examine volcanic aspects have particularly high Cl contents compared with samples collected for other purposes (9 out of 11 volcanic samples are affected, Figure 5). The reason for this is unclear.

The erosion rate will affect the ^{36}Cl exposure ages but is unmeasured on Jan Mayen. As an estimate, we therefore used an erosion rate of 1 ± 0.5 mm/ky, similar to what has previously been used in Iceland (1.11 mm/ky [39]). Changing the estimated erosion rates to either 0 ± 0 mm/ky or 2 ± 0.5 mm/ky changes the calculated ages by, on average, less than 1%. If a substantially higher erosion rate is used, e.g., 5 mm/ky, there is an average change of ~2%, although for a small number of samples, the effect is more pronounced (5 samples changed 5–7%).

The sample density will also have a small influence on the calculated ages. Although sample densities were not measured in this study, we assume densities measured by Licciardi et al. (2008) [35] for lava flows in Iceland, ranging from 1.76 to 2.44 g/cm^3, are likely to be similar to densities on Jan Mayen. We have therefore assumed a density of 2.1 ± 0.5 g/cm^3 for all samples, but the typically variable vesicularity of the sampled volcanic rocks indicates that a substantial spread in the true density of the samples is likely. To test the sensitivity of the reported ages to erosion, we calculated the ages using densities of 1.7 and 2.5 g/cm^3. The resulting ages changed by an average of <1%, well within uncertainties.

The production of cosmogenic nuclides can also be temporally influenced by local shielding effects, such as vegetation and snow or soil cover. The present-day vegetation on Jan Mayen is sparse (Figure 3) and not expected to influence the ages. Snow cover could be more important, but the maritime climate on Jan Mayen causes warm winters and generally thin snow cover, at least on lower elevations. Wind drift in the very open landscape could cause more substantial snow cover in shielded positions, but we consider the existing observations insufficient for estimating snow thickness for our samples and have, therefore, not corrected for snow shielding.

In addition, isostasy and topographic shielding will also influence the ages. The present-day topographic shielding was measured in the field, and although poor visibility limited the observations for some of the samples, we consider the actual influence of these uncertainties on the ages to be small, as most of the samples were taken in reasonably open terrain and during at least partial visibility. The possible exceptions are samples JM2017-36 and JM2018-101, for which no shielding measurements were possible. Isostatic changes will also have some influence on the calculated exposure ages as they may change the elevation of a sample. A recent study by Larsen et al. (2021) [2] identified a 14 m vertical tectonic displacement of southwest Beerenberg, following an eruption in 1732. So far, this is the only documented postglacial displacement, but we cannot exclude that there have been other instances of vertical displacements on the island due to both glacioisostasy and volcanic activity. As the extent and duration of such changes are difficult to quantify, no corrections are included here.

4.2. *Influence of Young Bedrock*

A unique problem for ^{36}Cl dating in a young volcanic landscape, such as Jan Mayen, is that the background production and decay have not necessarily reached equilibrium (Figure 2). The Jan Mayen samples provide additional evidence that the equilibrium assumption should not be applied to all samples. If equilibrium is (incorrectly) assumed for our samples, background production of ^{36}Cl is calculated to contribute to between 0.2 and 196.5% of the total measured ^{36}Cl in the samples. The latter value is clearly unreasonable, so we interpret it as an indication that these rocks have not reached equilibrium, which takes approximately 2 million years for ^{36}Cl. This is in agreement with the independent ages indicating that the bedrock is generally young. To estimate the potential influence of rock formation age, we considered five different scenarios for the rock formation age: equilibrium conditions, rock formation at 564, 250, or 50 ka, and assuming that rock formation age equaled the exposure age. In the following discussion, "equil." is used when referring to ages calculated assuming equilibrium conditions, and "young rock" is used to indicate ages calculated assuming that the exposure age equals the rock age.

Although all samples were calculated with multiple scenarios, the calculated age of most of the samples were relatively unaffected. However, as expected, there are a subset of samples that are more significantly affected by rock formation age assumptions (Figure 5). We observe, in detail, these samples to investigate these effects further. Specifically, we define "significantly affected" as samples where the compared ages differed by more than one standard deviation. For 25 samples, the age calculated assuming young rock conditions was more than one standard deviation from the age calculated using equilibrium conditions, with a maximum age difference of 6.1 ka in sample JM2017-39. Using a more realistic comparison between the oldest known rock formation age on Jan Mayen (564 ka) and the young rock scenario reduces the number of significantly affected samples to 22, with a maximum deviation of 4.4 ka.

The worst affected samples largely coincide with the samples classified as high Cl (>80 ppm, 19 out of 25 significantly affected samples), as would be expected since ^{35}Cl is required for background production. Although U and Th concentrations are not especially high (<10 ppm in all cases), the high native Cl concentrations and short exposure durations are sufficient to illustrate the issue. Older samples are less significantly affected because the potential difference between the equil. and nonequil. age becomes smaller through time due to the build-up of the background signal.

For volcanic samples such as lava flows, it is often reasonable to assume that the rock formation age is the same as the exposure age. The question is then whether or not the rock contains any ^{36}Cl formed before the lava solidified. Schimmelpfennig et al. (2009) [7] found that the total ^{36}Cl content in a fully shielded, high-Cl (1093 ppm), 400-year-old lava flow was close to the background values of ^{36}Cl found in the blanks, suggesting minimal contributions of ^{36}Cl formed before the lava solidified. Results from our only two samples collected from an independently-dated Jan Mayen lava flow support this. Samples JM2017-69 and JM2017-70 were dated to 1.1 ± 1 ka and 0.8 ± 1 ka when assuming equilibrium conditions, but to 3.8 ± 1.2 ka and 3.7 ± 1.3 ka when using the young rock age. Only the young rock age agrees with the minimum limiting age from radiocarbon dated plant material found directly overlying the same lava flow, which was dated to 2.761–2.747 cal. ka BP. Two additional samples, JM2014-09 and JM2018-115, had estimated equil. levels of background ^{36}Cl well above the measured total ^{36}Cl concentration (background + cosmogenic), indicating that those samples were not in equilibrium. We therefore conclude that the young rock age gives a better age estimate for the volcanic samples.

Even the oldest bedrock on the island (564 ± 6 ka [13]) has only reached a background production value of ~73% of the equilibrium value (Figure 2). For nonvolcanic samples, it is reasonable to assume that the rock formation age is unrelated to the exposure age, meaning that the true rock formation age lies somewhere between 564 ka and the young rock age. However, most of the glacial samples likely originate from the Nordvestkapp formation (Figure 1), which covers much of the land surface and which predates the deglaciation on south and central Jan Mayen. Most of the K/Ar and Ar/Ar dates from this formation fall between 50 and 250 ka [1,10,13]), which means that the background contribution has reached roughly 10–45% of the equilibrium values. We therefore suggest that the 50 ka and 250 ka formation age scenarios probably give the most accurate ages for these samples, and we use the 50 ka scenario as our best estimate for all the nonvolcanic samples. However, while this interpretation is reasonable on the group level, it does not necessarily hold true for a specific individual sample, and we therefore recommend that the 13 nonvolcanic samples with substantial (>1 standard deviation) difference between the 564 ka scenario age and the young rock age be interpreted with this potential variability in mind.

4.3. Deglaciation Pattern

Although an in-depth discussion of the deglaciation on Jan Mayen is outside the scope of this paper, some general conclusions can be drawn from the cosmogenic exposure ages alone. Note that as we consider the assumption of rock formation at 50 ka to be the most appropriate scenario for glacial samples, it has been used throughout this section. Choice of bedrock formation age model only slightly affects the general deglaciation pattern discussed here, with older formation ages leading to slightly younger exposure ages (Figure 5). For a more extensive discussion on the deglaciation on Jan Mayen, see Lyså et al. (2021) [1].

On south and central Jan Mayen, ages taken from glacial settings range from 27.7 to 2.6 ka, with a median value of 13.9 ka (n = 41, Figure 5). All but two of the ages are ~20 ka or younger, suggesting that deglaciation in this part started properly around 20 ka (Figure 5). In addition, 38 out of the 41 samples give ages in the interval from 20.5 to 6.8 ka, suggesting that most of south and central Jan Mayen became ice free during this time period. Only the final glacial age from this area, JM2015-100, with an age of 2.6 ± 3.2 ka, deviates from this general deglaciation pattern. On Nord-Jan and the slopes of Beerenberg, the deglaciation appears to begin somewhat later, as would be expected around a 2277 m high mountain, with deglaciation ages ranging from 17.2 to 5.7 ka (median 11.0 ka, n = 24), and all but two of them younger than 14 ka (Figure 5).

Apart from a single outlier age at 2.6 ka, none of the glacial samples taken outside of the LIA moraines is younger than 5.7 ka, and it seems reasonable to assume that the glaciers have been at or inside the LIA moraines for the last 5–6 thousand years. The eight samples from the LIA moraines range in age from 6.8 to 0.1 ka, with five of them giving ages younger than 1.5 ka (Figure 5).

5. Conclusions

- In this study, we present 89 cosmogenic exposure ages (^{36}Cl) from Jan Mayen, most of them dating either the deglaciation (n = 73) or postglacial volcanic events (n = 11) on the island.
- Based on the range of exposure ages at each location, large-scale deglaciation on Jan Mayen began ~20 ka and continued until 5.7 ka, after which the glaciers appear to have retreated inside the Little Ice Age moraines.
- The exposure ages on Jan Mayen were calculated using an updated version of CRONUScalc to account for the young bedrock formation ages at the site. Although the formation age assumption does not significantly affect most samples (n = 64), a number of exposure ages change substantially (n = 25) depending on the rock formation age assumed for the sample. On Jan Mayen, the most appropriate assumption for rock formation age varied by sample group: for samples dating volcanic activity, formation age should be assumed equal to the exposure age, whereas a rock formation age of 50 ka was used for the remaining samples.
- We recommend not assuming equilibrium conditions when calculating ^{36}Cl ages on rocks that meet the following criteria: (i) known young rock formation ages, and (ii) potentially susceptible composition, specifically high native Cl, or high U and/or Th concentrations that are likely to occur in volcanic rocks. Young exposure age samples will be particularly affected because of the large mismatch between expected equilibrium conditions and measured concentrations.

Supplementary Materials: The following are available online at https://www.mdpi.com/article/10.3390/geosciences11090390/s1: Supplementary Spreadsheet that includes Supplementary Table S1: Sample geochemistry results; Supplementary Table S2: All sample inputs for CRONUScalc calculations; Table S3: AMS data for all samples.

Author Contributions: A.L. and E.A.L. did most of the fieldwork, N.M. did most of the sample preparations under N.A.'s guidance, whereas C.V. performed the ^{36}Cl measurements. S.M. did most of the age calculations and adapted the CRONUScalc code for variable background production scenarios. J.A. participated in some of the fieldwork (2015), did some of the sample preparation and age calculations, and wrote most of the manuscript, with major contributions from S.M. and substantial contributions from the other coauthors. J.A., N.A., E.A.L., A.L., S.M., N.M. and C.V. all contributed to the interpretations and to the final version of the manuscript. Two anonymous reviewers gave constructive comments on the manuscript. All authors have read and agreed to the published version of the manuscript.

Funding: Financial support was given by the Geological Survey of Norway, The Research Council of Norway to A. Lyså (Grant No. 244135/E10 and Arctic Field Grant 2014), and to A. Lyså through The Royal Norwegian Society of Sciences and Letters (2019, DKNVS). Logistical support, and necessary permissions, were provided by the County Governor of Nordland and the Norwegian Defence Logistic Organisation (CYFOR, Jan Mayen).

Data Availability Statement: The complete geochemistry dataset and all the sample data used in the exposure age calculations are included in the supplementary material.

Conflicts of Interest: The authors declare no conflict of interest.

References

1. Lyså, A.; Larsen, E.A.; Anjar, J.; Akçar, N.; Ganerød, M.; Hiksdal, A.; Van Der Lelij, R.; Vockenhuber, C. The last glaciation of the Arctic volcanic island Jan Mayen. *Boreas* **2021**, *50*, 6–28. [CrossRef]
2. Larsen, E.; Lyså, A.; Höskuldsson, Á.; Davidsen, J.G.; Nadeau, M.J.; Power, M.; Tassis, G.; Wastegård, S. A dated volcano-tectonic deformation event in Jan Mayen causing landlocking of Arctic charr. *J. Quat. Sci.* **2021**, *36*, 180–190. [CrossRef]

3. Dallmann, W.K. Bergrunnsgeologi Jan Mayen. In *Natur og Kulturmiljøet på Jan Mayen—Med en Vurdering av Verneverdier, Kunnskapsbehov og Forvaltning*; Gabrielsen, G.W., Brekke, B., Alsos, I.G., Hansen, J.R., Eds.; Norsk Polarinstitutt Meddelelser: Tromso, Norway, 1997; Volume 144, p. 35.
4. Gosse, J.C.; Phillips, F.M. Terrestrial in situ cosmogenic nuclides: Theory and application. *Quat. Sci. Rev.* **2001**, *20*, 1475–1560. [CrossRef]
5. Marrero, S.M.; Phillips, F.M.; Borchers, B.; Lifton, N.; Aumer, R.; Balco, G. Cosmogenic nuclide systematics and the CRONUScalc program. *Quat. Geochronol.* **2016**, *31*, 160–187. [CrossRef]
6. Phillips, F.M.; Plummer, M.A. Chloe: A Program for Interpreting in-Situ Cosmogenic Nuclide Data for Surface Exposure Dating and Erosion Studies. *Radiocarbon* **1996**, *38*, 98–99.
7. Schimmelpfennig, I.; Benedetti, L.; Finkel, R.; Pik, R.; Blard, P.-H.; Bourlès, D.; Burnard, P.; Williams, A. Sources of in-situ 36Cl in basaltic rocks. Implications for calibration of production rates. *Quat. Geochronol.* **2009**, *4*, 441–461. [CrossRef]
8. Parmelee, D.E.; Kyle, P.R.; Kurz, M.D.; Marrero, S.M.; Phillips, F.M. A new Holocene eruptive history of Erebus volcano, Antarctica using cosmogenic 3He and 36Cl exposure ages. *Quat. Geochronol.* **2015**, *30*, 114–131. [CrossRef]
9. Marrero, S.; Phillips, F.M.; Caffee, M.W.; Gosse, J.C. CRONUS-Earth cosmogenic 36Cl calibration. *Quat. Geochronol.* **2016**, *31*, 199–219. [CrossRef]
10. Cromwell, G.; Tauxe, L.; Staudigel, H.; Constable, C.G.; Koppers, A.A.P.; Pedersen, R.-B. In search of long-term hemispheric asymmetry in the geomagnetic field: Results from high northern latitudes (Dataset). *Geochem. Geophys. Geosyst.* **2013**, *14*, 3234–3249. [CrossRef]
11. White, C.A. Petrology and Mineral Chemistry of Some Jan Mayen Volcanics. Master's Thesis, State University of New York at Albany, Albany, NY, USA, 1979.
12. Anda, E.O.O.; Mangerud, J.A.N. Late Holocene Glacier Variations and Climate at Jan Mayen. *Polar Res.* **1985**, *3*, 129–140. [CrossRef]
13. Larsen, E. (Geological Survey of Norway, Trondheim, Norway). Personal communication, 2021.
14. Imsland, P. The Volcanic Eruption on Jan Mayen, January 1985: Interaction between a Volcanic Island and a Fracture Zone. *J. Volcanol. Geotherm. Res.* **1986**, *28*, 45–53. [CrossRef]
15. Gjelsvik, T. Volcano on Jan Mayen Alive Again. *Nat. Cell Biol.* **1970**, *228*, 352. [CrossRef]
16. Sylvester, A.G. History and surveillance of volcanic activity on Jan Mayen island. *Bull. Volcanol.* **1975**, *39*, 313–335. [CrossRef]
17. Gjerløw, E. Holocene Volcanic Activity and Hazards of Jan Mayen, North-Atlantic. Ph.D. Thesis, University of Bergen, Bergen, Norway, 2018.
18. Maaloe, S.; Sorensen, I.; Hertogen, J. The Trachybasaltic Suite of Jan Mayen. *J. Pet.* **1986**, *27*, 439–466. [CrossRef]
19. Elmore, D.; Ma, X.; Miller, T.; Mueller, K.; Perry, M.; Rickey, F.; Sharma, P.; Simms, P.; Lipschütz, M.; Vogt, S. Status and plans for the PRIME Lab AMS facility. *Nucl. Instrum. Methods Phys. Res. Sect. B Beam Interact. Mater. Atoms* **1997**, *123*, 69–72. [CrossRef]
20. Ivy-Ochs, S.; Synal, H.-A.; Roth, C.; Schaller, M. Initial results from isotope dilution for Cl and 36Cl measurements at the PSI/ETH Zurich AMS facility. *Nucl. Instrum. Methods Phys. Res. Sect. B Beam Interact. Mater. Atoms* **2004**, *223–224*, 623–627. [CrossRef]
21. Desilets, D.; Zreda, M.; Almasi, P.F.; Elmore, D. Determination of cosmogenic 36Cl in rocks by isotope dilution: Innovations, validation and error propagation. *Chem. Geol.* **2006**, *233*, 185–195. [CrossRef]
22. Stone, J.; Allan, G.; Fifield, L.; Cresswell, R. Cosmogenic chlorine-36 from calcium spallation. *Geochim. Cosmochim. Acta* **1996**, *60*, 679–692. [CrossRef]
23. Ivy-Ochs, S.; Poschinger, A.; Synal, H.-A.; Maisch, M. Surface exposure dating of the Flims landslide, Graubünden, Switzerland. *Geomorphology* **2009**, *103*, 104–112. [CrossRef]
24. Zreda, M.G.; Phillips, F.M.; Elmore, D.; Kubik, P.W.; Sharma, P.; Dorn, R.I. Cosmogenic chlorine-36 production rates in terrestrial rocks. *Earth Planet. Sci. Lett.* **1991**, *105*, 94–109. [CrossRef]
25. Synal, H.-A.; Bonani, G.; Döbeli, M.; Ender, R.; Gartenmann, P.; Kubik, P.; Schnabel, C.; Suter, M. Status report of the PSI/ETH AMS facility. *Nucl. Instrum. Methods Phys. Res. Sect. B Beam Interact. Mater. Atoms* **1997**, *123*, 62–68. [CrossRef]
26. Christl, M.; Vockenhuber, C.; Kubik, P.; Wacker, L.; Lachner, J.; Alfimov, V.; Synal, H.-A. The ETH Zurich AMS facilities: Performance parameters and reference materials. *Nucl. Instrum. Methods Phys. Res. Sect. B Beam Interact. Mater. Atoms* **2013**, *294*, 29–38. [CrossRef]
27. Vockenhuber, C.; Miltenberger, K.-U.; Synal, H.-A. 36Cl measurements with a gas-filled magnet at 6 MV. *Nucl. Instrum. Nucl. Instrum. Methods Phys. Res. Sect. B Beam Interact. Mater. Atoms* **2019**, *455*, 190–194. [CrossRef]
28. Moore, A.K.; Granger, D.E. Calibration of the production rate of cosmogenic 36Cl from Fe. *Quat. Geochronol.* **2019**, *51*, 87–98. [CrossRef]
29. Stone, J.O.; Fifield, K.; Vasconcelos, P.M. Terrestrial Chlorine-36 Production from Spallation of Iron. In Proceedings of the 10th International Conference on Accelerator Mass Spectrometry, Berkeley, CA, USA, 5–10 September 2005.
30. Marrero, S.M.; Phillips, F.M.; Caffee, M.; Gosse, J.C. Corrigendum to "Cronus-Earth Cosmogenic 36Cl Calibration" [Quat. Geochronol. 31 (2016) 199–219]. *Quat. Geochronol.* **2021**, *61*, 101130. [CrossRef]
31. Leontaritis, A.D. The Late Quaternary Glacial History of Greece. Ph.D. Thesis, Harokopio University of Athens, Athens, Greece, 2021.
32. Moore, A. (Purdue University, West Lafayette, IN, USA). Personal communication, 2020.

33. Lal, D. Cosmic ray labeling of erosion surfaces: In situ nuclide production rates and erosion models. *Earth Planet. Sci. Lett.* **1991**, *104*, 424–439. [CrossRef]
34. Stone, J.O. Air pressure and cosmogenic isotope production. *J. Geophys. Res. Space Phys.* **2000**, *105*, 23753–23759. [CrossRef]
35. Licciardi, J.; Denoncourt, C.; Finkel, R. Cosmogenic 36Cl production rates from Ca spallation in Iceland. *Earth Planet. Sci. Lett.* **2008**, *267*, 365–377. [CrossRef]
36. Reimer, P.J.; Austin, W.E.N.; Bard, E.; Bayliss, A.; Blackwell, P.G.; Ramsey, C.B.; Butzin, M.; Cheng, H.; Edwards, R.L.; Friedrich, M.; et al. The Intcal20 Northern Hemisphere Radiocarbon Age Calibration Curve (0–55 Cal Kbp). *Radiocarbon* **2020**, *62*, 725–757. [CrossRef]
37. Le Maitre, R.; Streckeisen, A.; Zanettin, B.; Le Bas, M.; Bonin, B.; Bateman, P. (Eds.) *Igneous Rocks: A Classification and Glossary of Terms: Recommendations of the International Union of Geological Sciences Subcommission on the Systematics of Igneous Rocks*, 2nd ed.; Cambridge University Press: Cambridge, UK, 2002.
38. Phillips, F.M.; Stone, W.D.; Fabryka-Martin, J.T. An improved approach to calculating low-energy cosmic-ray neutron fluxes near the land/atmosphere interface. *Chem. Geol.* **2001**, *175*, 689–701. [CrossRef]
39. Principato, S.M.; Geirsdóttir, Á.; Jóhannsdóttir, G.E.; Andrews, J.T. Late Quaternary Glacial and Deglacial History of Eastern Vestfirdir, Iceland Using Cosmogenic Isotope (36cl) Exposure Ages and Marine Cores. *J. Quat. Sci.* **2006**, *21*, 271–285. [CrossRef]

geosciences

MDPI

Article

Glacial Erosion Rates Determined at Vorab Glacier: Implications for the Evolution of Limestone Plateaus

Olivia Steinemann [1,*], Alicia Martinez [2], Vincenzo Picotti [2], Christof Vockenhuber [1] and Susan Ivy-Ochs [1,2]

[1] Laboratory of Ion Beam Physics, ETH Zürich, Otto-Stern-Weg 5, 8093 Zürich, Switzerland; cvockenh@ethz.ch (C.V.); ivy@ethz.ch (S.I.-O.)
[2] Geological Institute, ETH Zürich, Sonneggstr. 5, 8092 Zürich, Switzerland; alicia.c.martinez.v@gmail.com (A.M.); vincenzo.picotti@erdw.ethz.ch (V.P.)
* Correspondence: okronig@phys.ethz.ch

Abstract: Understanding how fast glaciers erode their bedrock substrate is one of the key elements in reconstructing how the action of glaciers gives mountain ranges their shape. By combining cosmogenic nuclide concentrations determined in glacially abraded bedrock with a numerical model, we quantify glacial erosion rates over the last 15 ka. We measured cosmogenic ^{36}Cl in fourteen samples from the limestone forefield of the Vorab glacier (Eastern Alps, Switzerland). Determined glacial erosion rates range from 0.01 mm a^{-1} to 0.16 mm a^{-1}. These glacial abrasion rates differ quite markedly from rates measured on crystalline bedrock (>1 mm a^{-1}), but are similarly low to the rates determined on the only examined limestone plateau so far, the Tsanfleuron glacier forefield. Our data, congruent with field observations, suggest that the Vorab glacier planed off crystalline rock (Permian Verrucano) overlying the Glarus thrust. Upon reaching the underlying strongly karstified limestone the glacier virtually stopped eroding its bed. We attribute this to immediate drainage of meltwater into the karst passages below the glacier, which inhibits sliding. The determined glacial erosion rates underscore the relationship between geology and the resulting landscape that evolves, whether high elevation plateaus in limestone terrains or steep-walled valleys in granitic/ gneissic areas.

Keywords: glacial erosion rates; cosmogenic ^{36}Cl; Swiss Alps; limestone plateau; Bündnerbergjoch

Citation: Steinemann, O.; Martinez, A.; Picotti, V.; Vockenhuber, C.; Ivy-Ochs, S. Glacial Erosion Rates Determined at Vorab Glacier: Implications for the Evolution of Limestone Plateaus. *Geosciences* **2021**, *11*, 356. https://doi.org/10.3390/geosciences11090356

Academic Editors: Roger Urgeles Esclasans and Jesús Martínez-Frías

Received: 30 June 2021
Accepted: 20 August 2021
Published: 24 August 2021

1. Introduction

Many valleys and overdeepenings in forelands are the result of (sub) glacial erosion [1–7]. During the Quaternary, powerful ice masses were the main drivers in sculpting the Alpine landscape by bedrock erosion and production of large amounts of sediment [8–10]. Because there are numerous geological (e.g., lithology, structural geology, fractures and faults, amount of sediment) and glaciological factors (ice thickness, slope, orientation) that influence the rate of glacial erosion ([11,12] and references therein), and direct measurements are nearly impossible, knowledge about how fast glaciers are able to erode their underlying surface is still remarkably limited.

Over the previous century, different direct and indirect scientific approaches have been performed to determine subglacial erosion rates. In the past, researchers focused more on direct measurements [13,14] at the bedrock–ice interface or sediment yield measurements in meltwater streams [15–17]. Modern research concentrates more on numerical models based on glacier dynamics [18–22]. In the last decades, cosmogenic nuclide techniques have been shown to offer a unique possibility of determining glacial erosion rates directly on freshly exposed, glacially polished surfaces [23–26]. If two different cosmogenic nuclides (^{10}Be/^{26}Al or ^{10}Be/in-situ ^{14}C) are combined, this can even allow the possibility of determining the duration of the latest glacial burial history and, if glacial erosion was relatively small (e.g., marginal positions or beneath cold-based glaciers), it gives evidence on the glacial erosion rates [26]. If only one cosmogenic nuclide can be used, an independent archive to constrain the burial history of the sampled site is needed to determine the rate of glacial erosion [25]. This method, implementing cosmogenic ^{36}Cl, has the advantage of not being restricted to quartz-bearing lithologies (e.g., granites) but also allows the determination of glacial erosion rates on carbonates. In this study at Tsanfleuron glacier (Figure 1) [26], strikingly low glacial erosion rates (<0.08 mm/a) were determined on the flat-lying massive limestone bedrock.

In this study, we apply the method of using one cosmogenic nuclide, specifically ^{36}Cl, to measure how rapidly the Vorab glacier (Switzerland) erodes the underlying limestone bed. The fortuitous presence of the famous Glarus thrust allows the study of how a glacier responds when transitioning from an easy-to-erode rock, such as Permian schists, to a difficult-to-erode rock, such as limestone (cf. [25,26]). We examine some of the hypotheses presented in [25], in which we focussed on the Tsanfleuron glacier limestone forefield (Figure 1). Our results from the Vorab glacier site, in addition to providing direct measurements of how fast glaciers erode, also contribute to increasing the still vague understanding of how the high-elevation, low-relief limestone plateaus in the Alps have formed and persisted through numerous large-scale glaciations.

2. Study Area

Focus of the study is a high-elevated (2600 m a.s.l.), low-relief area located within the Little Ice Age (LIA) extent of the Vorab glacier, on the border between the political regions (cantons) of Glarus and Grisons in eastern Switzerland (Figure 1). The west-facing glacier (Figures 1 and 2) with a surface area of 1.20 km^2 (year 2014, [27]) stretches between the Bünder Vorab (3028 m a.s.l.) and the Gletscherhorn (2804 m a.s.l). During the LIA, it covered an area of 3.3 km^2 (Figure 2). During this time, the northern part of the glacier flowed down over the almost 400 m high, nearly vertical cliff to Glarus to the north, whereas the southern part extended about 1.7 km further southward than its present-day position, on a gently increasingly steep slope dipping towards the Grisons side. Ever since, the glacier has been retreating, with few years of stagnation or small brief advances [27]. A selection of glacier extents of different years, illustrated in Figure 2, shows the retreat history of the Vorab glacier.

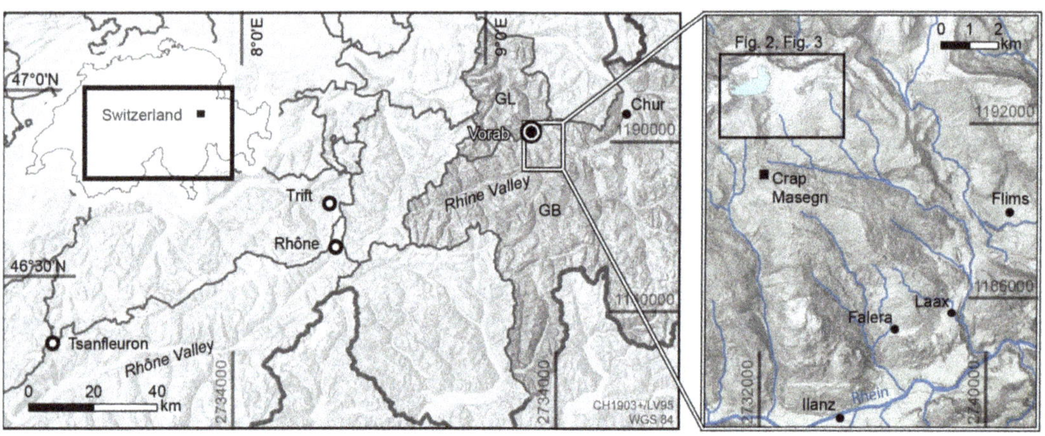

Figure 1. Overview maps. **Left:** Location of the Vorab glacier (black point) west of Chur at the border between the canton of Glarus (GL) and Grisons (GB). Extent of the map within Switzerland is indicated by the black rectangle on the inset map. Other locations, in the Swiss Alps, where glacial erosion rates were determined using cosmogenic nuclides are indicated as white points. **Right:** close-up of the study site area, with the Vorab glacier extent (2019) in light blue and the location of the weather station at Crap Masegn. The black rectangle shows the extent of Figures 2 and 3. Background maps and outlines of the cantons are reproduced with the authorization of the Swiss Federal Office of Topography (swisstopo).

Because of its unique geology, the area is part of the UNESCO World Heritage Tectonic Arena Sardona [28,29]. At several locations around the glacier front and likely also still below the glacier, the famous Glarus thrust plane [30,31] is exposed (Figure 2), where older (Permian) nappes overlie younger (lower Cretaceous) carbonates. In the study area, the Cretaceous carbonates are covering most of the glacier forefield. The area inside of the LIA extent is dominated by massive limestone beds and by siliceous limestone outside of the LIA extent (Figure 2). The hanging wall of the Glarus thrust is mainly composed of Permian Verrucano, visible as high peaks in the surrounding of the Vorab glacier (Gletscherhorn 2804 m a.s.l., Glarner Vorab 3018 m a.s.l. Bündner Vorab 3028 m a.s.l. and Laaxer Stöckli 2899 m a.s.l.).

During winter months the area is a ski resort (Flims-Laax-Falera, Figure 1), including the Vorab glacier itself. This is why certain areas in and outside the LIA extents were strongly affected and artificially modified, which has to be kept in mind during geomorphological mapping and interpretation.

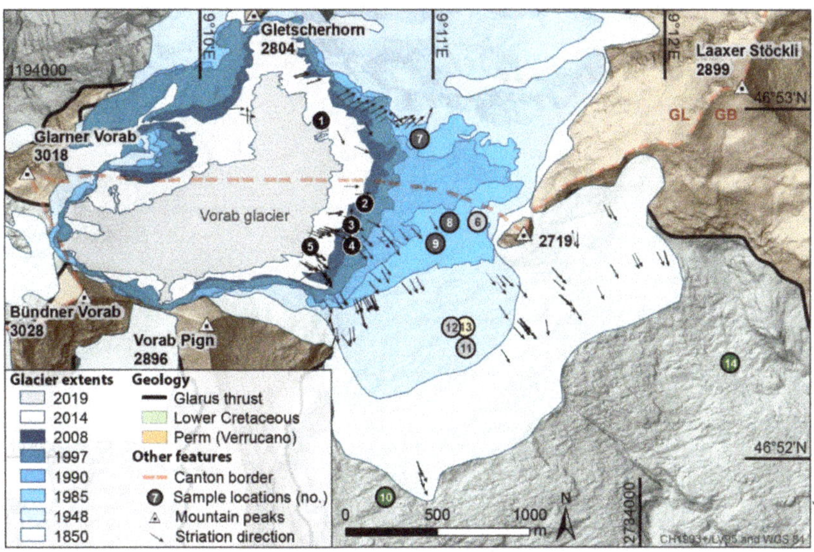

Figure 2. Geological map combined with the chronology of past glacier extents since the LIA. Circles show the location and number of the samples taken for ^{36}Cl analysis. Bedrock samples are grouped according to their position relative to the glacier extents: green: outside LIA extent (1850), light grey: inside 1948, dark grey: inside 1985, black: 2018; the only sampled boulder (Vorab-13) is yellow. Background map, geology and glacier extents are based on the swissAlti3D 2019, the geological 1:500,000 map and the topographic maps of the different years, reproduced with the authorization of the Swiss Federal Office of Topography (swisstopo). For location of the map see Figure 1.

3. Materials and Methods

3.1. Fieldwork

Fieldwork in the forefield of the Vorab glacier was restricted to the summer months in 2018, because of the continued presence of snow from the previous winter. Major focus was put on the geomorphology, e.g., the glacial landforms and deposits (moraine ridges, sediment stripes, striations). The area was mapped on a topographic map (1:10,000) supported by a tablet with GIS software (GIS Pro, Grafa 2014) that allowed to directly save striation measurements and the locations of mapped features, photographs and notes as layer files. The manually and digitally collected data, including knowledge from historical maps, were evaluated and combined to create a geomorphological map.

A total of 14 samples were taken for cosmogenic ^{36}Cl analysis (Figure 2). Eleven samples were collected from inside the LIA extend. We divided these into three sub-groups depending on their location relative to present and past ice margin positions (see Figure 2): (i) just in front of the 2018 glacier extent (Vorab-1-Vorab-5), (ii) near the extent of the glacier around 1985 (Vorab-7-Vorab-9), (iii) near the extent of the glacier of 1948 (Vorab-6, Vorab-11 and Vorab-12). Two samples were taken outside of the LIA extent (Vorab-10, Vorab-14). One sample (Vorab-13) was taken on a boulder located inside the LIA extent just next to Vorab-12. Only glacially polished surfaces were sampled, where clear striations were visible. Plucked areas were strictly avoided. An elevated location of the sample compared to its surrounding was favourable, to avoid shielding through micro topography. Additionally, it lowered the duration of snow cover due to frequent winds. About 500 g of the top few centimetres (<5 cm) were extracted with a battery-operated saw, hammer and chisel. The exact location of the sample, sample thickness, dip and dip direction and the topographic shielding were measured on site using a compass and clinometer.

3.2. ^{36}*Cl Sample Preparation*

Samples were crushed with a hydraulic press and sieved to a grain size of <400 µm. After leaching the samples in a weak HNO_3 solution, an aliquot of dried sample material (10 g) was sent to Actlabs S.A. (Ancaster, ON, Canada) to determine the composition of major and trace elements with ICP-MS (inductively coupled plasma mass spectrometry) (Table 1). Following this, ^{36}Cl sample preparation was undertaken at ETH Zürich (Laboratory of Ion Beam Physics) according to the sample preparation described by [32,33]. ^{36}Cl and total Cl concentrations were measured by AMS (accelerator mass spectrometry) (Tables 1 and 2) performed with the 6 MV TANDEM system at the Laboratory of Ion Beam Physics, ETH Zürich. Sample ratios were measured against the standard material K382/4N with a ^{36}Cl/Cl value of 17.36×10^{-12} [34]. The apparent exposure ages were calculated with an in-house MATLAB code (Table 2) using equations and constants described in [35] and the following ^{36}Cl production rates: a spallogenic production rate of calcium of 48.8 ± 3.4 atoms ${g_{Ca}}^{-1}$ a^{-1}, 5.3 ± 1.0 ^{36}Cl atoms ${g_{Ca}}^{-1}$ a^{-1} for muon capture in calcium and a neutron capture rate of 760 ± 150 neutrons ${g_{air}}^{-1}$. All production pathways were accounted for in the code (see also [23]).

3.3. *MECED Model*

The MECED model was programmed by C. Wirsig and V. Alfimov [23,35], then modified by O. Steinemann [25]. The code is based on equations, production rates and constants described in [35]. Additional input parameters are a glacier fluctuation history, i.e., when the sample location was covered by a glacier (no ^{36}Cl production) or was ice free (^{36}Cl production), and potential snow coverage (snow depth and duration in months), as well as the karst weathering rate during ice-free periods. The latter two lead to a reduced final ^{36}Cl concentration; the present maximal thickness of the Vorab glacier is 60–80 m (27). The glacier was thicker during the LIA and similar during late Holocene glacier advances. Production of ^{36}Cl, including production by muons, through this thickness of ice is 0.02–0.04% of total production on a fully exposed, ice-free surface. Already with an ice thickness of only 5 m, e.g., in marginal position, production is <5%.

With these inputs, the model gives two main outputs. The first is a sample specific diagram of ^{36}Cl concentration evolution over time based on the glacial fluctuation history (also see Section 5.2.1) including the final modelled (theoretical) nuclide concentration. This indicates what the concentration should be at that specific sample location presently with the used glacial fluctuation history, snow coverage input and karst weathering rate. The second output is a corresponding depth vs. concentration plot, showing the decrease of the ^{36}Cl concentration with depth into the limestone bedrock. By intersecting this modelled concentration-depth profile with the actual AMS measured ^{36}Cl concentration for that sample, one can determine how much of the bedrock surface must have been removed by the glacier at that specific sample location. By dividing this erosion depth by the total time of glacier coverage (glacier fluctuation history input), the glacial erosion rate is calculated. Detailed description of the MECED model and a thorough discussion of the influencing factors and limitations can be found in [36,37].

4. Overview of Geomorphology of Vorab Glacier Forefield

The most important observations made during the fieldwork will be described shortly; in this section the focus will be on glacial features such as moraine deposits, sediment stripes and striations. In the following section karst and bedrock features, such bedrock steps, erosional channels and swallow holes will be described. An overview of the identified landforms and their apparent exposure ages is illustrated in the geomorphological map (Figure 3).

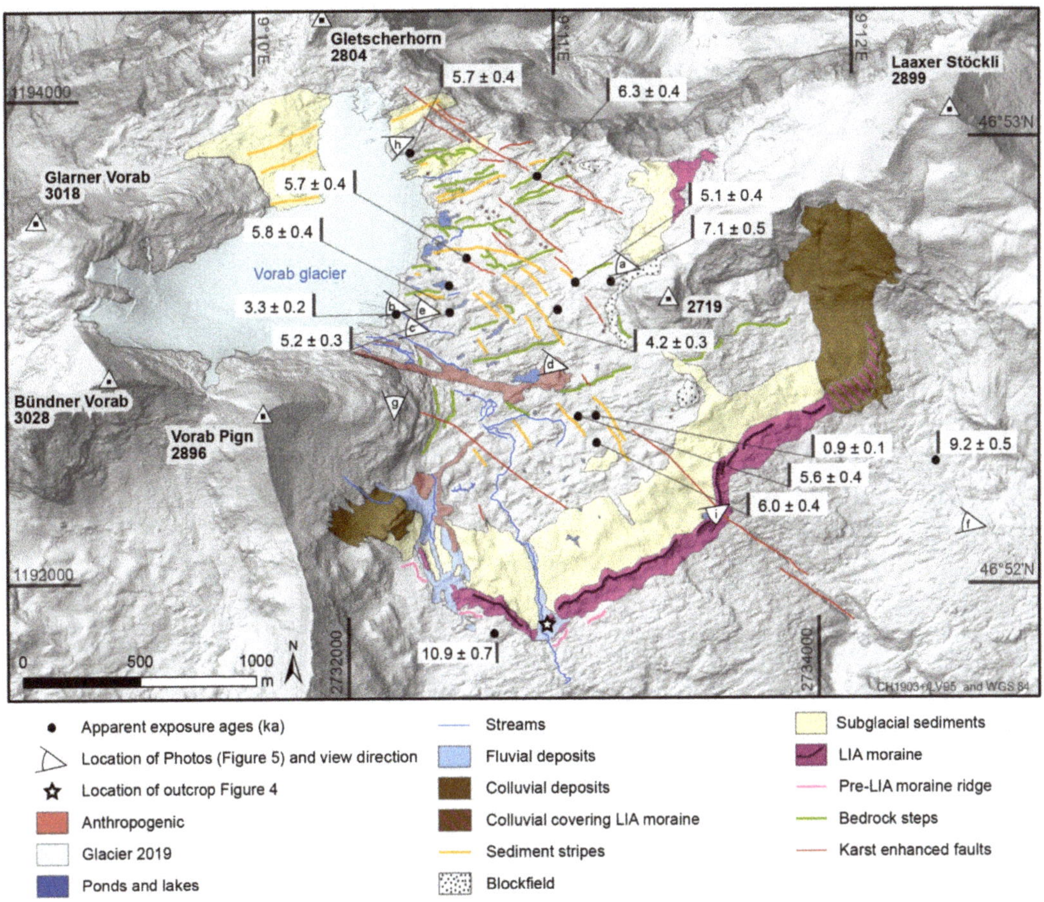

Figure 3. Geomorphological map of the Vorab glacier forefield showing the mapped features and the apparent ^{36}Cl exposure ages. Star shows the location of the described sedimentary outcrop (Figure 4). For location of the map see Figure 1.

4.1. Glacial Landforms and Sediments

The LIA margins were designated through field observations and according to the topographic map of 1895 [38], along with orthophoto and hillshade maps [39]. The largest, most continuous LIA moraine ridge is approximately 1 km long and up to 15 m high (Figure 3). The northernmost LIA moraine fragment has a length of about 200 m. These

LIA moraine ridges are not vegetated (or only patchily) and consist of angular clasts in a sandy to fine-grained matrix. Most large fragments (up to 15 cm) consist of Verrucano, larger components (0.5–1 m) are mainly limestones and angular. There are a few locations, especially in the northern moraine fragment, where larger Verrucano fragments are found more frequently.

The main meltwater stream crosscuts the LIA moraine in the southwest, creating a 2 m high and 6 m wide natural outcrop (Figures 3 and 4a). After the outcrop was cleaned with a shovel it was sketched in detail (Figure 4b) as described in [40] using the abbreviations from [41]. This investigation revealed the pushing and the gravitational reworking of sediment at the glacier front during the LIA because the folds show a clear forward movement toward the south. There is a wide variability in the size and lithology of the clasts. In the Dms (diamicton, matrix-supported, stratified) that clearly dominates the outcrop, there are interbedded layers that were deposited episodically (Figure 4b). These vary from clast-dominated layers of several centimetres thick that were deposited during periods of stronger energy and higher water flow to layers of very fine sand that were probably deposited during quieter periods. A repetitive sequence was not recognised. At the outermost edge, there is a hill structure consisting of GSc (clast-supported gravel and sand), which could be a remnant of a buried moraine (Figure 4b).

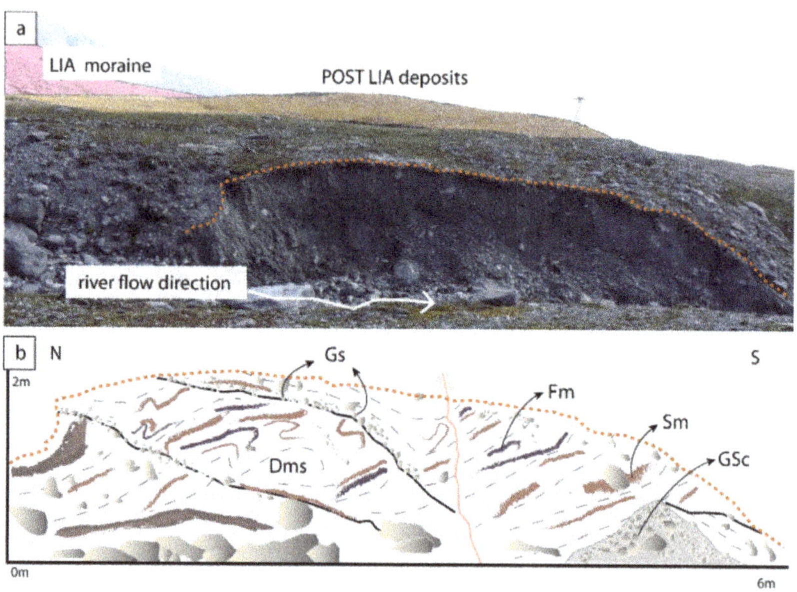

Figure 4. Photograph (**a**) and interpretation (**b**) of a river-cut outcrop through the LIA margin deposits. The interpretation shows that the advancing LIA glacier pushed the sediments. Dms: diamicton completely unsorted, matrix-supported, stratified, Gs: gravel stratified, Fm: fines massive, Sm: sand massive, GSc: gravel, sand, clast-supported [41]. LIA moraine coloured pink in the photograph, post-LIA deposits coloured in orange. See Figure 3 for location of the outcrop.

In contrast to the area just inside of the LIA moraine ridge, where secondary postglacial debris was deposited through fluvial and ice decay processes with a relatively fresh appearance, the pre-LIA moraines (outside LIA) are already strongly vegetated, and soil formation has started. These older moraines form clear 1–2 m high hills tens to hundreds of meters distal to the LIA moraines (Figure 3). The post-LIA moraines were recognized as elevated sediment accumulations, maximally 1 m high, perpendicular to the striation direction, which indicates formation of the material by the glacier. These faint ridges are mainly composed of cm- to dm-sized clast and very little matrix. It is noteworthy that the glacier forefield has no to very sparse subglacial till cover (Figure 5a), and if there are till deposits, for example just inside of the LIA moraines (Figure 5i), it lacks a fine matrix. However, some faint flute structures were observed.

In the forefield of the Vorab glacier, sediment stripes in different dimensions ranging from a few tens of centimetres up to several meters in width (Figure 5b,c) were observed and mapped (Figure 3). The clast size ranges from centimetre- up to metre-sized. It is striking that in the south-western part of the glacier forefield, the sediment stripes consist of much smaller and finer clasts; those in the north and northeast show larger as well as finer angular clasts. Based on field investigations, it can be concluded that these sediments were transported englacially; as Figure 5b shows, they emerged from within the ice and were not transported subglacially. As the glacier melts, the sediment remains. The trend of individual stripes is more or less straight, crossing over highs and lows in topography, indicating that they are not related to meltwater processes. Accordingly, they are excellent indicators of past ice-flow directions [25].

Over 200 striations were measured in the glacier forefield and are portrayed in Figure 2; an example is shown in Figure 5d. Most were measured on horizontal polished bedrock surfaces, where these typical glacier scratch marks could be seen easily. Others were measured based on the orientation of recrystallized calcite fibres formed on the lee side of small bedrock irregularities. Clearly, the glacier had two main directions; the more obvious one starting towards east and turning south. In the northern part, however, the traces point northwards, towards Glarus (Figures 1 and 2). In the centre, striations point toward the southeast, potentially locating the former ice divide. No crosscutting striations or other evidence was found that would indicate a recent change in ice-flow direction. The presence of fresh striations and the measured southerly direction on the southern side of the Verrucano ridge (between the 2719 m a.s.l. peak and the Laaxer Stöckli (Figure 3)) suggests that the glacier at least in part overtopped this ridge, as shown also by the historical maps (Figure 2). Outside the LIA moraine, no striations can be observed.

Figure 5. Photo panel, location of each photo and view direction is shown on Figure 3. (**a**) View across the flat limestone forefield of the Vorab glacier (in the background). Note the sparse sediment cover. (**b**) Sediment stripe with small clasts emerging englacially. (**c**) Sediment stripe with large clasts, see person for scale. (**d**) Polished and striated bedrock. (**e**) Bedrock step in front of the Vorab glacier, see person for scale. (**f**) Strongly karstified bedrock with deep channels and karren outside of the LIA extent. (**g**) Glarus thrust with Verrucano (phyllites) on top of Lower Cretaceous limestone. (**h**) Swallow hole where glacial meltwater disappears into the karst system. (**i**) LIA moraine ridge.

4.2. Bedrock, Karst and Hydrological Features

Bedrock steps are frequent features in the limestone forefield of the Vorab glacier. The steps are up to 5 m high and several are 100 m long (Figure 5e). In between the steps, which can be several meters apart, the surfaces are smooth and flat with an average dip of few degrees toward the southeast. Similar stepped limestone landforms have been described in the literature [42,43].

Karst weathering is an important factor influencing the appearance of the limestone, and it is also a significant parameter with respect to the modelling and calculation of the glacial erosion rates (see Section 5.2.1). Based on the presence and absence of glacial striations in the glacier forefield it can be concluded that the intensity varies across the limestone forefield. Striations were mainly observed on elevated crests of the bedrock, indicating that at these locations karst weathering rates have been rather low since the location was exposed. This is in contrast to the depressions and channels in between these local elevation highs. There the water has already found its way through the rock, often forming several dm-deep (subglacial) channels. These channels mainly follow the direction of the bedrock faults. Very few lie along the previously mentioned bedrock steps. At present, in spring and early summer, great amounts of meltwater flow out to the glacier forefield. Some of the meltwater is redirected through artificial pipes to prevent damage to infrastructure, like the gravel road. However, in many areas the (melt-)water disappears into swallow holes, especially in the north-western part of the glacier forefield (Figure 5h). Field observations indicate there are several metres-deep shafts and it is presumed that these connect to an extensive karst network, as has been shown to be the case in the Flims area just to the east [44]. Other swallow holes are filled with finer sediments, so little or no water seeps through and seasonal lakeletts form. Further meltwater lakes have formed in bedrock depressions. The size of the lakes ranges from 2 to 20 m^2 and they often have sandy-silty material on the ground and on the lake shores. Sometimes the ponds have superficial water in- and outlets. However, in most cases these superficial water streams are short and disappear into swallow holes, highlighting the very efficient karst drainage system. Outside of the LIA moraine ridges, water and creeks are less common, but the bedrock outcrops are strongly crossed by karren fields, which can be several meters deep (Figure 5f).

5. Glacial Erosion Rate Determinations

5.1. "Apparent" Exposure Ages

The AMS-measured ^{36}Cl concentrations and the apparent exposure ages calculated from them are shown in Table 1 and are plotted on the map of Figure 3. The ages listed in Table 1 are corrected for neither karst weathering erosion nor for snow cover, because they will not be used for chronological interpretations. The apparent exposure ages range from 3.3 ± 0.2 ka (Vorab-5) to 10.9 ± 0.7 ka (Vorab-10) excluding the boulder sample Vorab-13 with an exposure age of 0.9 ± 0.1 ka.

The samples located outside the LIA extent (Figures 2 and 3) Vorab-10 (10.9 ± 0.7 ka) and Vorab-14 (9.2 ± 0.5 ka) gave the oldest apparent exposure ages and conform relatively well to their presumable true exposure age, assuming they were last covered by the Egesen stadial glacier (12.7–11.5 ka). However, as shown in Figure 6, if snow cover and karst weathering were taken into account the modelled nuclide concentrations after 11.5 ka of exposure would be lower than the measured values (dashed green line in Figure 6a). Therefore, it seems likely that these samples were already exposed during the Bølling/Allerød (B/A) interstadial so that the bedrock surfaces could accumulate enough nuclides to account for the reduction or loss caused by the snow coverage and karst weathering. Including snow and karst correction, these ages are 13.1 ± 0.8 ka (Vorab-10) and 11.1 ± 0.6 ka (Vorab-14). These ages are in good agreement with the timing of the end of the Egesen stadial in the Alps at 11.7 ka, suggesting that excess ^{36}Cl from exposure prior to the last glacial maximum (LGM) is minimal. This question was explored in detail in the study at Tsanfleuron glacier using the same methodology [25]. There, a sample in a cave 6 m below the glacially polished surface had less than 2% of the surface concentration, in agreement with the muon-derived concentration in the calculated depth profile (see Figure 5 in [25]). Inheritance from before the LGM at Tsanfleuron was negligible. Although the same test was not performed at Vorab, the huge thickness of ice at Vorab (estimated to be several hundred meters) during the LGM [21,22] would, in principle, have similarly removed more than 2 m of bedrock and removed nearly all previously produced ^{36}Cl.

Table 1. Sample site information, AMS-measured ^{36}Cl concentrations and calculated apparent exposure ages.

Sample	Location	Latitude	Longitude	Elevation	Thickness	Topographic Shielding	^{36}Cl Concentration [1,2]	Apparent Exposure Age
		WGS 84		m a.s.l.	cm		10^6 at/g	ka
Vorab-1	2018	46.882	9.174	2653	2.5	0.9986	0.89 ± 0.05	5.7 ± 0.4
Vorab-2	2018	46.878	9.177	2635	1.0	0.9991	0.84 ± 0.05 *	5.7 ± 0.4
Vorab-3	2018	46.877	9.176	2639	2.0	0.9987	0.91 ± 0.05	5.8 ± 0.4
Vorab-4	2018	46.876	9.176	2618	1.5	0.9979	0.81 ± 0.03 *	5.2 ± 0.3
Vorab-5	2018	46.876	9.173	2640	3.0	0.9963	0.50 ± 0.03	3.3 ± 0.2
Vorab-6	1948	46.877	9.185	2623	1.5	0.9977	1.12 ± 0.06	7.1 ± 0.5
Vorab-7	1985	46.881	9.181	2616	2.0	0.9991	0.96 ± 0.05	6.3 ± 0.4
Vorab-8	1985	46.877	9.183	2609	1.5	0.9991	0.78 ± 0.05 *	5.1 ± 0.4
Vorab-9	1985	46.876	9.182	2600	1.5	0.9990	0.58 ± 0.03 *	4.2 ± 0.3
Vorab-10	outside LIA	46.864	9.178	2439	2.0	0.9962	1.26 ± 0.07	10.9 ± 0.7
Vorab-11	1948	46.871	9.184	2524	2.0	0.9988	0.77 ± 0.05 *	6.0 ± 0.4
Vorab-12	1948	46.872	9.183	2531	3.0	0.9985	0.80 ± 0.05	5.6 ± 0.4
Vorab-13	boulder	46.872	9.184	2540	6.0	0.9955	0.10 ± 0.01 *	0.9 ± 0.1
Vorab-14	outside LIA	46.870	9.203	2397	2.0	0.9955	1.23 ± 0.04 *	9.2 ± 0.5

[1] Measured against standard K382/4N (17.36 ± 0.35) × 10^{-12} [34,45]. [2] Corrected for processed blank of (1.2 ± 0.7) × 10^{-15} or (2.1 ± 0.4) × 10^{-15} if sample is marked with a *.

Table 2. Elemental composition of leached samples. Cl values are from AMS measurements.

Sample	Al_2O_3 %	CaO %	Fe_2O_3 %	K_2O %	MgO %	MnO %	Na_2O %	P_2O_5 %	SiO_2 %	TiO_2 %	Sm ppm	Gd ppm	U ppm	Th ppm	Cl ppm
Vorab-1	0.73	53.53	0.57	0.19	0.63	0.01	0.05	0.06	2.73	0.03	0.6	0.5	0.4	0.7	2.83 ± 0.08
Vorab-2	0.44	51.13	0.98	0.05	0.59	0.02	0.03	0.03	6.58	0.02	0.7	0.6	0.8	0.6	5.40 ± 0.16
Vorab-3	0.20	55.00	0.17	0.04	0.51	0.01	0.03	0.02	1.43	0.01	0.1	0.1	0.5	0.2	2.73 ± 0.10
Vorab-4	0.12	55.37	0.22	0.03	0.45	0.01	0.02	0.03	1.23	0.01	0.2	0.1	0.6	0.2	3.16 ± 0.06
Vorab-5	1.46	51.85	0.51	0.41	0.65	0.01	0.13	0.04	4.76	0.06	0.4	0.3	2.1	0.7	2.81 ± 0.10
Vorab-6	0.13	55.82	0.12	0.02	0.40	0.01	0.02	0.02	0.56	0.00	0.3	0.4	0.4	0.4	0.97 ± 0.03
Vorab-7	0.60	53.30	0.58	0.17	0.53	0.01	0.05	0.06	3.22	0.03	0.4	0.6	0.3	0.8	5.56 ± 0.21
Vorab-8	0.42	54.26	0.57	0.13	0.50	0.01	0.01	0.02	1.85	0.02	0.5	0.5	0.7	0.7	4.10 ± 0.14
Vorab-9	2.01	46.83	1.05	0.55	1.07	0.02	0.14	0.06	11.70	0.11	1.2	0.9	2.3	2.3	8.71 ± 0.15
Vorab-10	1.37	45.39	0.97	0.32	0.98	0.02	0.31	0.04	14.71	0.08	1.1	1	1.7	2	2.99 ± 0.11
Vorab-11	0.23	48.48	0.58	0.02	0.48	0.01	0.01	0.02	9.47	0.01	0.2	0.3	0.7	0.2	3.31 ± 0.05
Vorab-12	0.51	53.64	0.49	0.14	0.55	0.01	0.02	0.03	2.93	0.02	3.5	2.5	1.6	7.4	3.43 ± 0.09
Vorab-13	2.22	40.15	0.92	0.73	1.09	0.02	0.11	0.03	22.70	0.13	1.3	0.9	1.8	2.4	6.12 ± 0.06
Vorab-14	0.34	55.18	0.16	0.10	0.46	0.01	0.01	0.03	0.85	0.02	0.9	0.6	0.6	1.7	2.68 ± 0.03

All bedrock samples located inside the LIA extent, whose true exposure ages should range from about 170 years (just inside of the LIA moraine ridge) to zero years (samples which became ice-free only during the summer of sampling, 2018), have apparent exposure ages that are much too old (3.3–5.8 ka). These samples all contain considerable amounts of ^{36}Cl, which must have been acquired in previous ice-free periods during the Late Glacial and Holocene (shown graphically in Figures 6a and 7). This shows that glacial erosion was not deep enough (>2 m) to remove the cosmogenic signal built up during these ice-free periods [25]. Nevertheless, the apparent exposure ages give evidence of a total minimum exposure they must have experienced, which is an important boundary condition for the numerical modelling.

5.2. MECED Model Results

5.2.1. Evaluation and Definition of Input Parameters

The glacier fluctuation history (periods of coverage of the sampled bedrock by glacier vs. no cover by glacier), the depth of snow and duration of snow cover, as well as the karst weathering rate applicable to the ice-free periods are the three model input parameters. The most crucial input parameter for the numerical model is the local glacier fluctuation history. A variety of investigative methods have been applied to study and document fluctuations of glaciers in the Alps during the Holocene [46–48]. Local glacier histories have been reconstructed by applying dendrochronological methods and tracking past tree-line variations [49], looking at wood, peat and pollen [50–52] and lake sediment archives [53] and by interpreting historical pictorial data [54] and aerial photographs [55]. Since the 1990s, great inroads have been made into understanding past glacier variations in the Alps through the application of cosmogenic nuclide exposure dating [48,56–61]. This literature and local historic maps were used to construct a glacier fluctuation history for the Vorab glacier (further details given in [25,26]).

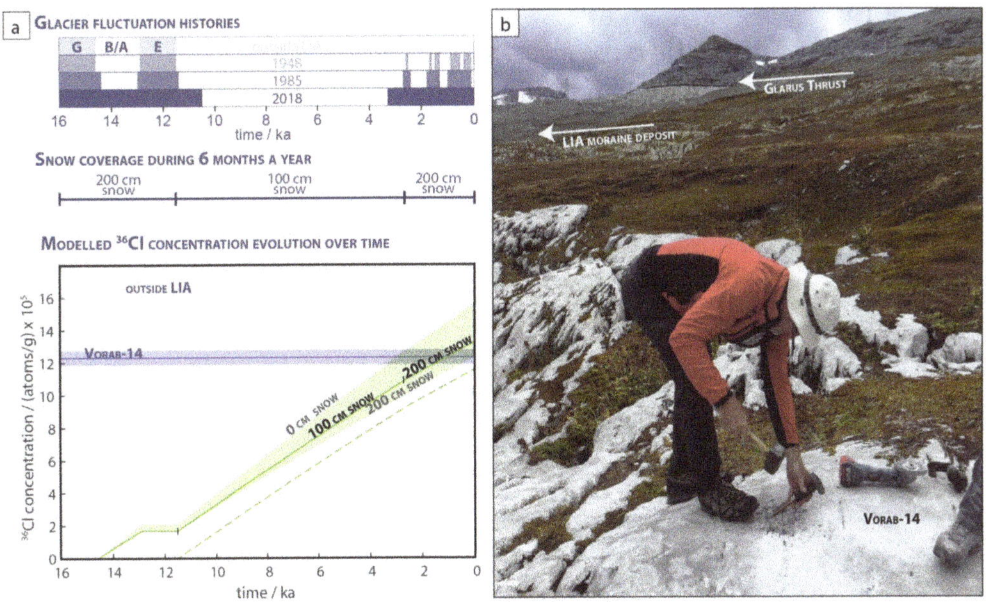

Figure 6. Inputs and output of the MECED model. (**a**) In the top bar diagram, the blue bars show periods of glacier coverage for each sample group (see Figure 2 and text for details). All sites were covered continuously during the last glacial maximum and Gschnitz stadial glaciers (G) of the Oldest Dryas. All groups except the 2018 group were exposed during the Bølling/Allerød interstadial (B/A) then covered again by the Egesen stadial glacier (E) during the Younger Dryas cold phase. This was followed by a longer ice-free phase before the late Holocene glacier advances. Just below, snow depth input shows that until the end of the Egesen and during the late Holocene 200 cm of snow for 6 months a year was used in the model and only 100 cm of snow during the Holocene warm period in between. The graph below shows the modelled ^{36}Cl concentration increase (solid green line) evolution with time (along the *x*-axis) for the sample location of Vorab-14 using the glacier fluctuation history for the group outside LIA, showing the increase in ^{36}Cl (green line) during first exposure during the B/A (14.6–12.9 ka), no production while covered by the Egesen stadial glacier (12.9–11.5 ka) and again an increase in ^{36}Cl in the exposed bedrock surface during the entire Holocene. The green band shows the effect of snow depth during ice-free times, where the upper boundary was modelled without any snow and the lower limit by using 200 cm for the entire period. The solid green line used reasonable snow values mentioned in the text. The blue horizontal line shows the AMS-measured ^{36}Cl nuclide concentration along with the uncertainty. The dashed green line shows the modelled concentration if the sample site only became ice-free after the Egesen glacier, highlighting that with this scenario the modelled concentration is lower than the one actually measured, demonstrating why an ice-free period with ^{36}Cl production of the B/A has to be included. (**b**) Sample location of Vorab-14 with the LIA moraine ridge and the Glarus thrust in the back. See Figures 2 and 3 for the location of samples and photos.

For each sample group (Figure 2) a separate glacier fluctuation history reflecting its location was defined (Figure 6a) based on the literature mentioned above. With these defined glacier histories, the model calculates for every sample the theoretical evolution of the cosmogenic nuclide concentration over time. A representative sample of each group is shown in Figure 7, where the effects of snow and karst weathering rate (see below) are also included. Glacial erosion rates are obtained by dividing the erosion depth by the total time of glacier coverage.

For the samples outside the LIA extent (Vorab-10, Vorab-14) we assume bedrock exposure starting at 14.5 ka with the beginning of the B/A until the Egesen glacier advanced and covered the area again from 12.9 ka to 11.5 ka [48] (Figure 6a). After that, the area outside the LIA was continuously exposed and never covered by a glacier again. The samples of the group 1948 (Vorab-6, Vorab-11, Vorab-12) experienced the same exposure during the B/A and coverage by the Egesen stadial glacier but were additionally repeatedly covered by late Holocene fluctuations of the Vorab glacier from 1.7–1.6 ka, 1.5–1.3 ka, 0.9–0.5 ka and 0.4–0.1 ka (total 2.5 ka of glacier coverage). The samples from the 1985 group (Vorab-7, Vorab-8, Vorab-9) are assumed to have a similar history as the 1950 group but with slightly longer intervals of glacier coverage because they are located at higher elevations and are closer to the present-day glacier front: 13.0–11.4 ka, 2.6–2.4 ka, 1.8–1.3 ka and 1.0–0.1 ka (total 3.3 ka of glacier coverage). It is assumed that the samples from the group 2018 were ice-covered until the early Holocene, when the Egesen stadial glacier retreated around 10.5 ka ago. During the late Holocene, it is likely that these areas were continuously glacier-covered from 3.3 ka until the summer of sampling (2018).

Snow depth data were derived by looking at records from the nearest station. According to the MeteoSchweiz [Schweizerische Eidgenossenschaft 2019] data, the station at Crap Masegn (Figure 1) (2480 m a.s.l.) is the nearest station. Precipitation records there span the past 30 years. The data show that at this elevation a constant snow cover of about 200 cm can be assumed for 6 months of the year. Knowing that during the middle Holocene warm phase [50,62] it is likely that there was less snow, we assume for this time period a reduced snow cover of 100 cm for 6 months a year. As shown in Figure 6, for our model we use 200 cm snow before the Holocene, 100 cm snow during the early and middle Holocene and 200 cm snow after 3.6 ka.

However, we are aware that these values are estimates. To visualise the effect of different snow depth values on the modelled ^{36}Cl concentration, minimal and maximal snow scenarios are indicated in Figures 6b and 7 as green bands. The upper limit of the green band tracks the modelled growth of ^{36}Cl in the bedrock with no snow cover at all and the lower limit of the green band tracks the modelled growth when a depth of 200 cm of snow is considered for all ice-free periods (including during the middle Holocene). The influence on the final modelled ^{36}Cl concentration in comparison to our chosen input parameters is about +15% (no snow) and -10% (200 cm of snow all of the time the forefield is ice-free). If the snow depth were further increased (>200 cm), for many locations it would not be possible to calculate erosion depth/rate values, as the modelled nuclide concentration would get lower than the actual AMS-measured one. The karst weathering rate of 5 mm ka^{-1}, a reasonable value for limestone surfaces in the Alps [63–65] causes a reduction on the modelled concentration of about 5%.

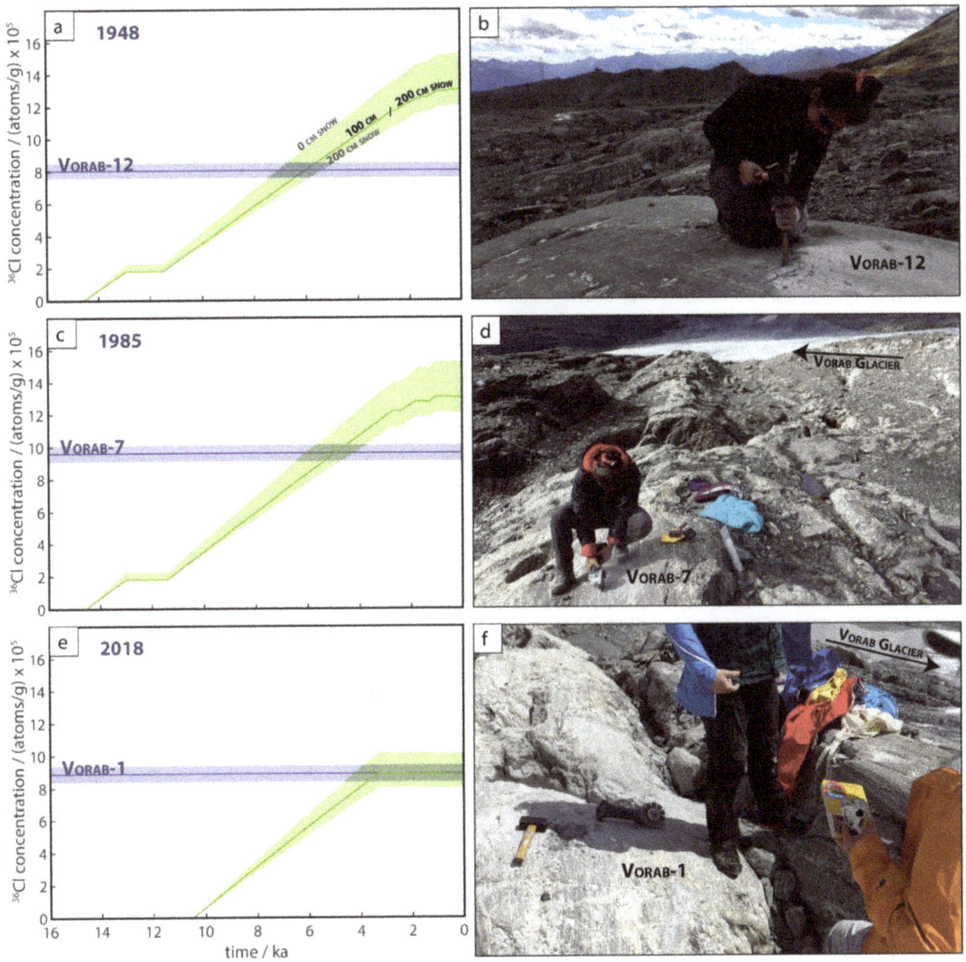

Figure 7. (**a**,**c**,**e**) MECED model output for a representative sample (Vorab-12, Vorab-7, Vorab1) of each group (1948, 1985, 2018) inside the LIA extent, showing the increase in ^{36}Cl during ice-free periods and no production during phases of glacier coverage (stays flat). (**b**,**d**,**f**) shows a photo of the corresponding sample. For a detailed description of the diagrams see caption of Figure 6. See Figures 2 and 3 for the location of samples and photos.

5.2.2. Glacial Erosion Depths and Rates

For every sample individually, the modelled depth profile of the nuclide concentration, including all the appropriate parameters described before, and the AMS measured ^{36}Cl concentrations were used to calculate the subglacial erosion depth [25] (Figure 8). Divided by the duration of the glacier cover we then obtained the average erosion rate values. An overview on the calculated erosion depths and erosion rates for all samples (excluding data from boulder Vorab-13) is given in Table 3 and Figure 9. Determined glacial erosion depths for the 2018 sample group are between 0 cm and 40 cm, with corresponding erosion rates of 0–0.12 mm a^{-1}. Erosion depth values for the group 1985 range from 17.8 cm to 52.7 cm resulting in erosion rates of 0.05–0.16 mm a^{-1}. For the group near the LIA extent (1948) erosion depths of 11.8–35.9 cm and erosion rates from 0.05–0.14 mm a^{-1} were determined. For outside the LIA extent, an erosion depth of 2.5–2.7 cm and an erosion rate of 0.02–0.05 mm a^{-1} could be calculated only for Vorab-14. The measured ^{36}Cl concentration of Vorab-10 was higher than the modelled one and thus it was not possible to determine erosion depth or rate. This means the data from this location, especially the measured ^{36}Cl concentration, do not allow the calculation of rock depth removed or glacial erosion rates. This would only be possible if the time of ice-free phases were increased. However, this seems rather unrealistic due to relatively high elevation of the location (2439 m a.s.l.) and considering that during the Younger Dryas the Vorab glacier reached several km further down-valley than its LIA position [66–68].

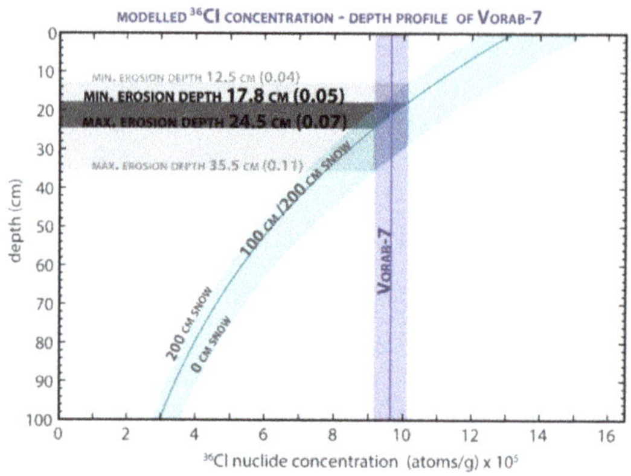

Figure 8. Modelled ^{36}Cl concentration-depth profile (light-blue) using the elemental and location data of Vorab-7. The solid line used the 200–100-200 cm snow scenario (see Figure 6 for details), the light blue band indicates the two extreme snow scenarios (0 cm snow and 200 cm snow). The AMS-measured concentration with its uncertainty is shown as vertical line in dark blue. The intersection of the two curves indicates the thickness of rock that had to be removed by the glacier to reach measured concentration (Table 1). The dark grey band highlights the erosion depth range (Table 3) with the intermediate snow scenario, the light grey indicates the minimal erosion depth when considering the 200 cm snow scenario and the maximal erosion depth with the zero snow scenario. By dividing this erosion depth by the total time of glacier coverage results in the average glacial erosion rates, given in parentheses.

Table 3. Determined glacial erosion depths and rates (sample locations in Figure 9).

Sample	Erosion Depth cm	Erosion Rate mm a^{-1}
Glacier front 2018		
Vorab-1	0.0–3.5	0.00–0.01
Vorab-2	0.0–4.1	0.00–0.01
Vorab-3	0.0–2.5	0.00–0.01
Vorab-4	3.1–8.8	0.01–0.03
Vorab-5	32.2–39.5	0.10–0.12
Glacier front 1985		
Vorab-7	17.8–24.5	0.05–0.07
Vorab-8	30.4–38.7	0.09–0.12
Vorab-9	46.0–53.7	0.14–0.16
Glacier front 1948		
Vorab-6	11.8–18.9	0.05–0.08
Vorab-11	24.3–32.6	0.10–0.13
Vorab-12	28.5–35.9	0.11–0.14
Outside LIA extent		
Vorab-10	nv–nv	nv–nv
Vorab-14	2.5–7.2	0.02–0.05

nv: no value.

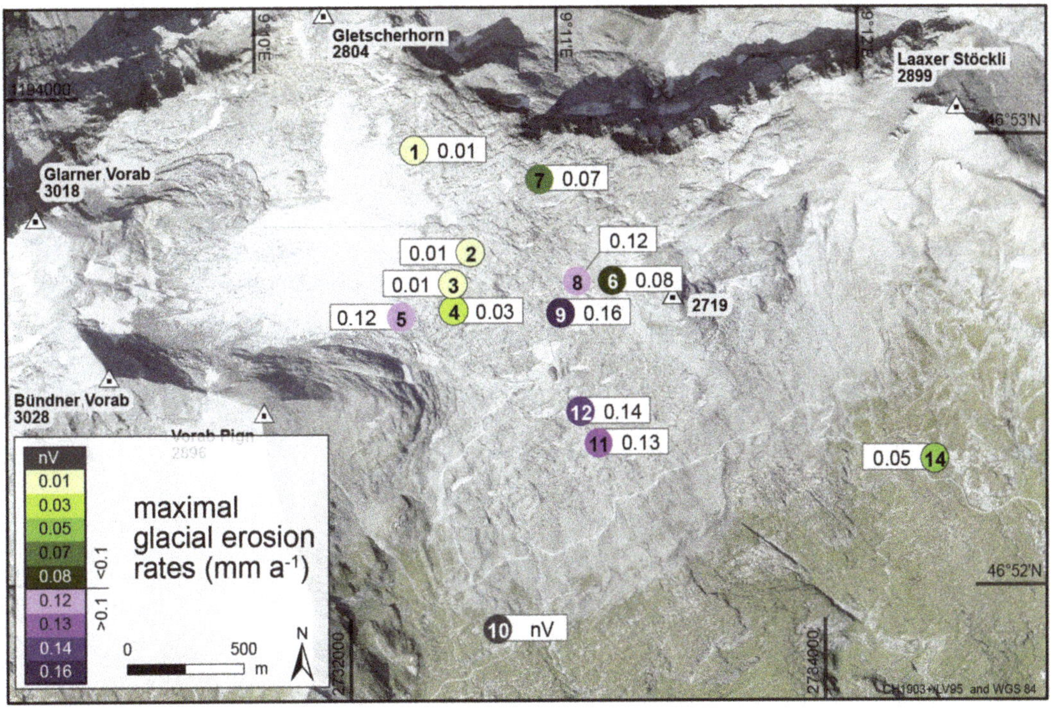

Figure 9. Determined maximal glacial erosion rates given in mm a^{-1} (Table 3). Green colours indicate glacial erosion rates less than and violet colours more than 0.1 mm a^{-1}. 3D view created with the swissIMAGE orthophoto 2019 and the SwissALTI3D 2019 (reproduced with authorization of the Swiss Federal Office of Topography, swisstopo).

6. Discussion and Conclusions

It is beyond question that the present-day shape of the Alps is the result of glacial erosion over numerous episodes of an extensive ice cap nearly completely covering the mountains and reaching to the forelands. However, actually determining glacial erosion rates remains challenging because conducting direct measurements is still extremely difficult [14]. Using cosmogenic nuclides in combination with a numerical model has been shown to be a strong tool for obtaining local glacial erosion rate values directly determined on glacially abraded bedrock surfaces [25,26]. As in the present investigation at the Vorab glacier, these studies give insight not only on local glacial and landscape forming processes but also increase understanding of the formation and evolution of the overall Alpine landscape.

At Vorab, the obtained high ^{36}Cl concentrations and calculated apparent exposure ages inside of the LIA extent between 7.1 ka to 3.3 ka provide clear evidence of the very small amount of erosion (<2 m deep) accomplished by the glacier during the late Holocene. The bedrock surfaces sampled were covered during the LIA as shown by historical maps (Figure 2). Coverage during the frequent cold phases of the late Holocene can be inferred from the numerous Holocene glacier variation studies carried out in the Alps (see Section 5.2.2). Accordingly, determined glacial erosion rates in the entire study area are low; all are less than <0.23 mm a^{-1}. We recognize a faint spatial pattern in the measured erosion rates across the glacier forefield as indicated by the colour coding in Figure 9. Erosion rates are lowest near the present-day glacier front (Vorab-1, Vorab-2, Vorab-3, Vorab-4, Vorab-5; green in Figure 9) and highest in the central part of the glacier forefield (Vorab-8: 0.12 mm a^{-1}, Vorab-9: 0.16 mm a^{-1}; purple in Figure 9). Vorab-6 is relatively close to these high values, with a rate of 0.08 mm a^{-1}. This slight local decrease in erosion efficiency could be due to its position. Vorab-6 is located where the slope slowly increases towards the approximately 100 m high bedrock ridge between Laaxer Stöckli and peak 2719. Here, the glacier flowed predominantly either to the left (northeast) or to the right (southward) as the bedrock ridge blocked its progress (see Figure 2). Our data also point to a slight increase in glacial erosion rate from north (Vorab-1: 0.01 mm a^{-1} and Vorab-7: 0.07 mm a^{-1}) to south (Vorab-5: 0.12 mm a^{-1} and Vorab-9: 0.23 mm a^{-1}). The glacial erosion rate determined from a bedrock sample outside of the LIA ice margin (Vorab-14) indicates that even during the Younger Dryas, during the Egesen stadial, the Vorab glacier was eroding at a very low rate of only 0.05 mm a^{-1}. However, it is important to keep the uncertainties of the input parameters of the MECED model in mind. Small changes in the glacier fluctuation history and the snow depth would slightly shift the erosion rate values. Nevertheless, because the input parameters were chosen conservatively to allow maximal erosion values within reasonable boundaries, the values should stay low or indeed get even lower. A thorough discussion of the effects of the input parameters is found in [25].

Our determined low glacial erosion rates are supported by field observations. On first sight, the forefield of the Vorab glacier appears to be strongly abraded by the glacier, as indicated by the heavily striated and highly polished bedrock surfaces. Taking a closer look, this only seems true for the previously present (i.e., before they were eroded) Verrucano and schistose lithologies that overlie the limestone. A crucial point is that in general, there is very little glacial sediment in the entire glacier forefield; most is bare limestone. Clasts are almost exclusively comprised Permian Verrucano, whereas limestone clasts are extremely rare and only found in the northern part of the study area. The same is true for the stone stripes, suggesting that the origin of these stone stripes is Permian bedrock obstacles below the glacier. In light of this isotopic and field evidence, we suggest that over many glacier advances, the glacier planed off the rock above the Glarus thrust (Permian Verrucano). As soon as the glacier eroded down to the Glarus thrust, where the lithology changes to limestone, its ability to erode dropped off decidedly. The Glarus thrust and the (remaining) overlying Verrucano units outcrop on both sides of the gap between Vorab Pign and peak 2719 (Figure 2). The glacier must have constantly widened instead of deepened this gap to head southward. In the passage through this gap and to the south, there are no remnants of Permian rock in the glacier forefield. The bedrock steps there indicate that plucking, not abrasion, is likely the dominant erosion process in the limestone units.

Field observations (numerous swallow holes and channels) are evidence that the limestone forefield is strongly karstified. Meltwater drains into a subsurface karst network to emanate in springs like those near Flims and Laax (Figure 1) [44,69]. The rapid loss of subglacial water into the karst system is likely the main factor for the low glacial erosion [25]. As a consequence of the lacking subglacial meltwater, the sliding capability of the glacier is reduced which directly influences the ability of the glacier to erode (abrade) the underlying bedrock [9,18,20,70,71]. This inhibition of sliding and low rate of erosion leads to low rates of subglacial sediment production. As noted above, there is almost no fine subglacial sediment in the glacier forefield. The limestones are massive and thick-bedded making them less susceptible to glacial erosion. This likely played a part in allowing the limestone plateau to develop as the Vorab glacier methodically shaved off the bedrock above the Glarus thrust (cf. [25]).

In this study, several factors were observed which are responsible for the formation of the limestone plateau in front of the Vorab glacier. These include the presence of: (i) the sharp geological contact (Glarus thrust), (ii) the massive, thick-bedded limestone below the thrust plane and (iii) the presence of a well-developed karst system. Similar findings were reported from the Lapis de Tsanfleuron (Figure 1) [25] and strengthen the hypothesis that high-elevation, low-relief limestone plateaus form due to limited glacial erosion because the glacier has a significantly reduced ability to slide as a consequence of losing water at the glacier bed into a well-developed karst system. At the Lapis de Tsanflueron, an even more extensive limestone plateau, determined glacial erosion rate values were similarly low or even lower (<0.08 mm a^{-1}) than those we determined at the Vorab glacier. The main difference between the two sites is the presence of the Glarus thrust plane right at the flat Vorab glacier forefield. Elevation (2600 m a.s.l.), aspect (east-facing) and precipitation values (150–200 m of snow) are comparable at both sites. Other striking similarities of the two sites are the absence of abundant glacial sediment, the presence of sediment stripes in glacier flow direction and the bedrock steps. Such clean (sediment-free) limestone plateaus at high elevations with associated underlying karst are common landscapes in the Alps, e.g., Hallstätter glacier, Dachstein Massif, Steinernes Meer, Siebenhengste and Hölloch-Silberen.

Glacial erosion rates determined on limestone beds are one to two orders of magnitude lower than those determined at sites located on crystalline bedrock, e.g., Rhône glacier (up to 0.66 mm a^{-1}) [24], Trift glacier (>1.8 mm a^{-1}) [26] (locations in Figure 1). In both of these crystalline study sites, the authors found a clear trend of low erosion rates at the margin of the glacier, where rates become nearly zero, while the highest erosion rates are measured below the central flow line of the glacier. At Trift glacier, rates of several millimetres per year were calculated from sites at the centre of the overdeepened glacial valley. In contrast, at the Lapis de Tsanfleuron and at the Vorab glacier, no significant or similarly extreme spatial variations in erosion rates were observed. Intriguingly, the slight decrease in glacial erosion that we noted at Vorab (Figure 9) occurs along the central flowline, the direct opposite of what was observed at Trift and the Rhône glacier. These stark differences in glacial erosion rates reveal quantitatively how the respective glaciers are carving quite distinct topographies. Whereas in limestone-dominated areas glacier forefields are plateau-shaped (broad and flat with very little relief); in crystalline areas (granite/gneiss) deep steep-walled valleys and at many sites overdeepened basins, e.g., the Rhône glacier and Trift glacier (Figure 1), tend to develop. The formation of overdeepenings is complex and usually a combination of geological (fracture spacing, weak zones, change in lithology) [72–74], geographical (glacier confluences) [75,76] and glaciological (near equilibrium line altitude, rates of ice flux) [12,77–79] factors. Recent research at Trift glacier, where there is an exceptionally deep overdeepening, highlighted that the presence of a gorge seems to have a crucial effect on the formation and the apparent depth of an overdeepening [26].

Undoubtedly many factors like elevation, precipitation sums, aspect, slope and bedrock geology influence the glacier and its behaviour, making the situation very complex and general statements on glacial erosion rates difficult. Nevertheless, recent studies on glacial erosion efficiency [24–26,80] highlight that there seems to be a relatively simple correlation between lithology, hydrology and glacial erosion rates. In crystalline areas (compact, impermeable bedrock) with an intact subglacial hydrology (subglacial meltwater present), glaciers tend to erode the bedrock strongly along the central flowline but only slightly near the lateral ice margins. Over hundreds of thousands of years this leads to the development of long, narrow, often overdeepened, steep-walled valleys. In contrast, flat glacier forefields form in areas dominated by horizontally bedded massive limestone with a well-developed karst system. Through the effective drainage of subglacial meltwater into the karst system, the ability of the glacier to slide is sharply curtailed. This results in an inability to erode the underlying substrate, favouring the formation of broad and flat landscapes (plateaus). A positive feedback loop develops: the glacier planes off a flat plateau as it reaches the massive flat-bedded limestone; the gentler the slope the further that sliding is inhibited. Over time, as these flat landscape elements cannot be reduced by glaciers, they evolve into plateaus—high-elevation islands. Despite the interplay of numerous factors, the importance of glaciers in the construction of mountainous landscapes is due as much to their ability to erode as to their ability to not erode.

Author Contributions: Conceptualization of the study was done by S.I.-O. and O.S.; methodology by A.M., O.S. and C.V.; investigation by A.M., S.I.-O., V.P. and O.S.; formal analysis by A.M. and O.S.; data curation by C.V., A.M. and O.S.; writing—original draft preparation, O.S., S.I.-O. and A.M.; writing—review and editing O.S. and S.I.-O.; visualisation, O.S. and A.M.; supervision, S.I.-O. and

V.P.; funding acquisition by S.I.-O. All authors have read and agreed to the published version of the manuscript.

Funding: Funding for this project was provided by "Schweizerischer Nationalfonds zur Förderung der Wissenschaftlichen Forschung" (SNSF) Project 175794 and 156187.

Data Availability Statement: All data are published within the publication.

Acknowledgments: We thank Sarah Kamleitner, Basil Weber and Ueli Steinemann for their support during fieldwork and Stefano Casale for his assistance in the laboratory. We are grateful for the overall support of the Laboratory of Ion Beam physics group and their excellent AMS measurements. Special thanks to the community of Laax and to the Weisse Arena Gruppe for providing driving permission and suitable vehicles for comfortable access to the study area. Funding for this project was provided by "Schweizerischer Nationalfonds zur Förderung der Wissenschaftlichen Forschung" (SNSF) Project 175794 and 156187. MeteoSchweiz (the Swiss Federal Office of Meteorology and Climatology) and Swisstopo (Federal Office of Topography swisstopo, provided local weather and geo data.

Conflicts of Interest: The authors declare no conflict of interest. The funders had no role in the design of the study; in the collection, analyses or interpretation of data; in the writing of the manuscript, or in the decision to publish the results.

References

1. Penck, A.; Brückner, E. *Die Alpen im Eiszeitalter*; Tauchitz: Leipzig, Germany, 1909.
2. Wright, G.F. The Muir Glacier. *Sci. Am.* **1887**, *23*, 9252. [CrossRef]
3. Agassiz, L. On the polished and striated surfaces of the rocks which form the beds of glaciers in the Alps. In *Proceedings of the Geological Society London*; Geological Society of London: London, UK, 1838; Volume 3, pp. 321–322.
4. Davis, W.M. The sculpture of mountains by glaciers. *Scott. Geogr. Mag.* **1906**, *22*, 76–89. [CrossRef]
5. Jordan, P. Analysis of overdeepened valleys using the digital elevation model of the bedrock surface of Northern Switzerland. *Swiss J. Geosci.* **2010**, *103*, 375–384. [CrossRef]
6. Preusser, F.; Reitner, J.; Schlüchter, C. Distribution, geometry, age and origin of overdeepened valleys and basins in the Alps and their foreland. *Swiss J. Geosci.* **2010**, *103*, 407–426. [CrossRef]
7. Dürst Stucki, M.; Schlunegger, F. Identification of erosional mechanisms during past glaciations based on a bedrock surface model of the central European Alps. *Earth Planet. Sci. Lett.* **2013**, *384*, 57–70. [CrossRef]
8. Sternai, P.; Herman, F.; Champagnac, J.-D.; Fox, M.; Salcher, B.; Willett, S.D. Pre-glacial topography of the European Alps. *Geology* **2012**, *40*, 1067–1070. [CrossRef]
9. Herman, F.; Beaud, F.; Champagnac, J.D.; Lemieux, J.M.; Sternai, P. Glacial hydrology and erosion patterns: A mechanism for carving glacial valleys. *Earth Planet. Sci. Lett.* **2011**, *310*, 498–508. [CrossRef]
10. Hinderer, M.; Kastowski, M.; Kamelger, A.; Bartolini, C.; Schlunegger, F. River loads and modern denudation of the Alps—A review. *Earth-Sci. Rev.* **2012**, *118*, 11–44. [CrossRef]
11. Herman, F.; De Doncker, F.; Delaney, I.; Prasicek, G.; Koppes, M. The impact of glaciers on mountain erosion. *Nat. Rev. Earth Environ.* **2021**. [CrossRef]
12. Cook, S.J.; Swift, D.A. Subglacial basins: Their origin and importance in glacial systems and landscapes. *Earth-Sci. Rev.* **2012**, *115*, 332–372. [CrossRef]
13. Boulton, G.S. Processes and Patterns of Glacial Erosion. In *Glacial Geomorphology*; Coates, D.R., Ed.; Springer: Dordrecht, The Netherland, 1982; pp. 41–87. [CrossRef]
14. Lütschg-Loetscher, O. 4. Kapitel. Die Bedeutung und Bewertung der Vorratsänderungen im Wasserhaushalt der Gletscher im Schweizer Hochgebirge, 5. Kapitel. Beobachtungen über das Verhalten des vorstossenden Obern Grindelwaldgletschers im Berner Oberland. In *Zum Wasserhaushalt Des Schweizer Hohgebrigers*; Kümmerly & Frey: Bern, Switzerland, 1944.
15. Bezinge, A.; Clark, M.J.; Gurnell, A.M.; Warburton, J. The management of sediment transported by glacial melt-water streams and its significance for the estimation of sediment yield. *Ann. Glaciol.* **1989**, *13*, 1–5. [CrossRef]
16. Bogen, J. Erosion rates and sediment yields of glaciers. *Ann. Glaciol.* **1996**, *22*, 48–52. [CrossRef]
17. Beaud, F.; Flowers, G.E.; Venditti, J.G. Efficacy of bedrock erosion by subglacial water flow. *Earth Surf. Dyn.* **2016**, *4*, 125–145. [CrossRef]
18. Cook, S.J.; Swift, D.A.; Kirkbride, M.P.; Knight, P.G.; Waller, R.I. The empirical basis for modelling glacial erosion rates. *Nat. Commun.* **2020**, *11*, 1–7. [CrossRef]
19. MacGregor, K.R.; Anderson, R.S.; Waddington, E.D. Numerical modeling of glacial erosion and headwall processes in alpine valleys. *Geomorphology* **2009**, *103*, 189–204. [CrossRef]
20. Koppes, M.; Hallet, B.; Rignot, E.; Mouginot, J.; Wellner, J.S.; Boldt, K. Observed latitudinal variations in erosion as a function of glacier dynamics. *Nature* **2015**, *526*, 100. [CrossRef]

21. Cohen, D.; Gillet-Chaulet, F.; Haeberli, W.; Machguth, H.; Fischer, U.H. Numerical reconstructions of the flow and basal conditions of the Rhine glacier, European Central Alps, at the Last Glacial Maximum. *Cryosphere* **2018**, *12*, 2515–2544. [CrossRef]
22. Seguinot, J.; Delaney, I. Last glacial cycle glacier erosion potential in the Alps. *Earth Surf. Dyn. Discuss.* **2021**, *9*, 923–935. [CrossRef]
23. Wirsig, C.; Ivy-Ochs, S.; Reitner, J.M.; Christl, M.; Vockenhuber, C.; Bichler, M.; Reindl, M. Subglacial abrasion rates at Goldbergkees, Hohe Tauern, Austria, determined from cosmogenic ^{10}Be and ^{36}Cl concentrations. *Earth Surf. Process. Landf.* **2017**, *42*, 1119–1131. [CrossRef]
24. Goehring, B.M.; Schaefer, J.M.; Schlüchter, C.; Lifton, N.A.; Finkel, R.C.; Jull, A.T.; Akçar, N.; Alley, R.B. The Rhone Glacier was smaller than today for most of the Holocene. *Geology* **2011**, *39*, 679–682. [CrossRef]
25. Steinemann, O.; Ivy-Ochs, S.; Grazioli, S.; Luetscher, M.; Fischer, U.H.; Vockenhuber, C.; Synal, H.A. Quantifying glacial erosion on a limestone bed and the relevance for landscape development in the Alps. *Earth Surf. Process. Landf.* **2020**, *45*, 1401–1417. [CrossRef]
26. Steinemann, O.; Ivy-Ochs, S.; Hippe, K.; Christl, M.; Haghipour, N.; Synal, H.A. Glacial erosion by the Trift glacier (Switzerland): Deciphering the development of riegels, rock basins and gorges. *Geomorphology* **2021**, *375*, 107533. [CrossRef]
27. GLAMOS The Swiss glaciers 1880-2016/17, Glaciological reports No 1-138. In *Yearbooks of the Cryospheric Commission of the Swiss Academy of Sciences (SCNAT) 1881–2019*; Published since 1964; VAW/ETH: Zürich, Switzerland; Available online: https://www.glamos.ch (accessed on 30 June 2021). [CrossRef]
28. Garwood, E.J. Features of Alpine Scenery Due to Glacial Protection. *Geogr. J.* **1910**, *36*, 310–336. [CrossRef]
29. Buckingham, T.; Pfiffner, O.A. Mountain Building and Valley Formation in the UNESCO World Heritage Tectonic Arena Sardona Region. In *Landscapes and Landforms of Switzerland*; Reynard, E., Ed.; Springer: Cham, Switzerland, 2021; pp. 173–186. [CrossRef]
30. Pfiffner, O.A. Structural map of the Helvetic Zone of the Swiss Alps: Including Vorarlberg (Austria) and Haute Savoie (France). *Geological Special Map 128/1-7* **2010**.
31. Pfiffner, O.A. Tektonische Untersuchungen im Infrahelvetikum der Ostschweiz. Ph.D. Thesis, ETH Zürich und Universität Zürich, Zürich, Switzerland, 1977.
32. Ivy-Ochs, S.; Synal, H.A.; Roth, C.; Schaller, M. Initial results from isotope dilution for Cl and ^{36}Cl measurements at the PSI/ETH Zurich AMS facility. *Nucl. Instrum. Methods Phys. Res. Sect B Beam Interact. Mater. At.* **2004**, *223*, 623–627. [CrossRef]
33. Stone, J.O.; Allan, G.L.; Fifield, L.K.; Cresswell, R.G. Cosmogenic chlorine-36 from calcium spallation. *Geochim. Cosmochim. Acta* **1996**, *60*, 679–692. [CrossRef]
34. Vockenhuber, C.; Miltenberger, K.U.; Synal, H.A. ^{36}Cl measurements with a gas-filled magnet at 6 MV. *Nucl. Instrum. Methods Phys. Res. Sect B Beam Interact. Mater. At.* **2019**, *455*, 190–194. [CrossRef]
35. Alfimov, V.; Ivy-Ochs, S. How well do we understand production of ^{36}Cl in limestone and dolomite? *Quat. Geochronol.* **2009**, *4*, 462–474. [CrossRef]
36. Steinemann, O. Using cosmogenic nuclides to understand rockglacier dynamics during the Lateglacial and to quantify glacial erosion on limestone and gneiss in the Alps. Ph.D. Thesis, ETH Zürich, Zürich, Switzerland, 2020.
37. Wirsig, C. Constraining the Timing of Deglaciation of the High Alps and Rates of Subglacial Erosion with Cosmogenic Nuclides. Ph.D. Thesis, ETH Zürich, Zürich, Switzerland, 2015.
38. Dufour, G.H.; Müllhaupt, H.; Stempelmann, J.; Kögel, H. *Altdorf, Chur*; Eidg. Topographisches Bureau: Genf, Switzerland, 1859.
39. Online Karten viewer. Available online: www.maps.geo.admin.ch (accessed on 6 June 2021).
40. Lukas, S.; Rother, H. Moränen versus Till: Empfehlungen für die Beschreibung, Interpretation und Klassifikation glazialer Landformen und Sedimente. *E&G Quat. Sci. J.* **2016**, *65*, 95–112. [CrossRef]
41. Keller, B. Lithofazies-Codes für die Klassifikation von Lockergesteinen. *Mitteilungen Schweiz. Ges. Boden- u. Felsmechanik* **1996**, *132*, 5–12.
42. Bögli, A. *Karst Hydrology and Physical Speleology*; Springer: Berlin, Germany, 1980. [CrossRef]
43. Veress, M.; Telbisz, T.; Tóth, G.; Lóczy, D.; Ruban, D.A.; Gutak, J.M. *Glaciokarsts*; Springer: Cham, Switzerland, 2019. [CrossRef]
44. Häuselmann, P.; Jeannin, P.Y. Wasserwege der Gemeinde Flims und ihre Einflüsse auf den Caumasee. *Schweiz. Inst. Speläologie und Karstforschung* **2009**, *24*, 1–24.
45. Christl, M.; Vockenhuber, C.; Kubik, P.W.; Wacker, L.; Lachner, J.; Alfimov, V.; Synal, H.A. The ETH Zurich AMS facilities: Performance parameters and reference materials. *Nucl. Instrum. Methods Phys. Res. Sect B Beam Interact. Mater. At.* **2013**, *294*, 29–38. [CrossRef]
46. Schneebeli, W.; Röthlisberger, F. 8000 Jahre Walliser Gletschergeschichte ein Beitrag zur Erforschung des Klimaverlaufs in der Nacheiszeit. Ph.D. Dissertation, University of Zürich, Verlag Schweizer Alpen-Club, Bern, Switzerland, 1976.
47. Holzhauser, H. Gletscherschwankungen innerhalb der letzten 3200 Jahre am Beispiel des grossen Aletsch- und des Gornergletschers. Neue Ergebnisse. In *Gletscher im Ständigen Wandel*; vdf Hochschulverlag AG an der ETH Zürich: Zürich, Switzerland, 1995; pp. 101–123.
48. Ivy-Ochs, S.; Kerschner, H.; Maisch, M.; Christl, M.; Kubik, P.W.; Schlüchter, C. Latest Pleistocene and Holocene glacier variations in the European Alps. *Quat. Sci. Rev.* **2009**, *28*, 2137–2149. [CrossRef]
49. Holzhauser, H. *Auf dem Holzweg zur Gletschergeschichte*; Mitteilungen Naturforschende Gesellschaft in Bern: Bern, Switzerland, 2009; Volume 66, pp. 173–208.

50. Hormes, A.; Müller, B.U.; Schlüchter, C. The Alps with little ice: Evidence for eight Holocene phases of reduced glacier extent in the Central Swiss Alps. *Holocene* **2001**, *11*, 255–265. [CrossRef]
51. Joerin, U.E.; Stocker, T.F.; Schlüchter, C. Multicentury glacier fluctuations in the Swiss Alps during the Holocene. *Holocene* **2006**, *16*, 697–704. [CrossRef]
52. Wick, L. Vegetational response to climatic changes recorded in Swiss Late Glacial lake sediments. *Palaeogeogr. Palaeoclimatol. Palaeoecol.* **2000**, *159*, 231–250. [CrossRef]
53. Leemann, A.; Niessen, F. Holocene glacial activity and climatic variations in the Swiss Alps: Reconstructing a continuous record from proglacial lake sediments. *Holocene* **1994**, *4*, 259–268. [CrossRef]
54. Zumbühl, H.J.; Nussbaumer, S.U. Little Ice Age glacier history of the Central and Western Alps from pictorial documents. *Cuad. Investig. Geogr.* **2018**, *44*, 115–136. [CrossRef]
55. Bauder, A.; Funk, M.; Huss, M. Ice-volume changes of selected glaciers in the Swiss Alps since the end of the 19th century. *Ann. Glaciol.* **2007**, *46*. [CrossRef]
56. Le Roy, M.; Deline, P.; Carcaillet, J.; Schimmelpfennig, I.; Ermini, M. ^{10}Be exposure dating of the timing of Neoglacial glacier advances in the Ecrins-Pelvoux massif, southern French Alps. *Quat. Sci. Rev.* **2017**, *178*, 118–138. [CrossRef]
57. Le Roy, M.; Nicolussi, K.; Deline, P.; Astrade, L.; Edouard, J.L.; Miramont, C.; Arnaud, F. Calendar-dated glacier variations in the western European Alps during the Neoglacial: The Mer de Glace record, Mont Blanc massif. *Quat. Sci. Rev.* **2015**, *108*, 1–22. [CrossRef]
58. Schimmelpfennig, I.; Schaefer, J.M.; Akçar, N.; Ivy-Ochs, S.; Finkel, R.C.; Schlüchter, C. Holocene glacier culminations in the Western Alps and their hemispheric relevance. *Geology* **2012**, *40*, 891–894. [CrossRef]
59. Schimmelpfennig, I.; Schaefer, J.M.; Akçar, N.; Koffman, T.; Ivy-Ochs, S.; Schwartz, R.; Finkel, R.C.; Zimmerman, S.; Schlüchter, C. A chronology of Holocene and Little Ice Age glacier culminations of the Steingletscher, Central Alps, Switzerland, based on high-sensitivity beryllium-10 moraine dating. *Earth Planet. Sci. Lett.* **2014**, *393*, 220–230. [CrossRef]
60. Kronig, O.; Ivy-Ochs, S.; Hajdas, I.; Christl, M.; Wirsig, C.; Schlüchter, C. Holocene evolution of the Triftje- and the Oberseegletscher (Swiss Alps) constrained with ^{10}Be exposure and radiocarbon dating. *Swiss J. Geosci.* **2018**, *111*, 117–131. [CrossRef]
61. Protin, M.; Schimmelpfennig, I.; Mugnier, J.-L.; Buoncristiani, J.-F.; Le Roy, M.; Pohl, B.; Moreau, L.; ASTER Team; Aumaître, G.; Bourlès, D.; et al. Millennial-scale deglaciation across the European Alps at the transition between the Younger Dryas and the Early Holocene—Evidence from a new cosmogenic nuclide chronology. *Boreas* **2021**, *50*, 671–685. [CrossRef]
62. Joerin, U.E.; Nicolussi, K.; Fischer, A.; Stocker, T.F.; Schlüchter, C. Holocene optimum events inferred from subglacial sediments at Tschierva Glacier, Eastern Swiss Alps. *Quat. Sci. Rev.* **2008**, *27*, 337–350. [CrossRef]
63. Plan, L. Factors controlling carbonate dissolution rates quantified in a field test in the Austrian alps. *Geomorphology* **2005**, *68*, 201–212. [CrossRef]
64. Häuselmann, P. Surface corrosion of an Alpine karren field: Recent measures at Innerbergli (Siebenhengste, Switzerland). *Int. J. Speleol.* **2008**, *37*, 107–111. [CrossRef]
65. Krklec, K.; Domínguez-Villar, D.; Braucher, R.; Perica, D.; Mrak, I. Morphometric comparison of weathering features on side by side carbonate rock surfaces with different exposure ages—A case from the Croatian coast. *Quat. Int.* **2017**, *494*, 275–285. [CrossRef]
66. Hantke, R. *Eiszeitalter*; Ott Verlag: Thun, Switzerland, 1980; Volume 2, p. 703.
67. Maisch, M. Zur Gletscher- und Klimageschichte des alpinen Spätglazials. *Geogr. Helv.* **1982**, *37*, 93–104. [CrossRef]
68. Jordi, U. *Glazialmorphologische und Gletschergeschichtliche Untersuchungen im Taminatal und im Rheintalabschnitt zwischen Flims und Feldkirch (Ostschweiz/Vorarlberg)*; Geographisches Institut der Universität Bern: Bern, Switzerland, 1986.
69. Weber, E.; Jordan, F.; Jeannin, P.-Y.; Vouillamoz, J.; Malard, A. Swisskarst project (NRP61): Towards a pragmatic simulation of karst spring discharge with conceptual semi-distributed model. The Flims case study (Eastern Swiss Alps). In Proceedings of the 9th Conference Limestone Hydrogeology, Besançon, France, 1–3 September 2011.
70. Hallet, B. A theoretical model of glacial abrasion. *J. Glaciol.* **1979**, *23*, 39–50. [CrossRef]
71. Ugelvig, S.V.; Egholm, D.L.; Anderson, R.S.; Iverson, N.R. Glacial Erosion Driven by Variations in Meltwater Drainage. *J. Geophys. Res. Earth Surf.* **2018**, *123*, 2863–2877. [CrossRef]
72. Dühnforth, M.; Anderson, R.S.; Ward, D.; Stock, G.M. Bedrock fracture control of glacial erosion processes and rates. *Geology* **2010**, *38*, 423–426. [CrossRef]
73. Krabbendam, M.; Glasser, N.F. Glacial erosion and bedrock properties in NW Scotland: Abrasion and plucking, hardness and joint spacing. *Geomorphology* **2011**, *130*, 374–383. [CrossRef]
74. Lane, T.P.; Roberts, D.H.; Rea, B.R.; Cofaigh, C.Ó.; Vieli, A. Controls on bedrock bedform development beneath the Uummannaq Ice Stream onset zone, West Greenland. *Geomorphology* **2015**, *231*, 301–313. [CrossRef]
75. Haeberli, W.; Linsbauer, A.; Cochachin, A.; Salazar, C.; Fischer, U.H. On the morphological characteristics of overdeepenings in high-mountain glacier beds. *Earth Surf. Process. Landf.* **2016**, *41*, 1980–1990. [CrossRef]
76. MacGregor, K.R.; Anderson, R.S.; Anderson, S.P.; Waddington, E.D. Numerical simulations of glacial-valley longitudinal profile evolution. *Geology* **2000**, *28*, 1031–1034. [CrossRef]
77. Sugden, D.E.; John, B.S. *Glaciers and Landscape: A Geomorphological Approach*; Arnold: London, UK, 1976.
78. Evans, I.S. *Glacial Erosional Processes and Forms Mountain Glaciation and Glacier Geography*; The Geological Society: London, UK, 2008.

79. Evans, I.S. Glaciers, rock avalanches and the 'buzzsaw' in cirque development: Why mountain cirques are of mainly glacial origin. *Earth Surf. Process. Landf.* **2021**, *46*, 24–46. [CrossRef]
80. Wirsig, C.; Ivy-Ochs, S.; Akçar, N.; Lupker, M.; Hippe, K.; Wacker, L.; Vockenhuber, C.; Schlüchter, C. Combined cosmogenic ^{10}Be, in situ ^{14}C and ^{36}Cl concentrations constrain Holocene history and erosion depth of Grueben glacier (CH). *Swiss J. Geosci.* **2016**, *109*, 379–388. [CrossRef]

geosciences MDPI

Article

Cosmogenic and Geological Evidence for the Occurrence of a Ma-Long Feedback between Uplift and Denudation, Chur Region, Swiss Alps

Ewerton da Silva Guimarães [1,2,*], Romain Delunel [1,3], Fritz Schlunegger [1], Naki Akçar [1], Laura Stutenbecker [1,4] and Marcus Christl [5]

1 Institute of Geological Sciences, University of Bern, Baltzerstrasse 3, 3012 Bern, Switzerland; romain.delunel@univ-lyon2.fr (R.D.); fritz.schlunegger@geo.unibe.ch (F.S.); akcar@geo.unibe.ch (N.A.); stutenbecker@geo.tu-darmstadt.de (L.S.)
2 Department of Earth Sciences, Vrije Universiteit Amsterdam, De Boelelaan 1085, 1081 HV Amsterdam, The Netherlands
3 French National Centre for Scientific Research (CNRS), UMR 5600 EVS/IRG-Lyon 2, 69676 BRON, France
4 Institute of Applied Geosciences, Technical University of Darmstadt, Karolinenplatz 5, 64289 Darmstadt, Germany
5 Laboratory of Ion Beam Physics, Swiss Federal Institute of Technology Zurich (ETHZ), Otto-Stern-Weg 5, 8093 Zurich, Switzerland; mchristl@ethz.ch
* Correspondence: e.dasilvaguimaraes@vu.nl

Citation: da Silva Guimarães, E.; Delunel, R.; Schlunegger, F.; Akçar, N.; Stutenbecker, L.; Christl, M. Cosmogenic and Geological Evidence for the Occurrence of a Ma-Long Feedback between Uplift and Denudation, Chur Region, Swiss Alps. *Geosciences* **2021**, *11*, 339. https://doi.org/10.3390/geosciences11080339

Academic Editors: Ángel Rodés and Jesus Martinez-Frias

Received: 9 June 2021
Accepted: 27 July 2021
Published: 12 August 2021

Abstract: We used concentrations of in situ cosmogenic ^{10}Be from riverine sediment to quantify the basin-averaged denudation rates and sediment fluxes in the Plessur Basin, Eastern Swiss Alps, which is a tributary stream to the Alpine Rhine, one of the largest streams in Europe. We complement the cosmogenic dataset with the results of morphometric analyses, geomorphic mapping, and sediment fingerprinting techniques. The results reveal that the Plessur Basin is still adjusting to the landscape perturbation caused by the glacial carving during the Last Glacial Maximum c. 20,000 years ago. This adjustment has been most efficient in the downstream part where the bedrock comprises high erodibility North Penninic flysch and Bündnerschist, whereas glacial landforms are still prominently preserved in the upstream region, comprising low erodibility South Penninic and Austroalpine bedrock. This geomorphic observation is supported by the ^{10}Be based denudation rate and sediment provenance analysis, which indicate a much faster sediment production in the flysch and schist lithologies. Interestingly, the reach of fast denudation has experienced the highest exhumation and rock uplift rates. This suggests that lithologic and glacial conditioning have substantially contributed to the local uplift and denudation as some of the driving forces of a positive feedback system.

Keywords: cosmogenic nuclides; sediment fingerprinting; geomorphometric analysis; positive feedback; Prättigau half-window

1. Introduction

In mountainous areas, the shape of a landscape is the expression of a complex interaction between tectonic and erosional processes over multiple temporal scales [1–4]. Tectonic forces create topography through the vertical upward-directed advection of crustal material, resulting in rock uplift, whereas erosional processes are mainly driven by gravitational forces and climate, and result in the downwearing of the accreted material. One particular expression of the interaction between tectonic and denudation is a positive feedback, where erosion-driven unloading has the potential to initiate an isostatic response of the lithosphere in the form of crustal uplift [5]. Such a mechanism at work has, for instance, been proposed for the Central European Alps [6,7]. However, for the Alps of Eastern Switzerland (Figure 1), several authors have suggested that the high uplift rates are rather a long-lived consequence of neotectonic shortening [8] than a feedback response to erosion. Furthermore, the occurrence of high uplift rates was explained within the context of a shorter timescale of observations, whereby the retreat and melting of glaciers of the Last Glacial Maximum (LGM), ca. 20 ka ago, was considered to have induced an isostatic rebound in response to unloading [9–11].

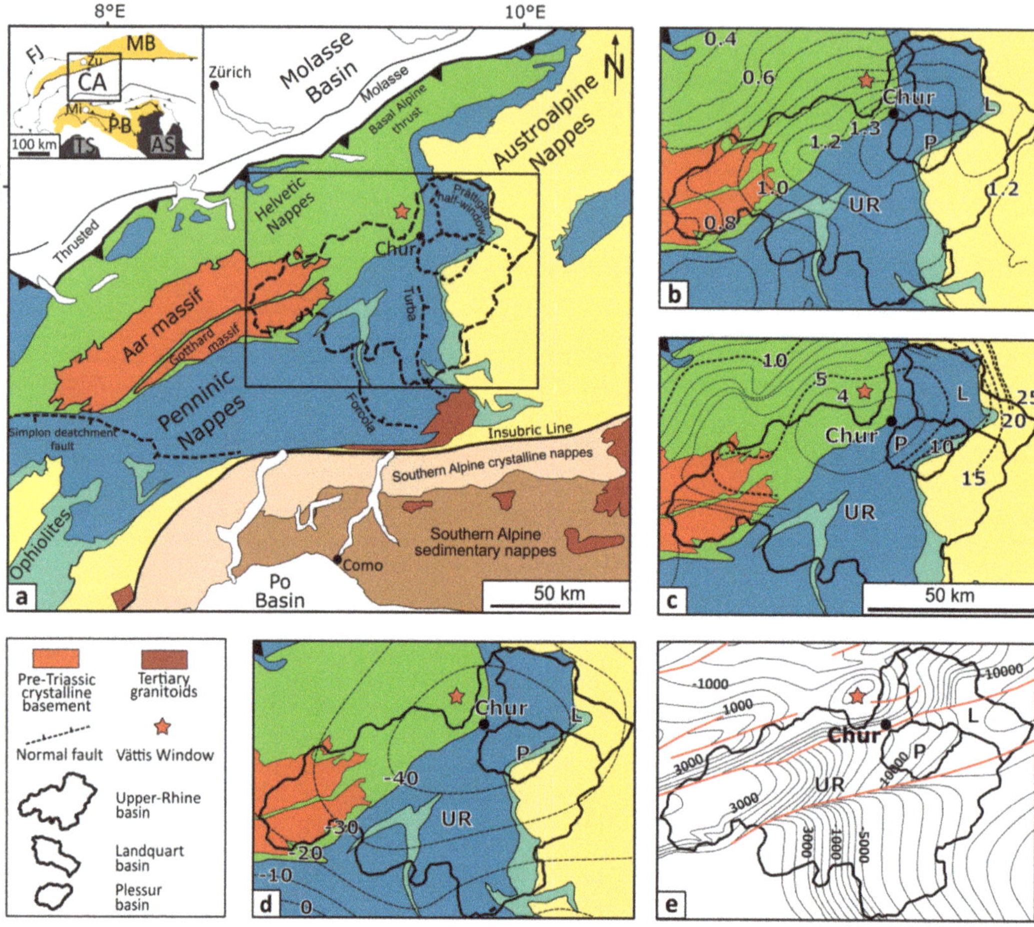

Figure 1. (**a**) Geological and tectonic framework of the Alps and the studied drainage basins (modified after [12]). FJ: Folded Jura, MB: Molasse Basin, CA: Central Alps, PB: Po Basin, TS: Tyrrhenian Sea, AS: Adriatic Sea, Zu: Zurich, Mi: Milan; (**b**) geodetic uplift rate in mm a^{-1} [13,14]; (**c**) apatite fission track pattern in Ma [15]; (**d**) isostatic anomaly in mGal [16]; (**e**) top crystalline basement in m [17].

It has recently been documented that surface denudation rates over the past hundreds to thousands of years are lower than geodetically measured rock uplift rates, particularly in the Central Alps of Europe [18,19]. This observation was based on a compilation of published and new ^{10}Be concentrations in riverine quartz in >350 rivers. Thus, an active tectonic driving force has to be invoked to explain the occurrence of a surface uplift component at the scale of the Central European Alps [19]. The authors of the aforementioned study also emphasized that the identifications of distinct driving forces on surface erosion and landscape shape at the regional scale and at that of individual catchments can be hampered because of the stochastic and site-specific nature of erosion.

The aim of this paper is to explore whether or not a positive feedback response to erosion can be invoked to explain the high uplift and exhumation rates in the Eastern Alps of Switzerland (Figure 1). To this end, we focused on the erosional mechanisms and related denudation rates in the Prättigau half-window near Chur, situated in the Eastern Alps of Switzerland (Figure 1a). This region exposes North Penninic flysch and Bündnerschist with a high bedrock erodibility in the core of the half-window, and these units are overlain by Penninic rocks and the Austroalpine orogenic lid that comprise limestones, gneiss, and schists with a lower erodibility [20]. We focused on this area because of previously published evidence for a tectonic control on erosion in the region and for the occurrence of an inferred, yet contested, positive feedback between denudation and uplift at least since the Pliocene [21,22] and possibly the Holocene [7]. We focused on the Plessur Basin (labeled as P in Figure 1b–e), situated in the southern part of this window where geodetically measured rock uplift rates are highest in the region (Figure 1b) and where the youngest apatite fission track ages have been reported (Figure 1c). We measured the denudation flux in this basin using in situ ^{10}Be. We identified the sediment source areas through provenance tracing, and we related these data to the lithotectonic architecture of the Plessur Basin and to the topographic imprint caused by the LGM and possibly previous glaciers. These data are combined with the results of previous work [22] conducted in the Landquart Basin (labeled as L in Figure 1b) situated in the northern part of the Prättigau half-window. We then used the combined dataset to reconstruct a picture about the pattern of erosion at the regional scale during the Holocene and over millennia. It additionally allows us to re-address the problem of whether or not a positive feedback response to erosion can be invoked to explain the uplift and exhumation pattern in the region.

2. Setting

2.1. Lithotectonic Architecture

The tectonic architecture in the area is characterized by a stack of Helvetic, Penninic, and Austroalpine nappes, which generally show a regional dip toward the SE (Figure 1a). During Mesozoic times, the Helvetic units were part of the stretched margin of the European plate and formed the transition toward the Valais Ocean. These rocks crop out to the NW of the study area and are mainly composed of limestones with marl interbeds. Currently, the Helvetic thrust nappes overlay the Aar massif, a pre-Triassic crystalline basement unit, which is largely exposed west of our study area where this massif forms major topographic peaks. The top of the crystalline rocks then plunges toward the NNE, where it currently lies at a depth of 10 km beneath the study area (Figure 1e). To the north of Chur, however, a small outcrop made up of basement crystalline rocks can be found in the Vättis window ([12,17]; red star on Figure 1).

The Penninic units in the region, which overlay the Helvetic thrust sheets, are divided into Lower, Middle, and Upper Penninic thrust nappes (Figure 2). The Lower Penninic units comprise Mesozoic–Cenozoic hemipelagic sediments that were deposited within the Valais trough to the south of the Helvetic sedimentary realm. In the study area, the Lower Penninic rocks are represented by the Mesozoic Bündnerschist and the Upper Cretaceous to Eocene North Penninic flysch that are exposed in the Prättigau half-window (Figure 2).

The only Middle Penninic unit that is found in the area is the Falknis nappe, which is mainly composed of Mesozoic limestones that were deposited on the northern margin of the Briançonnais zone. During Mesozoic times, this zone was a submarine swell that separated the Valais Ocean from the Piemont Ocean farther north and south, respectively [12,15,17,23]. On the southern margin of the Briançonnais swell, ophiolitic sequences and fragments of continental crustal rocks (mélange) were formed during the Jurassic phase of spreading within the Piemont Ocean, which are now found in the Arosa Zone of the Upper Penninic unit ([17,24,25]; Figure 2). The Austroalpine units, situated on top of the Penninic thrust sheets, are divided into the Lower and Upper Austroalpine nappes and comprise rocks of the Adriatic continental plate (upper Austroalpine rocks) and its northern margin ([24]; lower Austroalpine units). The Lower Austroalpine nappes are mainly composed of Mesozoic limestones and are represented in the study area by the Rothorn and Tschirpen nappes [17,26] (Figure 2). The Upper Austroalpine units are represented by Triassic dolomites that make up the Languard nappe, and by Paleozoic sediments of the basement in the Silvretta nappe [27].

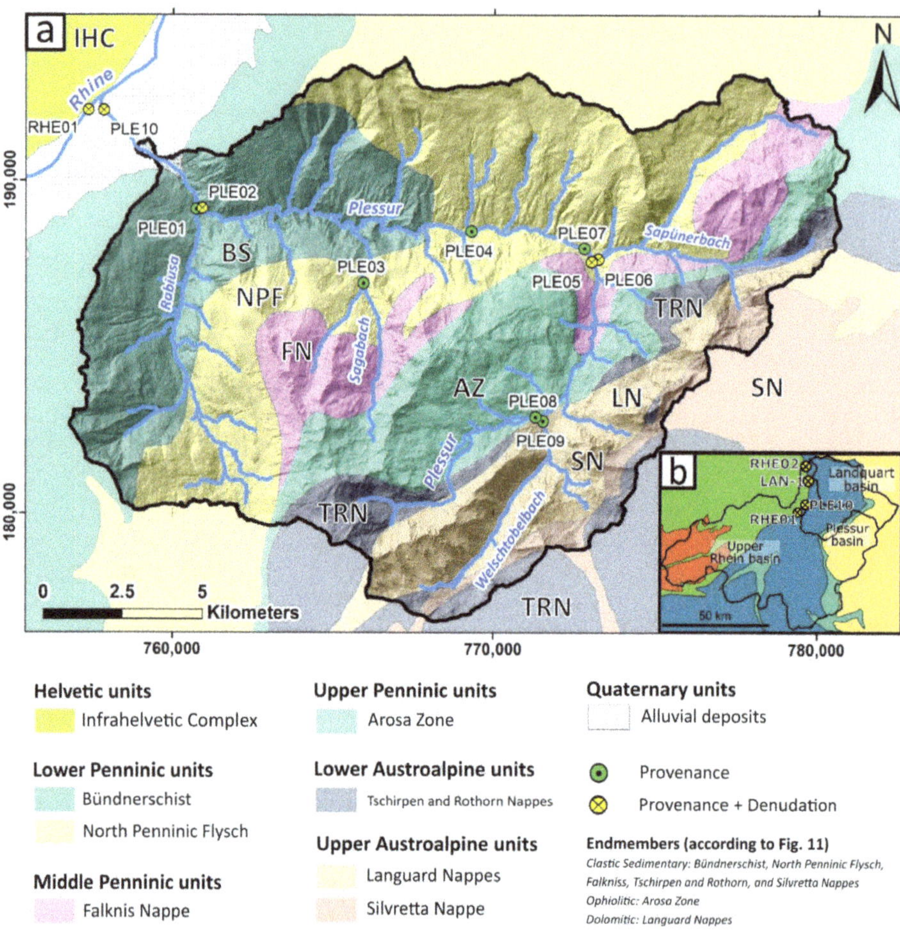

Figure 2. (**a**) Lithotectonic framework of the Plessur drainage basin and its main tributaries, and locations of collected samples. IHC: Infrahelvetic Complex; BS: Bündnerschist; NPF: North Penninic Flysch; FN: Falknis Nappe; AZ: Arosa zone; TRN: Tschirpen and Rothorn Nappes; LN: Languard Nappe; SN: Silvretta Nappe; Q: Quaternary deposits; (**b**) Location of samples used in this study but collected outside of the Plessur Basin. Samples RHE01 and RHE02 were collected in the Rhine River and are part of this work's dataset, whereas sample Lan-1 was collected at the outlet of the Landquart River and is part of the work by [22]. Green dots indicate sites where samples have been processed for provenance tracing purposes only, whereas yellow dots represent sites where the samples have been analyzed for both provenance and concentrations of ^{10}Be. Refer to Figure 1 for the explanation of colors. The numbers at the border of the figure indicate the Swiss coordinates in meters.

In summary, the lowermost part of the Plessur Basin is underlain by Lower Penninic Bündnerschist and North Penninic flysch, both of which are part of the Prättigau half-window (Figure 1a) and have a high erodibility [20]. The headwater part of the Plessur Basin, however, comprises ophiolites, limestones, and basement rocks of Penninic and

Austroalpine origin with a substantially lower bedrock erodibility [20]. As will be shown below, the differences in bedrock erodibilities are seen in the pattern of surface denudation and sediment fluxes.

2.2. Rock Uplift and Exhumation

As previously mentioned, the study area has experienced one of the highest uplift rates in the Swiss Alps during the past decades, reaching up to 1.3 mm a^{-1} [13,14] (Figure 1b). The contour lines of rock uplift rates display a similar shape as those of apatite fission track cooling ages (Figure 1c), where the occurrence of highest uplift and lowest apatite fission track ages of c. 4 Ma are centered near Chur [15,23,28]. These young ages point to one of the fastest exhumation rates in the Central European Alps [17,21,29], and the contour lines follow the boundaries of the Prättigau half-window (Figure 1c). Interestingly, the location with the highest exhumation rates also corresponds to the region with the largest negative Bouguer anomalies near Chur [16] (Figure 1d), which suggests that the loads related to the surface topography are overcompensated in the region where the study area is situated [30–32].

2.3. Geomorphologic Framework

The Plessur Basin is a tributary of the Alpine segment of the Rhine River (here named Upper Rhine), located in the southeastern Swiss Alps. The basin covers an area of ca. 265 km^2, and its trunk, the Plessur River, is ca. 30 km long (Figure 2). The lowermost section of the Plessur Basin is located within the Prättigau half-window. The Plessur Basin hosts four main tributaries streams: (i) the Rabiusa, which entirely drains Bündnerschist and North Penninic flysch rocks; (ii) the Sagabach, flowing through Penninic units; (iii) the Sapünerbach, which drains mainly Penninic rocks and a small part of Austroalpine units; and (iv) the Welschtobelbach, which is entirely sourced in Austroalpine units (Figure 2). The catchment areas of these tributary streams vary from ca. 25 to 60 km^2.

The cross-sectional geometries of the major tributary basins are generally V-shaped (Figure 3, transects EE′ and FF′) and reflect the ongoing dissection of the streams during the Holocene [33]. This observation is often verified in the lowermost part of the Plessur Basin, in the Rabiusa sub-catchment, and in the lowermost part of the Sagabach sub-catchment. In contrast, in the headwater areas of the Plessur Basin, which also includes the Sapünerbach and Welschtobelbach sub-catchments, U-shaped cross-sectional geometries and multiple convex-concave segments along the course of the valley are more common (Figure 3, transects AA′ and BB′, and Figure 4), both of which are an indication for glacial carving during the LGM and previous glaciations [34,35].

3. Methods

Following the scopes of the paper, we assembled geomorphic, provenance, and cosmogenic data to reveal that the Bündnerschist and North Penninic flysch units have been the most important sediment sources at least during the Holocene, and that the material contribution from the Austroalpine cover nappes has been less. We will relate the high denudation rates in the flysch and schist units to the high erodibility of these lithologies. We will then combine this information with data on long-term exhumation rates, offered by published fission track ages [15], and with the results of geodetic surveys where modern rock uplift rates has been measured [13]. These data suggest that uplift and exhumation have occurred at the highest rates in the Prättigau half-window that exposes Bündnerschist and Penninic flysch units with high bedrock erodibilities. We will use the combined information about (i) the long-term exhumation pattern; (ii) the geodetically measured uplift rates; (iii) the provenance of the material; (iv) the cosmogenic data; and (v) information on the landscape's properties to propose that (a) erosion has occurred at the highest rates where bedrock with high erodibilities is exposed, (b) this pattern of erosion has occurred during the past millions of years including the Holocene, and that (c) the area of highest denudation rates coincides with the region of highest rock uplift and exhumation rates.

Because of this spatial relationship between erosion and uplift over multiple time scales, we will propose that uplift and denudation could have been accelerated through a positive feedback mechanism. Following this concept, we will first describe the methods and the results of the geomorphic investigations, and we proceed with a presentation of how the sources of the material have been determined, and how, in combination with cosmogenic data, the pattern of sediment generation is quantified through budgeting. In doing so, we will also consider the dataset by Glaus et al. [22], which includes seven cosmogenic data so that our inferences will be based on a dataset with a total of 13 concentrations of cosmogenic ^{10}Be.

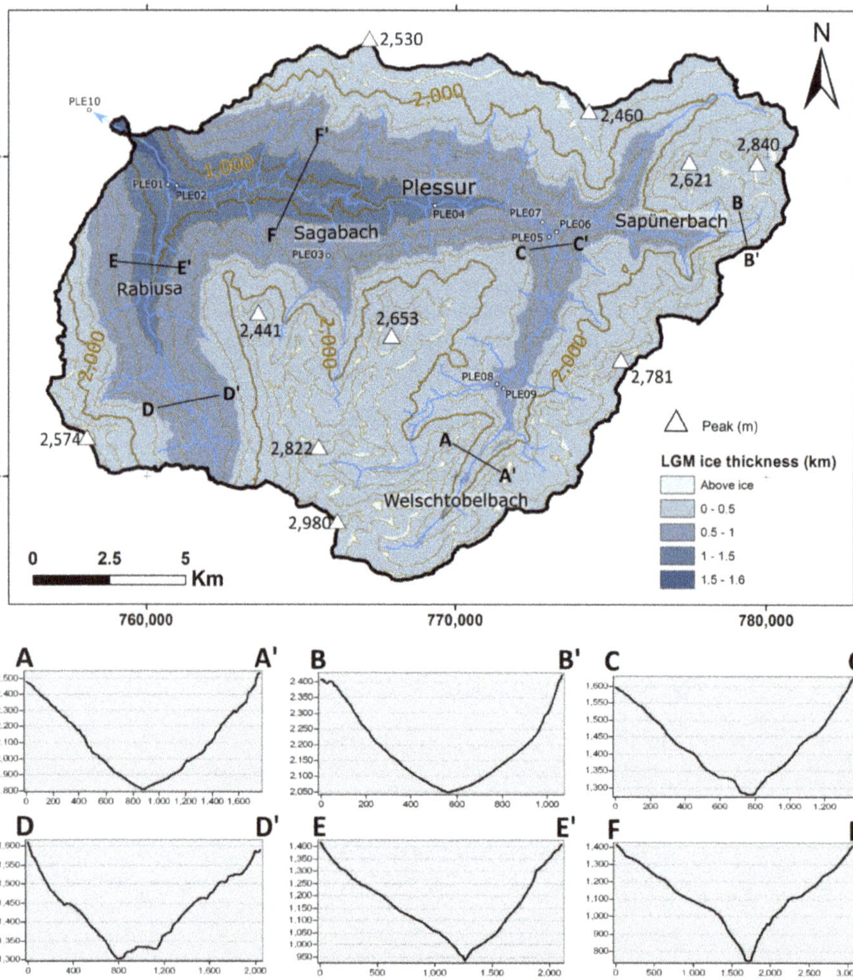

Figure 3. Map of the ice thickness in the Plessur basin during the LGM (from [35]) and main geomorphologic features. The contour lines of the ice thickness map display the elevations of the ice surface in meters above sea level. The map also shows the highest peaks in the study area. The transects from AA′ to FF′ (from up- to downriver) displays an ongoing transformation of an originally U-shaped valley to a V-shaped valley, reflecting the response to more prominent fluvial incision in the lowermost part of the Plessur basin. The numbers at the border of the figure indicate the Swiss coordinates in meters.

Figure 4. Picture of a U-shaped valley and an associate hanging valley in the Welschtobelbach sub-catchment, nearby the PLE09 site (Figures 2 and 3).

3.1. Morphometric Analysis

3.1.1. Knickzones and Steepness Indices

Topographic variables provide evidence for how the landscape has been shaped in response to tectonic and surface driving forces (e.g., [1,7,36,37]). Among the various morphometric parameters, longitudinal river profiles offer the most diagnostic information on how streams have adjusted to changes in uplift, erosional mechanisms, and rates [38].

River profiles ideally have a concave-up shape, and deviations thereof in general and knickzones in particular indicate the occurrence of perturbations such as a modification in erosion and/or uplift rates and patterns (e.g., [37,39,40]). Knickzones are generally initiated in the lowermost section of a river profile, from where these steps migrate upward through the channel network [38,40–42]. In this study we identified the occurrence of knickzones in the Plessur Basin using a 50 m resolution digital elevation model (DEM) that was resampled from a 2 m resolution LiDAR DEM provided by Swisstopo®. The resampling was done in order to reduce the computing time. We extracted the longitudinal stream profiles of the main channel and the main tributaries from this digital dataset within an open-source GIS environment. The data were then exported into a spreadsheet and included information on profile length, altitude, slope, and drainage area. From these data, we calculated the steepness indices for these river profiles following Flint's law [43] using TopoToolbox, a MATLAB script [44]. We normalized the steepness index (k_{sn}) through a constant concavity value of 0.45 [45], which allows for a better comparison of river profiles with varying drainage areas. Knickzones in the longitudinal stream profiles were then identified along reaches with high k_{sn} values. Their locations were also verified in the 2 m resolution LiDAR DEM.

3.1.2. Hypsometry, Slope Distribution, and Landscape Shape

The distribution of elevations within a basin bears relevant information on the extent of which a basin has been dissected by fluvial processes after the retreat of glaciers at the end of the LGM (e.g., [46]). Such data are commonly illustrated with a hypsographic curve where the cumulative area is plotted versus the relative elevation [47,48]. Additionally, the area under the curve, also known as the hypsometric integral (HI), can be calculated and normalized. This value represents the proportion of a basin that lies below a given elevation [49]. In addition, conceptual investigations have shown that a convex curve (HI > 0.5) is indicative for a young and thus immature stage in the development of a fluvially controlled landscape, whereas a concave curve (HI < 0.5) points toward a more mature stage [47,50]. Here, we reproduced such diagrams and extracted elevation data for the major tributaries and the trunk stream from a 2 m resolution DEM within an open source GIS environment.

Slope has been shown to exert one of the most important controls on erosion and sediment production [19,51,52]. In addition, hillslope steepness and length can positively influence runoff [53]. Moreover, it has been shown that denudation rates increase with steeper slopes until a threshold hillslope angle of c. 32°–35° (angle of repose) has been reached [19,52,54,55]. For landscapes where the hillslopes are steeper than this threshold, correlations between slope and denudation rates decouple [19,52]. We thus analyzed the slope distribution and the average slope of the Plessur and surrounding basins in an attempt to identify possible differences in sediment production and denudation between these basins. The analysis was based on a 50 m resolution DEM.

Finally, we mapped V-shaped versus U-shaped valley bottoms on the 2 m LiDAR-DEM for the Plessur Basin. These correspond to areas where streams have dissected into the previously glacially shaped landscape (e.g., cross section FF′ on Figure 3), or alternatively, these can be considered as reaches where the original glacial landscape is still preserved (e.g., Figure 4).

3.2. Sediment Provenance

We determined the bulk geochemistry of selected river samples to allocate the sources of the clastic riverine sediment. The samples were collected along the main stream as well as along tributaries of the Plessur River (Figure 2a, green dotted and yellow crossed circles, Figure S1 and Table S1), aiming for a good representation of the entire basin and its different lithotectonic units. From this detrital material, c. 5 g of the fine-grained sediment fraction of each sample (<63 µm) was separated, which was then analyzed through ICP-MS at the Activation Laboratories Ltd., in Ontario, Canada. This sediment fraction size was selected because silt grains are readily transported through the channel network, and since it was considered to adequately represent the different lithologies in the study area [56]. Therefore, the resulting sediment provenance model was only based on this sediment fraction, and it did not include the contribution of, for example, solute loads [57]. The results were expressed as major oxides and as selected trace elements (Sc, Cr, Ni, Sr, Y, Zr, Nb, and Ba). These were corrected for the loss of ignition and normalized to 100%. Values below the detection limit were changed to half of the detection limit.

The composition of all samples was investigated through principal component analysis (PCA). This method allows one to reduce the dimensionality of the dataset and, thus, facilitates the discrimination between the different samples according to their chemical composition [58,59]. Before conducting PCAs, we ran statistical tests to define the best combination of tracer elements that permits a discrimination between different compositional endmembers. For this purpose, we employed R-code fingerPro® [60] and conducted a three-step statistical test [61]: (i) in the first test, referred to as the range test, we explored whether the tracer elements are mass-conservative along the entire source to sink path; (ii) the second test, the Kruskal–Wallis H-test, allows one to exclude single tracer elements that do not vary significantly from the different samples and that, thus, do not provide a solid basis for a further discrimination; and (iii) we employed the Wilk's lambda test as the third step. This particular test is a stepwise linear discriminant analysis, which allows one to identify the ensemble of tracers that yield a maximum variation between the samples. This was used to enhance the reliability of the discrimination. As a following up task, we applied the routine fingerPro® code to assess the relative contribution of the potential sediment sources (endmembers) for each in-stream sample. The results of this analysis will finally build the basis to reconstruct a sediment provenance model.

3.3. *Catchment-Wide Denudation Rates Inferred from In Situ Produced Cosmogenic ^{10}Be*

3.3.1. Sediment Collection

We used concentrations of in situ ^{10}Be in detrital quartz grains of river-born sediments to estimate the catchment-averaged denudation rates [62]. A total of four riverine sediment samples were collected in the main channel and in selected tributaries (Figure 2a, yellow crossed circles). We also collected a sediment sample in the Rhine River upstream of the confluence with the Plessur River and downstream after the confluence with the Landquart River (RHE01 and RHE02, respectively; inset Figure 2b) to complete the budgeting of the material. We collected 2.5 kg of sand material for each sample location because high carbonate contents were expected in the stream sediment samples as the upstream basin comprises Bündnerschist and North Penninic flysch with a high calcite content [22]. Finally, we also included the results of [22] from the Landquart basin, represented by the cosmogenic sample site Lan-1 (inset Figure 2b) in our analysis.

3.3.2. Laboratory Work

In the laboratory, the sediment samples were processed using state-of-the art techniques established at the University of Bern [63]. Accordingly, we sieved the samples to the size-fraction of 0.25–0.5 mm, from which the non-magnetic fraction was separated using a Franz isodynamic magnetic separator. The remaining material was treated to gain pure quartz grain void of atmospheric ^{10}Be and other impurities. This includes (i) leaching with 5% hydrochloric acid (HCl) to dissolve the carbonates and organic components; (ii) three times treatment with 5% hydrofluoric acid (HF); and (iii) three treatment steps using 2.5% HF. As a last step of quartz purification, we used Aqua Regia in order to dissolve the remaining metallic components as well as residual carbonate and organic materials.

The chemical separation of ^{10}Be was then performed using the lab protocol of [63] including: (i) Addition of ca. 0.2 mg of a 1g/L ^{9}Be carrier to the purified quartz samples, which were then dissolved in concentrated HF; and (ii) completion of the evaporation of the solution. The sample was then fumed with HNO_3, Aqua Regia, and HCl. The separation was followed by ion-chromatography columns. Beryllium and iron were then co-precipitated as hydroxides at a pH of around 8. The precipitates were dried and baked in a furnace at 675 °C before the resulting beryllium-iron oxide was finally pressed into copper targets. ^{10}Be/^{9}Be ratios were measured at the 500 kV TANDY AMS facility at the ETH Zurich [64] and normalized to the ETH in-house standard S2007N [65] using the ^{10}Be half-life of 1.387 ± 0.012 Ma [66,67]. The full process blank ratio of $(2.41 \pm 0.13) \times 10^{-15}$ was then subtracted from the measured ratios in order to calculate the ^{10}Be concentrations for each sample.

3.3.3. Calculation of Denudation Rates, Scaling, and Sediment Fluxes

Calculation of the denudation rates was accomplished using the CAIRN® algorithm [68]. It calculates ^{10}Be production rates and shielding factors on a pixel-by-pixel basis and propagates the uncertainty in AMS measurements and production rates. Based on this calculation, the software estimates the ^{10}Be concentration of the basin considering a spatially homogeneous denudation rate. Finally, the software computes, through a Newton's iteration method, the best denudation rate that fits the measured ^{10}Be concentration. A 50 m-resolution DEM (resampled from a 2 m LiDAR DEM) was applied in the calculations as well as default parameters such as the SLHL ^{10}Be production rate of 4.30 at g^{-1} a^{-1} (e.g., [19]). The calculations of the topographic shielding were based on inferred values of 8° and 5° for zenith ($\Delta\varphi$) and azimuth ($\Delta\theta$), respectively [68]. A rock density of 2.65 g cm^{-3} was used. Snow shielding factors were estimated based on a combination of Swiss and French snow-data records [19]. We then used the calculated denudation rates to infer the sediment fluxes for specific areas. This was accomplished by multiplying the denudation rate of a specific sample by the drainage area upstream of this sample.

4. Results

4.1. Geomorphometric Analysis

4.1.1. Knickzones, Steepness Indices, and Hypsometries

Knickzones and their respective steepness index plots are found in Figure 5 and Table 1. The location of the knickzones is shown in Figure 6. In the Plessur River and Sagabach tributary channel, knickzones were found close to the lithological border between the Bündnerschist and the North Penninic flysch, having propagated horizontally 35% along the Plessur River and 14% along the Sagabach River, and vertically ca. 20% in both streams (Table 1). In the Sapünerbach and Welschtobelbach, the knickzones have propagated horizontally by 48% and 31%, respectively, and vertically, by 49% and 24%, respectively. The Rabiusa tributary stream does not present any knickzone. The steepness index plots reveal that the segments above the identified knickzones are consistently flatter (k_{sn} varying from 126 to 213 $m^{0.9}$) than the segments below the knickzones (k_{sn} varying from 128 to 347 $m^{0.9}$).

The hypsographic curves and hypsometric integrals (HI) of the Plessur River and its tributaries are shown in Figure 7. The Rabiusa tributary presents the lowest HI (0.48) of the Plessur Basin and a concave-shaped hypsographic curve, whereas the Sagabach has the highest HI (0.60) and a convex-shaped hypsometric curve. The Sapünerbach and Welschtobelbach have a HI of 0.55 and 0.51, respectively. The average is represented by the Plessur curve with a HI of 0.53.

Table 1. Normalized steepness index of the Plessur River and main tributaries and their upriver propagation.

		k_{sn}	Upstream/ Downstream k_{sn} Increase Factor	Knickzone Elevation (m)	Relative Knickzone (KZ) Distance Propagation (towards Upstream) (%)	Relative Knickzone (KZ) Relief Propagation (Relative Vertical Propagation) (%)
Plessur	Upstream (Up)	176	1.0	965	35	22
	Downstream (Down)	180				
Rabiusa	Up	172	-	-	-	-
	Down	-				
Sagabach	Up	213	1.6	1154	14	23
	Down	347				
Sapünerbach	Up	137	1.6	1613	48	49
	Down	214				
Welschtobelbach	Up	126	1.0	1814	31	24
	Down	128				

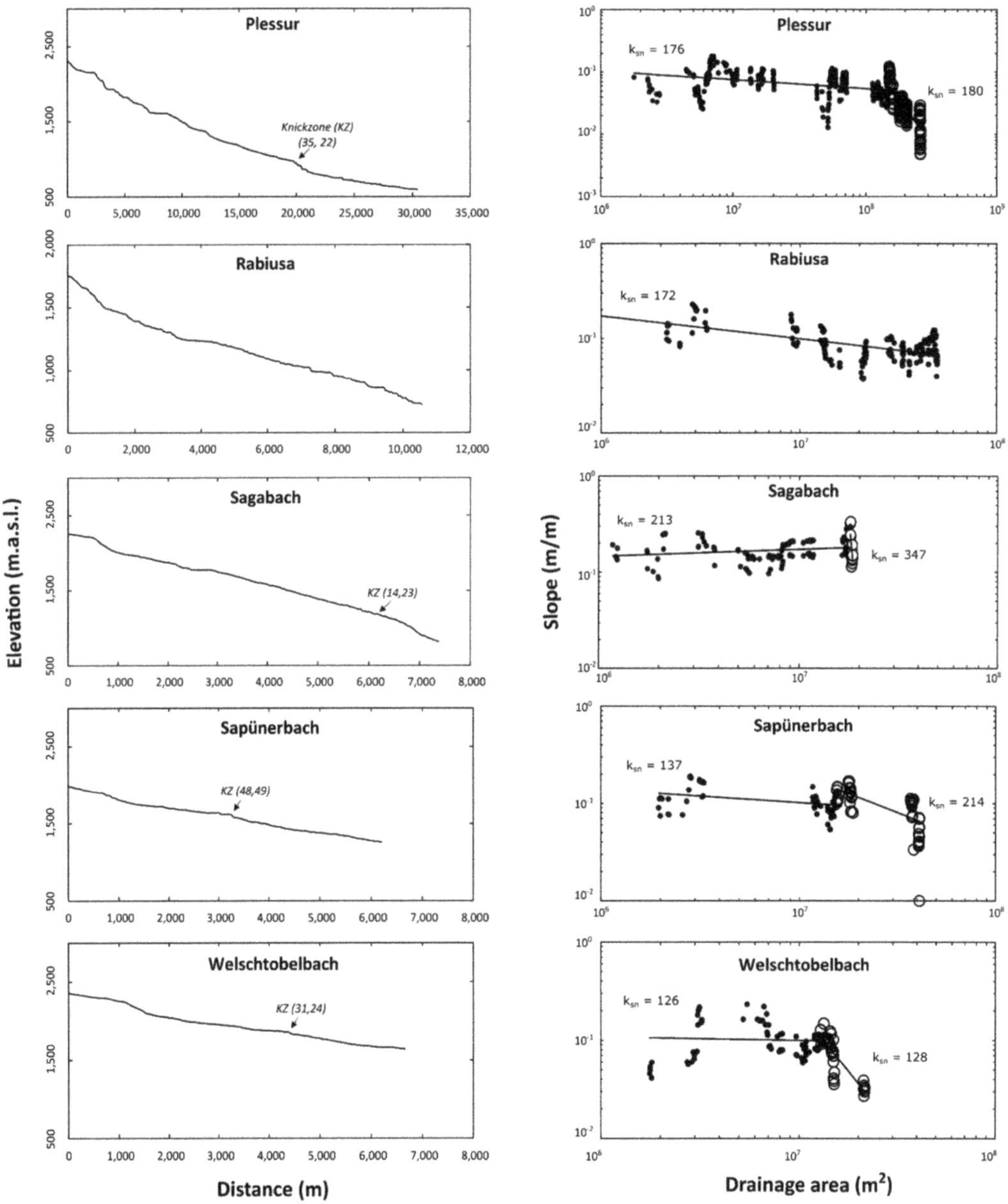

Figure 5. Longitudinal stream profiles of the Plessur River and main tributaries and respective normalized steepness index plots. Knickzones (KZ) are identified in the stream profiles. The KZ horizontal and vertical relative propagation are indicated as percentages, respectively. Refer to Table 1 for the dataset.

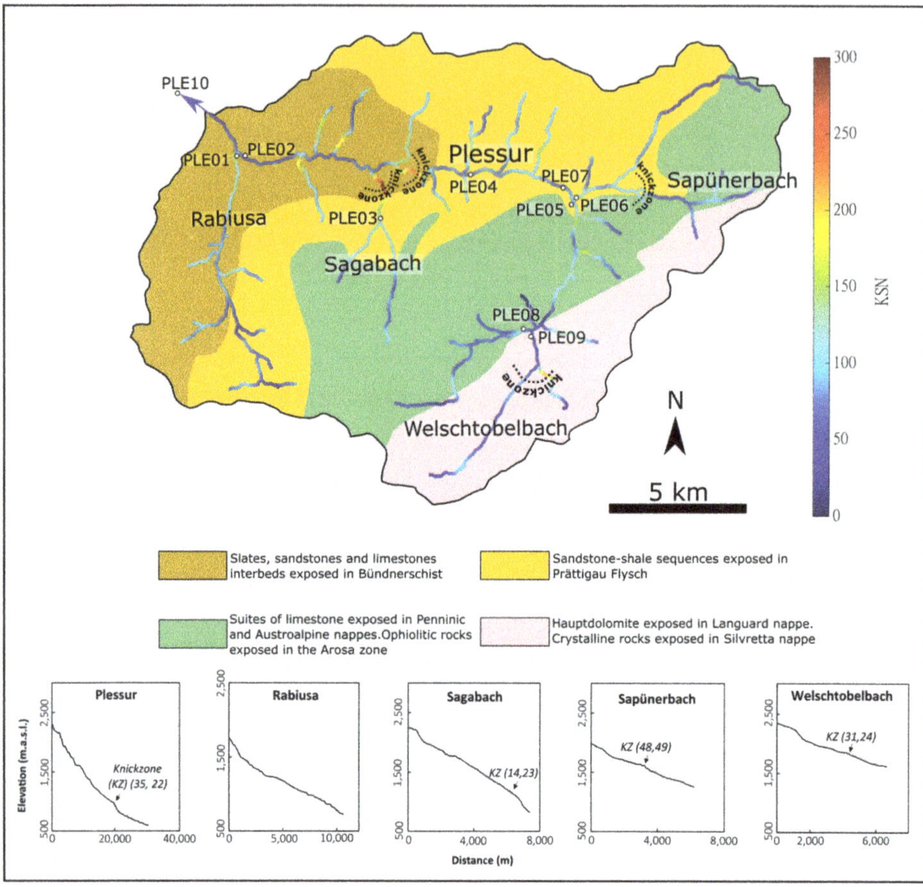

Figure 6. Map of normalized steepness indices (k_{sn}) displayed in a simplified lithotectonic framework for the Plessur Basin. The knickzones are identified in the map and in the stream profiles (see also Figure 5).

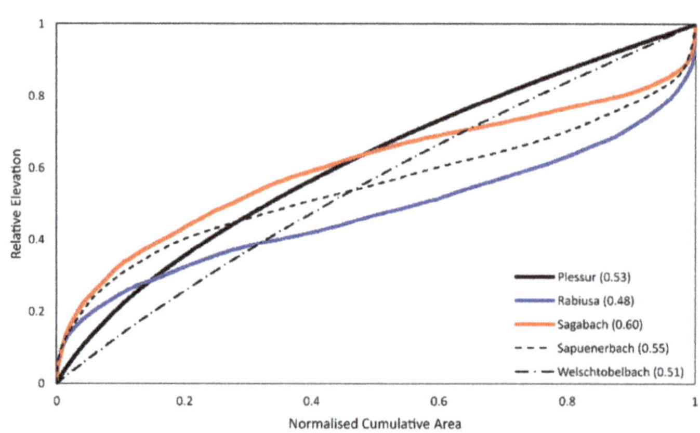

Figure 7. Comparison between the hypsographic curves of the Plessur Basin and its main tributaries. The resulting hypsometric integral (HI) is shown in the legend between parentheses.

The slope distribution within the Plessur, Landquart, and Upper Rhine Basins are displayed in frequency plots (Figure 8). The three basins present a very similar situation, with a normal slope distribution and none of them display average hillslope angles beyond the commonly inferred threshold (angle of repose) of c. 32°–35° (e.g., [19,52]). The Plessur has an average slope of 23.6° ± 9.9°, the Landquart is c. 24.8° ± 10.7° steep, and the Upper Rhine has hillslope angles in the range of 24.1° ± 11.2°.

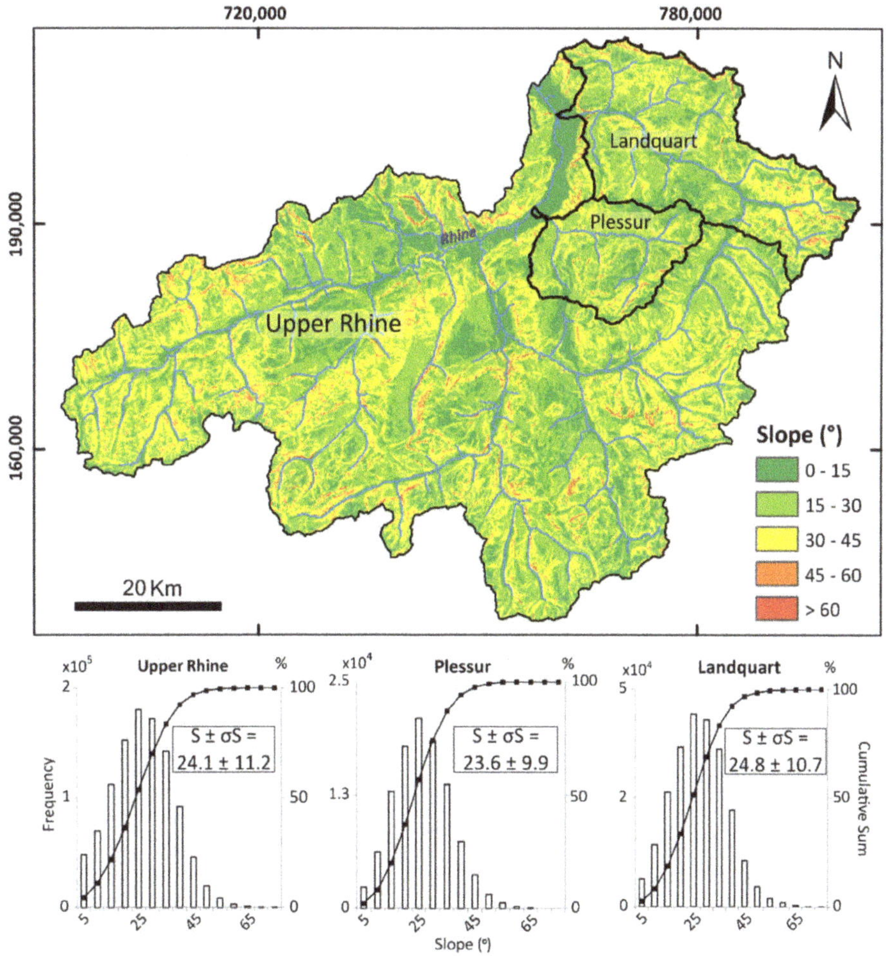

Figure 8. Slope distribution of the Plessur, Landquart, and Upper Rhine basin. The slope histograms of each basin display the slope averages in degrees (°), the associated uncertainties, and the cumulative sums in percentage (%). The numbers at the border of the figure indicate the Swiss coordinates in meters.

4.1.2. Landscape Patterns

The main glacial and fluvial patterns of the Plessur Basin are presented in Figure 9. Fluvial patterns (labelled as FP on Figure 9b,c) are predominantly represented by a rough aspect and channel dissection, whereas glacial patterns (labelled as GP on Figure 9b) have a smoother aspect with multiple concavities and convexities along the thalweg. For example, contrasts between glacial and fluvial patterns are found in the Sagabach sub-catchment (Figure 9b), whilst fluvial dissection predominates in the Rabiusa (Figure 9c). In general, the Plessur Basin presents clear fluvial dissection patterns in its lowermost sections where Lower Penninic Bündnerschist and North Penninic flysch predominates. However, it also displays glacial features in its uppermost areas, which are made up of Middle and Upper Penninic as well as Austroalpine units. The separation between both geomorphic domains is indicated by the dashed lines in Figure 9a,b.

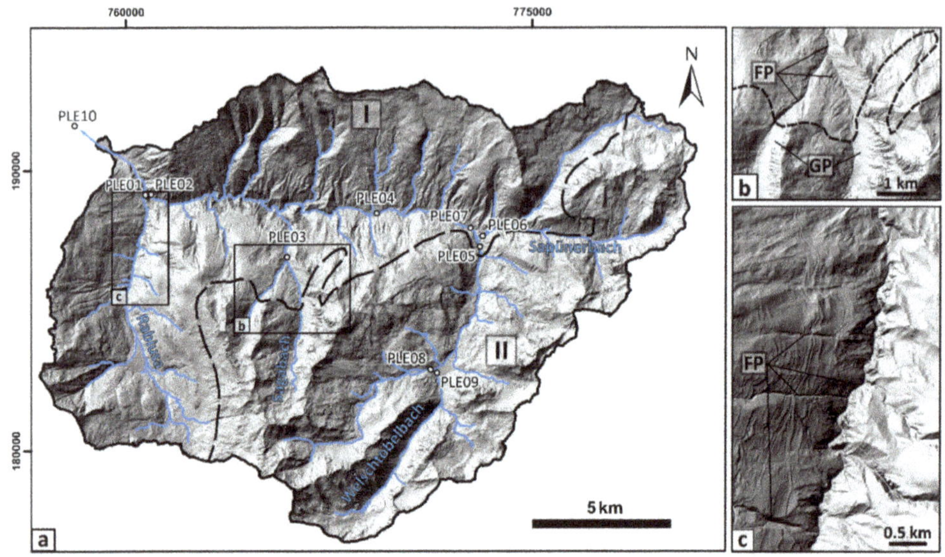

Figure 9. (**a**) Identification of fluvial patterns (FP) and glacial pattern (GP) in a digital elevation model of the Plessur Basin. The dashed line marks the division between lithologies with high and low erodibilities, and it separates the downstream sector I of the basin, which is dominated by fluvial processes and highly dissected, from the upstream sector II, which preserves glacial patterns, displayed by a smoother landscape. (**b**) Details of the Sagabach drainage area displaying the contrasts between the rougher fluvially controlled landscape (FP) and the topography that has still preserved features related to glacial erosion (GP). (**c**) Details of the Rabiusa drainage area showing landscape patterns indicative for fluvial erosion (FP) and associated hillslope erosion. The numbers at the border of the figure indicate the Swiss coordinates in meters.

4.2. Sediment Fingerprinting

4.2.1. Discrimination of Endmembers

The bulk geochemistry of each collected sample is displayed as the concentrations of major elements in the form of oxides (Table 2 and Table S2 for complete results) and of trace elements (Table 3 and Table S2 for complete results). The three statistical tests were then applied to this dataset. In the first test (range test), four tracers were excluded (P_2O_5, Sr, Y, and Zr). None of the remaining tracers passed the Kruskal–Wallis H-test and no patterns resulted from the Wilk's Lambda test. This might be due to the low number of samples compiled in the dataset. The results of the PCA are displayed in Figure 10a, which indicates that, despite the low sample quantity, the samples can be attributed to three endmembers: ophiolitic, dolomitic, and clastic sedimentary.

Table 2. Content of oxides (%).

Sample	Lithology	SiO_2	Al_2O_3	Fe_2O_3(T)	MnO	MgO	CaO	Na_2O	K_2O	TiO_2	P_2O_5
PLE01	Sedimentary	55.95	9.46	3.83	0.074	2.47	25.09	0.79	1.66	0.48	0.19
PLE02		49.54	8.65	4.48	0.088	6.54	27.81	0.71	1.49	0.524	0.17
PLE03	Sedimentary	45.79	10.18	4.35	0.112	9.67	26.51	0.65	2.06	0.547	0.13
PLE04		45.28	10.01	4.95	0.088	10.2	26.24	0.73	1.71	0.64	0.16
PLE05		34.04	7.39	4.48	0.122	21.43	29.96	0.48	1.53	0.468	0.1
PLE06	Sedimentary	46.71	5.74	3.87	0.092	10.01	31.12	0.74	0.93	0.6	0.18
PLE07		41.2	7.63	4.35	0.101	13.68	30.2	0.69	1.42	0.587	0.14
PLE08	Ophiolitic	52.17	13.52	7.82	0.192	12.28	9.65	0.95	2.43	0.81	0.17
PLE09	Dolomitic	7.93	2.12	1.24	0.103	35.06	52.68	0.09	0.63	0.108	0.04
PLE10		46.31	7.4	4.39	0.087	6.82	32.27	0.65	1.21	0.621	0.22

Table 3. Content of trace elements (ppm).

Sample	Lithology	Sr	Zr	Ba	Cr	Ni	Y	Nb	Sc
PLE01	Sedimentary	647	247	197	70	50	19.5	7.4	7
PLE02		614	227	182	160	90	18.4	8.1	7
PLE03	Sedimentary	316	134	224	80	60	14.7	8.4	7
PLE04		457	218	229	180	110	15.6	9.3	8
PLE05		152	129	144	280	200	11.6	6.2	6
PLE06	Sedimentary	531	361	124	160	100	17.1	8	6
PLE07		387	292	158	200	150	15	8.4	7
PLE08	Ophiolitic	127	179	310	550	450	21.7	11.3	13
PLE09	Dolomitic	112	51	75	10	10	3.1	1.2	1
PLE10		712	1039	153	150	80	28	9.1	7

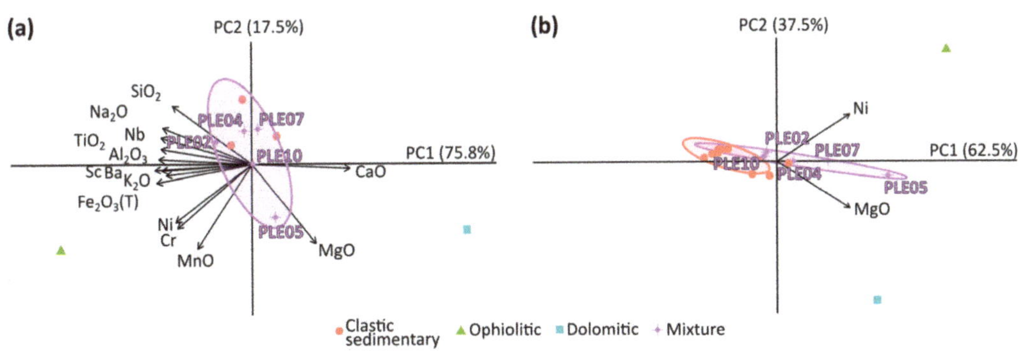

Figure 10. (**a**) Principal component analysis (PCA) of this work's dataset. In this case, only the range test was applied due to the low number of samples. The sum of the two principal components (PC1 and PC2) was 93.3%. (**b**) Results of PCA using a combination of the dataset presented in this work and [22]. The figure displays the results that were achieved after running the three tests mentioned in the text and with the combination of oxides/elements as suggested by the results of the Wilk's lambda test (MgO and Ni). The sum of PC1 and PC2 was 100%.

The ophiolitic endmember is characterized by higher contents of Fe_2O_3, MnO, Ni, and Cr, whereas the dolomitic endmember is defined by the higher concentration of CaO and MgO. The clastic sedimentary endmember displays a more diffuse but still consistent signal, which is characterized by an intermediate content of all the oxides and trace elements used in the analysis. We note, however, that the sedimentary endmember comprises both the North Penninic flysch and Bündnerschist, because they cannot be discriminated by the PCA due to compositional similarities. The mixture samples are the ones collected in the main trunk (Plessur River). They displayed an average content of all the oxides and trace elements, which showed a composition similar to the clastic sedimentary endmembers. In the next step, the samples representing the clastic sedimentary endmember of the Landquart Basin (Tables 3 and 4 in [22]) were combined with this work's dataset, making the analysis more robust, since the Plessur endmembers are derived from identical lithotectonic units as those from the neighboring Landquart Basin (Figure 1). The range test excluded P_2O_5, Y, and Zr, whereas no tracer passed the Kruskal–Wallis H-test. The results of the Wilk's Lambda test suggest that a combination of MgO and Ni is most suited to discriminate between the endmembers. The results are displayed in Figure 10b, in which Ni and MgO characterize the ophiolitic and dolomitic endmembers, respectively. Intermediate values between Ni and MgO represent the series of the clastic sedimentary endmember, where the data are well clustered near the center of the plot.

Table 4. Comparison between two datasets used in the calculation of the relative contribution of the three selected endmembers. In-stream locations are displayed in Figure 2. The uncertainties of the relative contribution are given as standard deviation. Goodness of fit (GOF) expresses the quality of the model. varying from 0 to 100.

Model Run.	In-Stream Location	GOF	Sedimentary Contribution (%)	Ophiolitic Contribution (%)	Dolomitic Contribution (%)
1 (all elements)	PLE02	93	72 ± 5	16 ± 3	12 ± 3
	PLE04	93	57 ± 5	29 ± 3	13 ± 2
	PLE05	95	10 ± 1	41 ± 1	49 ± 1
	PLE07	93	45 ± 5	28 ± 3	27 ± 2
	PLE10	82	84 ± 7	10 ± 4	5 ± 3
2 (MgO, Ni)	PLE02	99	83 ± 3	12 ± 1	5 ± 3
	PLE04	99	67 ± 3	17 ± 1	15 ± 3
	PLE05	99	13 ± 1	42 ± 1	45 ± 1
	PLE07	99	48 ± 2	28 ± 1	24 ± 2
	PLE10	99	83 ± 4	10 ± 2	6 ± 3

4.2.2. Sediment Provenance

Using the endmembers identified above, we built a sediment provenance model to quantify the relative amount of sediment produced in the different lithotectonic units. We used the merged dataset (this work's dataset and the Glaus et al. [22] dataset) to build the provenance model. We then compared the results before and after the application of the tests to the dataset (Table 4). Despite the similarities of the results in both runs, we selected the results of run 2 for our model because of the higher goodness of fit (GOF). In order to test whether the model is geologically consistent, we compared the model results with the lithological architecture of the Plessur Basin. Figure 11 displays a consistent pattern that is characterized, in the downstream direction, by an increase in the clastic sedimentary endmember contribution (from 13% ± 1% to 83% ± 4%) and a decrease in the ophiolitic and dolomitic contribution (from 42% ± 1% and 45% ± 1% to 10% ± 2% and 6% ± 3%, respectively). Furthermore, as also seen in Figure 11, the material composition of sample PLE05 displays a good agreement with the local lithology in the sense that it shows a relatively low contribution of the clastic sedimentary endmember (13% ± 1%) and relatively high contribution of both the ophiolitic (42% ± 1%) and dolomitic (45% ± 1%) endmembers.

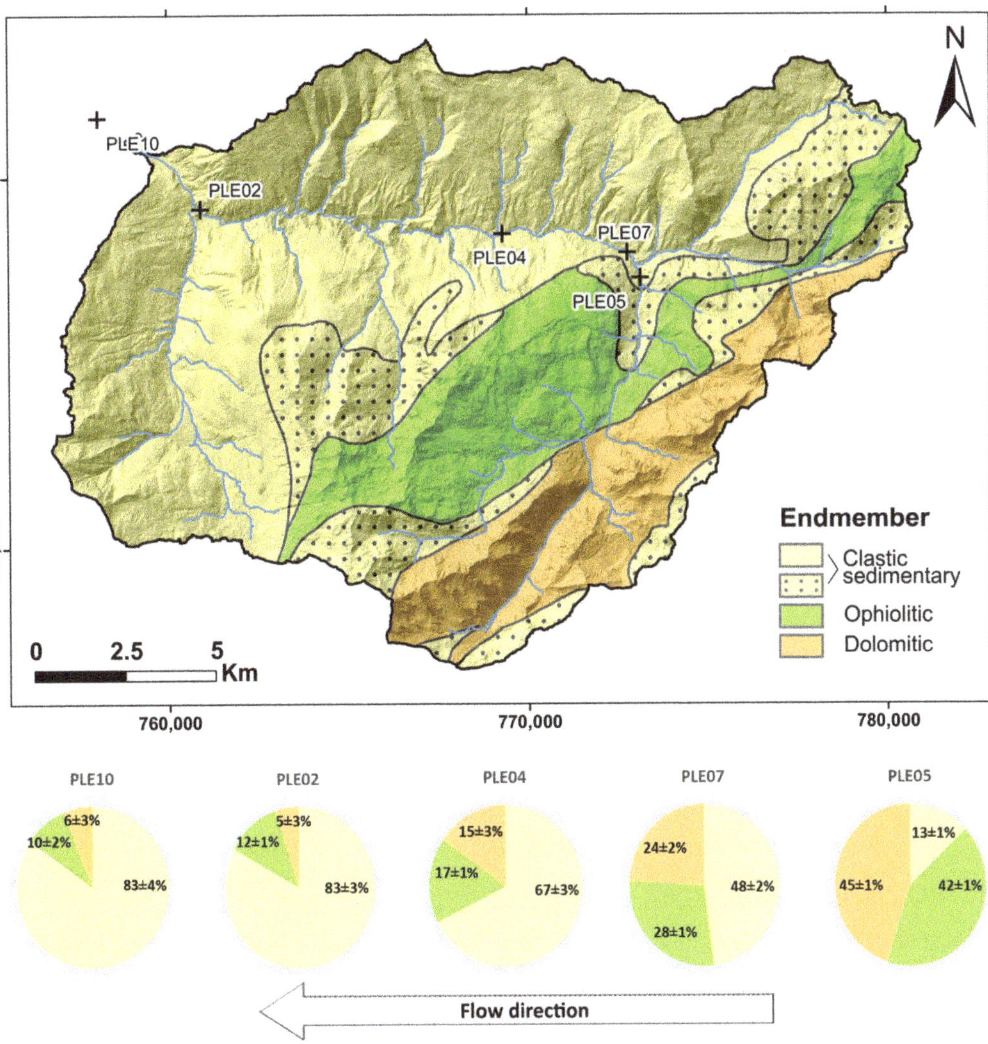

Figure 11. Sediment provenance considering three endmembers (Clastic sedimentary, Ophiolitic, and Dolomitic) and calculated for five in-stream samples (PLE02, PLE04, PLE05, PLE07, and PLE10). The results were obtained using the model run 2 of Table 4. The numbers at the border of the figure indicate the Swiss coordinates in meters.

4.3. Cosmogenic Data and Denudation Rates

The measured ^{10}Be concentrations (Table 5) vary from 1.53×10^4 atoms g^{-1} (PLE02) to 3.86×10^4 atoms g^{-1} (PLE06) for the Plessur Basin, whereas for the Upper Rhine segment, they vary from 1.48×10^4 atoms g^{-1} (RHE02) to 1.78×10^4 atoms g^{-1} (RHE01). The calculated spatially-averaged denudation rates (Table 6 and Figure 12) range from 0.34 ± 0.07 mm a^{-1} (PLE06) to 0.77 ± 0.16 mm a^{-1} (PLE02) for the Plessur Basin, whereas for the Upper Rhine segment, they vary from 0.70 ± 0.14 mm a^{-1} at RHE01 to 0.81 ± 0.16 mm a^{-1} at RHE02 after the confluence of the Plessur and Landquart Rivers. These results are in line with local denudation rates of similar lithotectonic units [22,69,70] as well as with denudation rates reported for the Central European Alps [18,19,71]. The sediment budget of the Plessur and Landquart Basins were then compared with that of the Upper Rhine

Basin in order to estimate the relative contribution of these tributary basins to the total Upper Rhine sediment budget (Table 7; Figure 13). The results show that the total sediment flux of the Upper Rhine is c. 3494 ± 677 × 10^3 m^3 a^{-1}. The contribution of the Plessur Basin to the Upper Rhine basin is approximately 181 ± 35 × 10^3 m^3 a^{-1}, which corresponds to c. 5.2% of the total budget while it covers a relative area of 6.1%. The Landquart Basin has a sediment flux of c. 704 ± 190 × 10^3 m^3 a^{-1}, which corresponds to 20.1% of the total sediment budget and represents a relative area of 14.3%.

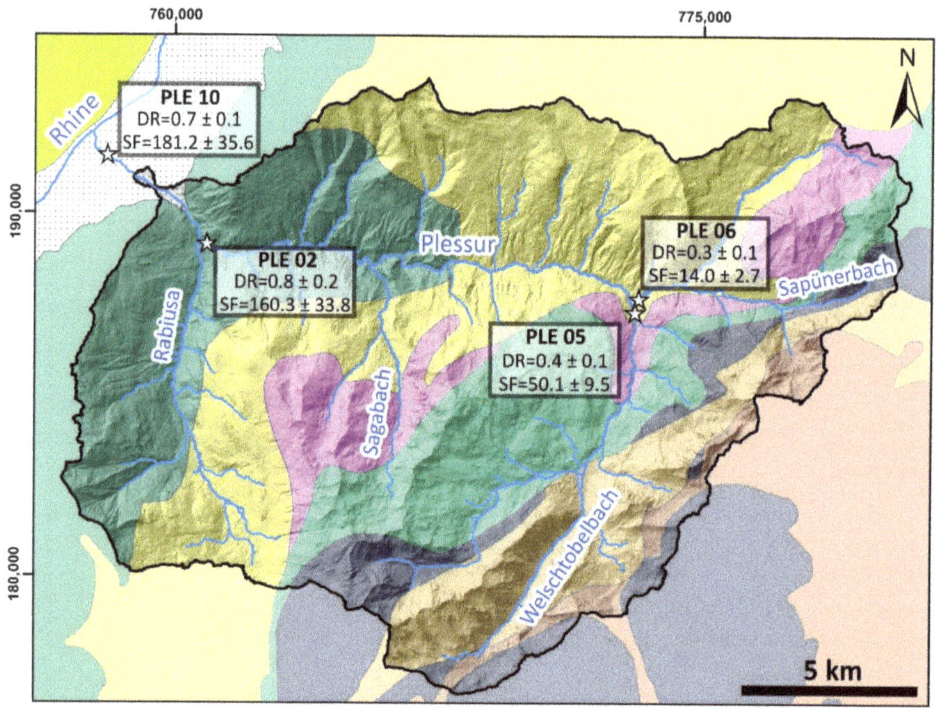

Figure 12. A lithotectonic map of the Plessur Basin (refer to Figure 2 for legend) with main tributaries, denudation rates (DR) in mm a^{-1}, and sediment flux (SF) in 10^3 m^3 a^{-1}. The numbers at the border of the figure indicate the Swiss coordinates in meters.

Table 5. Cosmogenic nuclide data.

Sample	Sample Weight (g)	^{9}Be Spike (mg)	AMS Ratio ($\times 10^{-12}$)	Uncertainty in AMS (%)	^{10}Be Concentration ($\times 10^4$ atoms g^{-1})
PLE02	37.03	0.195	0.044	9.9	1.53 ± 0.15
PLE05	44.54	0.188	0.106	4.3	2.99 ± 0.12
PLE06	49.55	0.195	0.147	5.4	3.86 ± 0.21
PLE10	50.47	0.192	0.066	6.2	1.68 ± 0.10
RHE01	50.17	0.190	0.070	7.1	1.78 ± 0.13
RHE02	50.13	0.198	0.056	5.6	1.48 ± 0.08
Lan-1	24.44	0.199	0.02	15.9	0.95 ± 0.17

AMS ratio uncertainty is at the 1σ level. The weighted average ^{9}Be/^{10}Be full process blank ratio is (2.59 ± 0.45) × 10^{-15}.

Table 6. ^{10}Be derived denudation rates.

Sample	Topographic Shielding [a]	Snow Shielding [b]	Denudation Rate (mm a^{-1}) [c]
PLE02	0.96	0.89	0.77 ± 0.16
PLE05	0.96	0.88	0.44 ± 0.08
PLE06	0.96	0.88	0.34 ± 0.07
PLE10	0.96	0.90	0.68 ± 0.13
RHE01	0.95	0.89	0.70 ± 0.14
RHE02	0.96	0.89	0.81 ± 0.16
Lan-1	0.97	0.90	1.14 ± 0.30

[a] Topographic shielding factor calculated after [68]. [b] Snow shielding factor calculated after [19]. [c] Denudation rates calculated by the CAIRN routine [68].

Table 7. Sediment budget calculated according to ^{10}Be derived denudation rates.

Sample	Area (Km^2)	Relative Area (%)	Denudation Rate (mm a^{-1})	Sediment Flux (10^3 m^3 a^{-1})	Relative Contribution (%)
PLE02	207.3	4.8	0.77 ± 0.16	160.28 ± 33.82	4.59 ± 1.31
PLE05	112.4	2.6	0.44 ± 0.08	50.12 ± 9.47	1.43 ± 0.04
PLE06	40.9	< 1	0.34 ± 0.07	14.03 ± 2.69	0.04 ± 0.01
PLE10	265.4	6.1	0.68 ± 0.13	181.16 ± 35.57	5.18 ± 1.43
RHE01	3270.9	75.8	0.70 ± 0.14	2297.26 ± 455.65	65.74 ± 18.23
RHE02	4311.7	100	0.81 ± 0.16	3494.07 ± 676.83	100.00 ± 27.39
Lan-1 *	616.1	14.3	1.14 ± 0.30	703.93 ± 186.23	20.15 ± 6.60

* Extracted from [22]. The sediment flux uncertainty was calculated in this study.

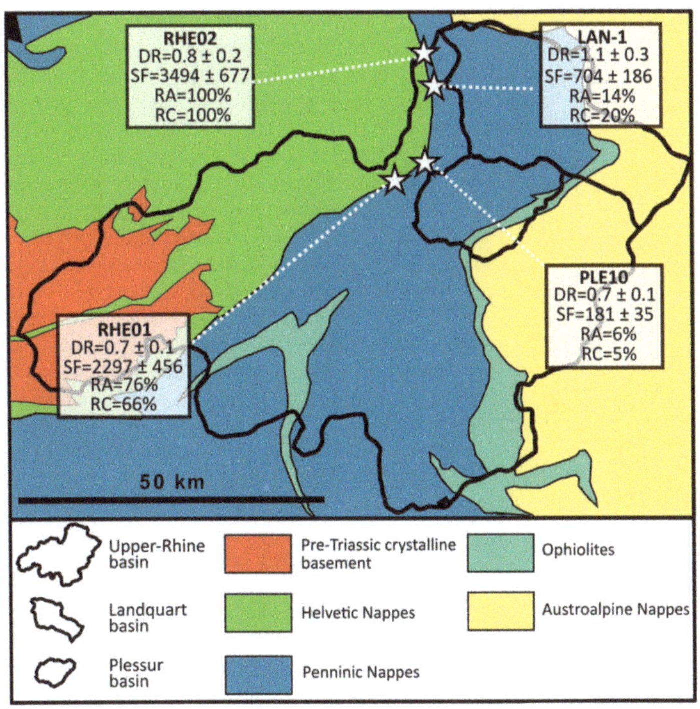

Figure 13. A simplified lithotectonic map of the Upper Rhine Basin including the Plessur and Landquart Basins, and their respective denudation rates (DR) in mm a^{-1}, sediment flux (SF) in 10^3 m^3 a^{-1}, relative area (RA), and relative contribution (RC).

5. Discussion

5.1. Topographic Parameters Reveal a Transient Stage

The combination of the data extracted from stream profiling facilitates the understanding of the landscape development. The normalized steepness index patterns show that the Plessur River, and those tributaries where knickzones were identified, have a relatively steeper downstream profile when compared with their upstream river segments (Table 1, Figure 5). This implies that the downstream part of the Plessur Basin, where normalized steepness indices are generally higher, has already undergone a rejuvenation process after the retreat of the LGM glaciers. In contrast, the areas where steepness indices are lower and where LGM-inherited glacial imprints are still preserved, the landscape shape is still in an immature stage with respect to fluvial processes rejuvenating a glacially conditioned landscape (e.g., [22,38,69]). In the Plessur Basin, evidence for fluvial carving is better developed in the mechanically weak lithologies such as the Bündnerschist and the North Penninic flysch, especially in the areas near the main trunk and within the Rabiusa drainage area (Figure 9). For instance, evidence for fluvial dissection can be seen in the Sagabach tributary sub-catchment. There, glacial features in the landforms underlain by rocks of the Falknis nappe contrast with the V-shaped fluvial landscape farther downstream where the bedrock is made up of the North Penninic flysch (Figure 9a). The Rabiusa sub-catchment and the lowermost areas of the Plessur Basin, which are underlain by Lower Penninic units, also display well-developed V-shaped cross-sectional valley geometries, thus pointing toward the occurrence of fluvial processes dissecting the landscape (Figure 9b). Interestingly, the Rabiusa tributary stream does not show any evidence for a knickzone (Figure 6). This suggests that this stream has not had any major steps in the long-stream profile after the retreat of the LGM glaciers, or more likely that a possible knickzone has already fully propagated through the entire channel network by headward retreat after the LGM. We support this interpretation with ample evidence for fluvial processes at work in the entire Rabiusa sub-catchment (V-shaped incision, Figure 9c), pointing toward an incised and rejuvenated landscape. In contrast to these morphologies, the uppermost areas of the Plessur Basin, where the bedrock comprises Austroalpine and Upper Penninic units, glacial landforms including concavities and convexities along the thalweg and U-shaped cross-sectional geometries occur more frequently. This is shown in Figure 9a for the uppermost part of the Plessur Basin (upstream of the dashed line), where all the other lithotectonic units are found, except for the Penninic flysch and Bündnerschist (Figure 2).

The hypsometric analysis (Figure 7) supports the aforementioned interpretations. The Plessur River presents a hypsometric integral of 0.53, which implies that the basin is in a late immature stage, almost reaching a denudation equilibrium [47]. Similarly, the Rabiusa sub-catchment has a hypsometric integral of 0.48, which implies a slightly more mature but yet an unequilibrated stage, whereas the Sagabach sub-catchment, with a hypsometric integral of 0.60, reveals a rather young stage of fluvial development [47]. The Sapünerbach and Welschtobelbach (HI of 0.55 and 0.51, respectively) are considered to be in a late immature stage.

Overall, considering the combination of topographic parameters, we propose that the Plessur Basin is in a transient stage, as evidenced by the on-going upstream migration of knickzones and the contrast between fluvial (downstream) and glacial patterns (upstream) in the landscape. As outlined by different authors [33,69,72], this transient state most likely reflects that the Alpine streams are still adjusting their geometries to the perturbation caused by the carving of the LGM glaciers.

5.2. The Relationship between Sediment Provenance and Denudation Rate

We built a sediment provenance model according to the three compositional endmembers suggested by the PCA (Figure 11). This was accomplished considering the <63 μm sediment fractions only, which limits a holistic analysis of the sediment production in the Plessur basin, but still yields meaningful results (e.g., [18,22]). Indeed, the data show that the clastic sedimentary contribution abruptly increases from 13% ± 1% to 48% ± 2% from

PLE05 to PLE07 farther downstream (Figure 11). This change could be related to a significant contribution of the sedimentary endmember component provided by the Sapünerbach sub-catchment. Another explanation could be attributed to a difference in the mechanical properties of the sedimentary rocks exposed in the Middle Penninic and Austroalpine units (dotted yellow, Figure 11), if compared with the Bündnerschist and flysch sediments of the Valais Ocean that make up the Lower Penninic units (yellow, Figure 11). In particular, the Middle Penninic and Austroalpine units (ophiolites and dolomites) upstream of PLE05 are considered to have a lower erodibility than the Lower Penninic units (Bündnerschist and flysch; [20]) that are exposed downstream of PLE05. We thus explain the relatively low contribution of the sedimentary endmember in the PLE05 sample by the low areal extent at which Bündnerschist and flysch lithologies are exposed upstream of that sample site. Farther downstream, a persistent increase of the clastic sedimentary endmember contribution along the Plessur River, rising from 13% $\pm$ 1% at the PLE05 site to 83% $\pm$ 4% at the PLE10 sample location, reflects the combined effect of a widespread exposure of the Bündnerschist and flysch source rocks and the high erodibility of these lithologies. Furthermore, although the sedimentary endmember covers an area of around 185 km^2 and thus ca. 69% of the total basin area, the relative abundance of the sedimentary endmember is almost 85% at the end of the Plessur River. In contrast, at the downstream end of the basin, dolomite and ophiolitic lithologies contribute up to 15% to the sediment budget, whilst covering an area of around 80 km^2 (ca. 31% of the total basin area). Therefore, we consider the area surrounding the Lower Penninic units made up of Bündnerschist and North Penninic flysch as a denudation hotspot within the Plessur Basin. However, we note that only the fine sediment fraction was analyzed, which could have resulted in an overestimation of the clastic sedimentary endmember contribution because it is possible that schist and flysch tend to break down to smaller grains more rapidly than limestones and ophiolites.

Similar to the material composition of the riverine material, the denudation rate pattern calculated for the Plessur Basin reflects the differences in the bedrock erodibilities of the underlying rocks. This is shown by an increase in the basin-averaged denudation rates in the downstream direction. In particular, material of the samples PLE05 and PLE06 (Figure 12) records the lowest catchment-averaged denudation rate (0.44 $\pm$ 0.08 mm a^{-1} and 0.34 $\pm$ 0.07 mm a^{-1}, respectively) in the Plessur Basin, whereas the PLE02 material yielded the highest values (0.77 $\pm$ 0.16 mm a^{-1}). We note that ^{10}Be concentrations in sediments at the lowermost PLE10 site yield a denudation rate of 0.68 $\pm$ 0.13 mm a^{-1}, which might be perceived as lower than the records at site PLE02 farther upstream. However, considering the uncertainties on the denudation rate estimates, the difference in the inferred catchment-averaged denudation rates becomes non-significant.

5.3. Upscaling and Including the Material Flux of the Rhine River

We upscale the results of the Plessur Basin analysis and include the material flux of the Upper Rhine through measuring concentrations of in situ ^{10}Be in samples collected upstream (RHE01) and downstream (REH02) of the confluences of the Plessur and Landquart Rivers (Figure 13). The scope of this task is to estimate the relative importance of the Plessur and Landquart Basins on the sediment budget in the region. Accordingly, sample RHE01 yields a basin averaged denudation rate of the Upper Rhine River before the Rhine receives the material of the Plessur and Landquart Rivers (Figure 13, Tables 6 and 7). Similarly, with sample RHE02, we estimate the average denudation rate of the entire Upper Rhine Basin including the contribution of the Plessur and Landquart Basins (Figure 13 and Table 7). Considering that the sediment flux calculated at RHE02 and the related upstream drainage area is 100%, then the Plessur River (drainage area of 6.1%) contributes up to 5.2% to the sediment flux at RHE02 (Table 7), whereas the Landquart River (drainage area of 14.3%) contributes up to 20.1% to this material budget (Table 7, data taken from [22]). Therefore, the sediment contribution of both the Landquart and Plessur Basins is, at the basin scale, in the same range as the other parts of the Upper Rhine Basin. However, it was documented for the Landquart Basin that a small downstream portion of the catchment that is underlain by Bündnerschist contributes between 60 to 70% of the basin's sediment flux, although it represents only <30% of the entire Landquart drainage area [22]. A comparable picture, although not as prominent as in the Landquart, arises if the denudation rate and sediment flux patterns in the Plessur Basin are considered. In particular, in the upper part of the Plessur Basin, which is underlain by rocks of the Austroalpine and Penninic units, the catchment-averaged denudation rates are ca. 0.3–0.4 mm a^{-1}. The rates nearly double to almost 0.8 mm a^{-1} as the stream flows through the Lower Penninic units where Bündnerschist and flysch are exposed, and the relative abundance of sedimentary material increases accordingly (Figure 11). Because the relative area of this downstream portion makes up to around 50% of the entire basin, then a doubling of the catchment-averaged denudation rates requires a three times larger contribution of material from the lower part, as calculated through mass balancing (see calculations in Equation S1 in the Supplementary Material). This implies that the Lower Penninic area, similar to the Landquart Basin, could be considered as hosting a denudation hotspot of the region (see also [22,70]).

5.4. Possible Controls on the Local Uplift Rates and Feedbacks with Denudation

The fast exhumation rates in the Landquart-Chur region are the response of a long-lived tectonic forcing ([15]; Figure 1). The updoming of the region was considered to have started ca. 4–5 Ma ago and might have exposed the Bündnerschist and North Penninic flysch to the surface, thereby forming the Prättigau half-window [15,21,29]. The site of updoming coincides in space with a peak of a negative isostatic anomaly (Figure 1d), caused by the almost-60-km-deep crustal root underneath the Central European Alps [30,32]. The related rocks, which are most likely buoyant [30], possibly contribute, or at least facilitate a possible positive feedback mechanism where rapid denudation of the Bündnerschist and North Penninic flysch with high erodibilities promote an isostatic response, which is likely reflected by the high exhumation rates over Ma. Thus, this isostatic anomaly not only has the potential to support a long-lived uplift signal, but it also points to the role of how deep crustal and geodynamic processes have the potential to exert a control on denudation at the surface (e.g., through contributing to the controls on the high uplift rates). In line with previous studies [7,22], we therefore propose that the exposure of mechanically weak lithologies has not only promoted surface denudation in the region where flysch and Bündnerschist are exposed, but also, through a positive feedback, the high modern uplift, and long-term exhumation rates. Equally relevant is the surface response to glacial carving, which could destabilize the local basin's equilibrium [11,69], possibly altering the local base level. The topographic parameters analyzed in this work indeed support the interpretation that the transient stage of the local basins is also due to the perturbation caused by the carving of the LGM glaciers. The rate of fluvial adjustments has then been controlled by the erodibility of the underlying bedrock. Furthermore, the melting of the LGM glaciers was likely to have initiated an isostatic rebound [11], thereby contributing to the maintenance of a possible positive feedback between uplift and lithology-controlled denudation. However, because the pattern of fission track ages comprises an exhumation record, which goes beyond the timescales of Alpine glaciations, and since the Prättigau half-window is characterized by the largest negative isostatic anomaly in the region, we infer a scenario where an active tectonic driving force has sustained the high uplift and denudation rates over millennia. This uplift, however, has most likely been accelerated in response to a denudation feedback, which in turn, could have been conditioned and controlled by lithology contrasts and possibly by glacial perturbations, at least since the Pleistocene.

6. Conclusions

The topographic variables reveal that the Plessur Basin is still adjusting to the perturbation caused by the termination of the LGM. This can be observed in the distribution of glacial and fluvial patterns in the landscape. Fluvial patterns are more often found in the lowermost areas of the basin where the thalweg of the trunk stream is steepest, whereas glacial patterns are more commonly observed in the uppermost and flatter areas of the basin. The presence of knickzones in the basin also supports the observations that a transient stage is still prevailing, most likely due to the last deglaciation event. Besides, the knickzones' locations disclose the role of the differential erosion mechanisms in the Plessur Basin. Apparently, the Bündnerschist is mechanically weaker than the North Penninic flysch, given that the two most prominent knickzones of the Plessur Basin are located between Bündnerschist and North Penninic flysch. Nonetheless, both units are considerably more erodible if compared with all the other units in the study site. Such high erodibility is reflected in the contribution of the mechanically weak units to the total sediment budget of the Plessur Basin. The data show an abrupt increase in the contribution of the Bündnerschist and flysch material to the Plessur sediment budget, rising from 13% $\pm$ 1% to 48% $\pm$ 2% within a reach of ca. 1 km where the corresponding bedrock changes. Furthermore, the highest catchment averaged-denudation rates of the Plessur Basin are found in locations where Bündnerschist and North Penninic flysch are predominant. At the regional scale, the long-lived uplift rates recorded by apatite fission track ages and geodetic surveys are possibly a consequence of the combination of two mechanisms: (i) sustainment of uplift rates by tectonic and deep crustal processes; and (ii) amplification of uplift rates caused by denudation unloading of the highly erodible Lower Penninic units (Bündnerschist and North Penninic flysch) of the Prättigau half-window. We thus conclude that lithologic, glacial, and geodynamic conditioning have substantially contributed to the local uplift and erosion mechanism as some of the driving forces of a positive feedback system.

Supplementary Materials: The following are available online at https://www.mdpi.com/article/10.3390/geosciences11080339/s1, Figure S1: Pictures of the sample sites. The location of samples PLE08, RHE01, and RHE02 were not photographed; Equation (S1): Mass balance; Table S1: Location, day, and time of sample collection; Table S2: Complete results of the ICP-MS analysis showing the composition of each sample.

Author Contributions: F.S. and R.D. designed the study, together with E.d.S.G. E.d.S.G. conducted all analyses with support from R.D., N.A., L.S. and M.C. E.d.S.G. wrote the text and drafted the figures with support by F.S. and R.D. All authors contributed to the discussion of the article. All authors have read and agreed to the published version of the manuscript.

Funding: This research was funded through various grants offered by the University of Bern, with major contributions by the Swiss National Science Foundation (project 147689 awarded to Fritz Schlunegger) and some support from the European Union's Horizon 2020 research and innovation programme under the Marie Sklodowska-Curie grant agreement No 860383 (awarded to Ronald van Balen).

Data Availability Statement: All data used in this paper are presented as tables or Supplementary Materials. The swisstopo DEMs are openly accessible.

Acknowledgments: The authors would like to thank Julijana Krbanjevic, Catharina Dieleman, and Serdar Yesilyurt for their technical support during the samples preparation. We are also grateful to the Laboratory of Ion Beam Physics accelerator mass spectrometry facility operated by the Swiss Federal Institute of Technology, Zurich, Switzerland. The quality of the paper has benefited from constructive reviews by three anonymous reviewers, which are especially acknowledged.

Conflicts of Interest: The authors declare no conflict of interest.

References

1. Willett, S.D.; Brandon, M.T. On steady states in mountain belts. *Geology* **2002**, *30*, 175. [CrossRef]
2. Burbank, D.W.; Anderson, R.S. *Tectonic Geomorphology*; John Wiley & Sons, Ltd.: Chichester, UK, 2011; Volume 13, ISBN 9781444345063.
3. Champagnac, J.-D.; Molnar, P.; Sue, C.; Herman, F. Tectonics, climate, and mountain topography. *J. Geophys. Res. Solid Earth* **2012**, *117*, B2. [CrossRef]
4. Whittaker, A.C. How do landscapes record tectonics and climate? *Lithosphere* **2012**, *4*, 160–164. [CrossRef]
5. Whipple, K.X. The influence of climate on the tectonic evolution of mountain belts. *Nat. Geosci.* **2009**, *2*, 97–104. [CrossRef]

6. Champagnac, J.-D.; Schlunegger, F.; Norton, K.; von Blanckenburg, F.; Abbühl, L.M.; Schwab, M. Erosion-driven uplift of the modern Central Alps. *Tectonophysics* **2009**, *474*, 236–249. [CrossRef]
7. Korup, O.; Schlunegger, F. Rock-type control on erosion-induced uplift, eastern Swiss Alps. *Earth Planet. Sci. Lett.* **2009**, *278*, 278–285. [CrossRef]
8. Persaud, M.; Pfiffner, O. Active deformation in the eastern Swiss Alps: Post-glacial faults, seismicity and surface uplift. *Tectonophysics* **2004**, *385*, 59–84. [CrossRef]
9. Gudmundsson, G.H. An order-of-magnitude estimate of the current uplift-rates in Switzerland caused by the Würm Alpine deglaciation. *Eclogae Geol. Helv.* **1994**, *87*, 545–557.
10. Barletta, V.R.; Ferrari, C.; Diolaiuti, G.; Carnielli, T.; Sabadini, R.; Smiraglia, C. Glacier shrinkage and modeled uplift of the Alps. *Geophys. Res. Lett.* **2006**, *33*, L14307. [CrossRef]
11. Mey, J.; Scherler, D.; Wickert, A.D.; Egholm, D.L.; Tesauro, M.; Schildgen, T.F.; Strecker, M.R. Glacial isostatic uplift of the European Alps. *Nat. Commun.* **2016**, *7*, 13382. [CrossRef]
12. Schmid, S.M.; Pfiffner, O.A.; Froitzheim, N.; Schönborn, G.; Kissling, E. Geophysical-geological transect and tectonic evolution of the Swiss-Italian Alps. *Tectonics* **1996**, *15*, 1036–1064. [CrossRef]
13. Kahle, H.; Marti, U.; Geiger, A.; Wirth, B.; Gubler, E. Recent crustal movements, geoid and density determination: Contribution from integrated satellite and terrestrial measurements. In *Deep Structure of the Swiss Alps: Results of NRP 20*; Pfiffner, O.A., Lehner, P., Heitzmann, P., Mueller, S., Steck, A., Eds.; Birkhäuser: Basel, Switzerland, 1997; pp. 251–259.
14. Schlatter, A.; Schneider, D.; Geiger, A.; Kahle, H.-G. Recent vertical movements from precise levelling in the vicinity of the city of Basel, Switzerland. *Int. J. Earth Sci.* **2005**, *94*, 507–514. [CrossRef]
15. Weh, M. *Tektonische Entwicklung der Penninischen Sediment-Decken in Graubünden (Prättigau bis Oberhalbstein)*; Universitaet Basel: Basel, Switzerland, 1998.
16. Klingelé, E.; Olivier, R. *Schwerekarte der Schweiz (Bouguer-Anomalien). Carte Gravimetrique de la Suisse (Anomalies de Bouguer). (Karte 4 der Geophysikalischen Landeskarten 1:500,000)*; Bundesamt für Landestopografie Swisstopo: Wabern/Bern, Switzerland, 1979.
17. Pfiffner, O. *Geology of the Alps*; John Wiley & Sons, Ltd.: Chichester, UK, 2014.
18. Stutenbecker, L.; Delunel, R.; Schlunegger, F.; Silva, T.A.; Šegvić, B.; Girardclos, S.; Bakker, M.; Costa, A.; Lane, S.N.; Loizeau, J.-L.; et al. Reduced sediment supply in a fast eroding landscape? A multi-proxy sediment budget of the upper Rhône basin, Central Alps. *Sediment. Geol.* **2018**, *375*, 105–119. [CrossRef]
19. Delunel, R.; Schlunegger, F.; Valla, P.G.; Dixon, J.; Glotzbach, C.; Hippe, K.; Kober, F.; Molliex, S.; Norton, K.P.; Salcher, B.; et al. Late-Pleistocene catchment-wide denudation patterns across the European Alps. *Earth Sci. Rev.* **2020**, *211*, 103407. [CrossRef]
20. Kühni, A.; Pfiffner, O. The relief of the Swiss Alps and adjacent areas and its relation to lithology and structure: Topographic analysis from a 250-m DEM. *Geomorphology* **2001**, *41*, 285–307. [CrossRef]
21. Vernon, A.; van der Beek, P.A.; Sinclair, H.D.; Rahn, M.K. Increase in late Neogene denudation of the European Alps confirmed by analysis of a fission-track thermochronology database. *Earth Planet. Sci. Lett.* **2008**, *270*, 316–329. [CrossRef]
22. Glaus, G.; Delunel, R.; Stutenbecker, L.; Akçar, N.; Christl, M.; Schlunegger, F. Differential erosion and sediment fluxes in the Landquart basin and possible relationships to lithology and tectonic controls. *Swiss J. Geosci.* **2019**, *112*, 453–473. [CrossRef]
23. Weh, M.; Froitzheim, N. Penninic cover nappes in the Prattigau half-window (Eastern Switzerland): Structure and tectonic evolution. *Eclogae Geol. Helv.* **2001**, *94*, 237–252. [CrossRef]
24. Trümpy, R. Penninic-Austroalpine boundary in the Swiss Alps: A presumed former continental margin and its problems. *Am. J. Sci.* **1975**, *275*, 209–238.
25. Ring, U.; Ratschbacher, L.; Frisch, W.; DÜrr, S.; Borchert, S. The internal structure of the Arosa Zone (Swiss-Austrian Alps). *Geol. Rundsch.* **1990**, *79*, 725–739. [CrossRef]
26. Ferreiro Mählmann, R.; Giger, M. The Arosa zone in Eastern Switzerland: Oceanic, sedimentary burial, accretional and orogenic very low- to low grade patterns in a tectono-metamorphic mélange. *Swiss J. Geosci.* **2012**, *105*, 203–233. [CrossRef]
27. Mohn, G.; Manatschal, G.; Masini, E.; Müntener, O. Rift-related inheritance in orogens: A case study from the Austroalpine nappes in Central Alps (SE-Switzerland and N-Italy). *Int. J. Earth Sci.* **2011**, *100*, 937–961. [CrossRef]
28. Wagner, G.; Reimer, G.M. Fission track tectonics: The tectonic interpretation of fission track apatite ages. *Earth Planet. Sci. Lett.* **1972**, *14*, 263–268. [CrossRef]
29. Fox, M.; Herman, F.; Willett, S.D.; Schmid, S.M. The Exhumation history of the European Alps inferred from linear inversion of thermochronometric data. *Am. J. Sci.* **2016**, *316*, 505–541. [CrossRef]
30. Kissling, E. Deep structure of the Alps—What do we really know? *Phys. Earth Planet. Inter.* **1993**, *79*, 87–112. [CrossRef]
31. Schlunegger, F.; Kissling, E. Slab rollback orogeny in the Alps and evolution of the Swiss Molasse basin. *Nat. Commun.* **2015**, *6*, 8605. [CrossRef] [PubMed]
32. Kissling, E.; Schlunegger, F. Rollback Orogeny Model for the Evolution of the Swiss Alps. *Tectonics* **2018**, *37*, 1097–1115. [CrossRef]
33. Salcher, B.C.; Kober, F.; Kissling, E.; Willett, S.D. Glacial impact on short-wavelength topography and long-lasting effects on the denudation of a deglaciated mountain range. *Glob. Planet. Chang.* **2014**, *115*, 59–70. [CrossRef]
34. Ivy-Ochs, S.; Kerschner, H.; Reuther, A.; Preusser, F.; Heine, K.; Maisch, M.; Kubik, P.W.; Schlüchter, C. Chronology of the last glacial cycle in the European Alps. *J. Quat. Sci.* **2008**, *23*, 559–573. [CrossRef]

35. Bini, A.; Buoncristiani, J.; Couterrand, S.; Ellwanger, D.; Felber, M.; Florineth, D.; Graf, H.; Keller, O.; Kelly, M.; Schluchter, C.; et al. *Die Schweiz Während des Letzteiszeitlichen Maximums (LGM)*; Bundesamt für Landestopografie Swisstopo: Wabern/Bern, Switzerland, 2009.
36. Ahnert, F. Functional relationships between denudation, relief, and uplift in large, mid-latitude drainage basins. *Am. J. Sci.* **1970**, *268*, 243–263. [CrossRef]
37. Tucker, G.E.; Slingerland, R. Drainage basin responses to climate change. *Water Resour. Res.* **1997**, *33*, 2031–2047. [CrossRef]
38. Whipple, K.X.; Tucker, G.E. Dynamics of the stream-power river incision model: Implications for height limits of mountain ranges, landscape response timescales, and research needs. *J. Geophys. Res. Solid Earth* **1999**, *104*, 17661–17674. [CrossRef]
39. Radaideh, O.M.A.; Mosar, J. Tectonics controls on fluvial landscapes and drainage development in the westernmost part of Switzerland: Insights from DEM-derived geomorphic indices. *Tectonophysics* **2019**, *768*, 228179. [CrossRef]
40. Korup, O.; Montgomery, D.R. Tibetan plateau river incision inhibited by glacial stabilization of the Tsangpo gorge. *Nature* **2008**, *455*, 786–789. [CrossRef] [PubMed]
41. Schlunegger, F.; Norton, K.P.; Zeilinger, G. Climatic Forcing on Channel Profiles in the Eastern Cordillera of the Coroico Region, Bolivia. *J. Geol.* **2011**, *119*, 97–107. [CrossRef]
42. Walsh, L.S.; Martin, A.J.; Ojha, T.P.; Fedenczuk, T. Correlations of fluvial knickzones with landslide dams, lithologic contacts, and faults in the southwestern Annapurna Range, central Nepalese Himalaya. *J. Geophys. Res. Earth Surf.* **2012**, *117*, F01012. [CrossRef]
43. Flint, J.J. Stream gradient as a function of order, magnitude, and discharge. *Water Resour. Res.* **1974**, *10*, 969–973. [CrossRef]
44. Schwanghart, W.; Kuhn, N.J. TopoToolbox: A set of Matlab functions for topographic analysis. *Environ. Model. Softw.* **2010**, *25*, 770–781. [CrossRef]
45. Wobus, C.; Whipple, K.X.; Kirby, E.; Snyder, N.; Johnson, J.; Spyropolou, K.; Crosby, B.; Sheehan, D. Tectonics from topography: Procedures, promise, and pitfalls. In *Tectonics, Climate, and Landscape Evolution*; Geological Society of America: Boulder, CO, USA, 2006.
46. Stutenbecker, L.; Costa, A.; Schlunegger, F. Lithological control on the landscape form of the upper Rhône Basin, Central Swiss Alps. *Earth Surf. Dyn.* **2016**, *4*, 253–272. [CrossRef]
47. Strahler, A.N. Hypsometric (Area-Altitude) Analysis of Erosional Topography. *Geol. Soc. Am. Bull.* **1952**, *63*, 1117–1142. [CrossRef]
48. Brozovic, N.; Burbank, D.W.; Meigs, A.J. Climatic Limits on Landscape Development in the Northwestern Himalaya. *Science* **1997**, *276*, 571–574. [CrossRef]
49. Hurtrez, J.-E.; Sol, C.; Lucazeau, F. Effect of drainage area on hypsometry from an analysis of small-scale drainage basins in the Siwalik Hills (Central Nepal). *Earth Surf. Process. Landforms* **1999**, *24*, 799–808. [CrossRef]
50. Brocklehurst, S.H.; Whipple, K.X. Hypsometry of glaciated landscapes. *Earth Surf. Process. Landf. J. Br. Geomorphol. Res. Group* **2004**, *29*, 907–926. [CrossRef]
51. Woodward, D. Method to predict cropland ephemeral gully erosion. *Catena* **1999**, *37*, 393–399. [CrossRef]
52. Ouimet, W.B.; Whipple, K.X.; Granger, D.E. Beyond threshold hillslopes: Channel adjustment to base-level fall in tectonically active mountain ranges. *Geology* **2009**, *37*, 579–582. [CrossRef]
53. Strahler, A.N. Quantitative Slope Analysis. *Geol. Soc. Am. Bull.* **1956**, *67*, 571–596. [CrossRef]
54. Strahler, A.N. Equilibrium theory of erosional slopes approached by frequency distribution analysis; Part II. *Am. J. Sci.* **1950**, *248*, 800–814. [CrossRef]
55. Burbank, D.W.; Leland, J.; Fielding, E.; Anderson, R.S.; Brozovic, N.; Reid, M.R.; Duncan, C. Bedrock incision, rock uplift and threshold hillslopes in the northwestern Himalayas. *Nature* **1996**, *379*, 505–510. [CrossRef]
56. Owens, P.; Blake, W.; Gaspar, L.; Gateuille, D.; Koiter, A.J.; Lobb, D.A.; Petticrew, E.L.; Reiffarth, D.G.; Smith, H.G.; Woodward, J.C. Fingerprinting and tracing the sources of soils and sediments: Earth and ocean science, geoarchaeological, forensic, and human health applications. *Earth Sci. Rev.* **2016**, *162*, 1–23. [CrossRef]
57. Hinderer, M.; Kastowski, M.; Kamelger, A.; Bartolini, C.; Schlunegger, F. River loads and modern denudation of the Alps—A review. *Earth Sci. Rev.* **2013**, *118*, 11–44. [CrossRef]
58. Aitchison, J. Principal component analysis of compositional data. *Biometrika* **1983**, *70*, 57–65. [CrossRef]
59. Vermeesch, P.; Garzanti, E. Making geological sense of 'Big Data' in sedimentary provenance analysis. *Chem. Geol.* **2015**, *409*, 20–27. [CrossRef]
60. Lizaga, I.; Latorre, B.; Gaspar, L.; Navas, A. FingerPro: An R Package for Tracking the Provenance of Sediment. *Water Resour. Manag.* **2020**, *34*, 3879–3894. [CrossRef]
61. Collins, A.; Walling, D. Selecting fingerprint properties for discriminating potential suspended sediment sources in river basins. *J. Hydrol.* **2002**, *261*, 218–244. [CrossRef]
62. Von Blanckenburg, F. The control mechanisms of erosion and weathering at basin scale from cosmogenic nuclides in river sediment. *Earth Planet. Sci. Lett.* **2005**, *237*, 462–479. [CrossRef]
63. Akçar, N.; Ivy-Ochs, S.; Alfimov, V.; Schlunegger, F.; Claude, A.; Reber, R.; Christl, M.; Vockenhuber, C.; Dehnert, A.; Rahn, M.; et al. Isochron-burial dating of glaciofluvial deposits: First results from the Swiss Alps. *Earth Surf. Process. Landf.* **2017**, *42*, 2414–2425. [CrossRef]

64. Christl, M.; Vockenhuber, C.; Kubik, P.; Wacker, L.; Lachner, J.; Alfimov, V.; Synal, H.-A. The ETH Zurich AMS facilities: Performance parameters and reference materials. *Nucl. Instrum. Methods Phys. Res. Sect. B Beam Interact. Mater. At.* **2013**, *294*, 29–38. [CrossRef]
65. Kubik, P.W.; Christl, M. 10Be and 26Al measurements at the Zurich 6MV Tandem AMS facility. *Nucl. Instrum. Methods Phys. Res. Sect. B Beam Interact. Mater. At.* **2010**, *268*, 880–883. [CrossRef]
66. Korschinek, G.; Bergmaier, A.; Faestermann, T.; Gerstmann, U.C.; Knie, K.; Rugel, G.; Wallner, A.; Dillmann, I.; Dollinger, G.; von Gostomski, C.L.; et al. A new value for the half-life of 10Be by Heavy-Ion Elastic Recoil Detection and liquid scintillation counting. *Nucl. Instrum. Methods Phys. Res. Sect. B Beam Interact. Mater. At.* **2010**, *268*, 187–191. [CrossRef]
67. Chmeleff, J.; von Blanckenburg, F.; Kossert, K.; Jakob, D. Determination of the 10Be half-life by multicollector ICP-MS and liquid scintillation counting. *Nucl. Instrum. Methods Phys. Res. Sect. B Beam Interact. Mater. At.* **2010**, *268*, 192–199. [CrossRef]
68. Mudd, S.M.; Harel, M.-A.; Hurst, M.D.; Grieve, S.W.D.; Marrero, S.M. The CAIRN method: Automated, reproducible calculation of catchment-averaged denudation rates from cosmogenic nuclide concentrations. *Earth Surf. Dyn.* **2016**, *4*, 655–674. [CrossRef]
69. Norton, K.P.; Abbühl, L.M.; Schlunegger, F. Glacial conditioning as an erosional driving force in the Central Alps. *Geology* **2010**, *38*, 655–658. [CrossRef]
70. Cruz Nunes, F.; Delunel, R.; Schlunegger, F.; Akçar, N.; Kubik, P.W. Bedrock bedding, landsliding and erosional budgets in the Central European Alps. *Terra Nova* **2015**, *27*, 370–378. [CrossRef]
71. Wittmann, H.; von Blanckenburg, F.; Kruesmann, T.; Norton, K.P.; Kubik, P.W. Relation between rock uplift and denudation from cosmogenic nuclides in river sediment in the Central Alps of Switzerland. *J. Geophys. Res. Earth Surf.* **2007**, *112*, F04010. [CrossRef]
72. Chittenden, H.; Delunel, R.; Schlunegger, F.; Akçar, N.; Kubik, P. The influence of bedrock orientation on the landscape evolution, surface morphology and denudation (10 Be) at the Niesen, Switzerland. *Earth Surf. Process. Landf.* **2014**, *39*, 1153–1166. [CrossRef]

geosciences

Article

Soil Formation and Mass Redistribution during the Holocene Using Meteoric ^{10}Be, Soil Chemistry and Mineralogy

Alessandra Musso [1,*], Dmitry Tikhomirov [1], Michael L. Plötze [2], Konrad Greinwald [3], Anne Hartmann [4], Clemens Geitner [5], Fabian Maier [1], Fanny Petibon [1] and Markus Egli [1]

1 Department of Geography, University of Zurich, Winterthurerstrasse 190, 8057 Zurich, Switzerland; dmitry.tikhomirov@geo.uzh.ch (D.T.); fabian.maier@geo.uzh.ch (F.M.); fanny.petibon@geo.uzh.ch (F.P.); markus.egli@geo.uzh.ch (M.E.)
2 Department of Civil, Environmental and Geomatic Engineering, ETH Zurich, Stefano-Franscini-Platz 3, 8093 Zurich, Switzerland; michael.ploetze@igt.baug.ethz.ch
3 Institute of Hydrology, Albert-Ludwig University Freiburg, Friedrichstraße 39, 79104 Freiburg, Germany; konrad.greinwald@hydrology.uni-freiburg.de
4 Section Hydrology, GFZ German Research Centre for Geosciences, Telegrafenberg, 14473 Potsdam, Germany; aha@gfz-potsdam.de
5 Institute of Geography, University of Innsbruck, Innrain 52, 6020 Innsbruck, Austria; clemens.geitner@uibk.ac.at
* Correspondence: alessandra.musso@geo.uzh.ch

Abstract: Soil development and erosion are important and opposing processes in the evolution of high-mountainous landscapes, though their dynamics are not fully understood. We compared soil development between a calcareous and a siliceous chronosequence in the central Swiss Alps at high altitudes, which both cover soil formation over the Holocene. We calculated element mass balances, long-term erosion rates based on meteoric ^{10}Be and we determined the rates of soil formation. We also analyzed the shifts in the mineralogical composition, weathering indices, the particle size distribution, carbon stocks and oxalate extractable Fe, Al, and Mn. The siliceous soils had high chemical weathering rates at the early stage of soil formation that strongly decreased after a few millennia. The development of calcareous soil was characterized by high carbonate losses and a shift to finer soil texture. Soil erosion hampered the upbuilding of soil horizons in the early stages of soil development, which led to a delay in soil and vegetation development. This study shows how soil formation drivers change over time. In the early stages of soil development, the parent material predominantly drives soil formation while at later stages the vegetation becomes more dominant as it influences surface stability, hydrological pathways, and chemical weathering that determine water drainage and retention.

Keywords: soil forming factors; cosmogenic nuclides; chronosequences; high-mountain soils; proglacial areas

Citation: Musso, A.; Tikhomirov, D.; Plötze, M.L.; Greinwald, K.; Hartmann, A.; Geitner, C.; Maier, F.; Petibon, F.; Egli, M. Soil Formation and Mass Redistribution during the Holocene Using Meteoric ^{10}Be, Soil Chemistry and Mineralogy. *Geosciences* **2022**, *12*, 99. https://doi.org/10.3390/geosciences12020099

Academic Editors: Jesús Ruiz-Fernández, Naki Akçar, Susan Ivy-Ochs, Fritz Schlunegger and Jesus Martinez-Frias

Received: 21 July 2021
Accepted: 13 February 2022
Published: 21 February 2022

1. Introduction

Soils and their formation, i.e., the physical and chemical alteration of the parent material, have been studied for more than a century. Dokuchaev [1] laid the foundation of soil science with his observations in the Russian steppe. He also laid the ground for the factorial model of soil development, which Jenny developed further in 1941 in his publication, Factors of Soil Formation [2]. Jenny described soil formation as the product of five independent factors: climate, organisms, relief, parent material, and time. Over the years, other models have been developed. Simonson [3] presented a model which looks at soils from a process-oriented point of view (i.e., additions, removals, translocations). Johnson and Watson-Stegner [4] introduced the concept of pathways. Their model allows soil formation to move in two directions and acknowledges soil formation as a highly dynamic process: soils experience progressive pedogenesis phases (characterized by horizonation, soil deepening, and developmental upbuilding) or regressive pedogenetic phases (halploidization, soil thinning, and retardant upbuilding). As the environmental conditions are never entirely stable, soil development may switch between progressive and regressive phases. The state of a soil at one point in time is merely the integration of all past soil development processes [4–6].

The formation of soils is in essence a balance between soil production (the conversion of bedrock or parent material to soil) and denudation (physical and chemical soil loss) [7]. This can be expressed as:

$$F_{soil} = P_{soil} - D_{soil} \quad (1)$$

F_{soil} is the soil formation rate, P_{soil} the soil production rate and D_{soil} the soil denudation rate.

Soil production (P_{soil}) is the sum of transformation of parent material to soil (TP_{soil}), atmospheric (dust) input (A), net organic matter input (O), and organic matter decay (G) (e.g., [8])

$$P_{soil} = TP_{soil} + A + (O - G) \quad (2)$$

Denudation (D_{soil}) is the sum of chemical weathering (W_{soil}) and erosion (E_{soil}):

$$D_{soil} = W_{soil} + E_{soil} \quad (3)$$

It has been shown that chemical weathering rates and physical erosion are closely linked [9]. Furthermore, the weathered layer acts as a protective layer which hinders weathering of the underlying parent material [10].

In glacial landscapes, glaciers shrink as a result of climate warming, exposing new surfaces and form fresh ground leading to significant amounts of sediment being eroded and accumulated downstream [11,12]. Periglacial areas are therefore useful to study processes related to early soil formation, chemical weathering, and erosion processes. They can provide good age constraints, and the till which generally forms the parent material can be regarded as well mixed. Based on the concept of Jenny [2] according to which the factor time and its influence on soil formation can be isolated when the other factors remain constant have become widely used approaches to trace and compare the transformation of soils over various timescales [13,14]. Chronosequences have been applied in various locations of the Alps for many decades. They show that most soil properties change very rapidly at early stages of soil development [15–17]. With time, this rate of alteration usually declines, reaching a plateau [14,18–22]. This is due to the high abundance of easily weatherable material in proglacial areas, because the glacial till has already undergone considerable physical weathering and provides, based on its grain size distribution, large surface areas [14]. Thus, weathering fluxes are usually highest in the beginning and taper off, usually after several millennia [23,24]. The rates depend on the composition and properties of the parent material. Granite or gneiss-based parent materials are expected to lose mainly silicon, iron, and aluminum, as they are the major components of micas and feldspars [25]. Micas have been observed to weather easily and be transformed into vermiculite and smectites, i.e., clay minerals [26]. Limestone-based soils will mainly loose calcium and varying amounts of magnesium through the dissolution of carbonates [23]. These shifts in the mineralogy will alter the physical and chemical soil properties, leading to finer texture and a higher proportion of fine pores which has a generally positive effect on the water retention capabilities of the soil [27,28].

When discussing soil development, it is important to consider the development of the biosphere as it influences the substrate considerably. Although Jenny defined it as independent factor, it is not. As both the soils and the vegetation influence each other simultaneously, their connection is not easy to quantify [29,30]. In the early stages of soil development, the processes of physical and chemical weathering improve the conditions for plant growth by providing the substrate for root growth and plant stability, water storage, and a steady supply of nutrients. At first, a combination of pioneer plants, which reproduce rapidly but are not very resilient, will add organic matter to the soils at a high rate via roots [31–33]. Roots are able to release large quantities of C to the soils as root detritus, exudates, and by transferring C to root symbionts [34,35]. Organic carbon and nitrogen can therefore be expected to accumulate rapidly in the first centuries after deglaciation [27,36,37]. The plants stabilize the soil surfaces and thus reduce soil erosion, and promote soil deepening [20,29,30]. As erosion is reduced [38] and the soil conditions provide more favorable and stable habitats, the composition of plant communities shift towards more resistant and competitive species. If the conditions are favorable, primary successions from pioneer species to trees can develop in less than two centuries in the Alps [39]. The subsequent accumulation of soil organic matter (SOM) [40] plays a crucial role in increasing soil water retention (by increasing the microporosity, [41]) and nutrient supply [42] which further improve the soil's qualities as a habitat.

The extent to which the vegetation influences the substrate depends not only on the degree of coverage but also on the functional plant traits [43]. In a species-rich plant community, the diversity of different traits is assumed to positively influence ecosystem functioning, e.g., by the complementarity of nutrient recycling (diversity hypothesis). In order to quantify these relationships, scientists often use the concept of functional diversity, which is calculated as the combination of the abundance of plant species and their functional traits [44]. In principle, functional diversity cannot be described by one single index. One prominent measure of functional diversity often related to soil processes is the functional richness (FRic) meaning the total amount of niche space filled by the traits of all species in a certain community [45].

Meteoric ^{10}Be is a new and novel method to obtain information about the balance between soil production and soil erosion [46]. Meteoric ^{10}Be is, among others, produced in the upper atmosphere through spallation of nitrogen and oxygen by cosmic radiation [47]. Precipitation washes it out of the atmosphere and deposits it in soils (hence it is referred to as 'meteoric' ^{10}Be, [48]). It is adsorbed strongly to soil particles which makes it a useful tracer for long-term soil erosion and sedimentation processes [46], soil residence times (e.g., [49]), and soil production (e.g., [50]). This method enables us to calculate the net soil loss during soil evolution. In contrast, fallout radionuclides (e.g., $^{239+240}$Pu) cover the last 60 years and enable the determination of erosion rates at a decadal scale. As soil erodibility generally decreases with (progressive) soil formation, the erosion rates are expected to decrease with increasing surface age. Vegetation and abiotic factors such as the topography and climatic conditions shape this temporal development. However, with increasing age, soils become less permeable and, thus, generate more surface runoff [51] which can induce more surface erosion. The roughness of the soil surface additionally influences the spatial distribution of the surface runoff. A rougher surface can lead to more channeling and, thus, increasing erodibility [52].

In order to improve our understanding of how high-mountainous (and, by extension, arctic) environments will develop under future climate warming, we must understand and quantify the processes that drive landscape evolution. This study explores soil formation in two proglacial areas with the aim to qualitatively and quantitatively assess the progression of chemical weathering and long-term erosion rates. We investigated soils on moraines of two periglacial chronosequences (one on granite/gneiss and the other one on carbonate parent material) which covered the Holocene and the Late Pleistocene. We used meteoric ^{10}Be to measure long-term soil erosion rates and related them with our data on weathering. We linked the results to short-term erosion rates [53], a vegetational survey [54], and hydrological properties of the soils [55,56], including a flow accumulation analysis, all of which were investigated in the same area as part of the same project.

2. Site Descriptions

2.1. Klausenpass

We studied the proglacial area of the Griess Glacier (2000–2200 m a.s.l.), a small glacier near the Klausenpass, central Switzerland (Figure 1). The glacial sediment is mostly composed of limestone from the Early Cretaceous and the Late Jurassic period with some addition of Flysch [57]. The mean annual temperature in the area is about +2 °C (annual averages for the 1961–1990 period, [58]) and the mean annual precipitation is ca. 2000 mm (1961–1990 period, [59]). The snowy season typically lasts from mid-October to mid-June.

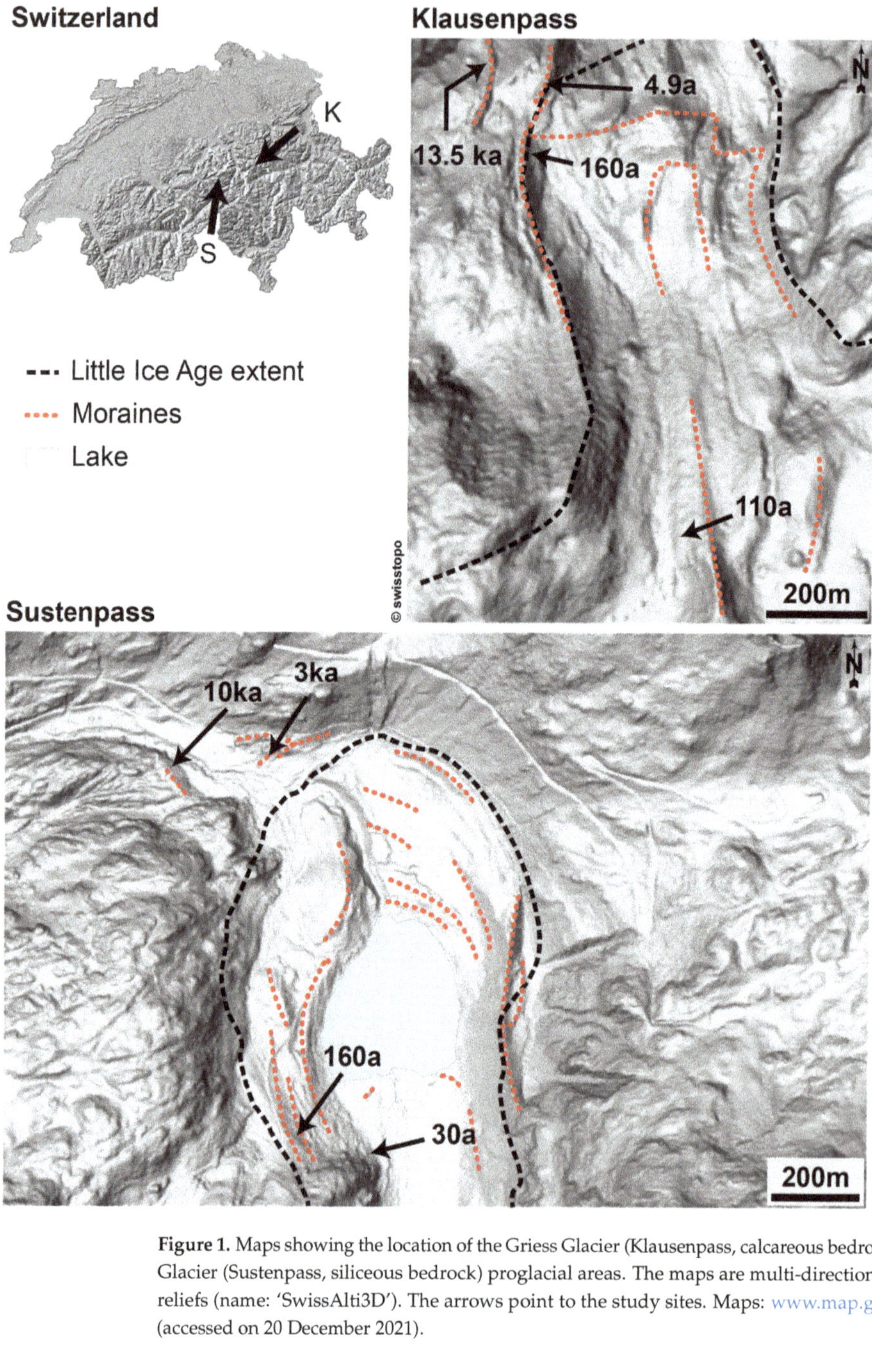

Figure 1. Maps showing the location of the Griess Glacier (Klausenpass, calcareous bedrock) and Stein Glacier (Sustenpass, siliceous bedrock) proglacial areas. The maps are multi-directional hill shade reliefs (name: 'SwissAlti3D'). The arrows point to the study sites. Maps: www.map.geo.admin.ch (accessed on 20 December 2021).

The chronosequence comprised four moraines: 110 a, 160 a, 4.9 ka, 13.5 ka (Table 1, Figures 2 and 3). We estimated the age of the 110 a site by tracking the glacier's retreat using historical maps [60]. The 160 a moraine corresponds to the Little Ice Age (LIA). We can identify LIA moraines with ease in the field and on aerial images as they are large and well preserved. Moreover, glaciers in the Alps advanced to almost the same extent as previously reached 3000 years ago, so that they form a prominent boundary between well-developed vegetation on the outside and the sparsely vegetated glacial foreland on the inside [61]. Approximations of the surface ages of the second oldest and the oldest moraines were obtained by radiocarbon dating of the H_2O_2 resistant fraction of soil organic matter in the top horizons [62]. Radiocarbon dating of soil organic matter is useful when no other dating methods are applicable, but there are limitations which need to be kept in mind. In general, the ages of the H_2O_2 resistant fraction indicate the oldest organic matter fraction in soils. These ages should rather be considered as minimum ages, because organic matter accumulation on moraines starts only then when plants are present which requires some time, especially when erosion rates are high (which may be the case in mountain soils, [53]). Musso et al. [62] obtained (2 σ) age ranges of 4969–4852 a cal BP for the second oldest and 14,001–13,760 and 13,287–13,120 a cal BP for the oldest moraine (Table 2 in [62]). They used the 'averages' 4900 a cal BP and 13,500 a cal BP as the respective surface ages. For consistency and simplicity, this paper will use the same ages and henceforth refer to these two moraine ages as '4.9 ka' and '13.5 ka'.

2.2. *Sustenpass*

The study site at Sustenpass is located in the proglacial area of the Stein Glacier (Figure 1). The parent material originates from the Aarmassif and consists of metagranitoids (variscan intrusion), gneiss, amphibolite, and mica-rich schist [57]. The mean annual temperature is ca 0–2 °C (1961–1990 period, [58]) and the mean annual precipitation ca 1700–2000 mm (1961–1990 period, [59]) in the study area. The snowy season typically lasts from mid-October to mid-June.

The Sustenpass chronosequence had four moraines: 30 a, 160 a, 3 ka, and 10 ka (Figures 2 and 3, Table 1). The age and location of the 30 a site was estimated by tracking the glacier's retreat after its advance in the mid-1980s using maps as well as aerial images. The soil barely shows any weathering and the vegetation is sparse and consists mostly of an initial grassland vegetation. The 160 a moraine (corresponding to the LIA) is well preserved in the east side of the valley, but it was not possible to sample there for logistical reasons. We therefore studied historical maps and located a suitable area on the opposite side of the valley which had the same surface age (Figures 1 and 2, [63–65]).

The proglacial area of Steingletscher was previously studied by King [66] and Heikkinen and Fogelberg [67]. King identified three ridges (Figure 1) which originated from different glacial advances. Two LIA moraines, one from the 17th century and one from the 19th century, are situated directly behind a distinctly older moraine. Schimmelpfennig et al. [68] provided ^{10}Be surface exposure ages on boulders and determined that the oldest of these three ridges was deposited nearly 3000 years ago. The 10 ka site was dated using the same method.

Figure 2. Photographs of the studied moraines at Klausenpass and Sustenpass (photos: A. Musso). Dashed lines have been added to highlight the moraine ridges or mark the outlines of the study sites.

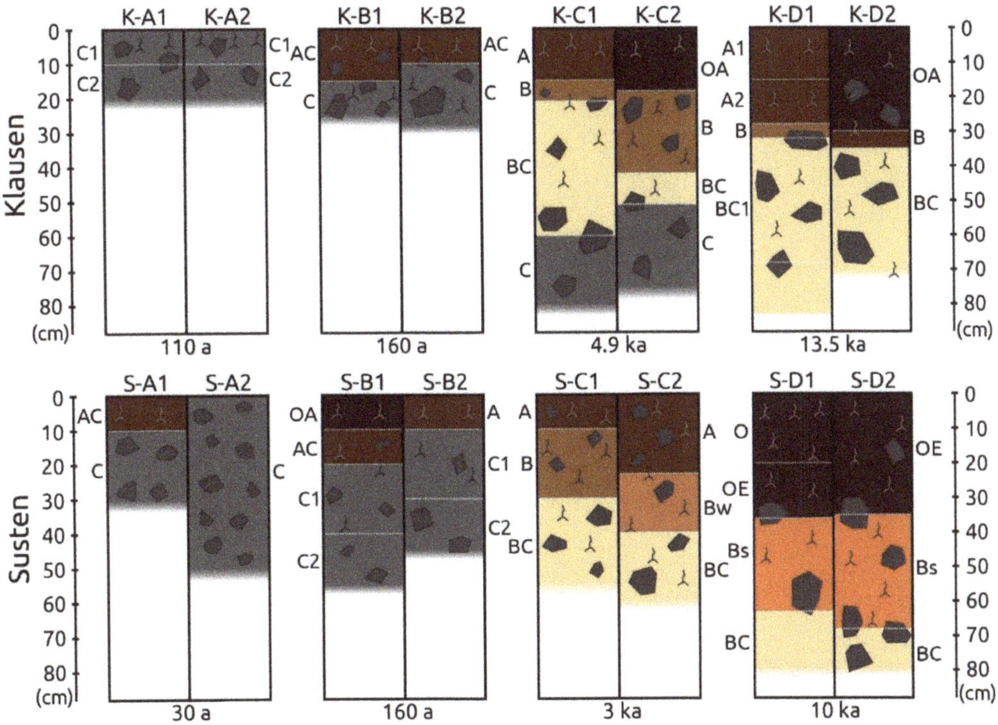

Figure 3. Soil profiles of both the Klausenpass and Sustenpass areas. The profile name codes include the location (K/S = Klausen/Susten), the age category (A = youngest, D = oldest) and the whether the functional richness of the vegetation at those soil pits' surroundings was high (=1) or low (=2). The functional richness is a measure encompassing plant species and their traits of a certain community. It is used to predict ecological functions and processes. The bottom depth of each profile represents the depth to which we were able to dig the profile (*y*-axis = depth in cm). Note that in the 30 years profiles at Sustenpass, we found both rounded and unrounded rocks. In general, the depicted amount and sizes of the rocks represent the amount and size of the largest rocks encountered in the soil pits.

Table 1. Site description for Klausenpass and Sustenpass with the surface ages (years) [62], aspect, slope (degrees), and elevation (m a.s.l.) and some basic soil parameters. The lowest depth of each profile refers to the depth that was reached during the dig (see also Figure 3). The soils were classified using the World Reference Base (WRB) classification [69]. The pH was measured with an absolute instrument error of ±0.003. LOI (Loss on ignition) represents the soil organic matter content of the fine earth fraction. The relative errors between lab replicates are 0.1%.

Soil Profile	Surface Age	Aspect	Slope	Elevation	Vegetation	Soil Type	Horizon	Depth	Bulk Density	Skeleton	pH ($CaCl_2$)	LOI
	(years)		(deg)	(m a.s.l.)	(Dominating Species)			(cm)	(g cm^{-3})	(wt. %)		(%)
K-A1	110	WNW	27	2200	*Saxifraga aizoides dominated stocks*	Hyperskeletic Leptosol	C1	0–10	1.8 ± 0.1	62	7.8	1.8
							C2	10–20+	1.4 ± 0.5	61.2	7.6	2
K-A2	110	WNW	27	2200	*Initial snowbed communities of type Arabidion caeruleae*	Hyperskeletic Leptosol	C1	0–10	1.5 ± 0.2	68.3	7.7	1.9
							C2	10–20+	1.3 ± 0.6	67.6	7.6	1.8
K-B1	160	NE	31	2030	*Thlaspietum rotundifolii*	Hyperskeletic Leptosol	AC	0–15	1.2 ± 0.4	71.6	7.3	8.4
							C	15–25+	0.7 ± 0.1	74.5	7.8	3.3
K-B2	160	NE	31	2030	*Dryadetum octopetalae*	Hyperskeletic Leptosol	AC	0–10	1.1 ± 0.7	68	7.5	8.9
							C	10–28+	0.8 ± 0	64.6	7.7	3.2
K-C1	4900 [1]	WNW	36	2020	*Caricetum ferruginei*	Calcaric Skeletic Cambisol	A	0–15	0.5 ± 0.1	11.3	5.8	18.7
							B	15–21	0.7 ± 0.1	64.5	7.3	10
							BC	21–60	1.5 ± 0.1	77.9	7.8	2.7
							C	60–80+	1.5 ± 0	85.3	7.7	2.2
K-C2	4900 [1]	N	37	2010	*Rhododendretum hirsuti*	Calcaric Skeletic Cambisol	OA	0–18	0.3 ± 0	3.8	4.5	34.4
							B	18–42	0.8 ± 0.1	22.7	6.7	14
							BC	42–51	1.5 ± 0	84.6	7.8	3.8
							C	51–75+	1.7 ± 0.1	80.8	7.7	2.7
K-D1	13,500 [2]	NW	44	2000	*Seslerio-Caricetum sempervirentis*	Calcaric Skeletic Cambisol	A1	0–15	0.3 ± 0.1	6.9	5.9	42.4
							A2	15–28	0.5 ± 0	40.8	6.8	20.5
							B	28–32	0.8 ± 0.2	73.5	7.6	5.8
							BC1	32–68	1.6 ± 0.1	75	7.8	1.64
							BC2	68–80+	1.7 ± 0.1	75	7.8	1.63
K-D2	13,500 [2]	ENE	32	2010	*Caricetum ferruginei*	Calcaric Skeletic Cambisol	OE	0–30	0.5 ± 0.1	22.8	6.6	24.5
							B	30–35	1.4 ± 0.1	74.3	7.8	2.3
							BC	35–70+	1.9 ± 0.1	82.1	7.8	1.7
S-A1	30	ENE	40	1950	*Initial grassland vegetation, rich in Trifolium badium/pallescens, Luzula alpino-pilosa*	Hyperskeletic Leptosol	AC	0–10	0.9 ± 0.6	53.5	6.1	1.1
							C	10–30+	1.5 ± 0.4	79.4	6.6	1.3

Table 1. *Cont.*

Soil Profile	Surface Age	Aspect	Slope	Elevation	Vegetation	Soil Type	Horizon	Depth	Bulk Density	Skeleton	pH ($CaCl_2$)	LOI
	(years)		(deg)	(m a.s.l.)	(Dominating Species)			(cm)	(g cm^{-3})	(wt. %)		(%)
S-A2	30	ENE	35	1950	*Pioneer vegetation at sandy sites, Epilobietum fleischeri subass rhacomitrietosum*	Hyperskeletic Leptosol	C	0–50+	0.7 ± 0.4	53.2	6.7	0.8
S-B1	160	ENE	31	1990	*Creeping dwarf shrup patches, high coverage of Salix retusa*	Hyperskeletic Leptosol	OA	0–10	1.7 ± 0.1	59.4	4.7	4.4
							AC	10–20	1.8 ± 0.1	57.4	4.8	1.7
							C1	20–40	1.9 ± 0	56	4.8	1.3
							C2	40–55+	1.6 ± 0.3	42.3	5.1	1.1
S-B2	160	NE	28	1990	*Grassland of the type Poion alpinae*	Hyperskeletic Leptosol	A	0–10	1.5 ± 0.3	42.7	5.5	38.4
							C1	10–30	1.9 ± 0.2	56.9	4.7	1.8
							C2	30–45+	1.8 ± 0.2	70.8	5	1.5
S-C1	3000	SW	25	1890	*Agrostio rupestris-Sempervivetum montani*	Skeletic Cambisol	A	0–10	1.3 ± 0.1	50.4	4.6	3.4
							B	10–30	1.5 ± 0.2	68.9	4.6	2.8
							BC	30–55+	1.9 ± 0	63.4	5.3	1.9
S-C2	3000	S	24	1910	*Carici sempervirentis*	Skeletic Cambisol	A	0–23	1 ± 0.5	72.2	5	22.5
							Bw	23–40	1.5 ± 0.3	81.2	5.6	13
							BC	40–60+	1.7 ± 0.1	79.5	6.1	2.4
S-D1	10,000	N	22	1880	*Geo montani-Nardetum*	Entic Podzol	O	0–20	0.7 ± 0.5	74.4	3.4	44
							OE	20–36	0.7 ± 0.2	47.9	3.6	28.7
							Bs	36–63	1.3 ± 0.1	54.8	4.4	5.2
							BC	63–80+	1.4 ± 0.1	41.7	4.4	6.1
S-D2	10,000	NNE	30	1880	*Rhododendro ferruginei-Vaccinietum*	Entic Podzol	OE	0–35	0.5 ± 0.2	12.8	3.6	38
							Bs	35–68	1 ± 0.2	34.1	4.1	10
							BC	68–80+	1.1 ± 0.1	73.9	4.3	16

[1] Radiocarbon 2 σ age range of soil organic matter: 4969–4852 a cal BC [61]. [2] Radiocarbon 2 σ age range of soil organic matter: 14,001–13,760, 13,287–13,120 a cal BC [62].

3. Materials and Methods

3.1. Sampling Strategy

This study is part of a series of studies which all used the same sampling scheme in order to ensure maximum comparability. Both the calcareous (Klausenpass) and the siliceous (Sustenpass) chronosequence comprised four moraines at which we excavated two soil pits each. We based the location of the pits on a vegetational mapping and the functional diversity of the vegetation. The functional diversity measures those components of plant diversity which influence the ecological functions and processes. The index used was the Functional Richness Index (FRI). It is based on the value, range, distribution, and the abundance of different plant traits in a vegetation community [70]. The parameters taken into account were: leaf dry matter content, specific leaf area, vegetative height [68], the plant life form, and the level of woodiness [71]. Soil pits were dug at two sites on each moraine where the FRI was highest and lowest, intending to cover the whole spectrum and see any effects, should they occur.

3.2. Physical and Chemical Properties

The skeleton (wt.%) is the weight percentage of the fraction >2 mm and was obtained by dry-sieving 1 kg of soil. Note that, for logistical reasons, we did not include rocks larger than roughly 5 cm. Therefore, the skeleton content presented here underestimates the actual situation in the field as there were many large cobbles and boulders in the soil profiles (Figures 3 and 4). The bulk density was determined by sampling soil with an Eijkelkamp soil corer (volume: 100 cm^3) and measuring its dry weight. Due to the size of the steel cylinders, rocks larger than a few centimeters were not included in the sampling, thus leading to a slight underestimation of the bulk density. The values presented (Table 1) are the means of 2–3 field replicates. The organic matter (OM) content was obtained via loss on ignition (LOI). 2 g of fine earth were dry-ashed at 550 °C for 6 h. Corrected for the skeleton content, the lost material represents the organic matter content of the soil [72]. The pH was determined by mixing 5 g of fine earth with 12.5 mL of 0.01 M $CaCl_2$, equating to a ratio of 1:2.5. We conducted the measurements on a Mehtrohm 692 pH/IonMeter with an absolute error of ±0.003. The carbonate content was calculated from total inorganic carbon (C_{inorg}) measurements using a MultiEA (Analytik Jena, Jena, Germany), which operates with phosphoric acid.

The total element contents were measured by X-ray fluorescence (XRF, Spectro Xepos, Spectro Analytical Instruments GmbH, Kleve, Germany). The instrument is calibrated using two glass beads (FLX-SP1, FLX-SP2, Fluxana, Bedburg-Hau, Germany). We used the certified reference soil sample SO-4 (Canada Centre for Mineral and Energy Technology, Canada) as a control standard during measurements to detect drifts. To measure the samples, 5 g of fine earth were finely milled and measured as powder samples. The instrument performs 6 repeat measurements per sample and outputs mean and standard error.

We calculated the oxide forms of the most abundant elements in the soils (Na, Mg, Al, Si, P, K, Ca, Ti, Mn, Fe). In order to differentiate the two Ca species ($CaCO_3$ and CaO), we subtracted Ca in carbonates ($CaCO_3$ contents obtained from C_{inorg}) from the total Ca content. The difference represents Ca in silicates (CaO). C_{tot} and N contents of the fine earth were obtained with EA-IRMS (Elemental Analyzer Isotope Ratio Mass Spectrometry; Flash HT Plus CNSOH, Thermo Scientific, Bremen, Germany) measuring 2 lab replicates.

We calculated three weathering indices for all soil profiles. All indices compare the molar ratios of easily weathered ('mobile') elements to more 'immobile' elements:

$$\frac{Ca + K}{Ti} \tag{4}$$

This ratio was introduced by Dorn [73] on rock varnishes and it has proven useful in soils as well, as Ti is mostly immobile in soils (alternatively, Zr can be used as the immobile element, [74]). Here, we calculated this ratio using the total Ca. It therefore also reflects the dissolution of carbonates. An alternative version to this ratio is:

$$\frac{\text{Na} + \text{K}}{\text{Ti}} \tag{5}$$

That only reflects weathering of silicates. We also used the Chemical Index of Alteration [75]. It is normally used in for igneous rocks or siliceous soils. There, an increase in the CIA (relative accumulation of Al) is indicative of stronger silicate weathering.

$$\text{CIA} = \frac{\text{Al}_2\text{O}_3}{\text{Al}_2\text{O}_3 + \text{CaO} + \text{Na}_2\text{O} + \text{K}_2\text{O}} \times 100 \tag{6}$$

The oxalate extractable Fe, Al, and Mn contents of the fine earth fraction were determined according to [76] with 2–3 replicates The Fe and Al measurements were conducted on a flame-AAS (atomic absorption spectrometer; ContrAA 700, Analytik Jena, Jena, Germany). The values presented are the averages of 2–3 replicates.

Figure 4. Profile pictures from the moraines of (**a**) Klausenpass and (**b**) Sustenpass (images: Musso et al. [61], reproduced with permission from Elsevier).

3.3. Mineralogical Analysis

The mineralogy of the samples was determined on randomly oriented powder specimens with X-ray diffraction (XRD) analysis. First, the samples were air-dried. The fine earth fraction (<2 mm) of the bulk soil was comminuted by hand using a mortar and sieved (400 μm). This fraction was subsequently Ca saturated. A representatively split aliquot of about 2 g was then milled in ethanol to a grain size below 20 μm with a McCrone micronizing mill and dried afterwards at 65 °C. For frontloading preparation, about 1 g of the powdered material was gently pressed in a sample holder for packing, sample-height adjustment, and forming a flat surface. Preferred orientation was avoided by using a blade for surface treatment [77]. For an improved identification of the clay mineralogy, appr. 1 g of the untreated <2 mm sample was sieved with a 20 μm sieve to enrich the clay minerals. From this <20 μm-fraction, we produced oriented specimens by carefully smearing a thin layer of the sample onto a glass slide. They were prepared for enhancement of the basal reflexes of layer silicates, thereby facilitating their identification. The textured samples were measured before and after treatment. The changes in the reflex positions in the XRD pattern by intercalation of different organic compounds (e.g., ethylene glycol) and after heating were used for the identification of expandable clay minerals.

X-ray diffraction measurements were made using a Bragg-Brentano X-ray diffractometer (D8 Advance, Bruker AXS, Karlsruhe, Germany) using CoKα (35 kV, 40 mA) radiation. The instrument worked with automatic beam optimization (equipped with an automatic theta compensating divergence slit and an automatic air scatter screen). The detector was an energy dispersive Lynx-Eye XE-T detector. The powder samples were step-scanned at room temperature from 2 to 80°2Theta (step width 0.02°2Theta, counting time 2 s per step).

The qualitative phase analysis was carried out with the software package DIFFRAC-plus (Bruker AXS, Karlsruhe, Germany). The phases were identified on the basis of the peak positions and relative intensities in the comparison to the PDF-2 database (International Centre for Diffraction Data). The quantitative amount of the mineral phases was determined with Rietveld analysis using the Rietveld program Profex/BGMN [78]. This full pattern-fitting method consists of the calculation of the X-ray diffraction pattern and its iterative adjustment to the measured diffractogram. In the refinements phase, specific parameters and the phase content were adapted to minimize the difference between the calculated and the measured X-ray diffractogram.

3.4. Meteoric ^{10}Be Sample Preparation and Analysis

Our method of extraction of ^{10}Be from soil is based on the methods described in [10,79]. A fine-milled soil sample is spiked with ^{9}Be standard solution and leached overnight twice using 16% HCl. The leachate undergoes a series of pH adjustments in order to separate metals, mainly iron, by precipitating as hydroxides. Thereafter, a fraction containing beryllium is isolated on ion-exchange resin Bio-Rad AG 50W-X8. A special grade of chemical reagents and water are used through all steps of sample preparation to ensure an ultra-low level of ^{10}B, which is an isobar of ^{10}Be.

Prior to the accelerator mass spectrometry (AMS) measurement, all the resulting BeO samples were mixed with Nb powder and pressed in a cathode insert. The measurements were performed with MILEA AMS system at ETH Zurich, Switzerland. The results were normalized to in-house standards S2007N and S2010N with nominal values of $^{10}Be/^{9}Be = 28.1 \times 10^{-12}$ and $^{10}Be/^{9}Be = 3.3 \times 10^{-12}$ and errors of 2.7% and 2.2%, respectively [80]. The in-house standards are calibrated to a primary standard ICN 01-5-1 [81]. The final ^{10}Be concentrations were corrected to preparation blank, while the final errors include a preparation blank error and an error of the in-house AMS standards.

3.5. Calculation of Weathering and Erosion Rates

The progression of chemical weathering was determined by comparing the differences in the chemical composition of the weathered and unweathered soil material. Using an immobile element (Ti), we calculated element-specific mass balances. As the soil ages were known, it was also possible to calculate long-term weathering rates (mass balances) [80,81]. We calculated these for Na, Mg, Al, Si, P, K, Ca, Mn, and Fe. We will refer to the sum of these elements as the 'total' mass balance, henceforth.

Following [82], we first calculated the volume change between the unweathered parent material and the weathered material, or strain coefficient ε, for each weathered horizon:

$$\varepsilon_w = \left(\frac{\rho_p C_{i,p}}{\rho_w C_{i,w}}\right) - 1 \quad (7)$$

where ρ_p and ρ_w are the bulk densities (kg m^{-3}) of the parent material (=lowermost horizon of each profile) and the weathered material, $C_{i,p}$ (g kg^{-1}) is the concentration of the immobile element (Ti was used) of the parent material, and $C_{i,w}$ (g kg^{-1}) is the concentration of the immobile element (Ti) of the weathered soil.

The open-system mass transport function $\tau_{j,w}$ [82] is defined as:

$$\tau_{j,w} = \left(\frac{\rho_p C_{j,w} C_{i,p}}{\rho_w C_{i,w} C_{i,p}}(\varepsilon_{i,w} + 1)\right) - 1 \quad (8)$$

with $C_{j,w}$ (g/kg) as the concentration of element j in the weathered material, and $C_{j,p}$ for the unweathered parent material (=lowermost horizon) and the strain coefficient ε_w. $C_{i,w}$ and $C_{i,p}$ (g kg^{-1}), the concentrations of the immobile element, were calculated with Ti. The change in mass for an element j ($m_{j,\,flux(z)}$ in kg m^{-2}) for n soil layers is defined by Egli and Fitze [83]:

$$m_{j,flux(z)} = \sum_{a=1}^{n} C_{j,p}\rho_p \left(\frac{1}{\varepsilon_{i,w} + 1}\right)\tau_{j,w}\Delta z_w \quad (9)$$

The long-term erosion rates were calculated using meteoric ^{10}Be in the soils. When the surface ages are known, soil erosion rates can be calculated by comparing the measured concentrations of ^{10}Be in the soils with the expected concentrations based on their age, as demonstrated by [84–86].

With no erosion, the surface age on a soil is defined as:

$$t = -\frac{1}{\lambda}\ln\left(1 - \lambda\frac{N_{exp}}{q}\right) \text{ and } N_{exp} = q\frac{e^{-\lambda t} - 1}{-\lambda} \quad (10)$$

with λ = the decay constant of ^{10}Be (4.997×10^{-7} y^{-1}), q (atoms cm^{-2} year^{-1}) = annual deposition rate of ^{10}Be (calculated according to [46,87,88]), and N_{exp} (atoms cm^{-2}) = the expected ^{10}Be inventory in the soil profile (i.e., the accumulated ^{10}Be in the soil due to atmospheric deposition, assuming no erosion or other disturbances). The deposition rates of ^{10}Be will have to be estimated for a specific area, as they are not known precisely. However, Maejima et al. [84] have shown that they are mostly dependent on the amount of precipitation.

With soil erosion, Equation (9) is rewritten as:

$$t = -\frac{1}{\lambda}\ln\left(1 - \lambda\frac{N_{exp}}{q - \rho C_{10Be} f E_{soil}}\right) \text{ and } E_{soil} = \frac{1}{\rho f C_{10Be}}\left(\frac{\lambda N}{e^{-\lambda t} - 1}\right) + q \quad (11)$$

where C_{10Be} (atoms g^{-1}) = concentration of ^{10}Be in the top eroding horizons (ca. the top 20 cm), E_{soil} (cm year^{-1}) = soil erosion rate, f = fine earth fraction, and ρ (g cm^{-3}) = bulk density of the top horizons, λ = decay constant of ^{10}Be, N = inventory of ^{10}Be. As C_{10Be} changes over time, it can be approximated by using the average between t = 0 and t, i.e., 0.5 $\times$ $C_{10Be(today)}$. We also assume that the erosion occurs mostly in the top 20 cm of the soil profile. We calculate the ^{10}Be inventory N using [86]:

$$N = \sum_{a=1}^{n}(z_w \rho_w C_w f_w) \quad (12)$$

for n horizons, and with the horizon thickness z, the bulk density ρ, the ^{10}Be concentration C and the fine earth fraction f of the weathered (w) material.

For comparison, we also used the method of Lal et al. [47]:

$$K_E = \frac{N_D}{N_S}\left[\frac{Q+q_a}{N_D} - \lambda\right] - \lambda \text{ and } E_{soil} = z_0 K_E \tag{13}$$

K_E is the first order rate constant for removal of soil, with N_S = ^{10}Be inventory in the S layer (comprising O and A horizons), N_D = ^{10}Be inventory in the D layer (below 20 cm), Q = flux of atmospheric ^{10}Be into topsoil (atoms cm^{-2} year^{-1}), q_a = flux of meteoric ^{10}Be (atoms cm^{-2} year^{-1}), and λ = decay constant of ^{10}Be. The soil erosion rate E_{soil} (cm year^{-1}) is then calculated using z_0 = thickness of topsoil horizons (comprising O and A horizons) multiplied by K_E.

In addition, the rates of soil formation were calculated. Most approaches are based on the assumption that the soil in question is in equilibrium, i.e., that the soil thickness remains stable because soil production and soil denudation are equal. As this is clearly not the case for such young and still intensely developing soils (see Figure 3), we used the changes in the profile thickness as an approximation for soil formation rates. The soil thickness function was first developed by [89] and further developed by Johnson et al. [90]. According to this model, the soil thickness (T) is a function of soil deepening factors (SD), upbuilding factors (U), and removals (R):

$$T = SD + U - R \tag{14}$$

We calculated the soil formation rates ($F = SD + U$) based on the changes in thickness (z) [91]:

$$F_{soil} = \frac{z_{profile}}{\text{age}} \tag{15}$$

3.6. Flow Accumulation Analysis

Digital surface models (DSM) were created using centimeter-resolution aerial drone images. The drone flights and the post-processing of the images were conducted by WWL Umweltplanung und Geoinformatik GbR [92]. Flow direction and flow accumulation analyses were conducted in ArcMap (ESRI). The flow direction analysis uses the DSM to compute the most likely flow path that water would take according to the steepness when moving from one grid cell to the next. The D-∞ ('D-infinity') method was used [93]. The flow accumulation shows the spatial distribution and intensity of the flow pathways, i.e., how much water is directed to each pathway. We used this output in combination with the microtopography measurements of Maier et al. [56] to better evaluate the results of our erosion analysis, as more bundled and therefore more intense flow paths have a higher potential for surface erosion than dispersed ones.

3.7. Uncertainty Calculation and Propagation

We defined the standard uncertainties for the different used parameters by calculating the standard deviation between field replicates (bulk density) and lab replicates (oxalate extractable Fe, Al, Mn; C_{org}, N contents) and by using the instrument measurement errors (elemental contents, mineralogy).

When using these datasets in further calculations (weathering indices, C_{org}, C_{org} stocks, C/N, element mass balances, ^{10}Be erosion rates), the standard uncertainties were accounted for by propagating them following the recommendations of the Guide of Expression of Uncertainty [94].

4. Results

4.1. Profile Descriptions

The soils of the 110 a moraine were classified as Hyperskeletic Leptosol. The moraine was covered in boulders and cobbles of varying sizes and was vegetated with initial snowbed communities of the type *Arabidion caeruleae* and *Saxifraga aizoides* (Figures 2–4, Table 1). The 160 a moraine was classified as Hyperskeletic Leptosol and had already accumulated a humus-rich A horizon. The vegetation was dominated by *Thlaspietum rotundifolii* and *Dryadetum octopetalae*. The 4.9 ka soils were Calcaric Skeletic Cambisols which reached a depth of approximately 60 cm. They were vegetated by dense grass, mainly *Caricetum ferruginei* and *Rhododendretum hirsuti* with roots to 40–50 cm. The 13.5 ka moraine had Calcaric Skeletic Cambisols as well and was also densely vegetated. The dominating species were *Seslerio-Caricetum sempervirentis* and *Caricetum ferruginei*.

The two soils at the 30 a moraine Hyperskeletic Leptosols, one of which had accumulated an A horizon (Figures 2–4). The surface was covered by pioneer species (mainly *Epilobium fleischeri*) and initial grassland vegetation rich in *Trifolium badium/pallescense* and *Luzula alpino-pilosa* (Table 1). The 160 a soils were also Hyperskeletic Leptosols with 10–20 cm thick A horizons. They were covered by a grassland vegetation of the type *Poion aplinae*. There were also creeping dwarf shrubs with a high coverage of Salix retusa present. The 3 ka soils were Skeletic Cambisols vegetated by grasses (mainly *Agrostio rupestris-Sempervivetum montani* and *Carici sempervirentis*). The 10 ka soils were Entic Podzols with thick, humus-rich topsoils that reached a depth of 30–40 cm. The surface was largely covered by shrubs (*Rhododendro ferruginei-Vaccinietum*) and some grasses (*Geo montani-Nardetum*) in between and underneath the shrubs.

4.2. Soil Properties

The bulk density is similar in both locations. At Klausenpass, it varied from 0.7 to 1.8 g cm^{-3} in the younger soils and from 0.3 to 1.9 g cm^{-3} in the older soils. The soils at Sustenpass display clear age trends, ranging from 1.2 to 1.7 g cm^{-3} in the 30 a, 160 a, and 3 ka soils and 0.7–1.4 in the 10 ka soils.

The soils had a high skeleton content throughout all sites, often over 50% (Table 1). At Klausenpass, it decreased with age, most strongly in the topsoil. There is also a strong increase in the skeleton fraction with depth. At Sustenpass, the skeleton content varied more strongly from moraine to moraine and the depth trend was less clear compared to Klausenpass.

The highest organic matter (OM) contents (measured as loss on ignition, Table 1) were found in the topsoil of the oldest soils, ranging from 19% to 42% at Klausenpass and 22% to 38% at Sustenpass. The OM in the young topsoils at Klausenpass (110 a, 160 a: 1.8–8.9%) was distinctly smaller compared to the young soils at Sustenpass (30 a, 160 a: 0.8–38.4%). The differences between the profiles of one moraine is also higher at Sustenpass than at Klausenpass.

The pH at Klausenpass was mostly in the neutral to weakly alkaline range (pH 6.7–7.8) except for the top horizons in the older soils where the pH was below 6. At Sustenpass, the pH was in the neutral to strongly acidic range (pH 3.4–6.7).

4.3. Weathering

4.3.1. Klausenpass

The elemental composition is presented in Table 2. The calculated strain coefficients and open-system mass transport functions for each Klausenpass soil profile are shown in Tables 3 and 4. The element mass fluxes of each moraine are summarized in Figure 5a. High losses were recorded for the mobile elements Mg and Ca, as well as P, while less-mobile elements passively accumulated. The total losses are highest at the 13.5 ka soil (-152 ± 49 kg m^{-2}), Figure 5b. When including the soil age to calculate rates of chemical weathering, the rates do not change very strongly over time (+28 $\pm$ 2 to -7 ± 1 m^{-2} $kyear^{-1}$ in the youngest soils to +1 $\pm$ 0 to -11 ± 4 m^{-2} $kyear^{-1}$ in the oldest soils).

Table 2. Elemental contents composition of the soil samples (fine earth fraction) in their oxidized form based on XRF measurements (Na_2O to ZrO_2). The Ca was split into $CaCO_3$ and CaO. The proportion was determined by separate C_{inorg} measurements.

Profile	Horizon	Depth	Na_2O	MgO	Al_2O_3	SiO_2	P_2O_5	SO_3	K_2O	$CaCO_3$	CaO	TiO_2	MnO	Fe_2O_3	ZrO_2
		cm	g/kg	g/kg	g/kg	g/kg	g/kg	g/kg	g/kg	g/kg	g/kg	g/kg	g/kg	g/kg	g/kg
			Klausenpass (Griess Glacier)												
110 years K-A1	C1	0–10	3.23 ± 0.08	33.6 ± 0.44	54.09 ± 0.18	289.58 ± 0.42	1.76 ± 0.01	0.8 ± 0.01	9.23 ± 0.02	555.49 ± 0.24	0 ± 0.14	3.46 ± 0.01	0.35 ± 0	30.75 ± 0.03	0.16 ± 0
	C2	10–20+	3.87 ± 0.09	34.73 ± 0.43	63.43 ± 0.22	283.27 ± 0.41	1.78 ± 0.01	0.74 ± 0.01	10.75 ± 0.02	545.3 ± 0.24	0 ± 0.13	3.7 ± 0.01	0.33 ± 0	32.14 ± 0.03	0.13 ± 0
K-A2	C1	0–10	3.19 ± 0.08	32.61 ± 0.43	53.18 ± 0.18	296.92 ± 0.41	1.68 ± 0.01	0.82 ± 0.01	8.93 ± 0.02	547.84 ± 0.24	0 ± 0.13	3.53 ± 0.01	0.41 ± 0	31.85 ± 0.03	0.12 ± 0
	C2	10–20+	2.36 ± 0.06	32.96 ± 0.43	54.54 ± 0.2	284.49 ± 0.41	1.79 ± 0.01	0.76 ± 0.01	9.23 ± 0.02	561.28 ± 0.24	0 ± 0.13	3.38 ± 0.01	0.35 ± 0	30.79 ± 0.03	0.15 ± 0
160 years K-B1	AC	0–15	8.88 ± 0.26	42.32 ± 0.49	113.04 ± 0.3	426.85 ± 0.43	1.05 ± 0.01	1.44 ± 0.02	21.57 ± 0.02	242 ± 0.13	0 ± 0.07	6.97 ± 0.01	0.58 ± 0	51.2 ± 0.03	0.26 ± 0
	C	15–25+	7.29 ± 0.2	41.53 ± 0.48	105.03 ± 0.28	389.75 ± 0.42	1.06 ± 0.01	1.18 ± 0.01	19.46 ± 0.02	347.14 ± 0.25	0 ± 0.14	6.28 ± 0.01	0.53 ± 0	47.1 ± 0.03	0.21 ± 0
K-B2	AC	0–10	7.79 ± 0.23	41.82 ± 0.49	107.59 ± 0.29	430.37 ± 0.43	1.02 ± 0.01	1.46 ± 0.02	20.48 ± 0.02	243.35 ± 0.13	0 ± 0.07	6.33 ± 0.01	0.64 ± 0	49.62 ± 0.03	0.23 ± 0
	C	10–28+	8.11 ± 0.16	40.15 ± 0.48	90.52 ± 0.26	365.38 ± 0.42	1.12 ± 0.01	1.03 ± 0.01	16.93 ± 0.02	396.89 ± 0.25	0 ± 0.14	5.15 ± 0.01	0.55 ± 0	41.85 ± 0.03	0.19 ± 0
4900 years K-C1	A	0–15	10.81 ± 0.39	35.6 ± 0.46	131.33 ± 0.33	521.18 ± 0.44	1.62 ± 0.02	2.82 ± 0.02	26.47 ± 0.02	0 ± 0.02	7.27 ± 0.01	7.65 ± 0.01	1.51 ± 0	66.41 ± 0.04	0.26 ± 0
	B	15–21	10.98 ± 0.36	40.28 ± 0.51	142.86 ± 0.35	503.99 ± 0.45	2.26 ± 0.02	1.62 ± 0.02	28.74 ± 0.03	83.96 ± 0.05	1.54 ± 0.03	8.43 ± 0.01	1.62 ± 0	73.79 ± 0.04	0.28 ± 0
	BC	21–60	0 ± 0	31.86 ± 0.38	48.67 ± 0.17	177.07 ± 0.26	2.38 ± 0.01	0.58 ± 0.01	8.39 ± 0.01	674.53 ± 0.23	0 ± 0.13	2.95 ± 0.01	0.37 ± 0	25.75 ± 0.03	0.09 ± 0
	C	60–80+	1.99 ± 0.04	33 ± 0.41	45.88 ± 0.17	163.54 ± 0.25	2.41 ± 0.01	0.54 ± 0.01	7.77 ± 0.02	696.6 ± 0.23	0 ± 0.13	2.76 ± 0.01	0.28 ± 0	22.84 ± 0.03	0.08 ± 0
K-C2	OA	0–18	10.44 ± 0.37	28.66 ± 0.42	108.43 ± 0.29	410.59 ± 0.44	2.14 ± 0.02	4.13 ± 0.02	23.24 ± 0.01	0 ± 0.02	7.18 ± 0.01	6.75 ± 0.01	0.89 ± 0	53.29 ± 0.03	0.23 ± 0
	B	18–42	11.16 ± 0.38	38.79 ± 0.49	150.88 ± 0.37	507.85 ± 0.65	2.2 ± 0.02	2.3 ± 0.02	31.7 ± 0.02	24.47 ± 0.03	6.16 ± 0.01	8.69 ± 0.01	1.68 ± 0	73.79 ± 0.04	0.27 ± 0
	BC	42–51	2.44 ± 0.06	33.49 ± 0.41	66.17 ± 0.21	202.09 ± 0.19	2.29 ± 0.01	0.66 ± 0.01	12.03 ± 0.02	608.08 ± 0.23	0 ± 0.13	4.11 ± 0.01	0.46 ± 0	30.46 ± 0.03	0.11 ± 0
	C	51–75+	2.05 ± 0.05	32.92 ± 0.42	51.27 ± 0.18	173.13 ± 0.26	2.33 ± 0.01	0.58 ± 0.01	9.51 ± 0.02	672.39 ± 0.23	0 ± 0.13	3.03 ± 0.01	0.37 ± 0	24.98 ± 0.03	0.1 ± 0
13,500 years K-D1	A1	0–15	11.83 ± 0.37	27.29 ± 0.41	80.3 ± 0.24	344.11 ± 0.46	2.77 ± 0.02	5.97 ± 0.03	18 ± 0.01	0 ± 0.03	18.94 ± 0.01	5.27 ± 0.01	1.26 ± 0	59.72 ± 0.03	0.19 ± 0
	A2	15–28	8.18 ± 0.3	32.59 ± 0.45	104.98 ± 0.3	466.5 ± 0.46	2.77 ± 0.02	3.53 ± 0.02	22.86 ± 0.01	52.94 ± 0.05	10.41 ± 0.03	6.8 ± 0.01	1.32 ± 0	81.98 ± 0.05	0.25 ± 0
	B	28–32	0.46 ± 0.01	29.08 ± 0.39	35.21 ± 0.14	159.18 ± 0.26	2.3 ± 0.01	0.95 ± 0.01	6.68 ± 0.01	673.8 ± 0.24	0 ± 0.13	2.16 ± 0.01	0.47 ± 0	31.69 ± 0.03	0.08 ± 0
	BC1	32–68	0 ± 0	29.62 ± 0.41	14.3 ± 0.08	41.8 ± 0.12	2.4 ± 0.01	0.49 ± 0.01	2.5 ± 0.01	880.17 ± 0.46	0 ± 0.26	0.95 ± 0.01	0.15 ± 0	11.32 ± 0.02	0.03 ± 0
	BC2	68–80+	0.9 ± 0.02	29.91 ± 0.41	18.72 ± 0.1	65.6 ± 0.16	2.4 ± 0.01	0.54 ± 0.01	3.48 ± 0.01	843.27 ± 0.46	0 ± 0.26	1.28 ± 0.01	0.24 ± 0	17.18 ± 0.01	0.05 ± 0
K-D2	O/OA	0–30	11.23 ± 0.39	36.16 ± 0.47	118.41 ± 0.31	468.5 ± 0.45	1.73 ± 0.02	3.84 ± 0.02	25.84 ± 0.03	0.19 ± 0.03	12.26 ± 0.01	6.81 ± 0.01	0.87 ± 0	69.25 ± 0.04	0.23 ± 0
	A	30–35	1.05 ± 0.02	29.68 ± 0.41	26.23 ± 0.12	94.61 ± 0.19	2.23 ± 0.01	0.62 ± 0.01	4.24 ± 0.01	799 ± 0.45	0 ± 0.25	1.69 ± 0.01	0.2 ± 0	17.43 ± 0.01	0.05 ± 0
	BC	35–70+	0.45 ± 0.01	30.24 ± 0.41	21.55 ± 0.1	80.35 ± 0.18	2.39 ± 0.01	0.51 ± 0.01	3.43 ± 0.01	826.37 ± 0.45	0 ± 0.25	1.44 ± 0.01	0.21 ± 0	16.28 ± 0.01	0.05 ± 0
		Sustenpass (Stein Glacier)													
30 years S-A1	AC	0–10	31.22 ± 0.67	24.83 ± 0.29	132.4 ± 0.35	705.99 ± 0.69	1.02 ± 0.02	0.43 ± 0.01	28.71 ± 0.03	0 ± 0	18.55 ± 0.02	5.86 ± 0.01	0.65 ± 0	38.83 ± 0.03	0.12 ± 0
	C	10–30+	32.15 ± 0.68	24.68 ± 0.28	139.66 ± 0.34	689.58 ± 0.68	0.73 ± 0.01	0.43 ± 0.01	31.57 ± 0.03	0 ± 0	17.98 ± 0.01	5.35 ± 0.01	0.7 ± 0	43.63 ± 0.03	0.08 ± 0
S-A2	C	0–50+	29.59 ± 0.67	24.25 ± 0.29	129.84 ± 0.35	714.21 ± 0.69	1.09 ± 0.02	0.53 ± 0.01	28.39 ± 0.03	0 ± 0	18.78 ± 0.02	6.03 ± 0.01	0.64 ± 0	38.12 ± 0.03	0.14 ± 0
160 years S-B1	OA	0–10	30.98 ± 0.65	19.05 ± 0.25	121.37 ± 0.32	699.74 ± 0.68	0.92 ± 0.01	1.01 ± 0.02	26.49 ± 0.03	0 ± 0	16.01 ± 0.01	5.08 ± 0.01	0.56 ± 0	34.34 ± 0.03	0.11 ± 0
	AC	10–20	29.44 ± 0.64	19.2 ± 0.25	124.68 ± 0.32	722.97 ± 0.68	0.77 ± 0.01	0.53 ± 0.01	28.46 ± 0.03	0 ± 0	15.91 ± 0.01	5.21 ± 0.01	0.56 ± 0	34.68 ± 0.03	0.11 ± 0
	C1	20–40	32.78 ± 0.62	18.13 ± 0.26	126.1 ± 0.32	722.13 ± 0.67	0.9 ± 0.02	0.81 ± 0.01	28.59 ± 0.03	0 ± 0	15.89 ± 0.01	5.39 ± 0.01	0.57 ± 0	35.56 ± 0.03	0.12 ± 0
	C2	40–55+	33.12 ± 0.6	17.4 ± 0.28	123.32 ± 0.31	733.17 ± 0.67	0.73 ± 0.01	0.38 ± 0.01	26.83 ± 0.02	0 ± 0	15.97 ± 0.01	4.87 ± 0.01	0.53 ± 0	32.62 ± 0.03	0.12 ± 0

Table 2. *Cont.*

Profile	Horizon	Depth	Na_2O	MgO	Al_2O_3	SiO_2	P_2O_5	SO_3	K_2O	$CaCO_3$	CaO	TiO_2	MnO	Fe_2O_3	ZrO_2
		cm	g/kg	g/kg	g/kg	g/kg	g/kg	g/kg	g/kg	g/kg	g/kg	g/kg	g/kg	g/kg	g/kg
S-B2	A	0–10	23.02 ± 0.55	13.95 ± 0.2	74.83 ± 0.23	427.05 ± 0.44	2.5 ± 0.02	4.29 ± 0.02	17.74 ± 0.01	0 ± 0	22.23 ± 0.01	3.64 ± 0.01	0.62 ± 0	25.82 ± 0.01	0.06 ± 0
	C1	10–30	30.85 ± 0.67	23.91 ± 0.28	131.28 ± 0.34	701.1 ± 0.68	0.88 ± 0.01	0.58 ± 0.01	28.97 ± 0.03	0 ± 0	17.32 ± 0.01	5.82 ± 0.01	0.67 ± 0	40.96 ± 0.03	0.12 ± 0
	C2	30–45+	31.08 ± 0.62	19.53 ± 0.28	125.94 ± 0.34	717.88 ± 0.67	0.9 ± 0.02	0.48 ± 0.01	28.24 ± 0.03	0 ± 0	17.39 ± 0.01	5.44 ± 0.01	0.63 ± 0	37.8 ± 0.03	0.1 ± 0
3000 years															
S-C1	A	0–10	36.68 ± 0.65	27.48 ± 0.34	133.38 ± 0.35	641.27 ± 0.66	1.42 ± 0.02	1.01 ± 0.02	21.02 ± 0.01	0 ± 0	33.17 ± 0.01	7.63 ± 0.01	1.1 ± 0	61.67 ± 0.04	0.08 ± 0
	B	10–30	30.05 ± 0.67	28.02 ± 0.31	138.55 ± 0.35	663.26 ± 0.7	0.93 ± 0.01	0.87 ± 0.01	27.23 ± 0.03	0 ± 0	17.45 ± 0.02	7.47 ± 0.01	0.83 ± 0	57.24 ± 0.03	0.13 ± 0
	BC	30–55+	31.92 ± 0.63	24.46 ± 0.31	131.42 ± 0.33	685.16 ± 0.67	0.93 ± 0.01	0.54 ± 0.01	25.27 ± 0.03	0 ± 0	18.33 ± 0.01	7.22 ± 0.01	0.8 ± 0	54.38 ± 0.03	0.11 ± 0
S-C2	A	0–23	29.09 ± 0.63	26.92 ± 0.3	108.08 ± 0.3	485.32 ± 0.45	2.89 ± 0.03	3.64 ± 0.03	16.97 ± 0.01	0 ± 0	30.35 ± 0.01	8.19 ± 0.01	1.37 ± 0	61.99 ± 0.03	0.07 ± 0
	Bw	23–40	27.15 ± 0.63	26.28 ± 0.28	119.1 ± 0.32	582.21 ± 0.68	1.38 ± 0.02	2.16 ± 0.02	22.65 ± 0.01	0 ± 0	21.8 ± 0.01	7.91 ± 0.01	0.89 ± 0	57.96 ± 0.03	0.12 ± 0
	BC	40–60+	35.37 ± 0.72	33.21 ± 0.32	141.44 ± 0.36	642.39 ± 0.69	1.24 ± 0.02	0.98 ± 0.02	23.41 ± 0.03	0 ± 0	31.51 ± 0.01	6.88 ± 0.01	1.06 ± 0	58.04 ± 0.03	0.08 ± 0
10,000 years			20.27 ± 0.53	10.34 ± 0.17	79.38 ± 0.25	383.81 ± 0.44	2.74 ± 0.02	5.95 ± 0.03	15.72 ± 0.01	0 ± 0	7.06 ± 0.01	7.02 ± 0.01	0.24 ± 0	27.52 ± 0.01	0.12 ± 0
S-D1	O	0–20	20.99 ± 0.53	14.12 ± 0.21	99.85 ± 0.28	491.92 ± 0.45	2.47 ± 0.02	4.1 ± 0.02	20.18 ± 0.01	0 ± 0	9.92 ± 0.01	9.2 ± 0.01	0.36 ± 0	39.3 ± 0.03	0.17 ± 0
	OE	20–36	27.88 ± 0.61	28.61 ± 0.31	135.28 ± 0.36	621.23 ± 0.67	2.09 ± 0.02	0.69 ± 0.01	26.98 ± 0.03	0 ± 0	17.59 ± 0.01	10.25 ± 0.01	1.23 ± 0	75.75 ± 0.04	0.17 ± 0
	Bs	36–63	25.75 ± 0.64	29.17 ± 0.3	140.03 ± 0.37	606.96 ± 0.69	2.41 ± 0.02	0.97 ± 0.02	27.74 ± 0.03	0 ± 0	16.98 ± 0.02	10.33 ± 0.01	1.22 ± 0	76.88 ± 0.05	0.18 ± 0
	BC	63–80+	21.31 ± 0.54	12.81 ± 0.19	90.74 ± 0.27	419.15 ± 0.44	2.62 ± 0.02	5.31 ± 0.03	16.64 ± 0.01	0 ± 0	7.59 ± 0.01	8.2 ± 0.01	0.32 ± 0	35.37 ± 0.03	0.13 ± 0
S-D2	OE	0–35	24.54 ± 0.61	28.56 ± 0.3	136.17 ± 0.36	576.97 ± 0.68	1.86 ± 0.02	1.23 ± 0.02	26.18 ± 0.03	0 ± 0	13.96 ± 0.01	10.75 ± 0.01	0.89 ± 0	78.95 ± 0.05	0.17 ± 0
	Bs	35–68	27.83 ± 0.64	41.91 ± 0.34	132.89 ± 0.35	504.05 ± 0.44	2.06 ± 0.03	1.42 ± 0.02	17.37 ± 0.01	0 ± 0	18.28 ± 0.01	8.59 ± 0.01	1.17 ± 0	84.82 ± 0.04	0.09 ± 0
	BC	68–80+	0 ± 0	0 ± 0	0 ± 0	0 ± 0	0 ± 0	0 ± 0	0 ± 0	0 ± 0	0 ± 0	0 ± 0	0 ± 0	0 ± 0	0 ± 0

Table 3. Strain coefficients (ε_{Ti}), open-system mass transport functions (τ) for each element at each soil profile. The lowermost horizon of each soil profile was used as the corresponding parent material in the calculations.

Profile	Horizon	Depth	ε	τ								
		(cm)	(Ti)	Na	Mg	Al	Si	P	K	Ca	Mn	Fe
110 years (hyperskeletic leptosol)												
K-A1	C1	0–10	−0.16 ± 40.5	−0.11 ± 0	0.03 ± 0	−0.09 ± 0	0.09 ± 0	0.06 ± 0	−0.08 ± 0	0.09 ± 0	0.11 ± 0	0.02 ± 0
	C2	10–20+	0 ± 0	0 ± 0	0 ± 0	0 ± 0	0 ± 0	0 ± 0	0 ± 0	0 ± 0	0 ± 0	0 ± 0
K-A2	C1	0–10	−0.15 ± 43.81	0.3 ± 0.01	−0.05 ± 0	−0.07 ± 0	0 ± 0	−0.1 ± 0	−0.07 ± 0	−0.07 ± 0	0.12 ± 0	−0.01 ± 0
	C2	10–20+	0 ± 0	0 ± 0	0 ± 0	0 ± 0	0 ± 0	0 ± 0	0 ± 0	0 ± 0	0 ± 0	0 ± 0
160 years (hyperskeletic leptosol)												
K-B1	AC	0–15	−0.53 ± 30.5	0.1 ± 0	−0.08 ± 0	−0.03 ± 0	−0.01 ± 0	−0.1 ± 0	0 ± 0	−0.37 ± 0	−0.02 ± 0	−0.02 ± 0
	C	15–25+	0 ± 0	0 ± 0	0 ± 0	0 ± 0	0 ± 0	0 ± 0	0 ± 0	0 ± 0	0 ± 0	0 ± 0
K-B2	AC	0–10	−0.4 ± 58.2	−0.22 ± 0.01	−0.15 ± 0	−0.03 ± 0	−0.04 ± 0	−0.25 ± 0	−0.02 ± 0	−0.5 ± 0	−0.04 ± 0	−0.04 ± 0
	C	10–28+	0 ± 0	0 ± 0	0 ± 0	0 ± 0	0 ± 0	0 ± 0	0 ± 0	0 ± 0	0 ± 0	0 ± 0
4900 years (calcaric skeletic cambisol)												
K-C1	A	0–15	−0.82 ± 29.83	0.96 ± 0.04	−0.61 ± 0.01	0.03 ± 0	0.15 ± 0	−0.76 ± 0.01	0.23 ± 0	−0.99 ± 0	0.96 ± 0.01	0.05 ± 0
	B	15–21	−0.73 ± 15.51	0.81 ± 0.03	−0.6 ± 0.01	0.02 ± 0	0.01 ± 0	−0.69 ± 0.01	0.21 ± 0	−0.96 ± 0	0.91 ± 0	0.06 ± 0
	BC	21–60	−0.4 ± 7.93	−1 ± 0.02	−0.1 ± 0	−0.01 ± 0	0.01 ± 0	−0.08 ± 0	0.01 ± 0	−0.09 ± 0	0.25 ± 0	0.05 ± 0
	C	60–80+	0 ± 2.85	0 ± 0	0 ± 0	0 ± 0	0 ± 0	0 ± 0	0 ± 0	0 ± 0	0 ± 0	0 ± 0

Table 3. *Cont.*

Profile	Horizon	Depth	ε	τ								
		(cm)	(Ti)	Na	Mg	Al	Si	P	K	Ca	Mn	Fe
K-C2	OA	0–18	−0.56 ± 6.73	1.28 ± 0.05	−0.61 ± 0.01	−0.05 ± 0	0.07 ± 0	−0.59 ± 0.01	0.1 ± 0	−0.99 ± 0	0.09 ± 0	−0.04 ± 0
	B	18–42	−0.81 ± 15.32	0.89 ± 0.04	−0.59 ± 0.01	0.03 ± 0	0.02 ± 0	−0.67 ± 0.01	0.16 ± 0	−0.98 ± 0	0.6 ± 0	0.03 ± 0
	BC	42–51	0.06 ± 6.25	−0.12 ± 0	−0.25 ± 0	−0.05 ± 0	−0.14 ± 0	−0.28 ± 0	−0.07 ± 0	−0.33 ± 0	−0.08 ± 0	−0.1 ± 0
	C	51–75+	0 ± 0	0 ± 0	0 ± 0	0 ± 0	0 ± 0	0 ± 0	0 ± 0	0 ± 0	0 ± 0	0 ± 0
13,500 years (calcaric skeletic cambisol)												
K-D1	A1	0–15	−0.67 ± 17.13	2.21 ± 0.11	−0.78 ± 0.02	0.04 ± 0	0.28 ± 0	−0.72 ± 0.01	0.26 ± 0	−0.99 ± 0	0.27 ± 0	−0.15 ± 0
	A2	15–28	−0.72 ± 11.91	0.72 ± 0.03	−0.79 ± 0.02	0.06 ± 0	0.34 ± 0	−0.78 ± 0.01	0.24 ± 0	−0.98 ± 0	0.03 ± 0	−0.1 ± 0
	B	28–32	0.15 ± 28.71	−0.69 ± 0.02	−0.42 ± 0.01	0.12 ± 0	0.44 ± 0	−0.43 ± 0	0.14 ± 0	−0.53 ± 0	0.14 ± 0	0.1 ± 0
	BC1	32–68	0.37 ± 10.85	−1 ± 0.02	0.34 ± 0	0.03 ± 0	−0.14 ± 0	0.35 ± 0	−0.03 ± 0	0.41 ± 0	−0.15 ± 0	−0.11 ± 0
	BC2	68–80+	0 ± 0	0 ± 0	0 ± 0	0 ± 0	0 ± 0	0 ± 0	0 ± 0	0 ± 0	0 ± 0	0 ± 0
K-D2	OA	0–30	−0.83 ± 22.49	4.29 ± 0.17	−0.75 ± 0.01	0.16 ± 0	0.23 ± 0	−0.85 ± 0.01	0.59 ± 0	−0.99 ± 0	−0.11 ± 0	−0.1 ± 0
	B	30–35	−0.2 ± 0	0.99 ± 0.03	−0.16 ± 0	0.04 ± 0	0 ± 0	−0.2 ± 0	0.05 ± 0	−0.18 ± 0	−0.18 ± 0	−0.09 ± 0
	BC	35–70+	0 ± 0	0 ± 0	0 ± 0	0 ± 0	0 ± 0	0 ± 0	0 ± 0	0 ± 0	0 ± 0	0 ± 0
30 years (hyperskeletic leptosol)												
S-A1	AC	0–10	−0.37 ± 0.24	−0.11 ± 0	−0.08 ± 0	−0.13 ± 0	−0.06 ± 0	0.29 ± 0.01	−0.17 ± 0	−0.06 ± 0	−0.14 ± 0	−0.19 ± 0
	C	10–30+	0 ± 0	0 ± 0	0 ± 0	0 ± 0	0 ± 0	0 ± 0	0 ± 0	0 ± 0	0 ± 0	0 ± 0
S-A2	C	0–50+	0 ± 0	0 ± 0	0 ± 0	0 ± 0	0 ± 0	0 ± 0	0 ± 0	0 ± 0	0 ± 0	0 ± 0
160 years (hyperskeletic leptosol)												
S-B1	OA	0–10	−0.08 ± 0.01	−0.07 ± 0	−0.02 ± 0	0 ± 0	0.01 ± 0	−0.18 ± 0	0.05 ± 0	−0.03 ± 0	−0.01 ± 0	−0.02 ± 0
	AC	10–20	−0.21 ± 0.04	0 ± 0	−0.1 ± 0	−0.02 ± 0	−0.03 ± 0	−0.08 ± 0	0.02 ± 0	−0.06 ± 0	−0.04 ± 0	−0.02 ± 0
	C1	20–40	−0.36 ± 0.06	0.11 ± 0	−0.05 ± 0	0.06 ± 0	0.09 ± 0	−0.17 ± 0	0.06 ± 0	0.04 ± 0	−0.01 ± 0	−0.01 ± 0
	C2	40–55+	0 ± 0	0 ± 0	0 ± 0	0 ± 0	0 ± 0	0 ± 0	0 ± 0	0 ± 0	0 ± 0	0 ± 0
S-B2	A	0–10	−0.06 ± 0.01	0.1 ± 0	0.07 ± 0	−0.11 ± 0	−0.11 ± 0	3.13 ± 0.06	−0.06 ± 0	0.91 ± 0	0.46 ± 0	0.02 ± 0
	C1	10–30	−0.38 ± 0.05	−0.07 ± 0	0.14 ± 0	−0.03 ± 0	−0.09 ± 0	−0.08 ± 0	−0.04 ± 0	−0.07 ± 0	−0.01 ± 0	0.01 ± 0
	C2	30–45+	0 ± 0	0 ± 0	0 ± 0	0 ± 0	0 ± 0	0 ± 0	0 ± 0	0 ± 0	0 ± 0	0 ± 0
3000 years (skeletic cambisol)												
S-C1	A	0–10	0.02 ± 0	0.09 ± 0	0.06 ± 0	−0.04 ± 0	−0.11 ± 0	0.43 ± 0.01	−0.21 ± 0	0.71 ± 0	0.3 ± 0	0.07 ± 0
	B	10–30	0.43 ± 0.07	−0.09 ± 0	0.11 ± 0	0.02 ± 0	−0.06 ± 0	−0.03 ± 0	0.04 ± 0	−0.08 ± 0	0 ± 0	0.02 ± 0
	BC	30–55+	0 ± 0	0 ± 0	0 ± 0	0 ± 0	0 ± 0	0 ± 0	0 ± 0	0 ± 0	0 ± 0	0 ± 0
S-C2	A	0–23	0.03 ± 0.01	−0.31 ± 0.01	−0.32 ± 0	−0.36 ± 0	−0.36 ± 0	0.96 ± 0.02	−0.39 ± 0	−0.19 ± 0	0.09 ± 0	−0.1 ± 0
	Bw	23–40	0.04 ± 0.01	−0.33 ± 0.01	−0.31 ± 0	−0.27 ± 0	−0.21 ± 0	−0.03 ± 0	−0.16 ± 0	−0.4 ± 0	−0.27 ± 0	−0.13 ± 0
	BC	40–60+	0 ± 0	0 ± 0	0 ± 0	0 ± 0	0 ± 0	0 ± 0	0 ± 0	0 ± 0	0 ± 0	0 ± 0
10,000 years (entic podsol)												
S-D1	O	0–20	5.18 ± 3.75	0.16 ± 0.01	−0.48 ± 0.01	−0.17 ± 0	−0.07 ± 0	0.67 ± 0.01	−0.17 ± 0	−0.39 ± 0	−0.71 ± 0	−0.47 ± 0
	OE	20–36	1.39 ± 0.38	−0.08 ± 0	−0.46 ± 0.01	−0.2 ± 0	−0.09 ± 0	0.15 ± 0	−0.18 ± 0	−0.34 ± 0	−0.67 ± 0	−0.43 ± 0
	Bs	36–63	0.36 ± 0.05	0.09 ± 0	−0.01 ± 0	−0.03 ± 0	0.03 ± 0	−0.12 ± 0	−0.02 ± 0	0.04 ± 0	0.02 ± 0	−0.01 ± 0
	BC	63–80+	0 ± 0	0 ± 0	0 ± 0	0 ± 0	0 ± 0	0 ± 0	0 ± 0	0 ± 0	0 ± 0	0 ± 0
S-D2	OE	0–35	−0.32 ± 0.11	−0.2 ± 0.01	−0.68 ± 0.01	−0.28 ± 0	−0.13 ± 0	0.33 ± 0.01	0 ± 0	−0.56 ± 0	−0.71 ± 0	−0.56 ± 0
	Bs	35–68	−0.66 ± 0.16	−0.3 ± 0.01	−0.46 ± 0.01	−0.18 ± 0	−0.08 ± 0	−0.28 ± 0	0.2 ± 0	−0.39 ± 0	−0.39 ± 0	−0.26 ± 0
	BC	68–80+	0 ± 0	0 ± 0	0 ± 0	0 ± 0	0 ± 0	0 ± 0	0 ± 0	0 ± 0	0 ± 0	0 ± 0

Table 4. Mass balances for each element for each soil profile. The lowermost horizon of each soil profile was used as the corresponding parent material for the calculations, therefore indicating a mass balance of zero.

Profile	Horizon	Depth	Mass Balance								
		cm	Na kg m^{-2}	Mg kg m^{-2}	Al kg m^{-2}	Si kg m^{-2}	P kg m^{-2}	K kg m^{-2}	Ca kg m^{-2}	Mn kg m^{-2}	Fe kg m^{-2}
110 years (hyperskeletic leptosol)											
K-A1	C1	0–10	−0.03 ± 0	0.04 ± 0	−0.32 ± 0.02	1.33 ± 0.07	0.01 ± 0	−0.08 ± 0	2.1 ± 0.11	0 ± 0	0.06 ± 0
	C2	10–20+	0 ± 0	0 ± 0	0 ± 0	0 ± 0	0 ± 0	0 ± 0	0 ± 0	0 ± 0	0 ± 0
K-A2	C1	0–10	0.02 ± 0	−0.02 ± 0	−0.08 ± 0.01	0 ± 0	0 ± 0	−0.02 ± 0	−0.62 ± 0.06	0 ± 0	−0.01 ± 0
	C2	10–20+	0 ± 0	0 ± 0	0 ± 0	0 ± 0	0 ± 0	0 ± 0	0 ± 0	0 ± 0	0 ± 0
160 years (hyperskeletic leptosol)											
K-B1	AC	0–15	0.02 ± 0	−0.03 ± 0.01	−0.05 ± 0.01	−0.06 ± 0.02	0 ± 0	0 ± 0	−1.52 ± 0.44	0 ± 0	−0.02 ± 0.01
	C	15–25+	0 ± 0	0 ± 0	0 ± 0	0 ± 0	0 ± 0	0 ± 0	0 ± 0	0 ± 0	0 ± 0
K-B2	AC	0–10	−0.04 ± 0.02	−0.05 ± 0.03	−0.04 ± 0.03	−0.19 ± 0.11	0 ± 0	−0.01 ± 0	−2.13 ± 1.24	0 ± 0	−0.03 ± 0.02
	C	10–28+	0 ± 0	0 ± 0	0 ± 0	0 ± 0	0 ± 0	0 ± 0	0 ± 0	0 ± 0	0 ± 0
4900 years (calcaric skeletic cambisol)											
K-C1	A	0–15	0.05 ± 0.01	−0.19 ± 0.06	0.02 ± 0.01	0.37 ± 0.11	−0.03 ± 0.01	0.05 ± 0.01	−9.02 ± 2.69	0.01 ± 0	0.02 ± 0.01
	B	15–21	0.02 ± 0	−0.09 ± 0.01	0.01 ± 0	0.01 ± 0	−0.01 ± 0	0.02 ± 0	−4.07 ± 0.63	0 ± 0	0.01 ± 0
	BC	21–60	−0.08 ± 0.01	−0.05 ± 0	−0.01 ± 0	0.05 ± 0	0 ± 0	0 ± 0	−1.51 ± 0.12	0 ± 0	0.05 ± 0
	C	60–80+	0 ± 0	0 ± 0	0 ± 0	0 ± 0	0 ± 0	0 ± 0	0 ± 0	0 ± 0	0 ± 0
K-C2	OA	0–18	0.12 ± 0.01	−0.34 ± 0.01	−0.08 ± 0	0.31 ± 0.01	−0.04 ± 0	0.05 ± 0	−15.8 ± 0.4	0 ± 0	−0.04 ± 0
	B	18–42	0.08 ± 0.01	−0.33 ± 0.05	0.04 ± 0.01	0.11 ± 0.02	−0.04 ± 0.01	0.08 ± 0.01	−15.55 ± 2.18	0.01 ± 0	0.03 ± 0
	BC	42–51	−0.01 ± 0	−0.09 ± 0	−0.05 ± 0	−0.43 ± 0	−0.01 ± 0	−0.02 ± 0	−3.45 ± 0.01	0 ± 0	−0.07 ± 0
	C	51–75+	0 ± 0	0 ± 0	0 ± 0	0 ± 0	0 ± 0	0 ± 0	0 ± 0	0 ± 0	0 ± 0
13,500 years (calcaric skeletic cambisol)											
K-D1	A1	0–15	0.01 ± 0	−0.07 ± 0.01	0 ± 0	0.08 ± 0.01	−0.01 ± 0	0.01 ± 0	−3.31 ± 0.53	0 ± 0	−0.02 ± 0
	A2	15–28	0.01 ± 0	−0.16 ± 0.02	0.01 ± 0	0.25 ± 0.03	−0.02 ± 0	0.02 ± 0	−7.88 ± 0.82	0 ± 0	−0.03 ± 0
	B	28–32	−0.01 ± 0	−0.04 ± 0.01	0.01 ± 0	0.16 ± 0.04	−0.01 ± 0	0 ± 0	−2.03 ± 0.57	0 ± 0	0.01 ± 0
	BC1	32–68	−0.1 ± 0.01	0.43 ± 0.04	0.04 ± 0	−0.63 ± 0.06	0.05 ± 0	−0.01 ± 0	20.22 ± 1.85	0 ± 0	−0.19 ± 0.02
	BC2	68–80+	0 ± 0	0 ± 0	0 ± 0	0 ± 0	0 ± 0	0 ± 0	0 ± 0	0 ± 0	0 ± 0
K-D2	OA	0–30	0.66 ± 0.15	−3.04 ± 0.68	0.84 ± 0.19	4.01 ± 0.89	−0.41 ± 0.09	0.78 ± 0.17	−153.33 ± 34.16	−0.01 ± 0	−0.55 ± 0.12
	B	30–35	0.01 ± 0	−0.02 ± 0	0.01 ± 0	0 ± 0	0 ± 0	0 ± 0	−1 ± 0.07	0 ± 0	−0.02 ± 0
	BC	35–70+	0 ± 0	0 ± 0	0 ± 0	0 ± 0	0 ± 0	0 ± 0	0 ± 0	0 ± 0	0 ± 0

Table 4. *Cont.*

Profile	Horizon	Depth	Mass Balance								
		cm	Na kg m^{-2}	Mg kg m^{-2}	Al kg m^{-2}	Si kg m^{-2}	P kg m^{-2}	K kg m^{-2}	Ca kg m^{-2}	Mn kg m^{-2}	Fe kg m^{-2}
30 years (hyperskeletic leptosol)											
S-A1	AC	0–10	−0.13 ± 0.09	−0.06 ± 0.04	−0.47 ± 0.33	−0.99 ± 0.7	0 ± 0	−0.21 ± 0.15	−0.04 ± 0.02	0 ± 0	−0.27 ± 0.19
	C	10–30+	0 ± 0	0 ± 0	0 ± 0	0 ± 0	0 ± 0	0 ± 0	0 ± 0	0 ± 0	0 ± 0
S-A2	C	0–50+	0 ± 0	0 ± 0	0 ± 0	0 ± 0	0 ± 0	0 ± 0	0 ± 0	0 ± 0	0 ± 0
160 years (hyperskeletic leptosol)											
S-B1	OA	0–10	−0.14 ± 0.03	−0.02 ± 0	0.01 ± 0	0.19 ± 0.05	−0.01 ± 0	0.09 ± 0.02	−0.03 ± 0.01	0 ± 0	−0.03 ± 0.01
	AC	10–20	−0.01 ± 0	−0.11 ± 0.03	−0.13 ± 0.03	−0.87 ± 0.22	0 ± 0	0.04 ± 0.01	−0.07 ± 0.02	0 ± 0	−0.06 ± 0.01
	C1	20–40	0.62 ± 0.15	−0.13 ± 0.03	0.9 ± 0.22	7.13 ± 1.74	−0.02 ± 0	0.29 ± 0.07	0.11 ± 0.03	0 ± 0	−0.05 ± 0.01
	C2	40–55+	0 ± 0	0 ± 0	0 ± 0	0 ± 0	0 ± 0	0 ± 0	0 ± 0	0 ± 0	0 ± 0
S-B2	A	0–10	0.14 ± 0.03	0.04 ± 0.01	−0.44 ± 0.1	−2.18 ± 0.49	0.07 ± 0.02	−0.08 ± 0.02	0.65 ± 0.15	0.01 ± 0	0.03 ± 0.01
	C1	10–30	−0.3 ± 0.05	0.29 ± 0.05	−0.32 ± 0.05	−5.2 ± 0.89	−0.01 ± 0	−0.17 ± 0.03	−0.15 ± 0.03	0 ± 0	0.05 ± 0.01
	C2	30–45+	0 ± 0	0 ± 0	0 ± 0	0 ± 0	0 ± 0	0 ± 0	0 ± 0	0 ± 0	0 ± 0
3000 years (skeletic cambisol)											
S-C1	A	0–10	0.14 ± 0.02	0.06 ± 0.01	−0.19 ± 0.02	−2.51 ± 0.29	0.01 ± 0	−0.31 ± 0.04	0.64 ± 0.07	0.01 ± 0	0.19 ± 0.02
	B	10–30	−0.21 ± 0.03	0.15 ± 0.02	0.12 ± 0.02	−2.03 ± 0.31	0 ± 0	0.08 ± 0.01	−0.1 ± 0.02	0 ± 0	0.06 ± 0.01
	BC	30–55+	0 ± 0	0 ± 0	0 ± 0	0 ± 0	0 ± 0	0 ± 0	0 ± 0	0 ± 0	0 ± 0
S-C2	A	0–23	−0.63 ± 0.34	−0.49 ± 0.27	−2.07 ± 1.13	−8.49 ± 4.61	0.04 ± 0.02	−0.59 ± 0.32	−0.33 ± 0.18	0.01 ± 0	−0.32 ± 0.17
	Bw	23–40	−0.49 ± 0.11	−0.35 ± 0.08	−1.13 ± 0.25	−3.59 ± 0.79	0 ± 0	−0.17 ± 0.04	−0.51 ± 0.11	−0.01 ± 0	−0.3 ± 0.07
	BC	40–60+	0 ± 0	0 ± 0	0 ± 0	0 ± 0	0 ± 0	0 ± 0	0 ± 0	0 ± 0	0 ± 0
10,000 years (entic podsol)											
S-D1	O	0–20	0.08 ± 0.06	−0.21 ± 0.16	−0.31 ± 0.23	−0.51 ± 0.37	0.02 ± 0.01	−0.1 ± 0.07	−0.12 ± 0.09	−0.02 ± 0.01	−0.65 ± 0.47
	OE	20–36	−0.09 ± 0.02	−0.42 ± 0.12	−0.78 ± 0.23	−1.34 ± 0.39	0.01 ± 0	−0.22 ± 0.06	−0.22 ± 0.06	−0.03 ± 0.01	−1.21 ± 0.35
	Bs	36–63	0.32 ± 0.06	−0.04 ± 0.01	−0.36 ± 0.06	1.67 ± 0.3	−0.02 ± 0	−0.08 ± 0.01	0.1 ± 0.02	0 ± 0	−0.07 ± 0.01
	BC	63–80+	0 ± 0	0 ± 0	0 ± 0	0 ± 0	0 ± 0	0 ± 0	0 ± 0	0 ± 0	0 ± 0
S-D2	OE	0–35	−0.62 ± 0.23	−2.6 ± 0.95	−3.03 ± 1.11	−4.58 ± 1.67	0.05 ± 0.02	0.01 ± 0	−1.12 ± 0.41	−0.1 ± 0.04	−5.06 ± 1.85
	Bs	35–68	−1.72 ± 0.49	−3.24 ± 0.91	−3.58 ± 1	−5.63 ± 1.58	−0.07 ± 0.02	0.83 ± 0.23	−1.43 ± 0.4	−0.1 ± 0.03	−4.28 ± 1.2
	BC	68–80+	0 ± 0	0 ± 0	0 ± 0	0 ± 0	0 ± 0	0 ± 0	0 ± 0	0 ± 0	0 ± 0

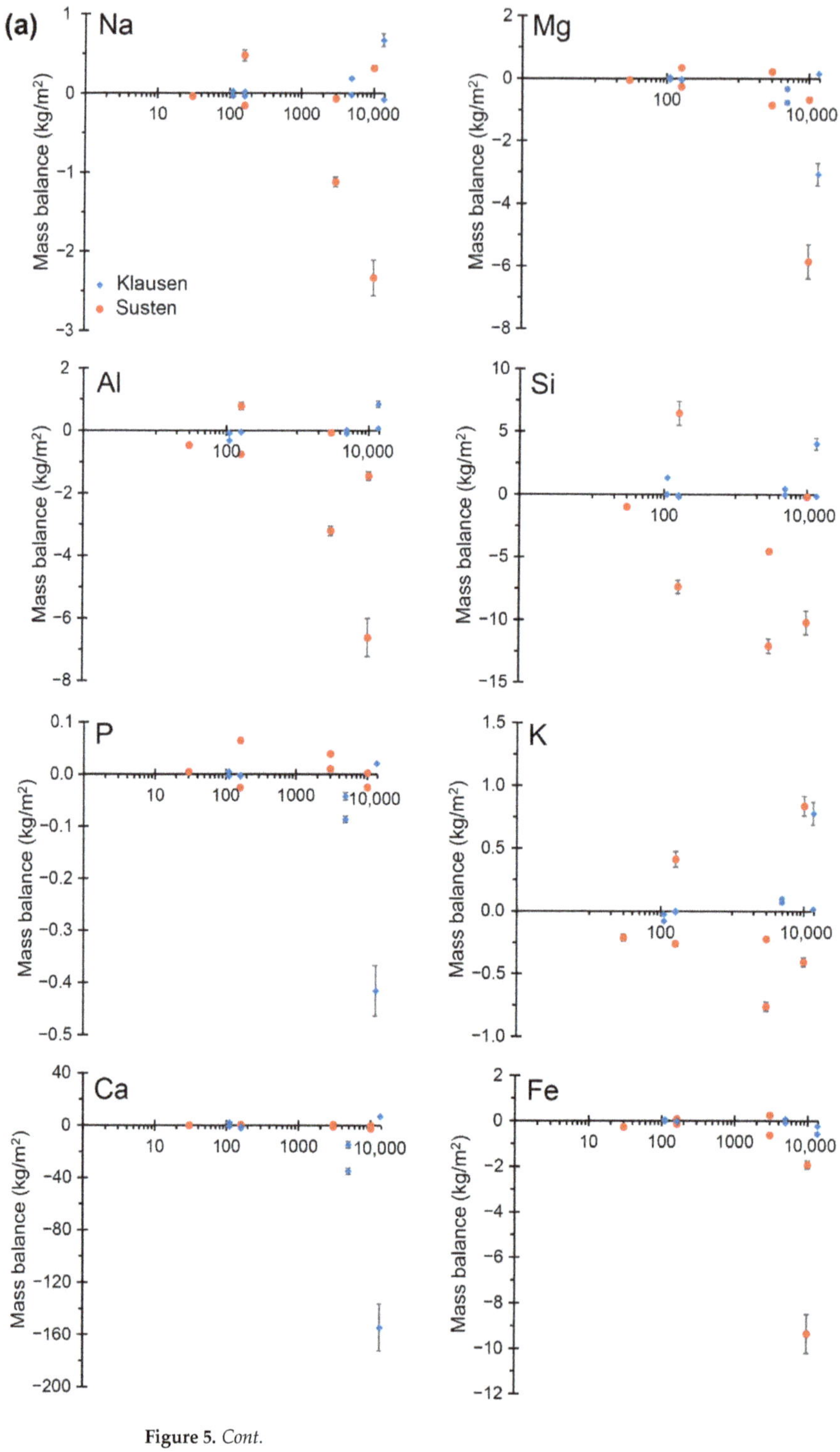

Figure 5. *Cont.*

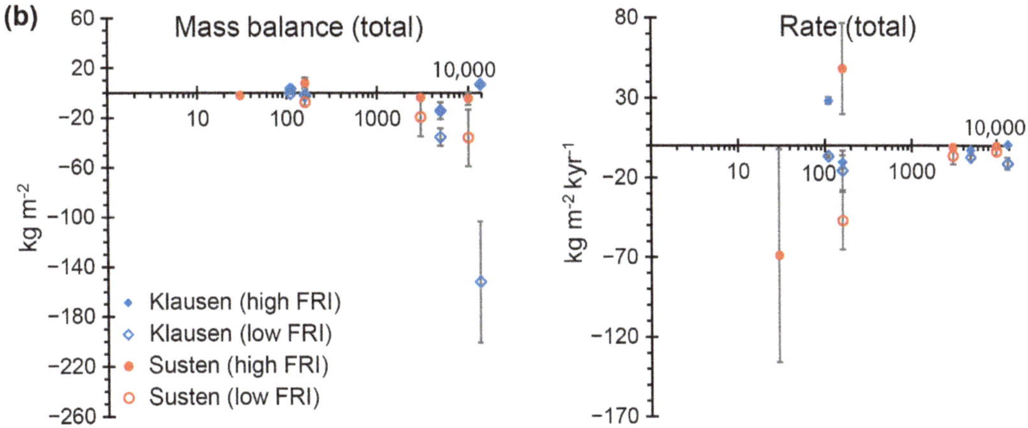

Figure 5. (**a**) Mass balances for Sustenpass and Klausenpass. The balances were calculated relative to the bottommost horizon of each soil profile. There are two soil profiles per moraine except for the 30 a moraine at Sustenpass, because profile S-A2 consisted of only one horizon and no mass balance could be calculated. (**b**) Total mass balances for Sustenpass and Klausenpass (includes Na, Mg, Al, Si, P, K, Ca, Mn, and Fe). The balances were calculated relative to bottommost horizon of each soil profile. There are two soil profiles per moraine except for the 30 a moraine at Sustenpass, because profile S-A2 consisted of only one horizon and no mass balance could be calculated.

The C_{org} stocks (Figure 6) at the youngest soils are 0.4 ± 0.0–0.8 ± 0.0 kg m^{-2} and increase to around 5.3 ± 0.1–15.9 ± 0.0 kg m^{-2} after 4.9 ka, where the stocks level off to 11.6 ± 0.0–12.5 ± 0.0 kg m^{-2} at the 13.5 ka moraine.

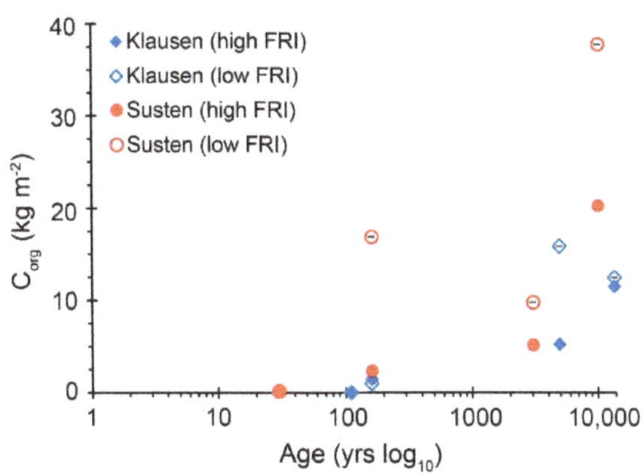

Figure 6. Stocks of organic carbon (kg m^{-2}) for the soil profiles at each moraine. FRI = functional richness index.

The weathering indices (Figure 7a) show strong alterations in the uppermost 20 cm of the soils. The variability between the soils at each moraine becomes slightly larger with age.

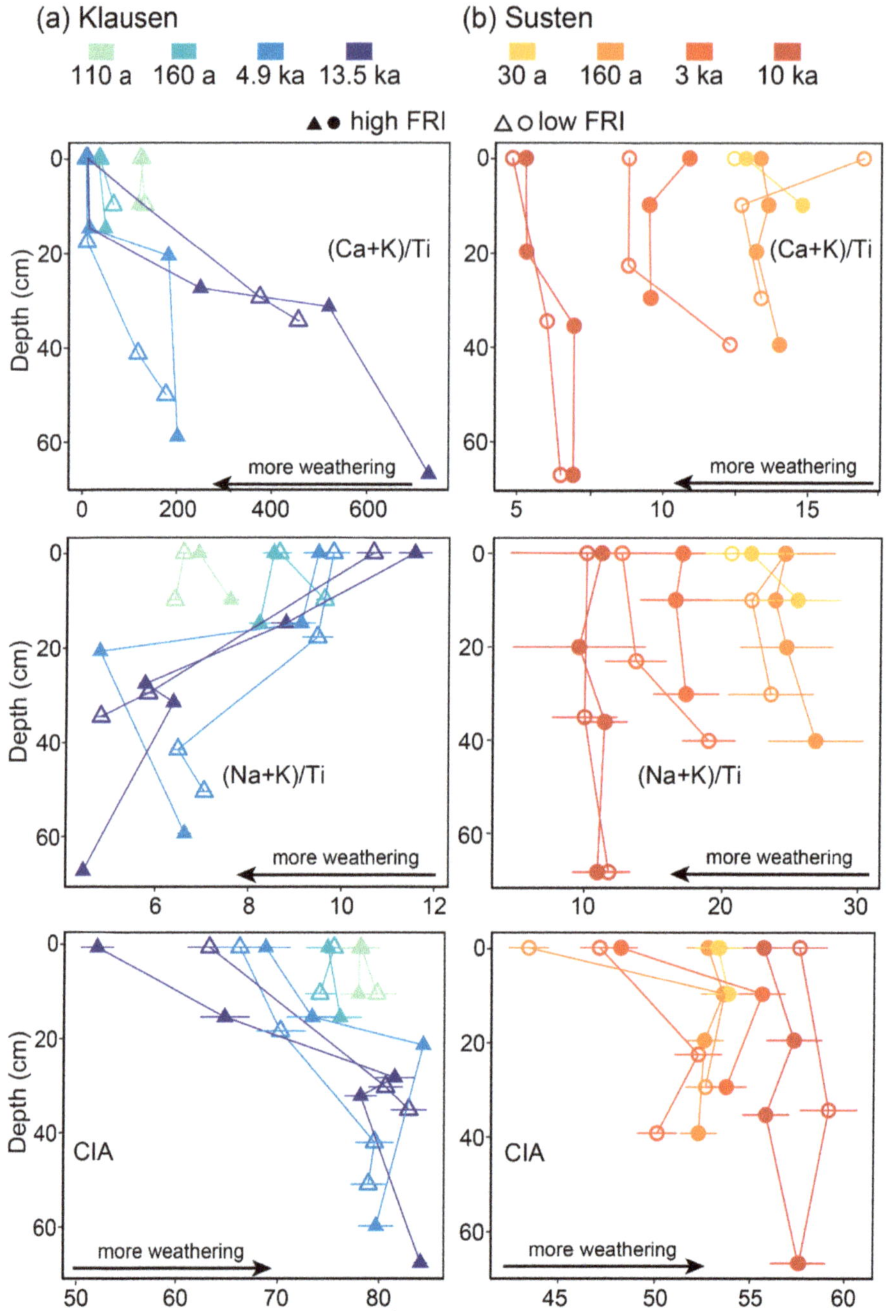

Figure 7. Weathering indices. The (Ca + K)/Ti ratio was calculated with total Ca in order to reflect the dissolution of calcium carbonate. In contrast, (Na + K)/Ti is not affected by carbonates. The Chemical Index of Weathering (CIA) reflects the loss of Al_2O_3, CaO, K_2O, and Na_2O relative to Al_2O_3.

The oxalate extractable forms of Fe, Al, and Mn increase throughout the chronosequence (Table 5, Figure 8a). The proportion of Fe_o/Fe_{tot} increases from ca 6% to7% in the young topsoils to 14 ± 1.8–22 ± 1.0% in the 4.9 ka soil, and decreases slightly in the 13.5 ka moraine (due to an increase of total Fe). The Mn_o/Mn_{tot} ratio increases sharply early on, reaching over 50% in the oldest soils. The Al_o/Al_{tot} ratio increases steadily to a maximum of around 4.8% in the older moraines.

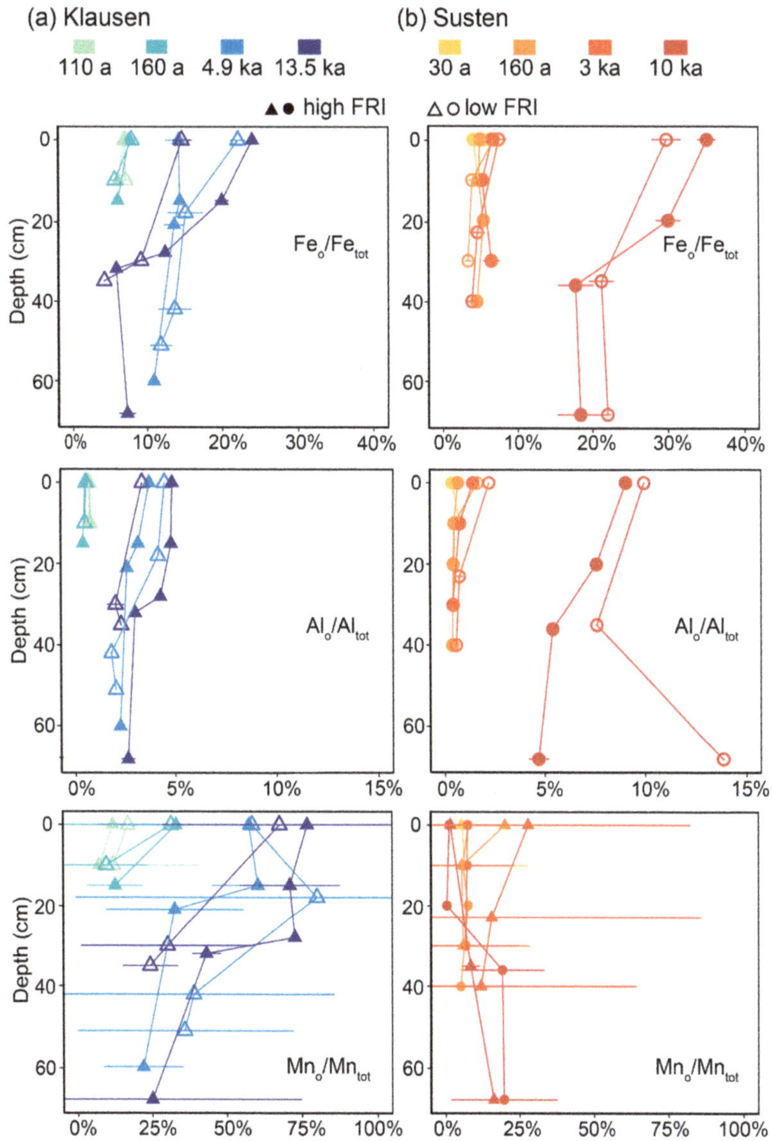

Figure 8. Oxalate extractable Fe_o, Al_o, and Mn_o (measured with atomic absorbance spectroscopy, AAS) as a fraction of the total contents, Fe_{tot}, Al_{tot}, and Mn_{tot} (measured with X-ray fluorescence, XRF). The error bars represent the measurement uncertainties.

Table 5. Organic matter and oxalate extractable Fe, Al, and Mn of the fine earth. OM (organic matter) content was derived from LOI measurements. C_{org}, N, and C/N are based on EA-IRMS measurements (C_{org} was calculated as C_{tot} minus C_{inorg}). The oxalate extractable fractions, Fe_o, Al_o, and Mn_o were measured with Flame-AAS.

Profile	Horizon	Depth	OM	C_{org}	N	C/N	Fe_o	Al_o	Mn_o
		cm	g kg^{-1}	g kg^{-1}	g kg^{-1}		mg kg^{-1}	mg kg^{-1}	mg kg^{-1}
Klausenpass (Griess Glacier)									
110 years									
K-A1	C1	0–10	17.51	0.96 ± 0	0.39 ± 0	2.87 ± 0	1494 ± 124	206 ± 0	32 ± 2
	C2	10–20+	19.83	0.29 ± 0	0.31 ± 0	1.11 ± 0	1390 ± 169	266 ± 0	19 ± 0
K-A2	C1	0–10	18.95	0.78 ± 0	0.35 ± 0.02	2.56 ± 0.04	1571 ± 121	191 ± 0	53 ± 2
	C2	10–20+	17.91	0	0.24 *	N/A	1545 ± 271	212 ± 0	32 ± 5
160 years									
K-B1	AC	0–15	83.85	26.99 ± 0.18	2.2 ± 0.05	14.33 ± 0.02	2751 ± 158	293 ± 0	149 ± 26
	C	15–25+	33.44	5.74 ± 0.06	0.68 ± 0.01	9.81 ± 0.01	2007 ± 91	215 ± 4	52 ± 14
K-B2	AC	0–10	89.29	22.07 ± 3.92	1.89 ± 0.33	13.62 ± 0.25	2767 ± 331	309 ± 48	157 ± 7
	C	10–28+	32.15	4.84 ± 0.02	0.67 ± 0.02	8.39 ± 0.02	1665 ± 137	229 ± 0	41 ± 6
4900 years									
K-C1	A	0–15	187.06	71.2 ± 0.02	6.71 ± 0.05	12.38 ± 0.01	6621 ± 831	2578 ± 183	671 ± 180
	B	15–21	99.65	25.61 ± 0.07	3.35 ± 0.13	8.93 ± 0.04	7416 ± 269	2379 ± 211	758 ± 282
	BC	21–60	27.37	0.94 ± 0	0.68 ± 0.01	1.62 ± 0.02	2451 ± 236	667 ± 0	93 ± 0
	C	60–80+	22.32	0	0.42 ± 0.01	N/A	1739 ± 115	544 ± 0	47 ± 0
K-C2	OA	0–18	344.02	150.89 ± 2.98	11.29 ± 0.16	15.59 ± 0.02	8225 ± 377	2564 ± 0	405 ± 57
	B	18–42	140.03	45.9 ± 0.17	5.07 ± 0.02	10.56 ± 0.01	7806 ± 1193	3317 ± 0	1040 ± 122
	BC	42–51	37.64	3.88 ± 0	0.92 ± 0.02	4.9 ± 0.02	2912 ± 475	630 ± 18	138 ± 11
	C	51–75+	27.33	2.12 ± 0	0.67 ± 0.03	3.68 ± 0.04	2065 ± 245	549 ± 6	102 ± 13
13,500 years									
K-D1	A1	0–15	424.36	185.86 ± 1.65	14.44 ± 0.19	15.01 ± 0.02	9993 ± 117	2059 ± 0	750 ± 88
	A2	15–28	204.90	78.91 ± 2.5	7.66 ± 0.15	12.03 ± 0.04	11,423 ± 528	2661 ± 132	724 ± 141
	B	28–32	57.96	13.82 ± 0.05	1.69 ± 0.02	9.52 ± 0.01	2748 ± 6	795 ± 0	262 ± 51
	BC1	32–68	16.44	0	0.44 ± 0	N/A	465 ± 3	226 ± 0	51 ± 10
	BC2	68–80+	16.36	0	0.42 ± 0.01	N/A	880 ± 128	260 ± 34	47 ± 5
K-D2	O/OA	0–30	244.66	99.66 ± 232.56	8.21 ± 0.09	14.17 ± 2.33	7092 ± 598	2102 ± 0	456 ± 2
	A	30–35	22.98	3.06 ± 3.37	0.63 ± 0.03	5.69 ± 1.1	1120 ± 99	279 ± 58	46 ± 2
	BC	35–70+	16.75	0 0	0.35 ± 0	N/A	487 ± 8	261 ± 0	39 ± 1

Table 5. *Cont.*

Profile	Horizon	Depth	OM	C_{org}	N	C/N	Fe_o	Al_o	Mn_o
		cm	g kg^{-1}	g kg^{-1}	g kg^{-1}		mg kg^{-1}	mg kg^{-1}	mg kg^{-1}
Sustenpass (Stein Glacier)									
30 years									
S-A1	AC	0–10	11.40	0.97 ± 0.19	0.11 ± 0	10.21 ± 0.2	1103 ± 147	266 ± 0	30 ± 0
	C	10–30+	13.46	0.26 ± 0.69	N/A	N/A	1436 ± 230	575 ± 0	35 ± 0
S-A2	C	0–50+	8.39	0.39 ± 3.49	N/A	N/A	1460 ± 243	413 ± 0	27 ± 1
160 years									
S-B1	OA	0–10	44.35	2.07 ± 2.95	0.3 ± 0.01	8.2 ± 1.42	1184 ± 107	433 ± 0	33 ± 0
	AC	10–20	17.48	1.4 *	0.19 *	8.73 *	1160 ± 58	308 ± 0	26 ± 0
	C1	20–40	13.04	1.21 *	0.11 *	12.43 *	1342 ± 15	290 ± 0	34 ± 0
	C2	40–55+	10.95	7.37 ± 19.26	1.24 ± 0.02	6.93 ± 2.61	1033 ± 90	256 ± 0	21 ± 1
S-B2	A	0–10	384.24	105.65 ± 12.63	12.48 ± 0.01	9.88 ± 0.12	1228 ± 140	653 ± 0	97 ± 4
	C1	10–30	17.53	1.96 *	0.29 *	7.9 *	1111 ± 138	337 ± 60	29 ± 3
	C2	30–45+	14.61	1.03 *	0.21 *	5.72 *	885 ± 112	286 ± 0	29 ± 1
3000 years									
S-C1	A	0–10	34.09	N/A	N/A	N/A	2829 ± 283	1009 ± 26	64 ± 2
	B	10–30	27.96	2.53 ± 0.71	0.57 ± 0	5.14 ± 0.28	2104 ± 217	549 ± 0	47 ± 0
	BC	30–55+	19.46	1.66 ± 5.55	0.21 ± 0.01	9.3 ± 3.35	2451 ± 414	283 ± 83	42 ± 2
S-C2	A	0–23	225.12	27.96 ± 4.43	8.31 ± 0.08	3.93 ± 0.16	3238 ± 469	1274 ± 0	297 ± 49
	Bw	23–40	130.40	10.6 ± 7.65	4 ± 0.03	3.09 ± 0.72	1862 ± 383	466 ± 100	108 ± 44
	BC	40–60+	24.39	1.34 ± 0.39	0.53 ± 0.01	2.95 ± 0.29	1565 ± 264	435 ± 0	98 ± 20
10,000 years									
S-D1	O	0–20	439.81	54.34 ± 142.12	13.71 ± 0.39	4.62 ± 2.62	6726 ± 232	3797 ± 110	3 ± 0
	OE	20–36	287.40	70.42 ± 65.93	9.26 ± 0.1	8.88 ± 0.94	8226 ± 438	4023 ± 0	1 ± 0
	Bs	36–63	52.24	6.51 *	0.71 *	10.78 *	9394 ± 1206	3874 ± 0	183 ± 18
	BC	63–80+	61.37	10.3 *	1.04 *	11.6 *	9887 ± 1655	3474 ± 362	186 ± 4
S-D2	OE	0–35	379.81	159.29 ± 25.03	11.83 ± 0.01	15.71 ± 0.16	7347 ± 469	4773 ± 0	4 ± 2
	Bs	35–68	99.77	21.03 *	2.09 *	11.75 *	11,650 ± 860	5502 ± 0	59 ± 2
	BC	68–80+	159.53	11.63 *	2.83 *	4.79 *	13,020 ± 230	9754 ± 0	146 ± 5

N/A = no data available; * no replicate measurements.

4.3.2. Sustenpass

The element contents are given in Table 2. As the pH of the Sustenpass soils was below 6.5 in most soil horizons (Table 1), total Ca corresponds to the silicate form CaO. The strains and the results of the open-system mass transfer function (τ) for the most abundant elements are shown in Tables 3 and 4. The total element range from +8 $\pm$ 5 to -36 ± 23 kg m^{-2}, resulting in high weathering rates at the youngest moraine (-69 ± 67 kg m^{-2} $kyear^{-1}$) which decrease strongly with age to 0 $\pm$ 1 to -4 ± 2 m^{-2} $kyear^{-1}$ (Figure 5a,b).

The C_{org} stocks of the soil profiles (Figure 6) increased at high rates with the youngest soils, starting from 0.17 $\pm$ 0.0–0.15 $\pm$ 0.0 kg m^{-2} (at the 30 a moraine) to 2.4 $\pm$ 0.0–16.9 $\pm$ 0.0 kg m^{-2} (160 a moraine). The highest C_{org} stocks occurred in the 10 ka moraine (20.3 $\pm$ 0.0–37.8 $\pm$ 0.0 kg m^{-2}).

Both the (Ca + K)/Ti and the (Na + K)/Ti ratios (Figure 7b) show similar patterns among the Sustenpass soils. The soils of each moraine show a homogenous profile and the soils are clearly grouped by age. The Chemical Index of Alteration (CIA) mostly increases with soil age. The variability is however high among the top horizons, particularly at the 160 a moraine site.

The ratio of Fe_o/Fe_{tot} only increased slowly; fluctuating at around 5–6 $\pm$ 0.5% in the first 3 ka of soil development. Between 3 ka and 10 ka, the ratio sharply increased to 29 $\pm$ 1.9–35 $\pm$ 1.2% in the topsoil (Figure 8b). Similarly, the Al_o/Al_{tot} ratio changed little during the first 3 ka of soil development (0.4–2.3 $\pm$ 0.2% in the topsoil) and then increased to 9–10% in the 10 ka soils. The Mn_o/Mn_{tot} ratio increased too with age, although the variability is very high.

4.4. *Mineralogy*

The mineral composition of the Klausenpass fine earth (Figure 9a) at the 110 a moraine consists mostly of calcite (59%), quartz (22–24%), mica (8–10%) chlorite (10%), and a small amount (3%) of feldspar. The 160 a soil contained 25–36% calcite, 28–33% quartz, 21–23% mica, and 10% chlorite. The 4.9 ka soil contained 0–79% calcite, 11–50% quartz, 7–33% mica, and 2–9% chlorite. Lastly, the 13.5 ka soil contained 0–95% calcite, 3–53% quartz, 2–31% mica, and 0–7% chlorite. The errors of the measurements were <1% throughout.

The mineral composition of the 30 a soil at Sustenpass (Figure 9a) was mainly quartz (30–33%), plagioclase (32–33%) and potassium feldspar (6%), mica (16–18%), and chlorite (5–7%). The 160 a moraine contained 34–36% quartz, 32–33% plagioclase and 6% potassium feldspar, 15–16% mica, and 5–6% chlorite. The 3 ka moraine contained 26–30% quartz, 28–31% plagioclase and 4–5% potassium feldspar, 16–19% mica, and 9–22% chlorite. The 10 ka moraine contained 25–38% quartz, 21–22% plagioclase and 4–5% potassium feldspar, 14–20% mica, and 3–29% chlorite. The errors of these measurements were <1%.

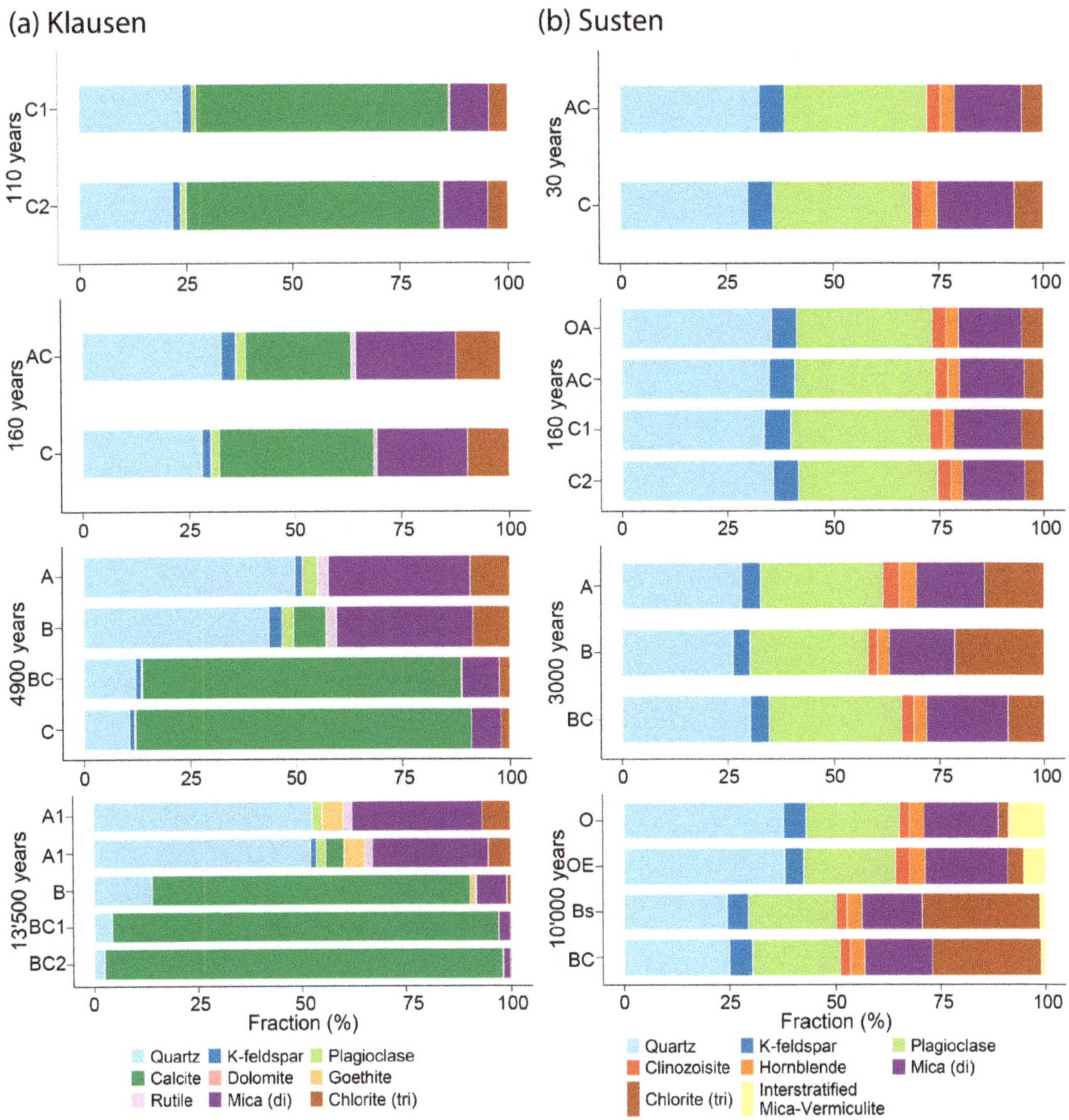

Figure 9. Bulk mineralogical composition of the fine earth at (**a**) Klausenpass and (**b**) Sustenpass for each horizon measured with XRD. (Based on one soil profile per moraine. For consistency, all soil profiles selected were from the high functional richness index (FRI) sites). di = dioctahedral, tri = trioctahedral. The XRD measurement uncertainties were <1% throughout.

4.5. Meteoric ^{10}Be Inventories and Erosion Rates

The ^{10}Be concentration in the fine earth at Klausenpass ranged from $0.63 \pm 0.01 \times 10^8$ to $13.32 \pm 0.31 \times 10^8$ at g^{-1} in the 4.9 ka soil and $1.46 \pm 0.02 \times 10^8$ to $17.39 \pm 0.46 \times 10^8$ at g^{-1} in the 13.5 ka soil (Figure 10, Table 6).

Table 6. Long-term erosion rates derived from meteoric ^{10}Be in soils. Different equations and q values from literature were used.

Profile	Horizon	Thickness	Bulk Density	Fine Earth	Skeleton	^{10}Be (conc.)	^{10}Be (per Horizon)	Erosion Rates Equation (10) [1]	Equation (10) [2]	Equation (10) [3]	Equation (13)
		(cm)	($g\ cm^{-3}$)	($g\ cm^{-2}$)	(wt. %)	($at\ g^{-1}\ 10^8$)	($at\ cm^{-2}\ 10^8$)	($t\ ha^{-1}\ year^{-1}$)	($t\ ha^{-1}\ year^{-1}$)	($t\ ha^{-1}\ year^{-1}$)	($t\ ha^{-1}\ year^{-1}$)
	Klausenpass 4900 years										
K-C1	A	15	0.50	7.45	11.31	13.32 ± 0.63	87.94 ± 4.15	0.26 ± 0.15	−0.30 ± 0.17	0.17 ± 0.10	−80 ± 0.10
	B	6	0.74	4.44	64.47	10.84 ± 0.48	17.11 ± 0.76				
	BC	39	1.53	59.62	77.93	1.12 ± 0.07	14.76 ± 0.88				
	C	20	1.48	29.53	85.30	0.63 ± 0.03	2.75 ± 0.12				
	Klausenpass 13,500										
K-D1	A1	15	0.33	4.94	6.88	15.29 ± 0.53	70.33 ± 2.44	0.01 ± 0.00	−0.38 ± 0.08	−0.05 ± 0.010	−0.15 ± 0.02
	A2	13	0.48	6.19	40.85	17.39 ± 0.91	63.73 ± 3.34				
	B	4	0.81	3.22	73.46	5.31 ± 0.15	4.54 ± 0.13				
	BC1	36	1.63	58.63	75.00	1.26 ± 0.04	18.5 ± 0.59				
	BC2	12	1.65	19.81	75.00	1.46 ± 0.04	7.24 ± 0.21				
	Sustenpass 3000 years										
S-C1	A	10	1.31	13.06	50.41	1.04 ± 0.03	6.76 ± 0.23	−2.64 ± 0.38	−14.11 ± 2.03	−4.58 ± 0.66	−11.26 ± 0.20
	B	20	1.51	30.28	68.90	0.3 ± 0.01	2.85 ± 0.12				
	BC	25	1.90	47.57	63.43	0.17 ± 0.01	2.89 ± 0.19				
	Sustenpass 10,000 years										
S-D1	O	20	0.73	14.62	74.37	3.55 ± 0.1	13.3 ± 0.39	−0.60 ± 0.20	−5.27 ± 1.75	−1.39 ± 0.46	−1.95 ± 0.27
	OE	16	0.71	11.36	47.94	3.31 ± 0.1	19.58 ± 0.57				
	Bs	32	1.29	41.16	54.76	1.31 ± 0.05	24.31 ± 0.88				
	BC	12	1.35	16.22	41.75	1.55 ± 0.06	14.68 ± 0.54				

Preparation blank (long-term average). [1] $q = 1.1 \times 10^6\ atm\ cm^{-2}\ year^{-1}$ [45]; [2] $q = 4.6 \times 106\ atm\ cm^{-2}\ year^{-1}$ [85]; [3] $q = 1.6 \times 106\ atm\ cm^{-2}\ year^{-1}$ [86].

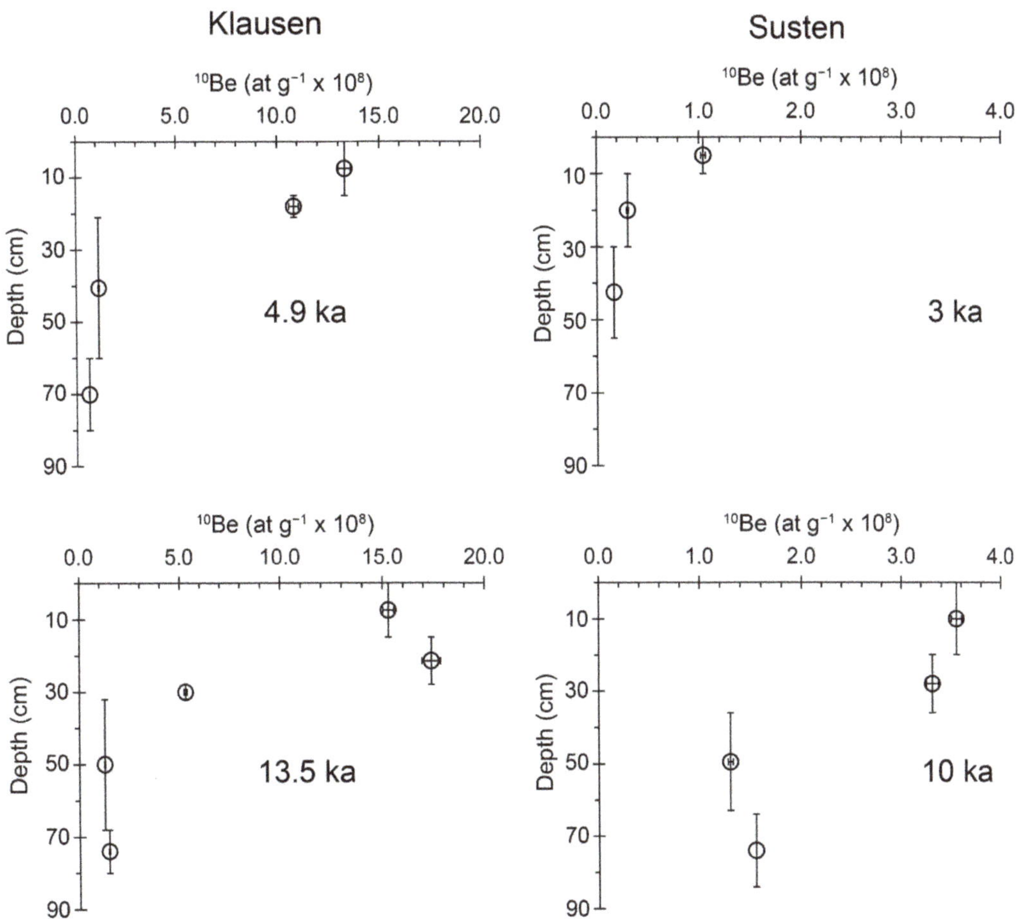

Figure 10. ^{10}Be concentrations at the oldest two moraines of Klausenpass and Sustenpass measured in atoms per gram of soil (fine earth). The horizontal error bars represent the measurement uncertainty of the ^{10}Be and the vertical errors represent the thickness of the horizon.

At Sustenpass, the ^{10}Be concentration ranged from 0.17 ± 0.00 to $1.04 \pm 0.02 \times 10^8$ at g^{-1} in the 3 ka soil and 1.31 ± 0.02 to 3.55 ± 0.05 at g^{-1} $\times$ 10^8 at g^{-1} in the 10 ka soil (Figure 10, Table 6).

4.6. Flow Accumulation

The GIS analysis to model the flow accumulation (Figure 11) produced a homogeneous pattern on the Klausenpass slopes. There, the surface runoff is dispersed with a low intensity. The Sustenpass slopes have a runoff pattern that indicates channeling and more intense flow accumulation (dark blue) of the surface water runoff. The same patterns also emerged in the younger moraines.

Figure 11. Output of the flow accumulation on top of high-resolution drone imagery (**a**) Klausenpass and (**b**) Sustenpass. North direction is upward in all images. The darker blue shades indicate stronger flow accumulation.

5. Discussion

5.1. Soil Development in Calcareous Parent Material

Weathering in calcareous soils is primarily characterized by carbonate leaching [95]. The values of carbonate leaching in the study area generally fit to the range of observations from other calcareous soils in the Alps [24,96]. The weathering front seems to advance in the soil column 'stepwise'. This leads to a divergence of top- and subsoil properties, where the topsoil is strongly altered, while the subsoil has changed very little or might have accumulated elements that were leached above (such as Ca). We observed this in the pH (drop of pH to <6 in the topsoil), mineralogy (dissolution of calcite), weathering indices and τ values (high element losses in the topsoil) as well as the oxalate extractable fractions (accumulation of Fe_o, Al_o and Mn_o). Soil texture data (Figures 12 and 13, obtained from [54] at the same moraines) showed that the topsoils at this location were altered very strongly with time, shifting from sandy loam to a silty clay loam texture.

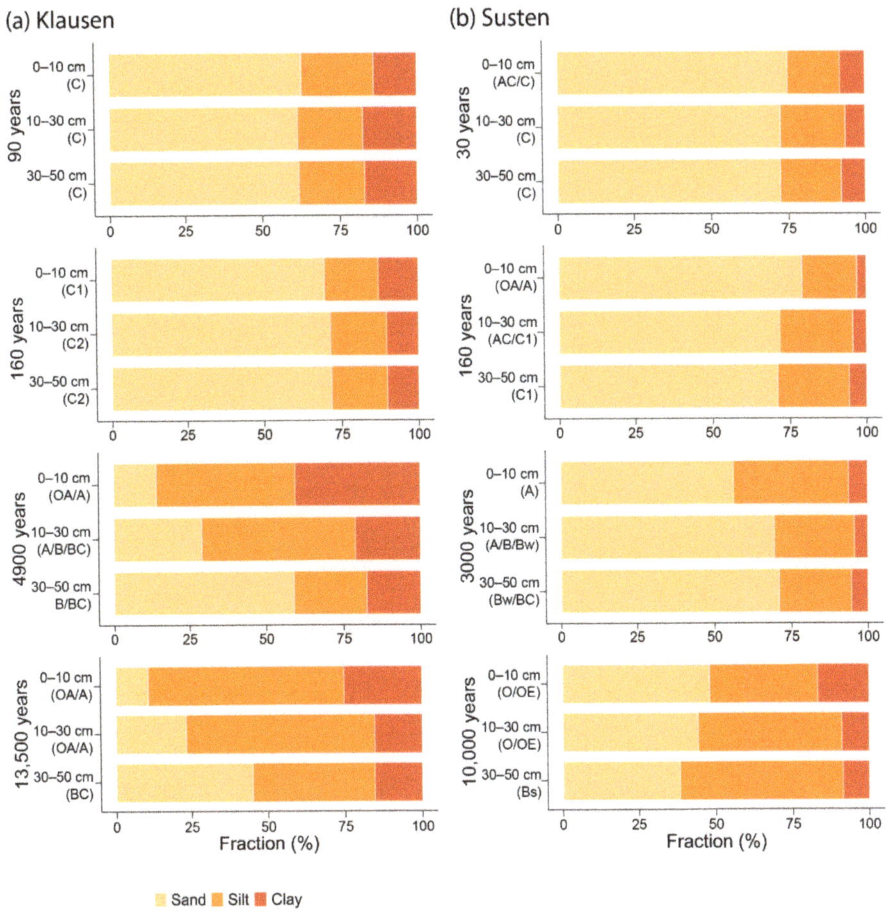

Figure 12. Particle size distributions for (**a**) Klausenpass and (**b**) Sustenpass. The particle size distributions were measured at fixed depths and with six replicates (here, the averages are shown [54]). The corresponding horizons are written in parentheses underneath (from Figure 3). Sand: 63–2000 μm, Silt: 2–63 μm, Clay: <2 μm. (Data reproduced with permission of [55]).

Despite 110 and 160 years of rapid soil carbonate leaching, the profiles of the youngest two moraines showed almost no carbonate losses. We calculated Ca mass balances between +2 and −2 kg m^{-2} which would translate to +6 to −6 kg m^{-2} $CaCO_3$. In comparison, [24] calculated losses of 15–30 kg m^{-2} in 193 a soils. The 4.9 and 13.5 ka soils had mass balances between +7 and −154 kg m^{-2}. Indeed, the rates of weathering (based in mass balances) remained roughly the same over the course of 13,500 years. Because the mass balances were calculated by comparing the soil column with the lowermost horizon, soil erosion may account for the small carbonate losses in the young soils. As soil erosion removes the weathered residue of the soil and exposes fresh material, it effectively leads to a rejuvenation of the soil.

5.2. Soil Development in Siliceous Parent Material

Soil development in siliceous parent materials is characterized by soil acidification and weathering of primary minerals. With weathering, elements are either rearranged into secondary minerals (i.e., clays) and remain in the soil profile [97], are used as nutrients by the vegetation, or lost through leaching into the groundwater. Overall, the progression of chemical weathering and the accumulation of soil organic carbon at Sustenpass is in agreement with other observations of mountainous soils of similar age and parent material [37,98–100]. Siliceous soils from other regions of the world show similar weathering and carbon accumulation rates [100–103]. The chemical weathering rates were high in the <160 a soils (losses of up to 69 kg m^{-2} $kyear^{-1}$) and decreased thereafter rapidly. These findings match the hydrological data on the same moraines by Maier et al. [51] and Hartmann et al [104]. The saturated hydraulic conductivity (K_{sat}) decreased significantly with soil age [51], leading to a less efficient draining of the soil column in the older moraines. Blue dye experiments by Hartmann et al. [104] also showed that the flow paths become more heterogeneous with age and that the macropore flow (via root channels) becomes more important in the older moraines. This concurs with the shift in soil texture from rather coarse, sandy soils to more silty compositions. The overall texture at Sustenpass is coarser compared to Klausenpass.

The elemental composition (Table 2) revealed some irregularities in the parent material composition. We found the Fe contents increased strongly with soil age and with depth. The concentrations of Fe in the oldest soil are roughly twice as high as in the youngest soils. This can be linked to the high chlorite concentrations in the 3 ka and 10 ka soils (Figure 9). As pedogenic chlorite forms very slowly, this must be primary chlorite, which suggests a fluctuation in the mineral composition of the glacial sediment. The distribution of the chlorite in the profiles of the 10 ka and 3 ka moraines is not uniform either. There is an accumulation of chlorite in lower horizons and an apparent depletion in the topsoil.

5.3. Soil Production vs. Erosion

The most challenging part in calculating erosion rates using meteoric ^{10}Be is the estimation of the meteoric ^{10}Be deposition rates. We applied a variety of approaches given in the literature [48,87,105]. Independent of the approach used, we found the expected inverse relationship between soil age and erosion at both locations. In addition, lower rates of long-term soil loss at the calcareous site (Klausenpass) were found when compared to the siliceous site (Sustenpass). This is in contrast to findings of the short-term erosion rates (averaged over decades). Short-term erosion rates using fallout plutonium showed similar erosion rates at both locations [53]. Overall, the long-term erosion rates measured at these two proglacial areas are in a similar range to findings of [106] for soils of the eastern Swiss Alps (up to 0.44 t ha^{-1} $year^{-1}$ of soil loss). Interestingly, ref. [107] found denudation rates that are twice as high at formerly glaciated slopes, which are now very active geomorphologically (760–2100 mm $kyear^{-1}$). They found comparable rates to the Sustenpass soils at slopes which were not influenced by glaciers (60–560 mm $kyear^{-1}$). In Southern Alpine soils (New Zealand), a range of 1.2–14 t ha^{-1} $year^{-1}$ soil erosion was found, which is on the higher end compared to the 0.1–10 ha^{-1} $year^{-1}$ for global soils [108].

The differences in the long-term erosion rates between the calcareous and the siliceous chronosequence are surprising. The sites at Klausenpass have a steeper slope and the material is siltier. These conditions usually would give rise to higher erosion rates [109,110]. Ref. [54] showed that these soils all have similar root tissue densities in the topsoil, from which we can derive that the soil surfaces experience similar stabilization by the plant roots. The SOM, which also greatly influences soil erodibility by acting as a binding agent for aggregation and protective layer for the mineral soil underneath [111,112], is higher at the 4.9 ka soil (187 g kg^{-1} vs. 37 g kg^{-1} at Sustenpass in the top horizon), and lower in the 13.5 ka soil (424 g kg^{-1} vs. 439 g kg^{-1}). The most important reason why the Klausenpass soils have lower erosion rates than expected may be related to the small-scale topography. Surface roughness determines the pathways of the surface water runoff and can, thus, have a large influence on the erodibility of a slope [52,113–115]. A flow direction and accumulation analysis revealed that the Sustenpass and Klausenpass slopes differ in their flow patterns (Figure 11). The Sustenpass slopes produced patterns that rather lead to a channeling of the surface water runoff, indicating a higher terrain roughness compared to Klausenpass, where the surface runoff is more spatially homogeneous and less intense. This is especially pronounced at the 10 ka moraine (Sustenpass), where the Rhododendron bushes promoted the formation of small-scale channels on the soil surface. The Klausenpass slopes, covered with grasses and having only few boulders interspersed on the surface (Figure 2), were able to retain a smoother surface. This GIS analysis has some limitations, as it is based on a digital surface model (DSM) which includes the volume and shapes of the vegetation, as it was produced from drone images, and not by using LiDAR (light detection and ranging, 'laser scanning'). However, we can support this hypothesis with the results of a detailed investigation of the terrain microtopography on the same slopes by Maier and van Meerveld [56]. The authors calculated a tortuosity index based on high-resolution transect measurements of the surface roughness of the moraine slopes. The tortuosity index at Sustenpass (10 ka: 0.64–1.14; 3 ka: 0.33–0.65) is indeed higher than at Klausenpass (13.5 ka: 0.25–0.30; 4.9 ka: 0.26–0.40), supporting the argument that the microtopography may be the factor causing lower erosion rates at Klausenpass [114,116].

We estimated the soil production rates (Equation (1)) using soil formation, weathering and erosion rates (Table 7). The soil production rates were considerably higher at the siliceous soils compared to the calcareous soils. Furthermore, the Klausenpass soils show clear signs of regressive soil formation. The denudation at the 110 a moraine greatly outstrips the soil production, preventing it from building up an organic top horizon (Figures 3 and 4). Such a delay in the soil development was also mirrored in the C_{org} stocks, which increased at a much smaller pace compared to the Sustenpass soils (Figure 6). It is further accompanied by a slower development of the vegetation: the Klausenpass moraines took more time to be covered in vegetation and to inhabit more complex species (i.e., grasses and woody species) compared to Sustenpass [54].

Table 7. Estimation of soil production rates. Profile thickening (Figure 3) was used for soil formation (F_{soil}), soil erosion rates from $^{239+240}$Pu and ^{10}Be were used as denudation rates (D_{soil}), and the total mass balances served as chemical weathering rates (W_{soil}). The meteoric ^{10}Be erosion rates (erosion = negative values) were used for the 3 ka, 4.9 ka, 10 ka, and 13.5 ka soils. The denudation rates of the young soils (30 a, 110 a, 160 a) were derived from the short-term ($^{239+240}$Pu) erosion rates.

Moraine	Age	Soil Formation		Soil Erosion [1]		Soil Weathering	Soil Production
	(years)	Equation (15) mm year^{-1}	t ha^{-1} year^{-1}	($^{239+240}$Pu) t ha^{-1} year^{-1}	(^{10}Be) t ha^{-1} year^{-1}	(Figure 5b) t ha^{-1} year^{-1}	Equation (1) [2] t ha^{-1} year^{-1}
Klausenpass (Griess Glacier)							
K-A	110	0.00	0.00	−10.23	N/A	0.11	10.34
K-B	160	0.78	6.24	−4.15	N/A	−0.13	10.52
K-C	4900	0.12	1.92	−1.04	0.05	−0.05	2.02
K-D	13,500	>0.06 *	1.08	−0.32	−0.14	−0.05	1.27 *
Sustenpass (Stein Glacier)							
S-A	30	1.65	28.05	N/A	N/A	−0.69	34.15 **
S-B	160	0.94	17.86	−5.41	N/A	0.00	23.27
S-C	3000	>0.19 *	>3.42 *	−2.65	−7.11	−0.04	10.57 *
S-D	10,000	>0.08 *	>1.04 *	−1.92	−2.42	−0.02	3.48 *

N/A = not available. [1] Short-term erosion rate (erosion = negative values) derived from $^{239+240}$Plutonium. It is the mean yearly soil loss over the past 54 years (data: [53]). [2] Equation (1): $P_{soil} = F_{soil} + D_{soil} = F_{soil} + W_{soil} + E_{soil}$.
* These are minimum values as the depth of the soil pits did not reach the bottom of the BC horizon (see Figure 3).
** Approximated using the erosion rate of S-B.

The soil production rates of the siliceous soils showed a strong decline with increasing age, which was to be expected. The 30 a moraine is estimated to have soil production rates of at least 34 t ha^{-1} $year^{-1}$ while the 10 ka moraine has 3.5 t $ha^{-1}year^{-1}$. This is three times as high as other Alpine soils (ca 0.4–10 t ha^{-1} $year^{-1}$ [107,117]). It is also high compared to global soil production and formation rates measured in soils and river sediments which are in the range of 1–30 t ha^{-1} $year^{-1}$ [8,108,118–120].

5.4. Interaction with Vegetation and Hydrology

In many areas of the Alps, the snow cover lasts for roughly six months or more and in summer, the day and night temperatures on the ground can vary strongly, as the shallow and debris-rich soils in the proglacial areas offer little moisture and little vegetation to buffer these fluctuations [121]. Therefore, in the earlier stages of soil development, the habitability for plants would be primarily given by the physical properties of the soil substrate. A loamy soil texture would have a higher water retention [122] but may also lead to a higher erodibility [123], which would slow down if not stop a deepening of the soil profile, thus inhibiting the progression of soil development. If pioneer plant species manage to inhabit the substrate, they will stabilize its surface with their roots and enhance soil aggregation by accumulating organic matter in the soils (via litter and root exudates).

These improved conditions for the soils would enable new species to inhabit these soils, thus setting in motion a succession of plant species and communities as well as a progressive soil development [124].

Overall, the texture at Klausenpass is consistently finer compared to the Sustenpass location, which remains low in clay-sized particles (Figures 12 and 13). Due to the dissolution of carbonates, the clay and oxide residues accumulate passively giving rise to fine-grained soils. Concurrent to the decrease in particle sizes, the water retention also increases with age at both sites [54]. What may have reinforced this effect of decreasing particle sizes in the topsoil, are loess inputs, which are common in the Alps [125].

Some data seem to indicate a noticeable influence of the vegetation on soil development. For example, the SOC stocks of the siliceous chronosequence increased consistently faster under the more functionally rich vegetation, while we could only see differences between some of the calcareous soil profiles in SOC accumulation (Figure 6). The total mass balances (Figure 5b) of the older soils, showed higher losses in the low FRI soils at both locations.

The weathering indices (Figure 7) showed stronger weathering in the low FRI profile at 3 ka. At the 10 ka moraine, this relationship is only detectable with the CIA. The oxalate-extractable fractions do not exhibit a clear and consistent pattern (Figure 8). Only the 10 ka soils show a difference between the two FRI types.

At later stages, however, the vegetation influences the site properties by increasing chemical weathering, reducing surface erosion and altering the hydrological properties of the substrate as well as the vertical water transport by causing heterogeneous infiltration patterns, leading to an increase in preferential flow paths [51,55,104]. The functional richness of the vegetation affects soil parameters that are directly related to the vegetation (e.g., SOC).

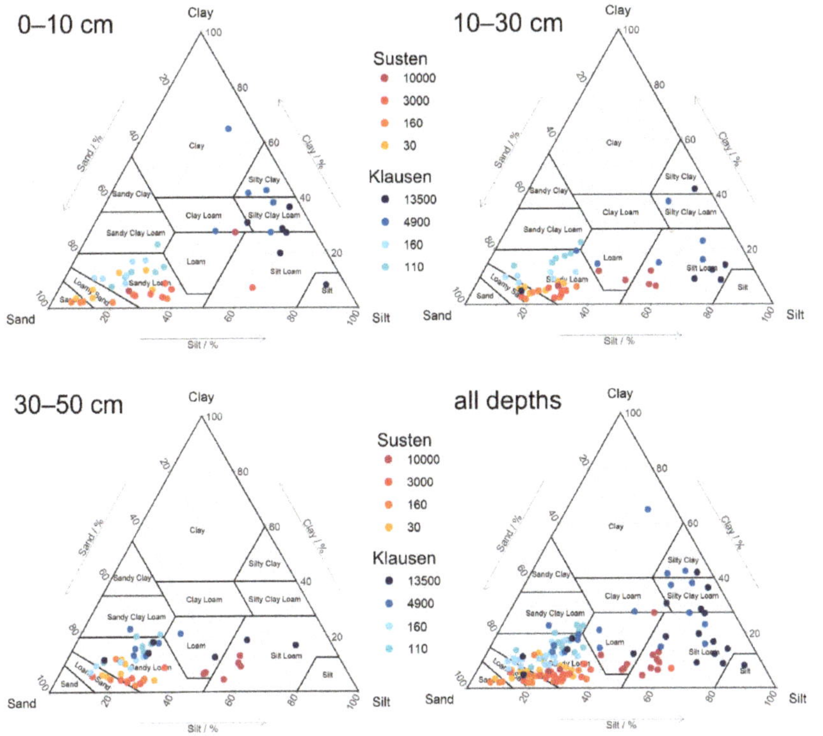

Figure 13. Particle size distribution on top of the USGS soil texture classification for each depth increment and for all depths together (same dataset as Figure 12 [55], plots were drawn in R [126]). The blue dots represent the Klausenpass and the red dots the Sustenpass soils. Light hues represent younger and the dark hues the older ages. Per age/moraine and depth, there are three replicates. (Data reproduced with permission of [55]). It therefore seems that the functional diversity of the vegetation has an effect on soil formation. The variability of functional diversity gives rise to a different litter input, root activity, and the microbial activity which in turn influence soil weathering and stabilization processes. How the FRI precisely relates to soil formation is not clear yet. The FRI, however, might be a useful parameter to better describe the natural variability of soils. Both chronosequences demonstrate that the FRI only started to have an effect on soil evolution when the vegetation was already well established, i.e., sometime between 160 and 3000 years of soil development. Before that, the physical site properties and the climate (mainly the precipitation, [127]) are the stronger drivers. The small-scale topography of the moraines (indentations offer protection from wind and erosion and have a higher retention of moisture [128]), inhomogeneity in the mineralogical and chemical composition of the parent material, and soil texture drive the soil formation and are therefore responsible for the high spatial variability of mountainous soils. These properties also influence the hydrological conditions of the soils [129] and have therefore a strong control on the early stages of vegetation settlement [39,90,130,131].

6. Conclusions

Calcareous and siliceous soil evolution in two proglacial areas in the Swiss Alps follow, not surprisingly, different traits of chemical weathering. Chemical weathering of the calcareous soils was characterized by high $CaCO_3$ losses over time, whereas the siliceous soils lost predominantly Si, Fe, and Al. By using meteoric ^{10}Be, we found higher long-term erosion rates along the siliceous chronosequence than along the calcareous chronosequence. A flow direction and flow accumulation analysis revealed that the lower terrain roughness at Klausenpass likely led to a lower erodibility. The temporal evolution of the erosion rates was completed by a combined approach of short-term ($^{239+240}Pu$) and long-term (^{10}Be) rates. The soil erosion rates counterbalanced the soil production rates at the calcareous location, delaying the development of both soil and vegetation in comparison to the siliceous soils. Overall, we found that the relative importance of the parent material decreased with increasing soil age and vegetation coverage. The vegetation is an important driver of soil development as it accelerates chemical weathering, increases surface stabilization, and changes the hydrological properties. Differences in the functional diversity are, however, only detectable with parameters that strongly depend on the vegetation such as SOC.

Author Contributions: Conceptualization, A.M. and M.E.; Methodology, A.M., M.E., M.L.P., D.T., F.P. and F.M.; Investigation, A.M., M.L.P., D.T., F.M. and F.P.; Resources, M.E.; Writing—Original Draft Preparation, A.M.; Writing—Review and Editing, A.M., M.E., K.G., A.H., C.G., D.T., M.L.P., F.P. and F.M.; Visualization, A.M.; Supervision, M.E.; Funding Acquisition, M.E. All authors have read and agreed to the published version of the manuscript.

Funding: This research is part of the HILLSCAPE project which was funded by the Swiss National Science Foundation (SNSF) and Deutsche Forschungsgemeinschaft (DFG) project grant no. 200021E-167563/1.

Data Availability Statement: The data that support the findings of this study will be made available in Pangaea (https://pangaea.de/, accessed on: 12 February 2022).

Acknowledgments: The authors would like to thank Gerald Raab, Raquel de Castro Portes and Diogo Spinola for their help in the field, and Tatjana Kraut, Yves Brügger (UZH), Annette Röthlisberger and Marion Rothaupt (ETH) for their assistance in the laboratory.

Conflicts of Interest: The authors declare no conflict of interest. The funders had no role in the design of the study; in the collection, analyses, or interpretation of data; in the writing of the manuscript, or in the decision to publish the results.

References

1. Dokuchaev, V.V. *Russkii Chernozem*; Moscow, Russia, 1883.
2. Jenny, H. *Factors of Soil Formation: A System of Quantitative Pedology*; McGraw-Hill: New York, NY, USA, 1941.
3. Simonson, R.W. Outline of a Generalized Theory of Soil Genesis. *Soil Sci. Soc. Am. J.* **1959**, *23*, 152–156. [CrossRef]
4. Johnson, D.L.; Watson-Stegner, D. Evolution Model of Pedogenesis. *Soil Sci.* **1987**, *143*, 349–366. [CrossRef]
5. Egli, M.; Poulenard, J. Soils of Mountainous Landscapes. In *International Encyclopedia of Geography: People, the Earth, Environment and Technology*; Richardson, D., Castree, N., Goodchild, M.F., Kobayashi, A., Liu, W., Marston, R.A., Eds.; John Wiley & Sons, Ltd.: Oxford, UK, 2016; pp. 1–10.
6. Sommer, M.; Gerke, H.H.; Deumlich, D. Modelling soil landscape genesis—A "time split" approach for hummocky agricultural landscapes. *Geoderma* **2008**, *145*, 480–493. [CrossRef]
7. Dosseto, A.; Buss, H.; Suresh, P. The delicate balance between soil production and erosion, and its role on landscape evolution. *Appl. Geochem.* **2011**, *26*, S24–S27. [CrossRef]
8. Heimsath, A.M.; Dietrich, W.E.; Nishiizumi, K.; Finkel, R.C. The soil production function and landscape equilibrium. *Nature* **1997**, *388*, 358–361. [CrossRef]
9. Riebe, C.S.; Kirchner, J.; Finkel, R.C. Long-term rates of chemical weathering and physical erosion from cosmogenic nuclides and geochemical mass balance. *Geochim. Cosmochim. Acta* **2003**, *67*, 4411–4427. [CrossRef]
10. von Blanckenburg, F.; Hewawasam, T.; Kubik, P.W. Cosmogenic nuclide evidence for low weathering and denudation in the wet, tropical highlands of Sri Lanka. *J. Geophys. Res. Earth Surf.* **2004**, *109*. [CrossRef]
11. Cannone, N.; Diolaiuti, G.; Guglielmin, M.; Smiraglia, C. Accelerating Climate Change Impacts on Alpine Glacier Forefield Ecosystems in the European Alps. *Ecol. Appl.* **2008**, *18*, 637–648. [CrossRef]
12. Delaney, I.; Bauder, A.; Huss, M.; Weidmann, Y. Proglacial erosion rates and processes in a glacierized catchment in the Swiss Alps. *Earth Surf. Process. Landf.* **2018**, *43*, 765–778. [CrossRef]
13. Huggett, R. Soil chronosequences, soil development, and soil evolution: A critical review. *CATENA* **1998**, *32*, 155–172. [CrossRef]
14. Sauer, D.; Schülli-Maurer, I.; Wagner, S.; Scarciglia, F.; Sperstad, R.; Svendgård-Stokke, S.; Sørensen, R.; Schellmann, G. Soil development over millennial timescales—A comparison of soil chronosequences of different climates and lithologies. *IOP Conf. Ser. Earth Environ. Sci.* **2015**, *25*, 12009. [CrossRef]
15. White, A.F.; Blum, A.E.; Schulz, M.; Bullen, T.D.; Harden, J.W.; Peterson, M.L. Chemical weathering rates of a soil chronosequence on granitic alluvium: I. Quantification of mineralogical and surface area changes and calculation of primary silicate reaction rates. *Geochim. Cosmochim. Acta* **1996**, *60*, 2533–2550. [CrossRef]
16. Bain, D.; Mellor, A.; Robertson-Rintoul, M.; Buckland, S. Variations in weathering processes and rates with time in a chronosequence of soils from Glen Feshie, Scotland. *Geoderma* **1993**, *57*, 275–293. [CrossRef]

17. Taylor, A.; Blum, J.D. Relation between soil age and silicate weathering rates determined from the chemical evolution of a glacial chronosequence. *Geology* **1995**, *23*, 979–982. [CrossRef]
18. Mavris, C.; Egli, M.; Plötze, M.; Blum, J.D.; Mirabella, A.; Giaccai, D.; Haeberli, W. Initial stages of weathering and soil formation in the Morteratsch proglacial area (Upper Engadine, Switzerland). *Geoderma* **2010**, *155*, 359–371. [CrossRef]
19. Vilmundardóttir, O.K.; Gísladóttir, G.; Lal, R. Between ice and ocean; soil development along an age chronosequence formed by the retreating Breiðamerkurjökull glacier, SE-Iceland. *Geoderma* **2015**, *259–260*, 310–320. [CrossRef]
20. D'Amico, M.E.; Freppaz, M.; Filippa, G.; Zanini, E. Vegetation influence on soil formation rate in a proglacial chronosequence (Lys Glacier, NW Italian Alps). *CATENA* **2014**, *113*, 122–137. [CrossRef]
21. Righi, D. Clay Formation and Podzol Development from Postglacial Moraines in Switzerland. *Clay Miner.* **1999**, *34*, 319–332. [CrossRef]
22. Bockheim, J.G. Solution and use of chronofunctions in studying soil development. *Geoderma* **1980**, *24*, 71–85. [CrossRef]
23. Egli, M.; Fitze, P.; Mirabella, A. Weathering and evolution of soils formed on granitic, glacial deposits: Results from chronosequences of Swiss alpine environments. *CATENA* **2001**, *45*, 19–47. [CrossRef]
24. Egli, M.; Fitze, P. Quantitative aspects of carbonate leaching of soils with differing ages and climates. *CATENA* **2001**, *46*, 35–62. [CrossRef]
25. Egli, M.; Mirabella, A.; Fitze, P. Clay mineral formation in soils of two different chronosequences in the Swiss Alps. *Geoderma* **2001**, *104*, 145–175. [CrossRef]
26. Föllmi, K.B.; Arn, K.; Hosein, R.; Adatte, T.; Steinmann, P. Biogeochemical weathering in sedimentary chronosequences of the Rhône and Oberaar Glaciers (Swiss Alps): Rates and mechanisms of biotite weathering. *Geoderma* **2009**, *151*, 270–281. [CrossRef]
27. Mavris, C.; Plötze, M.; Mirabella, A.; Giaccai, D.; Valboa, G.; Egli, M. Clay mineral evolution along a soil chronosequence in an Alpine proglacial area. *Geoderma* **2011**, *165*, 106–117. [CrossRef]
28. Birkeland, P.W. *Soils and Geomorphology*; Oxford University Press Inc.: New York, NY, USA; Oxford, UK, 1984.
29. Haselberger, S.; Ohler, L.-M.; Junker, R.R.; Otto, J.-C.; Glade, T.; Kraushaar, S. Quantification of biogeomorphic interactions between small-scale sediment transport and primary vegetation succession on proglacial slopes of the Gepatschferner, Austria. *Earth Surf. Process. Landf.* **2021**, *46*, 1941–1952. [CrossRef]
30. Eichel, J.; Corenblit, D.; Dikau, R. Conditions for feedbacks between geomorphic and vegetation dynamics on lateral moraine slopes: A biogeomorphic feedback window. *Earth Surf. Process. Landf.* **2016**, *41*, 406–419. [CrossRef]
31. Dümig, A.; Smittenberg, R.; Kögel-Knabner, I. Concurrent evolution of organic and mineral components during initial soil development after retreat of the Damma glacier, Switzerland. *Geoderma* **2011**, *163*, 83–94. [CrossRef]
32. Burga, C.A. Vegetation development on the glacier forefield Morteratsch (Switzerland). *Appl. Veg. Sci.* **1999**, *2*, 17–24. [CrossRef]
33. Hudek, C.; Stanchi, S.; D'Amico, M.; Freppaz, M. Quantifying the contribution of the root system of alpine vegetation in the soil aggregate stability of moraine. *Int. Soil Water Conserv. Res.* **2017**, *5*, 36–42. [CrossRef]
34. Kumar, R.; Pandey, S.; Pandey, A. Plant roots and carbon sequestration. *Curr. Sci.* **2006**, *91*, 885–890.
35. Tresder, K.K.; Morris, S.J.; Allen, M.F. The Contribution of Root Exudates, Symbionts, and Detritus to Carbon Sequestration in the Soil. In *Roots and Soil Management: Interactions between Roots and the Soil*; Zobel, R.W., Wright, S.F., Eds.; Agronomy No. 48; American Society of Agronomy: Madison, WI, USA, 2005; pp. 145–162.
36. Khedim, N.; Cécillon, L.; Poulenard, J.; Barré, P.; Baudin, F.; Marta, S.; Rabatel, A.; Dentant, C.; Cauvy-Fraunié, S.; Anthelme, F.; et al. Topsoil organic matter build-up in glacier forelands around the world. *Glob. Chang. Biol.* **2021**, *27*, 1662–1677. [CrossRef] [PubMed]
37. Montagnani, L.; Badraghi, A.; Speak, A.F.; Wellstein, C.; Borruso, L.; Zerbe, S.; Zanotelli, D. Evidence for a non-linear carbon accumulation pattern along an Alpine glacier retreat chronosequence in Northern Italy. *PeerJ* **2019**, *7*, e7703. [CrossRef] [PubMed]
38. Zhang, B.; Yang, Y.-S.; Zepp, H. Effect of vegetation restoration on soil and water erosion and nutrient losses of a severely eroded clayey Plinthudult in southeastern China. *CATENA* **2004**, *57*, 77–90. [CrossRef]
39. Burga, C.A.; Krüsi, B.; Egli, M.; Wernli, M.; Elsener, S.; Ziefle, M.; Fischer, T.; Mavris, C. Plant succession and soil development on the foreland of the Morteratsch glacier (Pontresina, Switzerland): Straight forward or chaotic? *Flora Morphol. Distrib. Funct. Ecol. Plants* **2010**, *205*, 561–576. [CrossRef]
40. Hinsinger, P. Plant-induced changes in soil processes and properties. In *Soil Conditions and Plant Growth*; Gregory, P.J., Nortcliff, S., Eds.; Wiley-Blackwell: Chichester, West Sussex, UK; Ames, IA, USA, 2013; pp. 323–365.
41. Walczak, R.; Rovdan, E.; Witkowska-Walczak, B. Water retention characteristics of peat and sand mixtures. *Int. Agrophys.* **2002**, *16*, 161–165.
42. Yang, F.; Zhang, G.-L.; Yang, J.-L.; Li, D.-C.; Zhao, Y.-G.; Liu, F.; Yang, R.-M.; Yang, F. Organic matter controls of soil water retention in an alpine grassland and its significance for hydrological processes. *J. Hydrol.* **2014**, *519*, 3086–3093. [CrossRef]
43. Faucon, M.-P.; Houben, D.; Lambers, H. Plant Functional Traits: Soil and Ecosystem Services. *Trends Plant Sci.* **2017**, *22*, 385–394. [CrossRef]
44. Villéger, S.; Mason, N.W.H.; Mouillot, D. New Multidimensional Functional Diversity Indices for a Multifaceted Framework in Functional Ecology. *Ecology* **2008**, *89*, 2290–2301. [CrossRef]
45. Mason, N.W.H.; de Bello, F. Functional diversity: A tool for answering challenging ecological questions. *J. Veg. Sci.* **2013**, *24*, 777–780. [CrossRef]

46. Willenbring, J.K.; von Blanckenburg, F. Meteoric cosmogenic Beryllium-10 adsorbed to river sediment and soil: Applications for Earth-surface dynamics. *Earth-Sci. Rev.* **2010**, *98*, 105–122. [CrossRef]
47. Lal, D. New methods for studies of soil dynamics utilizing cosmic ray produced radionuclides. In *Sustaining the Global Farm. 10th International Soil Conservation Organization Meeting*; Stott, D.E., Mohtar, R.H., Steinhardt, G.C., Eds.; Purdue University and USDA-ARS National Soil Erosion Research Laboratory: Purdue, IN, USA, 2001; pp. 1044–1052.
48. Heikkilä, U.; Beer, J.; Feichter, J. Meridional transport and deposition of atmospheric 10Be. *Atmos. Chem. Phys.* **2009**, *9*, 515–527. [CrossRef]
49. Foster, M.; Anderson, R.S.; Wyshnytzky, C.E.; Ouimet, W.B.; Dethier, D.P. Hillslope loweing rates and mobile-regolith residence times from in situ and meteoric 10Be analysis, Boulder Creek Critical Zone Observatory, Colorado. *GSA Bull.* **2015**, *127*, 862–878. [CrossRef]
50. Schoonejans, J.; Vanacker, V.; Opfergelt, S.; Christl, M. Long-term soil erosion derived from in-situ 10Be and inventories of meteoric 10Be in deeply weathered soils in southern Brazil. *Chem. Geol.* **2017**, *466*, 380–388. [CrossRef]
51. Maier, F.; van Meerveld, I.; Greinwald, K.; Gebauer, T.; Lustenberger, F.; Hartmann, A.; Musso, A. Effects of soil and vegetation development on surface hydrological properties of moraines in the Swiss Alps. *CATENA* **2020**, *187*, 104353. [CrossRef]
52. Helming, K.; Römkens, M.J.M.; Prasad, S.N. Surface Roughness Related Processes of Runoff and Soil Loss: A Flume Study. *Soil Sci. Soc. Am. J.* **1998**, *62*, 243–250. [CrossRef]
53. Musso, A.; Ketterer, M.E.; Greinwald, K.; Geitner, C.; Egli, M. Rapid decrease of soil erosion rates with soil formation and vegetation development in periglacial areas. *Earth Surf. Process. Landf.* **2020**, *45*, 2824–2839. [CrossRef]
54. Greinwald, K.; Dieckmann, L.A.; Schipplick, C.; Hartmann, A.; Scherer-Lorenzen, M.; Gebauer, T. Vertical root distribution and biomass allocation along proglacial chronosequences in Central Switzerland. *Arct. Antarct. Alp. Res.* **2021**, *53*, 20–34. [CrossRef]
55. Hartmann, A.; Weiler, M.; Blume, T. The impact of landscape evolution on soil physics: Evolution of soil physical and hydraulic properties along two chronosequences of proglacial moraines. *Earth Syst. Sci. Data* **2020**, *12*, 3189–3204. [CrossRef]
56. Maier, F.; van Meerveld, I. Long-Term Changes in Runoff Generation Mechanisms for Two Proglacial Areas in the Swiss Alps I: Overland Flow. *Water Resour. Res.* **2021**, *57*, e2021WR030221. [CrossRef]
57. Bundesamt für Landestopographie (Swisstopo). Geologische Karte der Schweiz. 2008. Available online: https://s.geo.admin.ch/8f6671cce7 (accessed on 1 December 2021).
58. Meteoswiss. Climate Normals Temperature 1961–1990. 2019. Available online: https://s.geo.admin.ch/93c9144885 (accessed on 20 October 2021).
59. Meteoswiss. Climate Normals Precipitation 1961–1990. 2019. Available online: https://s.geo.admin.ch/93c91528d3 (accessed on 20 October 2021).
60. Bundesamt für Landestopographie (Swisstopo). Siegfriedkarte: TA50, Blatt 403. 1899. Available online: https://s.geo.admin.ch/8f679fc9f6 (accessed on 1 December 2021).
61. Ivy-Ochs, S.; Kerschner, H.; Maisch, M.; Christl, M.; Kubik, P.W.; Schlüchter, C. Latest Pleistocene and Holocene glacier variations in the European Alps. *Quat. Sci. Rev.* **2009**, *28*, 2137–2149. [CrossRef]
62. Musso, A.; Lamorski, K.; Sławiński, C.; Geitner, C.; Hunt, A.; Greinwald, K.; Egli, M. Evolution of soil pores and their characteristics in a siliceous and calcareous proglacial area. *CATENA* **2019**, *182*, 104154. [CrossRef]
63. Bundesamt für Landestopographie (Swisstopo). Aerial Image. 1988. Available online: https://s.geo.admin.ch/8804d2594d (accessed on 15 December 2019).
64. Bundesamt für Landestopographie (Swisstopo). Aerial Image. 1992. Available online: https://s.geo.admin.ch/8804d210f5 (accessed on 15 December 2019).
65. Bundesamt für Landestopographie (Swisstopo). Aerial Image. 1993. Available online: https://s.geo.admin.ch/8804cdd7ca (accessed on 15 December 2019).
66. King, L. *Studien zur postglazialen Gletscher- und Vegetationsgeschichte des Sustenpassgebietes*, 2nd ed.; Helbing und Lichtenhahn: Basel, Switzerland, 1974.
67. Heikkinen, O.; Fogelberg, P. Bodenentwicklung im Hochgebirge: Ein Beispiel vom Vorfeld des Steingletschers in der Schweiz. *Geogr. Helv.* **1980**, *35*, 107–112. [CrossRef]
68. Schimmelpfennig, I.; Schaefer, J.M.; Akçar, N.; Koffman, T.; Ivy-Ochs, S.; Schwartz, R.; Finkel, R.C.; Zimmerman, S.; Schlüchter, C. A chronology of Holocene and Little Ice Age glacier culminations of the Steingletscher, Central Alps, Switzerland, based on high-sensitivity beryllium-10 moraine dating. *Earth Planet. Sci. Lett.* **2014**, *393*, 220–230. [CrossRef]
69. FAO. World reference base for soil resources 2014. In *International Soil Classification System for Naming Soils and Creating Legends for Soil Maps*; World Soil Resources Reports 106; FAO: Rome, Italy, 2014.
70. Tilman, D. Functional diversity. In *Encyclopedia of Biodiversity*; Levin, S.A., Ed.; Academic Press: San Diego, CA, USA, 2001; pp. 109–120.
71. Lauber, K.; Wagner, G.; Gygax, A. (Eds.) *Flora Helvetica: Illustrierte Flora der Schweiz: Mit Artbeschreibungen und Verbreitungskarten von 3200 wild Wachsenden Farn- und Blütenpflanzen, Einschliesslich Wichtiger Kulturpflanzen*; Sechste, Vollständig Überarbeitete Auflage; Haupt Verlag: Bern, Switzerland, 2018.
72. Heiri, O.; Lotter, A.F.; Lemcke, G. Loss on ignition as a method for estimating organic and carbonate content in sediments: Reproducibility and comparability of results. *J. Paleolimnol.* **2001**, *25*, 101–110. [CrossRef]
73. Dorn, R. Cation-ratio dating: A new rock varnish age-de-termination technique. *Quat. Res.* **1983**, *20*, 49–73. [CrossRef]

74. Stiles, C.A.; Mora, C.I.; Driese, S.G. Pedogenic processes and domain boundaries in a Vertisol climosequence: Evidence from titanium and zirconium distribution and morphology. *Geoderma* **2003**, *116*, 279–299. [CrossRef]
75. Nesbitt, H.W.; Young, G.M. Early Proterozoic climates and plate motions inferred from major element chemistry of lutites. *Nature* **1982**, *299*, 715–717. [CrossRef]
76. McKeague, J.A.; Brydon, J.E.; Miles, N.M. Differentiation of Forms of Extractable Iron and Aluminum in Soils. *Soil Sci. Soc. Am. J.* **1971**, *35*, 33–38. [CrossRef]
77. Zhang, G.; Germaine, J.T.; Martin, T.; Whittle, A.J. A simple sample-mounting method for random powder X-ray diffraction. *Clays Clay Miner.* **2003**, *51*, 218–225. [CrossRef]
78. Döbelin, N.; Kleeberg, R. Profex: A graphical user interface for the Rietveld refinement program BGMN. *J. Appl. Crystallogr.* **2015**, *48*, 1573–1580. [CrossRef]
79. von Blanckenburg, F.; Belshaw, N.; O'Nions, R. Separation of 9Be and cosmogenic 10Be from environmental materials and SIMS isotope dilution analysis. *Chem. Geol.* **1996**, *129*, 93–99. [CrossRef]
80. Christl, M.; Vockenhuber, C.; Kubik, P.; Wacker, L.; Lachner, J.; Alfimov, V.; Synal, H.-A. The ETH Zurich AMS facilities: Performance parameters and reference materials. *Nucl. Instrum. Methods Phys. Res. Sect. B Beam Interact. Mater. At.* **2013**, *294*, 29–38. [CrossRef]
81. Nishiizumi, K.; Imamura, M.; Caffee, M.W.; Southon, J.R.; Finkel, R.C.; McAninch, J. Absolute calibration of 10Be AMS standards. *Nucl. Instrum. Methods Phys. Res. Sect. B Beam Interact. Mater. At.* **2007**, *258*, 403–413. [CrossRef]
82. Chadwick, O.A.; Brimhall, G.H.; Hendricks, D.M. From a black to a gray box—A mass balance interpretation of pedogenesis. *Geomorphology* **1990**, *3*, 369–390. [CrossRef]
83. Egli, M.; Fitze, P. Formulation of pedologic mass balance based on immobile elements: A revision. *Soil Sci.* **2000**, *165*, 437–443. [CrossRef]
84. Maejima, Y.; Matsuzaki, H.; Higashi, T. Application of cosmogenic 10Be to dating soils on the raised coral reef terraces of Kikai Island, southwest Japan. *Geoderma* **2005**, *126*, 389–399. [CrossRef]
85. Tsai, C.; Chen, Z.; Kao, C.; Ottner, F.; Kao, S.; Zehetner, F. Pedogenic development of volcanic ash soils along a climosequence in Northern Taiwan. *Geoderma* **2010**, *156*, 48–59. [CrossRef]
86. Egli, M.; Brandová, D.; Böhlert, R.; Favilli, F.; Kubik, P.W. 10Be inventories in Alpine soils and their potential for dating land surfaces. *Geomorphology* **2010**, *119*, 62–73. [CrossRef]
87. Graly, J.A.; Reusser, L.J.; Bierman, P.R. Short and long-term delivery rates of meteoric 10Be to terrestrial soils. *Earth Planet. Sci. Lett.* **2011**, *302*, 329–336. [CrossRef]
88. Deng, K.; Wittmann, H.; von Blanckenburg, F. The depositional flux of meteoric cosmogenic 10Be from modeling and observation. *Earth Planet. Sci. Lett.* **2020**, *550*, 116530. [CrossRef]
89. Phillips, J.D.; Marion, D.A.; Luckow, K.; Adams, K.R. Nonequilibrium Regolith Thickness in the Ouachita Mountains. *J. Geol.* **2005**, *113*, 325–340. [CrossRef]
90. Johnson, D.; Domier, J.; Johnson, D.N. Animating the biodynamics of soil thickness using process vector analysis: A dynamic denudation approach to soil formation. *Geomorphology* **2005**, *67*, 23–46. [CrossRef]
91. Egli, M.; Dahms, D.; Norton, K. Soil formation rates on silicate parent material in alpine environments: Different approaches–different results? *Geoderma* **2014**, *213*, 320–333. [CrossRef]
92. Lustenberger, F. Event-Based Surface Hydrological Connectivity and Sediment Transport on Young Moraines. Master's Thesis, University of Zurich, Zurich, Switzerland, 2019.
93. Tarboton, D.G. A new method for the determination of flow directions and upslope areas in grid digital elevation models. *Water Resour. Res.* **1997**, *33*, 309–319. [CrossRef]
94. Joint Committee for Guides in Metrology. *JCGM 100:2008: Guide to the Expression of Uncertainty in Measurement (JCGM)*; BIPM: Sevres, France, 2008.
95. Martignier, L.; Verrecchia, E.P. Weathering processes in superficial deposits (regolith) and their influence on pedogenesis: A case study in the Swiss Jura Mountains. *Geomorphology* **2013**, *189*, 26–40. [CrossRef]
96. Egli, M.; Merkli, C.; Sartori, G.; Mirabella, A.; Plötze, M. Weathering, mineralogical evolution and soil organic matter along a Holocene soil toposequence developed on carbonate-rich materials. *Geomorphology* **2008**, *97*, 675–696. [CrossRef]
97. Allen, C.; Darmody, R.; Thorn, C.; Dixon, J.; Schlyter, P. Clay mineralogy, chemical weathering and landscape evolution in Arctic–Alpine Sweden. *Geoderma* **2001**, *99*, 277–294. [CrossRef]
98. Egli, M.; Sartori, G.; Mirabella, A.; Giaccai, D. The effects of exposure and climate on the weathering of late Pleistocene and Holocene Alpine soils. *Geomorphology* **2010**, *114*, 466–482. [CrossRef]
99. Böhlert, R.; Mirabella, A.; Plötze, M.; Egli, M. Landscape evolution in Val Mulix, eastern Swiss Alps—Soil chemical and mineralogical analyses as age proxies. *Catena* **2011**, *87*, 313–325. [CrossRef]
100. Mahaney, W.; Kalm, V.; Kapran, B.; Milner, M.; Hancock, R. A soil chronosequence in Late Glacial and Neoglacial moraines, Humboldt Glacier, northwestern Venezuelan Andes. *Geomorphology* **2009**, *109*, 236–245. [CrossRef]
101. Earl-Goulet, J.R.; Mahaney, W.C.; Sanmugadas, K.; Kalm, V.; Hancock, R.G. Middle-Holocene timberline fuctuation: Influence on the genesis of Podzols (Spodosols), Norra Storfjället Massif, northern Sweden. *Holocene* **1998**, *8*, 705–718. [CrossRef]

102. Eger, A.; Almond, P.C.; Condron, L.M. Pedogenesis, soil mass balance, phosphorus dynamics and vegetation communities across a Holocene soil chronosequence in a super-humid climate, South Westland, New Zealand. *Geoderma* **2011**, *163*, 185–196. [CrossRef]
103. Dahlgren, R.; Boettinger, J.; Huntington, G.; Amundson, R. Soil development along an elevational transect in the western Sierra Nevada, California. *Geoderma* **1997**, *78*, 207–236. [CrossRef]
104. Hartmann, A.; Semenova, E.; Weiler, M.; Blume, T. Field observations of soil hydrological flow path evolution over 10 millennia. *Hydrol. Earth Syst. Sci.* **2020**, *24*, 3271–3288. [CrossRef]
105. Chen, P.; Yi, P.; Czymzik, M.; Aldahan, A.; Ljung, K.; Yu, Z.; Hou, X.; Zheng, M.; Chen, X.; Possnert, G. Relationship between precipitation and 10Be and impacts on soil dynamics. *CATENA* **2020**, *195*, 104748. [CrossRef]
106. Zollinger, B.; Alewell, C.; Kneisel, C.; Meusburger, K.; Brandová, D.; Kubik, P.; Schaller, M.; Ketterer, M.; Egli, M. The effect of permafrost on time-split soil erosion using radionuclides (137Cs, 239 + 240Pu, meteoric 10Be) and stable isotopes (δ 13C) in the eastern Swiss Alps. *J. Soils Sediments* **2015**, *15*, 1400–1419. [CrossRef]
107. Norton, K.P.; von Blanckenburg, F.; Kubik, P.W. Cosmogenic nuclide-derived rates of diffusive and episodic erosion in the glacially sculpted upper Rhone Valley, Swiss Alps. *Earth Surf. Process. Landf.* **2010**, *35*, 651–662. [CrossRef]
108. Larsen, I.J.; Almond, P.C.; Eger, A.; Stone, J.O.; Montgomery, D.R.; Malcolm, B. Rapid Soil Production and Weathering in the Southern Alps, New Zealand. *Science* **2014**, *343*, 637–640. [CrossRef]
109. Renard, K.; Foster, G.R.; Weesies, G.A.; McCool, D.K.; Yoder, D.C. (Eds.) *Predicting Soil Erosion by Water: A Guide to Conservation Planning with the Revised Universal Soil Loss Equation (RUSLE)*; USDA, Agricultural Research Service: Washington, DC, USA, 1997.
110. Saedi, T.; Shorafa, M.; Gorji, M.; Moghadam, B.K. Indirect and direct effects of soil properties on soil splash erosion rate in calcareous soils of the central Zagross, Iran: A laboratory study. *Geoderma* **2016**, *271*, 1–9. [CrossRef]
111. Pintaldi, E.; D'Amico, M.E.; Stanchi, S.; Catoni, M.; Freppaz, M.; Bonifacio, E. Humus forms affect soil susceptibility to water erosion in the Western Italian Alps. *Appl. Soil Ecol.* **2018**, *123*, 478–483. [CrossRef]
112. Stanchi, S.; Falsone, G.; Bonifacio, E. Soil aggregation, erodibility, and erosion rates in mountain soils (NW Alps, Italy). *Solid Earth* **2015**, *6*, 403–414. [CrossRef]
113. Darboux, F.; Huang, C. Does Soil Surface Roughness Increase or Decrease Water and Particle Transfers? *Soil Sci. Soc. Am. J.* **2005**, *69*, 748–756. [CrossRef]
114. Luo, J.; Zheng, Z.; Li, T.; He, S. Assessing the impacts of microtopography on soil erosion under simulated rainfall, using a multifractal approach. *Hydrol. Process.* **2018**, *32*, 2543–2556. [CrossRef]
115. Luo, J.; Zheng, Z.; Li, T.; He, S. Spatial variation of microtopography and its effect on temporal evolution of soil erosion during different erosive stages. *CATENA* **2020**, *190*, 104515. [CrossRef]
116. Luo, J.; Zheng, Z.; Li, T.; He, S.; Zhang, X.; Huang, H.; Wang, Y. Quantifying the contributions of soil surface microtopography and sediment concentration to rill erosion. *Sci. Total Environ.* **2021**, *752*, 141886. [CrossRef] [PubMed]
117. Alewell, C.; Egli, M.; Meusburger, K. An attempt to estimate tolerable soil erosion rates by matching soil formation with denudation in Alpine grasslands. *J. Soils Sediments* **2015**, *15*, 1383–1399. [CrossRef]
118. Riebe, C.S.; Kirchner, J.W.; Granger, D.E.; Finkel, R.C. Strong tectonic and weak climatic control of long-term chemical weathering rates. *Geology* **2001**, *29*, 511. [CrossRef]
119. Small, E.E.; Anderson, R.S.; Hancock, G.S. Estimates of the rate of regolith production using and from an alpine hillslope. *Geomorphology* **1999**, *27*, 131–150. [CrossRef]
120. Dixon, J.L.; von Blanckenburg, F. Soils as pacemakers and limiters of global silicate weathering. *Comptes Rendus Geosci.* **2012**, *344*, 597–609. [CrossRef]
121. Körner, C. Alpine plant life. In *Functional Plant Ecology of High Mountain Ecosystems*, 2nd ed.; Körner, C., Ed.; Springer: Heidelberg, Germany; New York, NY, USA, 2003.
122. Saxton, K.E.; Rawls, W.J.; Romberger, J.S.; Papendick, R.I. Estimating generalized soil-water characteristics from texture. *Soil Sci. Soc. Am. J.* **1986**, *50*, 1031–1036. [CrossRef]
123. Thorarinsdottir, E.F.; Arnalds, O. Wind erosion of volcanic materials in the Hekla area, South Iceland. *Aeolian Res.* **2012**, *4*, 39–50. [CrossRef]
124. Dontsova, K.; Balogh-Brunstad, Z.; Chorover, J. Plants as drivers of rock weathering. In *Biogeochemical Cycles: Ecological Drivers and Environmental Impact*; Dontsova, K., Balogh-Brunstad, Z., Le Roux, G., Eds.; Wiley & Sons Inc.: Hoboken, NJ, USA; Americal Geophysical Union: Washington, DC, USA, 2020; pp. 33–58.
125. Gild, C.; Geitner, C.; Sanders, D. Discovery of a landscape-wide drape of late-glacial aeolian silt in the western Northern Calcareous Alps (Austria): First results and implications. *Geomorphology* **2018**, *301*, 39–52. [CrossRef]
126. R Core Team. *R: A Language and Environment for Statistical Computing*; R Foundation for Statistical Computing: Vienna, Austria, 2020.
127. Egli, M.; Lessovaia, S.N.; Chistyakov, K.; Inozemzev, S.; Polekhovsky, Y.; Ganyushkin, D. Microclimate affects soil chemical and mineralogical properties of cold alpine soils of the Altai Mountains (Russia). *J. Soils Sediments* **2015**, *15*, 1420–1436. [CrossRef]
128. Holtmeier, F.-K.; Broll, G. The Influence of Tree Islands and Microtopography on Pedoecological Conditions in the Forest-Alpine Tundra Ecotone on Niwot Ridge, Colorado Front Range, U.S.A. *Artic Antarct. Alp. Res.* **1992**, *24*, 216. [CrossRef]
129. Lohse, K.A.; Dietrich, W.E. Contrasting effects of soil development on hydrological properties and flow paths. *Water Resour. Res.* **2005**, *41*, 1–17. [CrossRef]

130. Wondzell, S.M.; Cornelius, J.M.; Cunningham, G.L. Vegetation patterns, microtopography, and soils on a Chihuahuan desert playa. *J. Veg. Sci.* **1990**, *1*, 403–410. [CrossRef]
131. McGrath, G.S.; Paik, K.; Hinz, C. Microtopography alters self-organized vegetation patterns in water-limited ecosystems. *J. Geophys. Res. Biogeosci.* **2012**, *117*, 1–19. [CrossRef]

geosciences

Article

Seismic Activity of the Manisa Fault Zone in Western Turkey Constrained by Cosmogenic ^{36}Cl Dating

Nasim Mozafari [1,*], Çağlar Özkaymak [2,3], Dmitry Tikhomirov [1,4], Susan Ivy-Ochs [5], Vasily Alfimov [5], Hasan Sözbilir [6,7], Christian Schlüchter [1] and Naki Akçar [1]

1 Institute of Geological Sciences, University of Bern, Baltzerstrasse 1+3, 3012 Bern, Switzerland; dmitry.tikhomirov@geo.uzh.ch (D.T.); christian.schluechter@geo.unibe.ch (C.S.); akcar@geo.unibe.ch (N.A.)
2 Department of Geological Engineering, Afyon Kocatepe University, Ahmet Necdet Sezer Kampusü Gazligöl Yolu, Afyonkarahisar 03200, Turkey; caglarozkaymak@aku.edu.tr
3 Earthquake Implementation and Research Center, Afyon Kocatepe University, Afyonkarahisar 03200, Turkey
4 Department of Geography, University of Zurich, Winterthurerstrasse 190, 8057 Zürich, Switzerland
5 Institute for Particle Physics, ETH Hönggerberg, Schafmattstrasse 20, 8093 Zürich, Switzerland; ivy@phys.ethz.ch (S.I.-O.); vassili.alfimov@gmail.com (V.A.)
6 Geological Engineering Department, Dokuz Eylül University, Izmir 35160, Turkey; hasan.sozbilir@deu.edu.tr
7 Earthquake Research and Implementation Center, Dokuz Eylül University, Izmir 35160, Turkey
* Correspondence: nasim.mozafari@geo.unibe.ch

Abstract: This study reports on the cosmogenic ^{36}Cl dating of two normal fault scarps in western Turkey, that of the Manastır and Mugırtepe faults, beyond existing historical records. These faults are elements of the western Manisa Fault Zone (MFZ) in the seismically active Gediz Graben. Our modeling revealed that the Manastır fault underwent at least two surface ruptures at 3.5 ± 0.9 ka and 2.0 ± 0.5 ka, with vertical displacements of 3.3 ± 0.5 m and 3.6 ± 0.5 m, respectively. An event at 6.5 ± 1.6 ka with a vertical displacement of 2.7 ± 0.4 m was reconstructed on the Mugırtepe fault. We attribute these earthquakes to the recurring MFZ ruptures, when also the investigated faults slipped. We calculated average slip rates of 1.9 and 0.3 mm yr^{-1} for the Manastır and Mugırtepe faults, respectively.

Keywords: active tectonics; fault scarp dating; cosmogenic ^{36}Cl; Gediz Graben; western Anatolia; earthquake; Holocene

Citation: Mozafari, N.; Özkaymak, Ç.; Tikhomirov, D.; Ivy-Ochs, S.; Alfimov, V.; Sözbilir, H.; Schlüchter, C.; Akçar, N. Seismic Activity of the Manisa Fault Zone in Western Turkey Constrained by Cosmogenic ^{36}Cl Dating. *Geosciences* **2021**, *11*, 451. https://doi.org/10.3390/geosciences11110451

Academic Editors: Rosa Nappi and Jesus Martinez-Frias

Received: 23 September 2021
Accepted: 29 October 2021
Published: 31 October 2021

1. Introduction

Although earthquakes are one of the most hazardous natural disasters, seismic records from instrumental and historical earthquake data cover only a limited time frame [1–4]. Therefore, the forecasting of future earthquake events and disaster mitigation design are based upon short and incomplete seismic records (e.g., [1]). The dearth of such data limits our understanding of the spatial extent of deformation and magnitude of future earthquakes, which may lead to a misevaluation of high seismic risk areas [5–7].

Numerous fault studies have been conducted worldwide using different techniques (e.g., [8–14]). One of the possible tools for tracking the pace of earthquakes on individual faults over timeframes that exceed those included in the existing seismic records, is fault scarp dating. This is a valuable tool that directly date episodic exposures of normal fault scarps produced by large magnitude earthquakes and was first proposed by Zreda and Noller [15]. This tool has been used and progressively improved by many other researchers over the last two decades [15–38]. The investigation of fault scarp exposure using cosmogenic ^{36}Cl allows for the reconstruction of the timing, vertical displacement, recurrence interval, and magnitude of earthquakes as well as the fault slip rate. Thereby, the aforementioned technique offers the opportunity to extend the timeframe of slip histories on individual faults providing additional knowledge with respect to regional seismic behavior. The overall dating concept is as follows. On a fresh fault surface, exposed by a dip-slip component of rupture, cosmogenic ^{36}Cl begins to accumulate along the newly exposed segment at a uniform distribution and at a higher rate than the unexposed part of the scarp under the colluvium. Periods of earthquake activity are then disentangled based on: (1) cosmogenic ^{36}Cl concentrations measured along a continuous strip on the fault scarp; and (2) differences in ^{36}Cl accumulation rates on the exposed and covered surfaces during the quiescence times [15–19,33].

Strong earthquakes in extensional tectonic regimes may cause surface ruptures and deformation, principally documented as normal fault scarps that juxtapose Quaternary alluvium or colluvium against bedrock at a variety of scales [39,40]. An example of such a setting is Western Anatolia, Turkey, which includes approximately E-W-trending graben systems as a result of roughly N-S extension evidenced by the occurrence of large normal fault scarps occurring in the limestone bedrock (Figure 1).

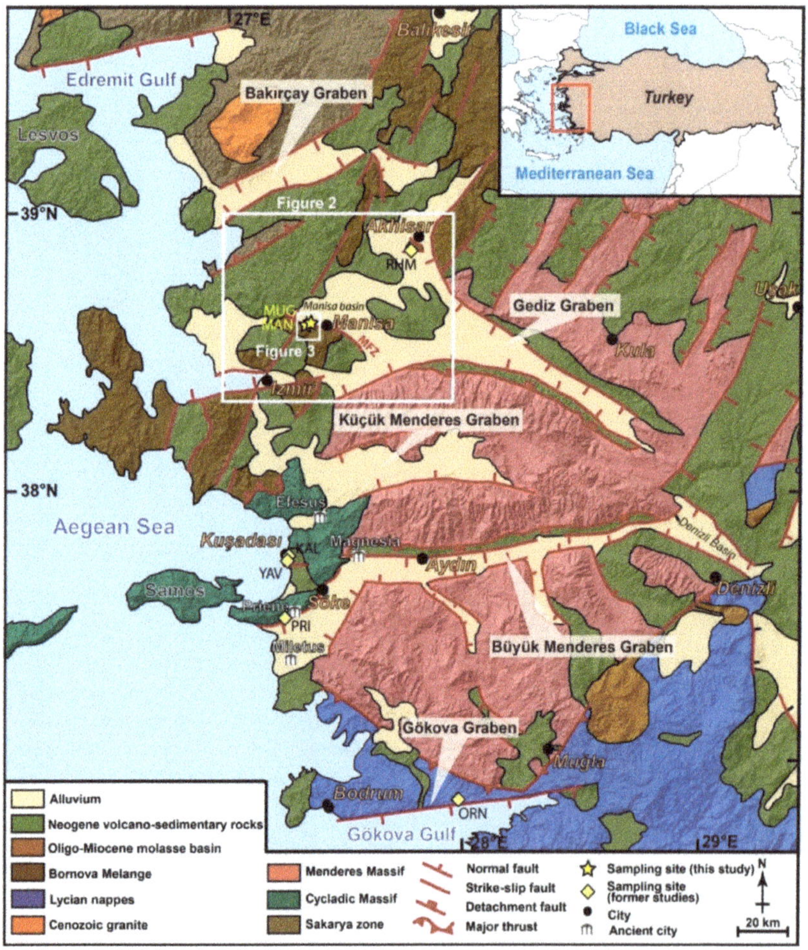

Figure 1. Simplified geological map of western Anatolia showing major structural elements and location of Manastır and Mugırtepe faults (modified from [24,33,34,41,42]). Inset white boxes mark the location of the Figures 2 and 3. Faults abbreviations: MFZ: Manisa Fault Zone, RHM: Rahmiye, MAN: Manastır, MUG: Mugırtepe, KAL: Kalafat, YAV: Yavansu, PRI: Priene-Sazlı, ORN: Ören.

In addition, the active zone of the Izmir-Balıkesir Transfer Zone (IBTZ) extended between Izmir and Balıkesir cities, generally consists of N-S and NNE-SSE trending strike slip faults, and acts as the western border of the E-W-trending grabens i.e., Gediz (Figure 1). IBTZ was demonstrated to be a deep crustal transform fault zone during Late Cretaceous, which acted as a transtensional transfer zone in the Neogene period ([43–45]). Recent seismicity with focal mechanism, Global Navigation Satellite System (GNSS) measurements and several geological studies indicate that IBTZ is undergoing an E-W shortening as well as N-S extension (e.g., [43,44,46–49]). Here, recent investigations confirm a very close connection between the normal surface ruptures and large magnitude 6 or higher earthquakes (e.g., [33–35,50–52]). Earthquakes are considered imminent in this intensively active region, with the most recent destructive event occurring offshore Samos Island (south of Izmir region) on 30th October 2020, with Mw 7.0 [53]. However, the association of historic earthquakes with individual normal faults in Western Anatolia continues to be very limited (e.g., [54–56]).

In this study, we focus on one of the fault zones in the western sector of the Gediz Graben (western Turkey) to obtain a broader insight into the seismic behavior of active faults beyond the historical and instrumental earthquake archives (Figure 2). We focus on two fault scarps within the western part of the Manisa Fault Zone (MFZ), documented as one of the most seismically-damaged regions in history [3,48,55,57,58]. Specifically, we applied the Fault Scarp Dating Tool (FSDT) computation code [37] to recover rupture histories of the Manastır and Mugırtepe faults (MAN and MUG in Figure 3, respectively) in the western segment of the active MFZ. We analyzed 87 samples from the Manastır fault surface and remodeled the cosmogenic ^{36}Cl concentrations already measured on the Mugırtepe fault surface [24]. We show that the MFZ experienced numerous ground-rupturing earthquakes during Holocene. We also provide a comparison and interpretation of our results with respect to paleoseismological data in our effort to better constrain the seismic history for the western MFZ. We find that this region experienced clustered earthquakes during late Holocene.

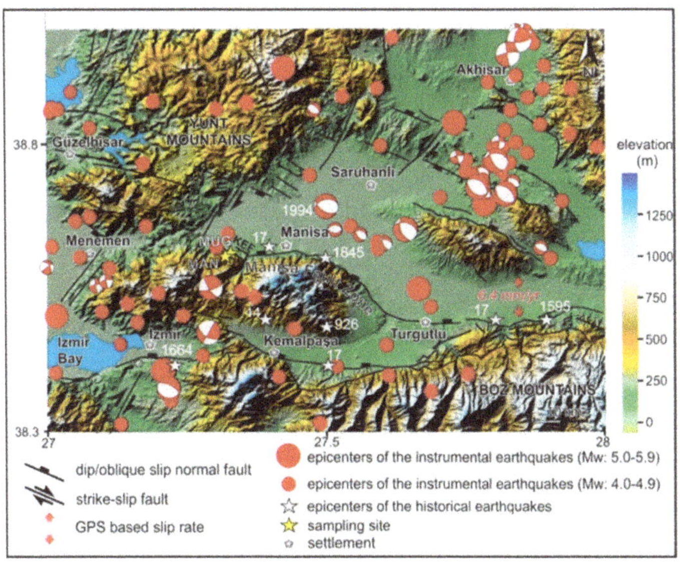

Figure 2. Seismotectonic map of the Izmir-Manisa region showing the epicenters of instrumental and historical earthquakes (modified from [59]). MAN: Manastır fault, MUG: Mugırtepe fault, KEF: Keçiliköy fault; GPS based slip rate is from [47].

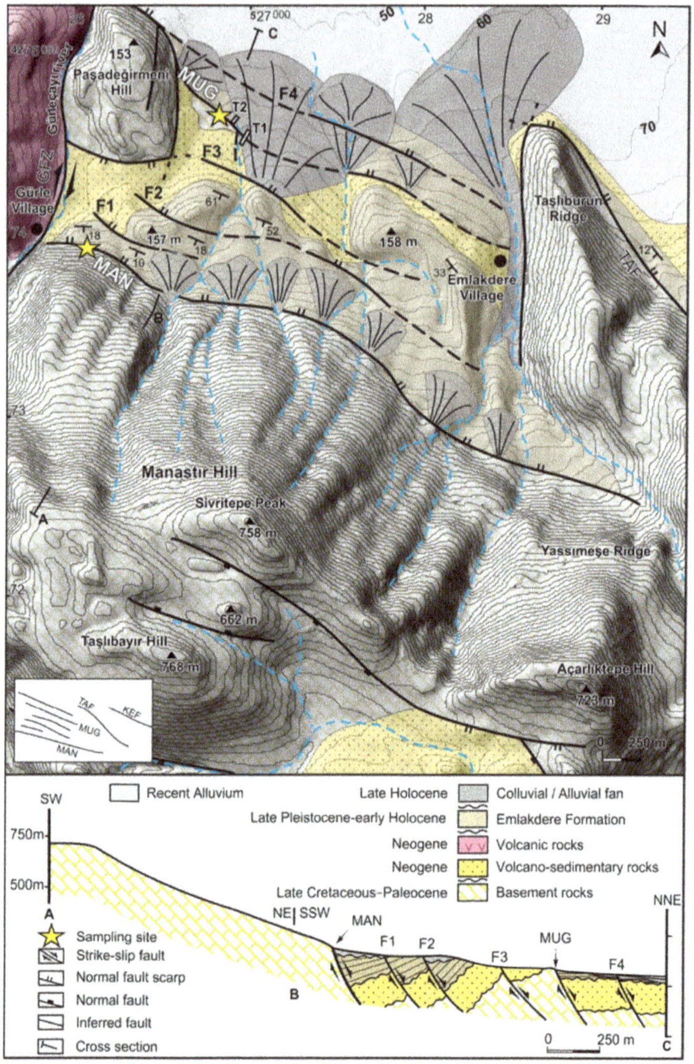

Figure 3. Detailed geological map of the study area showing NW–SE-trending active faults; location of paleoseismological trench sites from [59] along Mugırtepe fault are shown as T1 and T2. Yellow stars locate fault scarp dating sampling sites of Manastır (MAN) and Mugırtepe (MUG) faults, respectively. GFZ: Gürle Fault Zone, TAF: Taşlıburun fault, KEF: Keçiliköy fault; Inset is a simple sketch to show Taşlıburun fault, which its location is beyond the map frame. Geological cross section shows stratigraphic and structural relationships of the units. Note that offset of Quaternary deposits by several instances of synthetic Holocene faulting in the hangingwall of the Manastır fault is the most direct evidence for their activity (modified from [24,59]).

2. Study Area

The approximately WNW-ESE-trending MFZ within the Gediz Graben extends for 35 km at the southern margin of the Manisa Basin [44,51,60] and includes a large number of Quaternary fault scarps [41,44,51,61,62] (Figures 1 and 2). The MFZ is considered to be a northeastward-arcuate structure of the graben [59]. Different groups of kinematic indicators, including sinistral strike-slip, dextral strike-slip, and normal-slip denote three phases of activity in the MFZ since the early Miocene, respectively [44,51]. In the investigated area (westernmost part of the MFZ), at least six WNW-ESE-trending and NNE-dipping normal fault scarps displaced Late Cretaceous–Paleocene carbonate footwall, against the Late Pleistocene–early Holocene sediments of Emlakdere Formation, occasionally covered by colluvium deposits on the hangingwall (Figures 3 and 4). This group of faults comprise the Manastır, F1, F2, F3, Mugırtepe, and F4 form separate scarp terraces and defines the southwestern boundary of the Manisa Basin. The Manastır fault was probably initiated as a master fault with the onset of graben system formation during the Early-Miocene or later in western Anatolia. While the other faults are interpreted to be formed as a consequence of basinward migration evidenced by back-tilting of the Emlakdere Formation and the Neogene volcano-sedimentary rocks [59]. The strata are parallel in these rock units separated by unconformity, indicating synchronized tilting [59] (Figures 3 and 4). Gradual deposition and rotation of hangingwall deposits caused by slip on the Manastır master fault is evidenced by a clear angular unconformity recorded in the upper part of the Emlakdere Formation showing dissimilar dip of strata of similar lithology. Based on the radiocarbon ages, this syn-depositional tilting was considered to occur between ca. 19 and 9 cal kyr BP [59]. In an extensional tectonic setting, shallow antithetic and synthetic faults within the hangingwall of the larger master fault are typically prevalent; these parallel/subparallel faults are refractions of the master fault dip and maintain evolution towards the basin and can move in synchrony with the master fault they are linked to (e.g., [63–65]). Secondary faults are normally incapable of producing significant earthquakes with magnitudes exceeding 5.5 and are considered as non-seismogenic faults [64]. Among this set of faults, the approximately 4.5-km long and 140-m-high Manastır fault is considered as the master fault. The Manastır fault is connected to the approximately 3-km long Taşlıburun fault through an N-S-trending relay ramp (Figure 3). The Taşlıburun fault is, in turn, linked to the Keçiliköy fault with a similar length on its northeast side (Figures 2 and 3). These three faults constitute an en échelon structure linked to MFZ at its westernmost end [44].

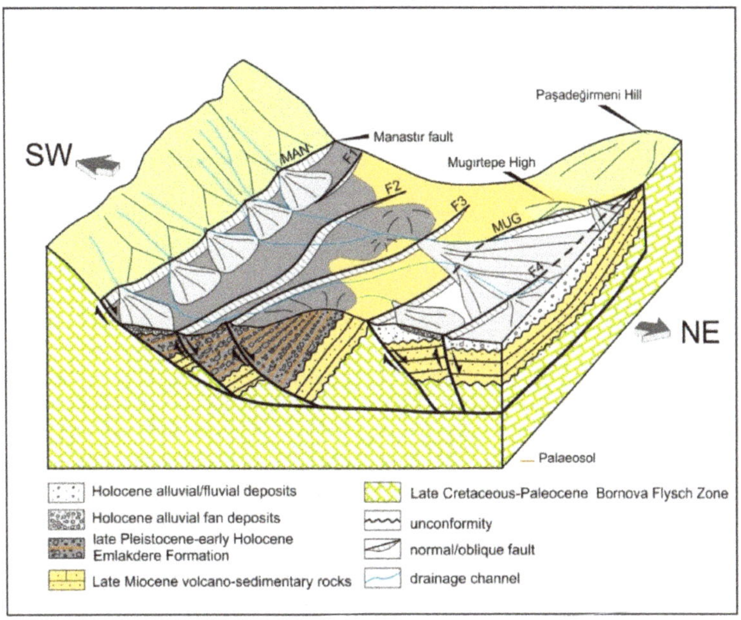

Figure 4. 3D view of western Manisa Fault Zone showing the staircase of fault scarps in the Manastır hangingwall. MAN: Manastır, MUG: Mugırtepe. Note that in the cartoons the exact horizontal and vertical scales, vertical displacement values as well as thickness of sedimentary layers are disregarded. However, MAN fault depth is known to be 5–10 km in about 10 km northeast of Manisa.

The Manastır fault activity is expressed by sets of screes, landslides, and at least two generations of triangular facets [59]. The overprinting of the strike-slip slickenlines by the dip-slip ones indicate reactivation of the Manastır fault. Accordingly, the exposure of the sampled fault surface is attributed to an approximately N-S trending extensional tectonic regime that started in Quaternary. The Manastır fault extends to the NE side of Manastır Hill and intersects the NW end of the approximately 1.5 km wide strike-slip Gürle Fault Zone [59] (Figure 3). Gürle Fault Zone, a segment of IBTZ, is characterized by segmented parallel-subparallel faults, which extends to the north of Paşadeğirmeni Hill and bounds MFZ on its west end. All the synthetic faults located at the western part of the MFZ are assumed to be linked in depth to the master Manastır fault and there are five secondary faults to the north (F1, F2, F3, Mugırtepe, and F4), which run parallel to the Manastır fault [59]. Faults F1, F2, and F3 are approximately 1, 3, and 2 km in length, respectively. Their dip ranges between 45° and 65° NNE with an average scarp height of 3 m, whereas the Mugırtepe fault has a maximum of 4 m height and is approximately 3 km long. To the northwest, the Mugırtepe and the similar-sized F4 faults merge at the foot of Paşadeğirmeni Hill and cut across the Gürle Fault Zone [59] (Figure 3).

Six strong earthquakes have been recorded in Manisa and the surrounding region historically (Figure 2). The oldest occurred in Lydia in 17 AD and had an intensity of IX. It caused significant damage in 13 or 16 ancient cities, mostly located in the Manisa Basin [1,48,55,57,58]. This earthquake is an example of the discrepancies in the geographic locations and intensity/magnitudes of ancient earthquakes recorded by different sources. The succeeding destructive earthquake dates back to 44 AD and likely damaged the ancient Greek cities of Magnesia, Samos, Militus and Ephesus, with an intensity of VIII [1,3,58,66]. In addition, [1] reported an earthquake in 926 (925) AD, in the province of the Thraceseans, caused traverse of the region by the Gediz and Menderes rivers. An earthquake in 1595 was documented ca. 60 km to the east of Manisa [1,55,67]. Another major earthquake with an intensity of VII was reported to have occurred in Izmir in 1664 [1,3,66,67], although [55] claim this event to have occurred near Izmir, and perhaps towards Manisa. The last recorded historical earthquake occurred in 1845 in Lesvos, and felt in Manisa [3,66], with a reconstructed intensity of VIII [3] or M = 6.7 [68]. The earthquake of 28th January 1994, with Mw 5.2 or 5.4, is the largest instrumental earthquake recorded close to Manisa with estimated focal depth of 5 to 10 km [48,53,69] and epicentered about ten kilometers northeast of Manisa (Figure 2).

The rupture history of the Mugırtepe fault is reconstructed [24] using a Matlab® code developed by Schlagenhauf et al. [22]. They proposed two scenarios. The first one yielded two seismic events at 13.7 ± 0.8 ka and 7.8 ± 0.5 ka with a displacement of 0.5 ± 0.2 m and 2.15 ± 0.35 m, respectively. The second scenario resulted in a single seismic event of 8.5 ± 0.6 ka with a vertical displacement of 2.65 ± 0.35 m. Here, we note that this event, similar to most of the events recovered by FSDT, consists of clusters of earthquakes that occurred close in time. In addition to the abovementioned documented and reconstructed earthquakes, three palaeoearthquakes defined during the last 1 kyr using radiocarbon dating of palaeosol samples collected inside two paleoseismic trenches dug across the Mugırtepe fault [59]. These three events were tentatively linked to the historical earthquakes of 926 AD, 1595/1664 AD, and 1845 AD.

3. Materials and Methods

3.1. Sampling

To select appropriate sampling sites, we explored the fault surfaces in several outcrops along the MFZ. We considered the most suitable site as the well-preserved surface with negligible evidence of weathering (Figure 3). The site is close to the Mugırtepe fault scarp studied [24,59]. The scarp of the Manastır fault was sampled in summer 2008 following a similar sampling strategy by Mitchell et al. [16] (Figure 5). Two parallel vertical slots spaced approximately 12 cm apart were cut to a depth of 3–4 cm into the scarp surface using a hand-held circular saw with a diamond blade. The rock strip was then divided into 5–10-cm slabs perpendicular to the vertical slots. Finally, the sample slabs were broken off with a chisel and hammer. Being vital factors for earthquake modeling, the scarp geometry elements including scarp dip, scarp height, top surface dip, and colluvium dip were determined in the field. Rock density and water content of the bedrock and colluvium were also estimated, and colluvium density was measured in the field. In addition, top and bottom positions of each sample were documented for modeling.

Figure 5. Manastır fault scarp, view towards SSW with three sampling strips of MAN-A, MAN-B and MAN-C. Note: lowest notch below MAN-B strip is not related to this study; Schematic sketch shows fault scarp with used parameters for modeling. White dashed line represents the sampled surface.

Along the Manastır fault scarp, three sampling strips (MAN-A, MAN-B, and MAN-C) that were spaced a few meters apart, were cut to cover the maximum height along the scarp surface (Figure 5). We collected 87 samples, which covered approximately 7 m of the 12-m scarp surface height from the ground level; 36 samples were obtained from both MAN-A and MAN-B, and 15 samples from MAN-C. In addition, the fault scarp geometry parameters (i.e., scarp dip, scarp height, top surface dip, and colluvium dip, Figure 5) were precisely measured in the field (e.g., [22,27,37]. In comparison, Akçar et al. [24] collected 44 samples along a ca. 2.7-m sampling profile of the 4-m high Mugırtepe fault scarp (Figure 6).

Figure 6. Mugırtepe fault scarp, view towards south with two sampling strips of MUG-A and MUG-B. The sampling has been done by [24]. Schematic sketch shows fault scarp with used parameters for modeling. White dashed line represents the sampled surface.

3.2. Cosmogenic ^{36}Cl Analysis

The samples collected from the Manastır fault were processed at the Surface Exposure Laboratory of the Institute of Geological Sciences, University of Bern, following the procedure reported by Stone et al. [70] and Ivy-Ochs et al. [71,72], and the isotope dilution method [72,73]. A full description of the standard protocol of the laboratory for ^{36}Cl extraction from limestone samples is presented in previous publications (cf. [24,33,34]). The total Cl and ^{36}Cl of the Manastır samples were measured at the TANDEM accelerator mass spectrometry (AMS) facility at ETH Zurich. The calcium concentrations of individual samples from Manastır as well as major and trace elements of five proxy samples were also measured using inductively coupled plasma mass spectrometry (ICP-MS) at SGS Mineral Services, Canada. In addition, we determined the calcium concentrations of the Mugırtepe fault scarp from the study conducted by Akçar et al. [24].

3.3. Fault Scarp Dating Tool

To analyze the distribution of ^{36}Cl concentrations accumulated on the Manastır fault scarp (MAN) and reanalyze the dataset for the Mugırtepe fault scarp (MUG), we applied the computation code based on the Monte-Carlo method, which allows the reconstructions of the time-slip histories of normal fault scarps through two separate stages of database building and data simulation [37]. In the database-building stage, the chemical composition of the bedrock, sample positions, shielding of the scarp and colluvium are considered to calculate the production of the isotope in every sample depending on fixed slip step and rock erosion. Creation of the database improves time efficiency via the approximation of pre-calculated isotope production during each round of simulation. The maximal erosion rate was set to 15 cm kyr^{-1} to provide some flexibility in the current analysis. In the simulation stage, exposure histories are generated within an earthquake scenario based on the number of earthquakes, earthquake ages, slip values, and erosion rates (cf. [33,34]). Simulated ^{36}Cl concentration of the samples are statistically compared with measured concentrations taking into account the measurement errors of ^{36}Cl, parent elements and production rates (Tables 1–4). Following the preliminary simulation of the fault dataset, we began our main simulations by entering an excessive number of earthquakes. After identifying the most accurate number of events in terms of the lowest statistical criteria, we modeled the time-slip histories using minimum 100,000 simulations to achieve the best fit scenario based on one and two earthquake scenarios for the Manastır fault and one to three earthquake scenarios for the Mugırtepe fault. In the FSDT code [37] "Beginning of exposure" indicates the time when the ^{36}Cl starts to accumulate in the analyzed section of the fault scarp at depth, but it does not refer to any exposure and/or any seismic event. Thus, the analyzed strip is assumed to be still underground (e.g., covered by the colluvium) at the beginning of cosmogenic ^{36}Cl accumulation. The thickness of the overburden can theoretically be in the order of several meters. To avoid any confusion, in this paper we use the term "beginning of accumulation". It is important to note that the fault scarp dating process only allows for the detection of large earthquakes with considerable displacement values, thereby yielding a lower estimate of earthquake frequency (cf. [20,21,33,34]). Furthermore, episodic earthquakes occurring within the uncertainty of the analysis are not identified as a series of earthquakes but rather as a single event, which cannot be disentangled by any code. This causes lower resolution of the older ages and larger slips [37]. The simulation output is given as a plot of measured ^{36}Cl concentrations against the sample height along the sampled profile. Following a comparison of the measured and modeled ^{36}Cl concentrations, the most realistic scenario is selected based on the lowest statistic criteria.

Table 1. Input parameters of the Manastır and Mugırtepe fault scarps for earthquake modeling.

	Manastır Fault	**Mugırtepe Fault**
Latitude	38° 36.729′ N	38° 37.101′ N
Longitude	27° 17.917′ E	27° 18.498′ E
Altitude	141 m	80 m
Scarp strike	N88° E	N65° W
Colluvium dip	5°	0°
Scarp dip	80°	52°
Top surface dip	30°	0°
Scarp height	1200 cm	415 cm
Scarp rock density	2.4 g/cm^3	2.4 g/cm^3
Colluvium density	1.5 g/cm^3	1.4 g/cm^3
Rock water content	0.1%	0.1%
Colluvium water content	1%	1%
Spallation on Ca: 48.8 ± 3.5 at g^{-1} yr^{-1} [70]		
Spallation on K of 170 ± 25 at g^{-1} yr^{-1} [74]		
Spallation on Ti of 13 ± 3 at g^{-1} yr^{-1} [75]		
Spallation on Fe of 1.9 ± 0.2 at g^{-1} yr^{-1} [76]		
Epithermal neutrons from fast neutrons: 760 ± 150 n/g^{-1} yr^{-1} [77]		
Scaling scheme [78]		

Table 2. Cosmogenic nuclide data of the Manastır Fault scarp.

Sample Name	Top Position (cm)	Bottom Position (cm)	Thickness (cm)	^{36}Cl * (105 at/g)	^{36}Cl Uncertainty * (105 at/g)	Cl Total * (ppm)	Cl Total Uncertainty * (ppm)	Ca † (ppm)	O (%)	C (%)
MAN-A02	645	638	2.0	1.056	0.099	9.3	0.09	372,143	48.78	11.16
MAN-A03	638	631	2.0	1.114	0.046	9.9	0.10	346,429	52.13	10.39
MAN-A04	631	624	2.0	1.086	0.054	7.9	0.08	386,429	46.93	11.59
MAN-A05	624	617	2.0	1.056	0.045	11.2	0.11	372,143	48.78	11.16
MAN-A06	617	610	2.0	1.144	0.049	10.6	0.11	375,714	48.32	11.27
MAN-A07	610	603	3.0	1.154	0.055	9.3	0.09	346,429	52.13	10.39
MAN-A09	596	589	2.0	0.934	0.043	9.2	0.09	346,429	52.13	10.39
MAN-A10	589	582	2.0	0.905	0.048	10.6	0.11	373,571	48.60	11.21
MAN-A11	582	575	2.0	1.021	0.056	9.7	0.10	346,429	52.13	10.39
MAN-A12	575	568	2.0	1.055	0.044	9.5	0.09	373,571	48.60	11.21
MAN-A13	568	561	3.0	0.974	0.043	7.2	0.07	368,571	49.25	11.06
MAN-A14	561	554	2.0	0.840	0.037	5.4	0.05	362,143	50.08	10.86
MAN-A15	554	547	3.0	1.005	0.052	7.9	0.08	350,714	51.57	10.52
MAN-A16	547	540	2.0	0.880	0.040	8.3	0.08	370,000	49.06	11.10
MAN-A17	540	533	2.0	0.782	0.038	6.8	0.07	346,429	52.13	10.39
MAN-A18	533	526	2.0	0.827	0.036	5.7	0.06	346,429	52.13	10.39
MAN-A19	526	519	2.0	0.769	0.034	3.9	0.04	346,429	52.13	10.39
MAN-A20	519	512	2.0	1.053	0.054	5.9	0.06	360,714	50.27	10.82
MAN-A21	512	505	2.0	0.843	0.037	4.9	0.05	346,429	52.13	10.39
MAN-A22	505	498	2.0	0.796	0.042	10.7	0.11	346,429	52.13	10.39
MAN-A23	498	491	2.0	0.749	0.047	8.2	0.08	346,429	52.13	10.39
MAN-A24	491	484	3.0	0.871	0.054	8.7	0.09	346,429	52.13	10.39
MAN-A25	484	477	2.0	0.815	0.040	7.9	0.08	346,429	52.13	10.39
MAN-A26	477	470	2.0	0.774	0.039	8.4	0.08	346,429	52.13	10.39
MAN-A27	470	463	2.0	0.792	0.037	9.8	0.10	346,429	52.13	10.39
MAN-A28	463	456	2.0	0.854	0.071	10.2	0.10	369,286	49.16	11.08
MAN-A29	456	449	2.0	0.781	0.038	11.7	0.12	341,429	52.78	10.24
MAN-A30	449	442	2.0	0.747	0.035	10.0	0.10	345,000	52.31	10.35
MAN-A31	442	435	2.0	0.721	0.038	9.4	0.09	370,714	48.97	11.12
MAN-A32	435	428	2.0	0.795	0.042	10.0	0.10	354,286	51.11	10.63
MAN-A33	428	421	2.0	0.694	0.036	8.9	0.09	353,571	51.20	10.61
MAN-A35	414	407	2.0	0.800	0.045	13.5	0.13	317,857	55.84	9.54
MAN-A36	407	400	2.0	0.713	0.038	25.1	0.25	321,429	55.38	9.64
MAN-A37	400	393	2.0	0.696	0.035	21.5	0.21	320,000	55.56	9.60
MAN-A38	393	386	2.0	0.654	0.045	30.8	0.31	298,571	58.35	8.96
MAN-A39	386	379	2.0	0.565	0.031	13.2	0.13	337,143	53.33	10.11
MAN-B01	379	372	3.0	0.777	0.047	10.7	0.05	345,000	52.31	10.35
MAN-B02	372	365.5	2.0	0.694	0.033	4.8	0.11	345,000	52.31	10.35
MAN-B03	365.5	357.5	2.0	0.714	0.038	11.3	0.13	338,571	53.15	10.16
MAN-B04	357.5	351	2.0	0.791	0.046	13.3	0.08	343,571	52.50	10.31
MAN-B05	351	343	2.0	0.604	0.033	7.6	0.07	347,143	52.03	10.41
MAN-B06	343	336.5	2.0	0.582	0.034	6.9	0.06	357,143	50.73	10.71
MAN-B07	336.5	329.5	2.0	0.752	0.045	5.9	0.12	325,000	54.91	9.75
MAN-B08	329.5	322	2.0	0.740	0.045	11.8	0.23	315,714	56.12	9.47
MAN-B09	322	314	2.0	0.664	0.040	23.1	0.18	315,714	56.12	9.47
MAN-B10	314	306	2.0	0.758	0.067	17.9	0.25	323,571	55.10	9.71
MAN-B11	306	298	2.0	0.504	0.036	25.2	0.21	317,857	55.84	9.54
MAN-B12	298	290.5	2.0	0.539	0.029	21.3	0.13	333,571	53.80	10.01
MAN-B13	305.5	296.5	2.0	0.572	0.031	13.1	0.09	341,429	52.78	10.24
MAN-B14	296.5	289.5	2.0	0.663	0.058	9.1	0.09	350,000	51.66	10.50
MAN-B15	289.5	282.5	2.0	0.516	0.029	8.7	0.08	325,000	54.91	9.75
MAN-B16	282.5	277.0	2.0	0.539	0.035	7.5	0.05	364,286	49.81	10.93
MAN-B17	277.0	270.0	2.0	0.659	0.055	4.9	0.10	352,143	51.38	10.56
MAN-B18	270.0	263.5	2.0	0.578	0.032	9.5	0.08	357,143	50.73	10.71
MAN-B19	263.5	257.5	2.0	0.596	0.033	8.0	0.10	346,429	52.13	10.39
MAN-B20	257.5	252.0	2.0	0.589	0.032	10.4	0.10	346,429	52.13	10.39
MAN-B21	252.0	230.0	2.0	0.614	0.042	10.2	0.10	349,286	51.76	10.48
MAN-B24	230.0	223.0	2.0	0.529	0.029	10.1	0.15	330,714	54.17	9.92
MAN-B25	223.0	216.0	2.0	0.546	0.045	15.4	0.12	350,714	51.57	10.52
MAN-B26	216.0	207.5	2.0	0.456	0.028	12.3	0.18	335,714	53.52	10.07
MAN-B27	207.5	201.0	2.0	0.499	0.029	17.7	0.14	324,286	55.00	9.73
MAN-B28	201.0	194.0	2.0	0.601	0.033	14.1	0.14	328,571	54.45	9.86

Table 2. *Cont.*

Sample Name	Top Position (cm)	Bottom Position (cm)	Thickness (cm)	^{36}Cl * (105 at/g)	^{36}Cl Uncertainty * (105 at/g)	Cl Total * (ppm)	Cl Total Uncertainty * (ppm)	Ca † (ppm)	O (%)	C (%)
MAN-B29	194.0	187.5	2.0	0.557	0.068	14.4	0.09	342,857	52.59	10.29
MAN-B30	187.5	180.5	2.0	0.483	0.028	9.3	0.08	353,571	51.20	10.61
MAN-B31	180.5	173.5	2.0	0.515	0.028	8.2	0.06	358,571	50.55	10.76
MAN-B32	173.5	166.5	2.0	0.713	0.071	6.1	0.11	343,571	52.50	10.31
MAN-B33	166.5	160.5	2.0	0.584	0.031	11.4	0.07	356,429	50.83	10.69
MAN-B34	160.5	153.0	2.0	0.497	0.026	7.0	0.03	378,571	47.95	11.36
MAN-B35	153.0	141.0	2.0	0.574	0.042	2.8	0.04	369,286	49.16	11.08
MAN-B37	141.0	135.0	2.0	0.612	0.053	4.2	0.09	342,143	52.68	10.26
MAN-B38	135.0	127.5	2.0	0.549	0.028	8.8	0.06	371,429	48.88	11.14
MAN-B39	127.5	120.0	2.0	0.498	0.027	6.1	0.08	365,000	49.71	10.95
MAN-C01	105.0	101.0	2.0	0.535	0.030	8.4	0.08	346,429	52.13	10.39
MAN-C02	101.0	94.0	2.0	0.739	0.047	4.3	0.04	357,143	50.73	10.71
MAN-C03	94.0	88.0	2.0	0.577	0.034	6.3	0.06	364,286	49.81	10.93
MAN-C04	88.0	81.0	2.0	0.515	0.039	12.5	0.12	344,286	52.40	10.33
MAN-C05	81.0	74.3	2.0	0.555	0.041	12.4	0.12	336,429	53.43	10.09
MAN-C06	74.3	67.5	2.0	0.245	0.020	6.7	0.07	346,429	52.13	10.39
MAN-C07	67.5	60.0	2.0	0.534	0.034	10.5	0.11	338,571	53.15	10.16
MAN-C08	60.0	52.0	2.0	0.540	0.043	5.4	0.05	368,571	49.25	11.06
MAN-C09	52.0	45.0	2.0	0.557	0.042	4.3	0.04	358,571	50.55	10.76
MAN-C10	45.0	36.5	2.0	0.544	0.044	9.1	0.09	354,286	51.11	10.63
MAN-C11	36.5	30.0	2.0	0.381	0.030	10.4	0.10	357,143	50.73	10.71
MAN-C12	30.0	21.0	2.0	0.511	0.035	12.1	0.12	355,000	51.01	10.65
MAN-C13	21.0	14.0	2.0	0.402	0.030	22.1	0.22	313,571	56.40	9.41
MAN-C14	14.0	6.5	2.0	0.399	0.033	26.0	0.26	297,143	58.53	8.91
MAN-C15	6.5	0.0	2.0	0.523	0.033	27.6	0.28	296,429	58.62	8.89

* Measured with accelerator mass spectrometry (AMS). † Measured with inductively coupled plasma mass spectrometry (ICP-MS).

Table 3. Cosmogenic nuclide data of the Mugırtepe scarp [24].

Sample Name	Top Position (cm)	Bottom Position (cm)	Thickness (cm)	^{36}Cl * (105 at/g)	^{36}Cl Uncertainty * (105 at/g)	Cl Total * (ppm)	Cl Total Uncertainty * (ppm)	Ca † (ppm)	O (%)	C (%)
MUG-B01	267.5	262.5	2.0	4.512	0.125	12.5	0.12	383,571	49.69	11.51
MUG-B02	262.5	257.5	2.0	4.743	0.120	14.9	0.15	390,714	48.76	11.72
MUG-B03	257.5	250.0	2.0	4.245	0.321	14.2	0.14	382,143	49.88	11.46
MUG-B04	250.0	240.0	2.0	3.986	0.089	13.7	0.14	374,286	50.90	11.23
MUG-B05	240.0	231.5	2.0	4.127	0.121	14.0	0.14	389,286	48.95	11.68
MUG-B06	231.5	226.0	2.0	4.121	0.088	13.0	0.13	352,857	53.68	10.59
MUG-B07	226.0	218.5	2.0	4.372	0.128	17.4	0.17	389,286	48.95	11.68
MUG-B08	218.5	210.0	2.0	4.114	0.104	11.0	0.11	400,000	47.55	12.00
MUG-B09	210.0	203.5	2.0	3.808	0.108	17.3	0.17	396,429	48.02	11.89
MUG-B10	203.5	196.5	2.0	3.649	0.133	16.6	0.17	383,571	49.69	11.51
MUG-B11	196.5	189.0	2.0	3.695	0.116	17.6	0.18	396,429	48.02	11.89
MUG-B12	189.0	181.5	2.0	3.626	0.125	14.0	0.14	396,429	48.02	11.89
MUG-B13	181.5	174.5	2.0	3.309	0.124	13.9	0.14	384286	49.60	11.53
MUG-B14	174.5	166.0	2.0	3.405	0.100	16.2	0.16	384,286	49.60	11.53
MUG-B15	166.0	157.0	2.0	3.328	0.125	17.7	0.18	384,286	49.60	11.53
MUG-B16	157.0	150.5	2.0	3.085	0.091	16.8	0.17	384,286	49.60	11.53
MUG-B17	150.5	144.0	2.0	3.626	0.123	19.5	0.19	384,286	49.60	11.53
MUG-B18	144.0	136.5	2.0	3.209	0.140	16.3	0.16	384,286	49.60	11.53
MUG-B19	136.5	130.0	2.0	3.100	0.129	15.3	0.15	384,286	49.60	11.53
MUG-B20	130.0	123.5	2.0	2.762	0.095	13.7	0.14	384,286	49.60	11.53
MUG-B21	123.5	116.5	2.0	3.024	0.129	13.5	0.13	384,286	49.60	11.53
MUG-B22	116.5	109.5	2.0	3.004	0.119	12.3	0.12	384,286	49.60	11.53
MUG-B23	109.5	103.5	2.0	2.826	0.087	11.7	0.12	384,286	49.60	11.53
MUG-A01	122.5	117.5	2.0	3.287	0.100	14.2	0.14	382,823	49.32	11.59
MUG-A02	117.5	112.5	2.0	3.039	0.108	12.0	0.12	386,429	50.06	11.42
MUG-A03	112.5	107.5	2.0	2.883	0.129	10.1	0.10	380,714	49.41	11.57
MUG-A04	107.5	101.5	2.0	2.902	0.085	12.3	0.12	385,714	48.39	11.81

Table 3. *Cont.*

Sample Name	Top Position (cm)	Bottom Position (cm)	Thickness (cm)	^{36}Cl * (105 at/g)	^{36}Cl Uncertainty * (105 at/g)	Cl Total * (ppm)	Cl Total Uncertainty * (ppm)	Ca † (ppm)	O (%)	C (%)
MUG-A05	101.5	96.0	2.0	2.899	0.109	10.8	0.11	393,571	48.67	11.74
MUG-A06	96.0	91.0	2.0	2.838	0.118	12.7	0.13	391,429	49.60	11.53
MUG-A07	91.0	85.0	2.0	2.856	0.113	11.8	0.12	384,286	49.41	11.57
MUG-A08	85.0	79.0	2.0	2.726	0.083	10.1	0.10	385,714	49.60	11.53
MUG-A09	79.0	73.0	2.0	2.803	0.108	11.3	0.11	384,286	49.88	11.46
MUG-A10	73.0	66.5	2.0	2.448	0.105	11.4	0.11	382,143	50.71	11.27
MUG-A11	66.5	59.0	2.0	2.598	0.080	11.4	0.11	375,714	49.60	11.53
MUG-A12	59.0	51.5	2.0	2.665	0.098	11.5	0.12	384,286	49.78	11.49
MUG-A13	51.5	45.0	2.0	2.507	0.075	11.4	0.11	382,857	49.60	11.53
MUG-A14	45.0	38.5	2.0	2.468	0.118	12.0	0.12	384,286	49.69	11.51
MUG-A15	38.5	31.0	2.0	2.531	0.098	11.7	0.12	383,571	49.78	11.49
MUG-A16	31.0	24.5	2.0	2.409	0.069	11.2	0.11	382,857	50.06	11.42
MUG-A17	24.5	19.0	2.0	2.563	0.073	11.3	0.11	380,714	50.15	11.40
MUG-A18	19.0	13.0	2.0	2.626	0.102	11.1	0.11	380,000	50.25	11.38
MUG-A19	13.0	7.5	2.0	2.259	0.075	10.8	0.11	379,286	49.60	11.53
MUG-A20	7.5	4.5	2.0	2.225	0.068	12.7	0.13	384,286	50.34	11.36
MUG-A21	4.5	0.0	1.0	2.540	0.091	12.7	0.13	378,571	51.64	11.06

* Measured with accelerator mass spectrometry (AMS). † Measured with inductively coupled plasma mass spectrometry (ICP-MS).

Table 4. Average chemical composition of the bedrock and colluvium of the Manastır (this study) and Mugırtepe (modified after [24]) fault scarps used in earthquake modeling.

Fault	Cl, ppm	O, ppm	C, ppm	Na, ppm	Mg, ppm	Al, ppm	Si, ppm	P, ppm	K, ppm
Manastır	10.8	520,796	104,037	445	20,718	1398	4051	218	398
Mugırtepe	13.4	496,184	115,237	371	2593	344	491	196	166
	Ca, ppm	Ti, ppm	Mn, ppm	Fe, ppm	B, ppm	Sm, ppm	Gd, ppm	U, ppm	Th, ppm
Manastır	346,790	72	39	1022	3.4	0.24	0.7	1.16	0.18
Mugırtepe	384,123	30	39	210	1.5	0.05	0.025	0.565	0.05

Note: Cl is measured with accelerator mass spectrometry (AMS), the rest are measured with inductively coupled plasma mass spectrometry (ICP-MS). Average chemical composition is determined from representative samples.

We reanalyzed the Mugırtepe fault data reported by Akçar et al. [24] using the FSDT [37] by applying relatively more precise bottom and top positions of the samples rather than their heights along the sampling profile. Moreover, the calcium concentrations of the individual samples were used in the simulation. In this study, as we remodeled the Mugırtepe fault, which was formerly examined by Akçar et al. [24] using the Schlagenhauf et al. [22] code, it is useful to outline the main differences between the two modeling strategies. With respect to cosmic ray shielding by the fault scarp, while the Schlagenhauf code [22] applies scarp shielding only to neutron spallation, the FSDT code considers all cosmogenic particles producing ^{36}Cl; that is neutron spallation, fast muons, and thermal and epithermal neutrons (cf. [26,37]). In addition, in the Schlagenhauf code, one exponential simplification of muon attenuation is considered, whereas the FSDT approach uses the full model by Heisinger et al. [79,80]. Moreover, considering the exact position of bottom and top of the samples along the fault surface in FSDT is required to obtain more accurate results in terms of distributions of particles at nodes of three-dimensional mesh. This provides coverage of all possible positions of the sample strip to calculate the theoretical ^{36}Cl, which might have been produced. Dimensions of mesh are considered as the depth of sample perpendicular to scarp surface, the position of the samples along the fault surface and the relative position of footwall and colluvium wedge. These differences affect the model outputs with differences of a few percentage points. The FSDT code applies a broad-ranging search for the optimal solution using the Monte-Carlo method. In addition, despite both codes apply forward modeling, the FSDT method uses a two-step modeling process whereby a database is created during the first step, which has the advantage of reducing the simulation running time.

4. Results

4.1. Cosmogenic ^{36}Cl Concentration Analysis

The fault scarp parameters used for the database and default rates of ^{36}Cl production are presented in Table 1. The samples positions, thicknesses, cosmogenic ^{36}Cl and natural Cl values and uncertainties, calcium, oxygen, and carbon contents are provided in Tables 2 and 3 for the Manastır and Mugırtepe faults, respectively. The average compositions of major and trace elements of the bedrock and colluvium are listed in Table 4.

4.2. Time-Slip Histories of the Manastır and Mugırtepe Fault Scarps

Our best-fit model for the Manastır and Mugırtepe faults yields two and one earthquake(s), respectively (Table 5). The best-fit solution resulting from the simulation of the Manastır dataset indicates seismic events at 3.5 ± 0.9 ka and 2.0 ± 0.5 ka, with the beginning of accumulation at 8.8 ka (Figure 7). The modeled slips for these events are 3.3 ± 0.5 m and 3.6 ± 0.5 m, respectively. The Akaike information criterion (AICc) of this simulation was 444.46, the weighted root-mean-square (RMSw) was 2.12, and the chi-square (χ2) value was 4.91 (Table 5). The incremental slip rate of 2.2 mm yr^{-1} is calculated for the time interval between the first and second modeled earthquakes, and 1.8 mm yr^{-1} between the second earthquake and the present. The average slip rate of 1.9 mm yr^{-1} is estimated based on a 6.7-m cumulative throw since the oldest modeled earthquake (Figure 8).

Table 5. Best fit results for the modeling of the Manastır and Mugırtepe fault scarps dataset.

Fault	Beginning of Accumulation (ka)	Age (ka)	Slip (cm)	Throw/Maximum Vertical Displacement (cm)	IncrementalSlip Rate (mm yr^{-1})	Average Slip Rate (mm yr^{-1})	X^2	AICc	RMSw
Manastır	8.8	2.0 ± 0.5 3.5 ± 0.9	3.6 ± 0.5 3.3 ± 0.5	3.5 ± 0.5 3.2 ± 0.5	2.2 1.8	1.9	4.91	444.46	2.12
Mugırtepe	27.0	6.5 ± 1.6	2.7 ± 0.4	2.1 ± 0.3	0.3	0.3	1.88	178.31	1.31

Note: Slip rates are calculated using the maximum vertical displacement (e.g., throw).

The re-analysis of the Mugırtepe fault data from Akçar et al. [24] showed a single seismic event at 6.5 ± 1.6 ka, with the beginning of exposure at 27 ka and a vertical slip of 2.7 ± 0.4 m (Figure 9). For this single-earthquake scenario, the best-fit (AICc) analysis yielded a value of 178.31, RMSw a value of 1.31, and χ^2 a value of 1.88 (Table 5). A slip rate of 0.3 mm yr^{-1} was calculated based on maximum vertical displacement divided by the age of the modeled seismic event (Figure 10). Because the upper parts of both the Manastır and Mugırtepe faults were not suitable for sampling, it should be noted that the number of reconstructed seismic events is minimum.

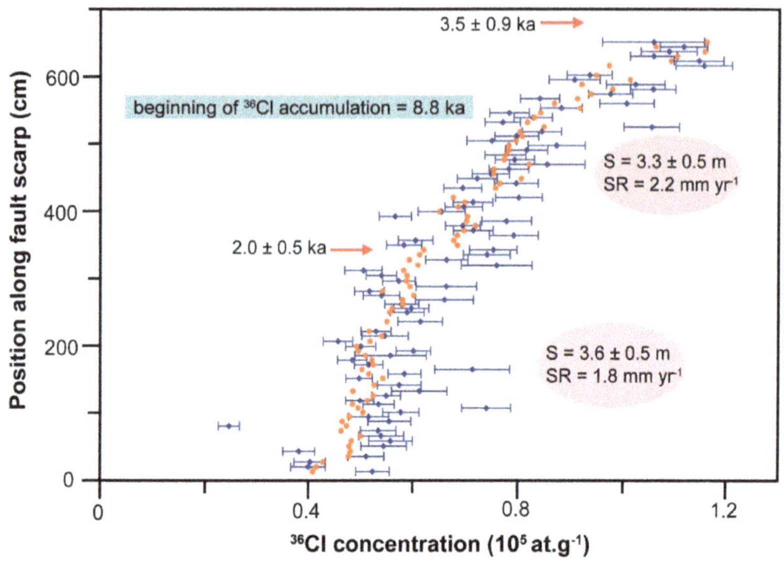

Figure 7. Best fit (filled circles) of the data for the samples from the Manastır fault scarp with a two seismic event model and beginning of ^{36}Cl accumulation ca. 8.8 ka. Dots with 2σ uncertainties are measured ^{36}Cl concentrations. The arrows mark the colluvium positions before the modeled seismic event. S and SR define the amount of slip and incremental slip rate, respectively.

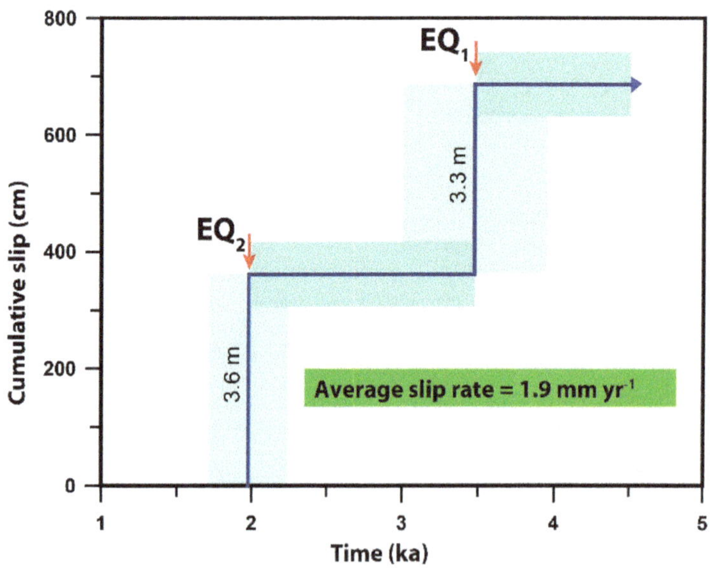

Figure 8. Cumulative slip versus time. Green bands indicate the uncertainties of seismic events ages and colluvium level obtained from modeling of Manastır fault; the average slip rate is 1.9 mm yr^{-1}.

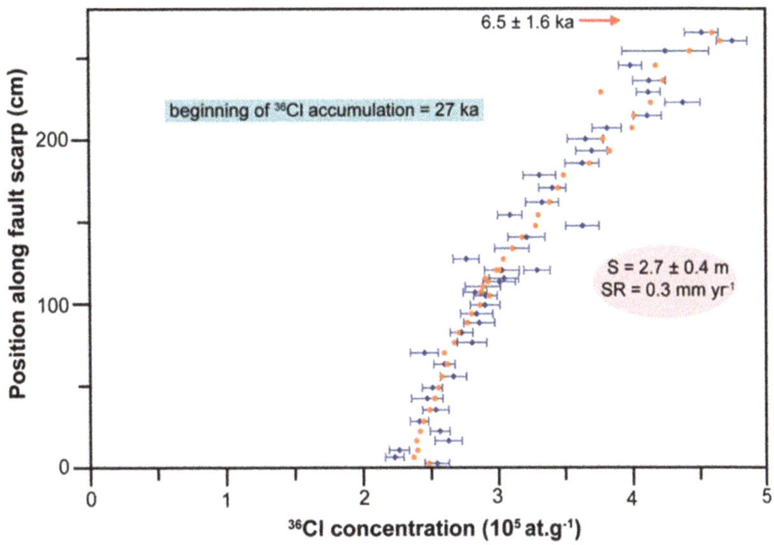

Figure 9. Best fit (filled circles) of the data for the samples from the Mugırtepe fault scarp with a single seismic event model and beginning of ^{36}Cl accumulation ca. 27 ka. Dots with 2σ uncertainties are measured ^{36}Cl concentrations. The arrow marks the colluvium positions before the modeled seismic event. S and SR define the amount of slip and incremental slip rate, respectively.

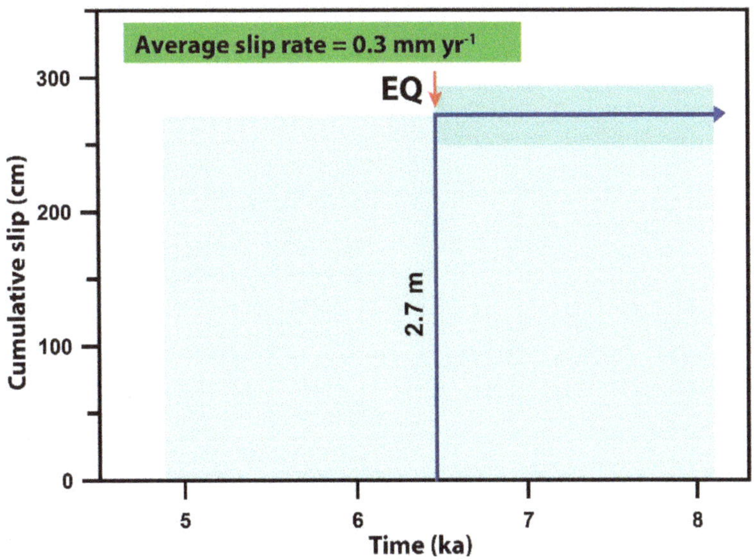

Figure 10. Cumulative slip versus time. Green bands show the uncertainties of seismic events ages and colluvium level obtained from modeling of Mugırtepe fault; the average slip rate is 0.3 mm yr^{-1}.

5. Discussion

5.1. Plausibility of Earthquake Modeling

The modeling of seismic events associated with the Mugırtepe and Manastır faults indicates that both faults slipped during the Holocene. As mentioned above, these faults are elements of the MFZ, therefore they must have moved in response to slip on this fault zone at 6.5 ± 1.6, 3.5 ± 0.9 and 2.0 ± 0.5 ka. The modeled seismic event at 6.5 ± 1.6 ka is close to that of the youngest earthquake modeled by Akçar et al. [24]. Based on the assumption that the Manastır fault is the principal slip surface of the fault zone, we suggest that this event (and probably the older one(s)) was recorded in the upper 5 m of the Manastır fault scarp. We support this argument by comparing measured the cosmogenic ^{36}Cl concentrations and timing of modeled seismic events of Manastır and Mugırtepe faults. The chemical compositions of Mugırtepe and Manastır samples, especially ^{40}Ca concentrations as the main target of cosmogenic ^{36}Cl production, are very similar (Tables 2 and 3) and the longer is the fault surface exposed, the more concentration of cosmogenic ^{36}Cl is expected. We assume that the accumulation pattern of the measured cosmogenic ^{36}Cl concentrations on the Manastır fault (Figure 7) is similar to the accumulation pattern on the Mugırtepe fault (Figure 9). Therefore, we expect that the cosmogenic ^{36}Cl concentrations on the higher unsampled surface of the Manastır fault (>6.5 m), if it was possible to measure, should increase upscarp surface and be relatively close to those on the Mugırtepe fault scarp. However, no chronology can be attributed to this unsampled section with a high degree of certainty, owing poor surface preservation for sampling.

Here, we discuss the fault parameters that arise from our modeling by applying empirical relationships (Table 6) that link the modeled earthquake magnitudes to the surface rupture lengths and displacements [50,81,82]. Theoretically, the instantaneous rupture of the entire 35-km-long MFZ would have required an earthquake with a magnitude of approximately 6.9 and an average slip amount of 1–1.7 m regardless of modeling [50,81,82] (Table 6). By considering Equation (6) in Table 6, the maximum slip of 3.1 m for MFZ was calculated, which fits to the lower bound of modeled slip, though is the most appropriate approach in this case. Therefore, our modeled slips of over 3 m can probably be explained by at least two large-magnitude earthquakes (>6) occurring over a short time span within the uncertainty of modeled ages. However, such concurrent earthquakes cannot be differentiated by the FSDT or any other code or recognized as separate events (cf. [22,37]). We assert the occurrence of clustered earthquake in a close time, because in addition to theoretical calculations above, basically the amount of displacement close to tips of the normal faults is smaller than that around the fault's center (e.g., [83–85]). We propose that the earthquakes that occurred in the MFZ triggered the synchronous displacement of all or some of the main segments of the fault zone, including the Manastır master fault, which in turn resulted in the exposure of secondary fault scarps such as the Mugırtepe.

Table 6. Regression of SRL (surface rupture length), magnitude (Ms/M) and vertical displacement (MVD/MD) values calculated for the Manastır and Mugırtepe faults and the Manisa Fault Zone.

	SRL/FL	**35 km**
Sin (θ) = Maximum Vertical Displacement/Slip		**Manisa Fault Zone (avg. θ = 60°)**
[50]	Ms = 0.9 × Log (SRL) + 5.48	6.9
	Log (MVD) = 1.14 × Ms − 7.82	MVD = 1.0; Slip ~ 1.3
[81]	M = 4.86 + 1.32 × log (SRL)	6.9
	Log (MD) = −5.90 + 0.89 × M	MD (Slip) = 1.7
[82]	Mw = 6.12 + 0.47 × log (SRL)	6.9
	Maximum Slip = 0.09 × SRL Average slip = 0.03 × SRL	Maximum slip = 3.1 Average slip = 1.0

Note: The unit of slip, MVD and MD is in meters. MVD (maximum vertical displacement) is converted to slip or MD (maximum displacement) by applying fault surface dip (sin (θ) = Maximum vertical displacement/slip).

The time span of the modeled seismic events covers a part of the activity of the MFZ during the Holocene, extending the seismic archives significantly beyond the historical records. Indeed, the youngest modeled seismic event at 2.0 ± 0.5 ka temporarily coincides with the most devastating historical earthquake in this region. Many ancient cities within the Manisa Basin and its environs were damaged or completely destroyed in 17 AD by an earthquake with an intensity of IX [3] and a reconstructed Mw of 7.4 (e.g., [57]). Shortly after, in 44 AD, another event with an intensity of VIII damaged the ancient cities of Magnesia and Ephesus [66]. Considering the rough magnitude value of an earthquake possibly sourced by the 35-km-long MFZ (Mw 6.9) and slip (1–3.1 m), we argue that the 17 AD earthquake is a reliable candidate for the event at 2.0 $\pm$ 0.5 ka. The earthquake of 44 AD is mainly attributed to rupture the southern faults close to Izmir and Kemalpaşa (Figure 2); Thus, the 3.6 $\pm$ 0.5 m rupture of the Manastır fault scarp is triggered by earthquake of the 17 AD event and probably a smaller unrecorded earthquake.

Our modeling did not yield any seismic event younger than 2.0 $\pm$ 0.5 ka. At least three additional historical destructive earthquakes, which caused damages in the region, have been reported: the 926 AD, 1595/1664 AD, and 1845 AD events (Figure 2). The epicenter location of the earthquake of 926 AD similar to that of 44 AD appears to be associated with the faults in the south. Moreover, the reconstructed epicenters of the 1595/1664 AD events are located on the eastern part of the southern main boundary of the Gediz Graben and the Izmir fault, respectively (Figure 2) [3,48,58]. In addition, the 1845 AD earthquake, with an epicenter to the east of Manisa is not a definitive earthquake and considered as extreme exaggeration of June 5, 1845 AD Izmir earthquake. The evidence of this event is missing in the Church Missonary Society archives for 1845–46 AD damages of the Izmir–Manisa region [1]. Ambraseys [1] states that this earthquake is only reported by Perrey [86], which claims several weeks before July 23, Manisa was completely destroyed by an earthquake, and this was accordingly reported in modern earthquake catalogues. The abovementioned earthquakes must have initiated significant rupturing of nearby faults but, most likely, negligible or zero rupturing of more distal faults, such as MFZ. However, these events left evidence of liquefaction and lateral spreading in the colluvium in front of the Mugırtepe fault; but their impact in terms of surface rupture or dip-slip displacement of the fault is unclear owing to a lack of field data [59]. These findings could be directly related to availability of organic material, however further discussion regarding the possibilities is beyond the scope of this study.

5.2. Evolution of the Western Manisa Fault Zone

There is a dearth of information about the timing of the initiation of surface rupture of the Manastır fault, which is the main and longest fault in the western MFZ. According to radiocarbon dating of charcoal samples in palaeosol and bulk sediment samples from the Emlakdere Formation, the progressive accumulation and tilting of the Emlakdere Formation in front of the Manastır fault in the hanging-wall began ca. 19 cal kyr BP [59], which can be considered as the lower bound for the initial surface rupture of the Manastır fault (Figure 11). Deposition and progressive tilting of hanging-wall deposits are indicated by diverse dip of bedding planes of the Emlakdere Formation. This syn-sedimentary faulting continued until ca. 9 cal kyr BP based on the ^{14}C age of a palaeosol sample collected from the uppermost part of the Emlakdere Formation within the hanging-wall of F1, where the dip-slip offset of Emlakdere block is a minimum of 12 m (Figure 11) [59]. The time span of sedimentation (from ca. 19 to 9 ka) in this area correlates with the timing of Last Glacial Maximum (LGM) and Termination-I in the northern hemisphere [87–89], when the rate of sedimentation is assumed to be rapid. This implies that the rupture of the secondary faults in the Manisa basin (F1 to F4 including the Mugirtepe Fault) should have initiated after ca. 9 ka. Based on these lines of evidence, we argue that activity of F1, similarly to all the other secondary faults, are younger than surface rupture of the Manastır fault.

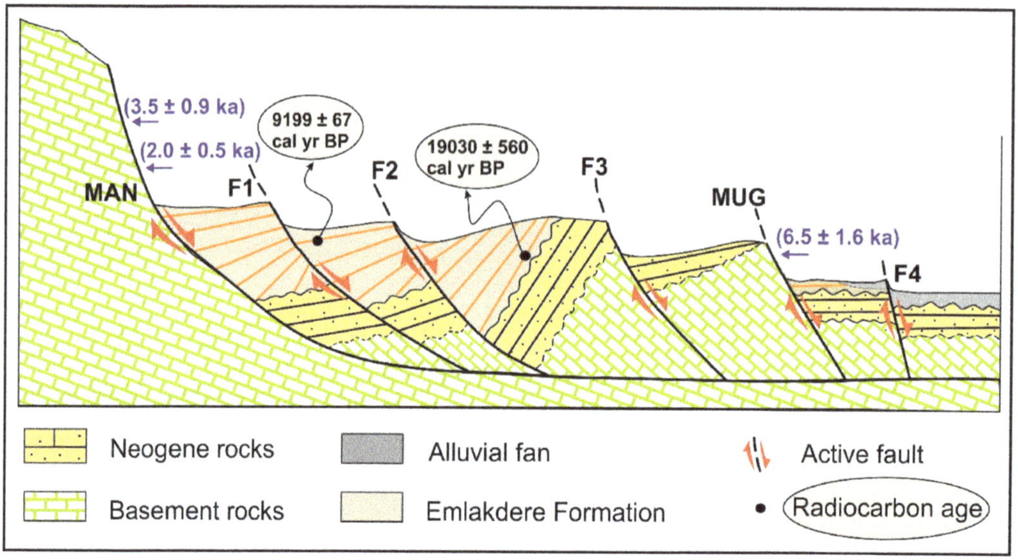

Figure 11. Simplified schematic sketch showing western Manisa Fault Zone and its synthetic faults displacing formations of various ages (Early-late Holocene). Note that in the cartoons the exact horizontal and vertical scales, vertical displacement values as well as thickness of sedimentary layers are disregarded. The approximate lateral distance between MAN and F4 is 400 m. MAN fault depth is 5–10 km at about 10 km northeast of Manisa. MAN: Manastır fault, MUG: Mugırtepe fault. Radiocarbon ages are taken from study of Özkaymak et al. [59], as ca. 19 and 9 cal kyr BP, representing the lowermost and uppermost paleosols within the Emlakdere formation, respectively.

Back-tilting of bedding planes in the hangingwall of the faults, in general decreases towards the basin. Accordingly, we plead that the closest faults to the Manastır fault experienced most likely more earthquakes and higher subsequent slip than those close to the basin. Among those, hangingwall of F2 accommodates the highest backtilting, while the MUG and F4 are characterized by sub-horizontal bedding planes in their hangingwalls. This reveals that most likely not all secondary faults are affected by seismic event simultaneously. Although F4 was not dated in this study, there is field evidence that a Late Holocene alluvial fan is displaced by F4 (Figures 3 and 4). Though, we interpret that F4 was broken by a younger activity than that of responsible for Mugırtepe fault rupture, presumably synchronized with Manastır fault activity either at ca. 3.5 or 2 ka. The deformation of the alluvial fan by F4, which overlies the Mugırtepe fault, assures that F4 ruptured later than the Mugırtepe fault. This sequence of events might be a hint for the basinward migration of the faulting. However, the evolution of the faulting in the Manisa Fault Zone remains obscure and needs to be explored by additional dating studies.

Nevertheless, we propose that the Manastır and Mugırtepe faults could underwent number of earthquakes between ca. 9 cal kyr BP and 6.5 $\pm$ 1.6 ka. These should have resulted in associated slips, which are obscured today in the poorly preserved upper 5 m and 1.3 m of these faults. The seismic event at 6.5 $\pm$ 1.6 ka displaced the Mugırtepe secondary fault by 2.7 $\pm$ 0.4 m, as revealed by our modeling (Figure 11). This event might have occurred as two clustered earthquakes in the Manastır master fault that caused the simultaneous displacement of the Mugırtepe fault, if this is true, these ruptures should presumably be recorded in the current upper 5 m of the Manastır fault. At 3.5 $\pm$ 0.9 ka, the Manastır fault moved by 3.3 $\pm$ 0.5 m as a result of several subsequent earthquakes, which appear not to cause any movement of the Mugırtepe fault. The Manastır fault experienced another seismic event at 2.0 $\pm$ 0.5 ka with a significant displacement of 3.6 $\pm$ 0.5 m (Figure 11), which we attribute to the destructive earthquakes of 17 AD and a probable smaller event missing in historical records (Figure 11).

5.3. Timing of Seismically Active Periods in Western Anatolia

Using fault scarp dating, we reconstructed the oldest discovered seismic event in MFZ at 6.5 ± 0.5 ka followed by the subsequent event at 3.5 ± 0.9 ka. The subsequent modeled earthquake at 2.0 ± 0.5 ka temporally coincides with the 17 destructive earthquakes recorded in the historic records.

We showed that MFZ was active during Holocene similar to other faults in the region (cf. [33,34]). The 2.0 ± 0.5 ka earthquake is highly concordant with the timing of the youngest earthquakes discovered using fault scarp dating on the Yavansu, Priene-Sazlı, and Ören faults (Figure 1). In addition to the Rahmiye fault, all of these faults are considered to have been activated in a close time by the modeled Manastır fault earthquake at 3.5 ± 0.9 ka. Similarly, the timing of the reconstructed earthquake at 6.5 ± 1.6 ka for the Mugırtepe fault is compatible with the age of the reconstructed earthquakes of the Priene-Sazlı and Ören faults. Overall, our fault scarp dating shows that regional seismic activity in Western Anatolia has a rhythmic pattern and is broadly characterized by clusters of surface rupturing earthquakes with phases of high seismic activities with a recurrence interval of ca. 2000 yr.

6. Conclusions

Fault scarp dating in the western MFZ has been observed to be a means of exploring major earthquake events. We documented the occurrence of two and one seismic events, respectively, for the Manastır and Mugırtepe faults as a component of the MFZ during the Holocene. Each of these events is considered to result from clustered earthquakes with the modeled displacements representing the cumulative slip due to these events. The youngest of these events coincides with earthquakes documented in the historic record at 17. The reconstructed earthquakes associated with the Mugırtepe fault are interpreted to have occurred as a consequence of activity on the Manastır fault. While both the Manastır and Mugırtepe faults are tectonic, the former is considered to be seismogenic and the latter non-seismogenic. Our results together with the geological and paleoseismological investigations [59] demonstrate that in the western MFZ, the hangingwall of the master Manastır fault experienced syn-depositional rotation during the Late Pleistocene-early Holocene. Thereafter, secondary faults developed during the Early–late Holocene as a consequence of repeated earthquakes. Our results can unfortunately not solve the growth of the secondary faults. Whether they display a migration pattern or irregular rupture pattern remains to be explored. Our findings are consistent with previous fault scarp dating results from western Turkey [33–35]. This demonstrates the significant potential of this method for deriving the critical parameters required for precise evaluations of seismic risk.

Supplementary Materials: The following are available online at https://www.mdpi.com/article/10.3390/geosciences11110451/s1.

Author Contributions: Conceptualization, N.M., Ç.Ö., D.T., V.A., S.I.-O., H.S., C.S. and N.A.; methodology, D.T. and N.A.; software, D.T.; validation, N.M. and N.A.; formal analysis, N.M.; investigation, N.M., D.T. and N.A.; resources, Ç.Ö., S.I.-O., V.A., H.S., C.S. and N.A.; data curation, N.A.; writing—original draft preparation, N.M. and D.T.; writing—review and editing, N.M. and N.A.; visualization, N.M. and Ç.Ö.; supervision, H.S. and N.A.; project administration, N.A.; funding acquisition, H.S. and N.A.; All authors have read and agreed to the published version of the manuscript.

Funding: This research was funded by Dokuz Eylul University, Turkey [Research Project No. 2007.KB.FEN.047 and 2008.KB.FEN.008]; the Surface Exposure Dating Laboratory, University of Bern, Switzerland; the Bern University Research Foundation and the Swiss National Science Foundation, Switzerland [Project No. 200021-147065].

Data Availability Statement: Data required for simulation including FSDT code, excel files, and databases for both faults are provided as Supplementary Material.

Acknowledgments: We would like to specially thank the Laboratory of Ion Beam Physics operated by the Swiss Federal Institute of Technology, Zurich, Switzerland. We are grateful to Bora Uzel and Onur Sarıoğlu for their help during field studies. We especially acknowledge the constructive comments and feedback given by three anonymous reviewers.

Conflicts of Interest: The authors declare no conflict of interest.

References

1. Ambraseys, N. *Earthquakes in the Mediterranean and Middle East: A Multidisciplinary Study of Seismicity up to 1900*; Cambridge University Press: Cambridge, UK, 2009; p. 947.
2. Shebalin, N.V.; Karnik, V.; Hadzievski, D. *Catalogue of Earthquakes of the Balkan Region. I, UNDP-UNESCO Survey of the Seismicity of the Balkan Region*; UNDP: Skopje, Yugoslavia, 1974; p. 600.
3. Soysal, H.; Sipahioğlu, S.; Kolçak, D.; Altınok, Y. *Historical Earthquake Catalogue of Turkey and Surrounding Area (2100 B.C.–1900 A.D.)*; Technical Report, TÜBİTAK, No. TBAG-341; The Scientific and Technological Research Council of Turkey: Ankara, Turkey, 1981.
4. Papadopoulos, G.A. A Seismic History of Crete: Earthquakes and Tsunamis 2000 B.C.–2011 A.D. *Ocelotos Publ.* **2011**, *65*, 415.
5. England, P.; Jackson, J.A. Uncharted seismic risk. *Nat. Geosci.* **2011**, *4*, 348–349. [CrossRef]
6. Stein, S.; Friedrich, A.M. How much can we clear the crystal ball? *Astron. Geophys.* **2014**, *55*, 2.11–2.17. [CrossRef]
7. Konstantinou, K.I.; Mouslopoulou, V.; Saltogianni, V. Seismicity and Active Faulting around the Metropolitan Area of Athens, Greece. *Bull. Seismol. Soc. Am.* **2020**, *110*, 1924–1941. [CrossRef]
8. Morley, C.K. Patterns of displacement along large normal faults; implications for basin evolution and fault propagation, based on examples from East Africa. *Am. Assoc. Pet. Geol. Bull.* **1999**, *83*, 613–634.
9. Biasi, G.P.; Weldon, R.J.; Fumal, T.E.; Seitz, G.G. Paleoseismic event dating and conditional probability of large earthquakes on the southern San Andreas fault, California. *Bull. Seismol. Soc. Am.* **2002**, *92*, 2761–2781. [CrossRef]
10. Nicol, A.; Walsh, J.; Berryman, K.; Villamor, P. Interdependence of fault displacement rates and paleoearthquakes in an active rift. *Geology* **2006**, *34*, 865–868. [CrossRef]
11. Nicol, A.; Walsh, J.J.; Villamor, P.; Seebeck, H.; Berryman, K.R. Normal fault interactions, paleoearthquakes and growth in an active rift. *J. Struct. Geol.* **2010**, *32*, 1101–1113. [CrossRef]
12. Berryman, K.R.; Villamor, P.; Nairn, I.A.; Van Dissen, R.J.; Begg, J.G.; Lee, J.M. Late Pleistocene surface rupture history of the Paeroa Fault, Taupo Rift, New Zealand. *N. Z. J. Geol. Geophys.* **2008**, *51*, 135–158. [CrossRef]
13. Friedrich, A.M.; Lee, J.; Wernicke, B.P.; Sieh, K. Geologic context of geodetic data across a Basin and Range normal fault, Crescent Valley, Nevada. *Tectonics* **2004**, *23*, TC2015. [CrossRef]
14. Mouslopoulou, V.; Nicol, A.; Walsh, J.J.; Begg, J.G.; Townsend, D.B.; Hristopulos, D.T. Fault-slip accumulation in an active rift over thousands to millions of years and the importance of paleoearthquake sampling. *J. Struct. Geol.* **2012**, *36*, 71–80. [CrossRef]
15. Zreda, M.; Noller, J.S. Ages of prehistoric earthquakes revealed by cosmogenic chlorine-36 in a bedrock fault scarp at Hebgen Lake. *Science* **1998**, *282*, 1097–1099. [CrossRef]
16. Mitchell, S.G.; Matmon, A.; Bierman, P.R.; Enzel, Y.; Caffee, M.; Rizzo, D. Displacement history of a limtesone normal fault scarp, northern Israel, from cosmogenic ^{36}Cl. *J. Geophys. Res.* **2001**, *106*, 4247–4264. [CrossRef]
17. Benedetti, L.; Finkel, R.; King, G.; Armijo, R.; Papanastassiou, D.; Ryerson, F.J.; Flerit, F.; Farber, D.; Stavrakakis, G. Motion on the Kaparelli fault (Greece) prior to the 1981 earthquake sequence determined from Cl-36 cosmogenic dating. *Terra Nova* **2003**, *15*, 118–124. [CrossRef]
18. Benedetti, L.; Finkel, R.; Papanastassiou, D.; King, G.; Armijo, R.; Ryerson, F.; Farber, D.; Flerit, F. Post-glacial slip history of the Sparta fault (Greece) determined by Cl-36 cosmogenic dating: Evidence for non-periodic earthquakes. *Geophys. Res. Lett.* **2002**, *29*, 87-1–87-4. [CrossRef]
19. Benedetti, L.; Manighetti, I.; Gaudemer, Y.; Finkel, R.; Malavieille, Y.; Pou, K.; Arnold, M.; Aumaître, G.; Bourlès, D.; Keddadouche, K. Earthquake synchrony and clustering on Fucino faults (Central Italy) as revealed from in situ ^{36}Cl exposure dating. *J. Geophys. Res. Solid Earth* **2013**, *118*, 4948–4974. [CrossRef]
20. Palumbo, L.; Benedetti, L.; Bourles, D.; Cinque, A.; Finkel, R. Slip history of the Magnola fault (Apennines, Central Italy) from ^{36}Cl surface exposure dating: Evidence for strong earthquake over the Holocene. *Earth Planet. Sci. Lett.* **2004**, *225*, 163–176. [CrossRef]
21. Carcaillet, J.; Manighetti, I.; Chauvel, C.; Schlagenhauf, A.; Nicole, J.M. Identifying past earthquakes on an active normal fault (Magnola, Italy) from the chemical analysis of its exhumed carbonate fault plane. *Earth Planet. Sci. Lett.* **2008**, *271*, 145–158. [CrossRef]
22. Schlagenhauf, A.; Gaudemer, Y.; Benedetti, L.; Manighetti, I.; Palumbo, L.; Schimmelpfennig, I.; Finkel, R.; Pou, K. Using in situ Chlorine-36 cosmonuclide to recover past earthquake histories on limestone normal fault scarps: A reappraisal of methodology and interpretations. *Geophys. J. Int.* **2010**, *182*, 36–72. [CrossRef]
23. Schlagenhauf, A.; Manighetti, I.; Benedetti, L.; Gaudemer, Y.; Finkel, R.; Malavieille, J.; Pou, K. Earthquake supercycles in Central Italy, inferred from ^{36}Cl exposure dating. *Earth Planet. Sci. Lett.* **2011**, *307*, 487–500. [CrossRef]

24. Akçar, N.; Tikhomirov, D.; Özkaymak, Ç.; Alfimov, V.; Ivy-Ochs, S.; Sözbilir, H.; Uzel, B.; Schlüchter, C. ^{36}Cl Exposure dating of paleoearthquakes in the Eastern Mediterranean: First results from Western Anatolian Extensional Province, Manisa Fault Zone, Turkey. *Geol. Soc. Am. Bull.* **2012**, *124*, 1724–1735. [CrossRef]
25. Mouslopoulou, V.; Moraetis, D.; Benedetti, L.; Guillou, V.; Bellier, O.; Hristopulos, D. Normal faulting in the forearc of the Hellenic subduction margin: Paleoearthquake history and kinematics of the Spili Fault, Crete, Greece. *J. Struct. Geol.* **2014**, *66*, 298–308. [CrossRef]
26. Tikhomirov, D.; Akçar, N.; Ivy-Ochs, S.; Alfimov, V.; Schlüchter, C. Calculation of shielding factors for production of cosmogenic nuclides in fault scarps. *Quat. Geochronol.* **2014**, *19*, 181–193. [CrossRef]
27. Tikhomirov, D. An Advanced Model for Fault Scarp Dating and Paleoearthquake Reconstruction, with a Case Study of the Gediz Graben Formation (Turkey). Ph.D. Thesis, University of Bern, Bern, Switzerland, 2014; p. 113.
28. Kong, P.; Fink, D.; Chunguang, N.; Xiao, W. Dip-slip rate determined by cosmogenic surface dating on a Holocene scarp of the Daju fault, Yunnan, China. *Tectonophysics* **2010**, *493*, 106–112. [CrossRef]
29. Tesson, J.; Pace, B.; Benedetti, L.; Visini, F.; Delli Rocioli, M.; Arnold, M.; Aumaître, G.; Bourlès, D.L.; Keddadouche, K. Seismic slip history of the Pizzalto fault (central Apennines, Italy) using in situ-produced ^{36}Cl cosmic ray exposure dating and rare earth element concentrations. *J. Geophys. Res. Solid Earth* **2016**, *121*, 1983–2003. [CrossRef]
30. Cowie, P.A.; Phillips, R.J.; Roberts, G.P.; McCaffrey, K.; Zijerveld, L.J.J.; Gregory, L.C.; Faure Walker, J.; Wedmore, L.N.J.; Dunai, T.J.; Binnie, S.A.; et al. Orogen-scale uplift in the central Italian Apennines drives episodic behaviour of earthquake faults. *Sci. Rep.* **2017**, *7*, 44858. [CrossRef]
31. Beck, J.; Wolfers, S.; Roberts, J.P. Bayesian earthquake dating and seismic hazard assessment using chlorine-36 measurements (BED v1). *Geosci. Model Dev.* **2018**, *11*, 4383–4397. [CrossRef]
32. Mechernich, S.; Schneiderwind, S.; Mason, J.; Papanikolaou, I.D.; Deligiannakis, G.; Pallikarakis, A.; Reicherter, K. The seismic history of the Pisia fault (eastern Corinth rift, Greece) from fault plane weathering features and cosmogenic ^{36}Cl dating. *J. Geophys. Res. Solid Earth* **2018**, *123*, 4266–4284. [CrossRef]
33. Mozafari, N.; Sümer, Ö.; Tikhomirov, D.; Ivy-Ochs, S.; Alfimov, V.; Vockenhuber, C.; İnci, U.; Sözbilir, H.; Akçar, N. Holocene seismic activity of the Priene-Sazlı Fault revealed by cosmogenic ^{36}Cl, western Anatolia, Turkey. *Turk. J. Earth Sci.* **2019**, *28*, 410–437. [CrossRef]
34. Mozafari, N.; Tikhomirov, D.; Sümer, Ö.; Özkaymak, Ç.; Uzel, B.; Yeşilyurt, S.; Ivy-Ochs, S.; Vockenhuber, C.; Sözbilir, H.; Akçar, N. Dating of active normal fault scarps in the Büyük Menderes Graben (western Anatolia) and its implications for seismic history. *Quat. Sci. Rev.* **2019**, *220*, 111–123. [CrossRef]
35. Mozafari, N.; Özkaymak, Ç.; Sümer, Ö.; Tikhomirov, D.; Uzel, B.; Yeşilyurt, S.; Ivy-Ochs, S.; Vockenhuber, C.; Sözbilir, H.; Akçar, N. Seismic history of western Anatolia during the last 16 kyr determined by cosmogenic ^{36}Cl dating. *Swiss J. Geosci.* **2021**. in revision.
36. Tesson, J.; Benedetti, L. Seismic history from in situ ^{36}Cl cosmogenic nuclide data on limestone fault scarps using Bayesian reversible jump Markov chain Monte Carlo. *Quat. Geochronol.* **2019**, *52*, 1–20. [CrossRef]
37. Tikhomirov, D.; Mozafari, N.; Ivy-Ochs, S.; Alfimov, V.; Vockenhuber, C.; Akçar, N. Fault Scarp Dating Tool—A MATLAB code for fault scarp dating with in-situ chlorine-36 along with datasets of Yavansu and Kalafat faults. *Data Brief* **2019**, *26*, 104476. [CrossRef]
38. Goodall, H.J.; Gregory, L.C.; Wedmore, L.N.J.; McCaffrey, K.J.W.; Amey, R.M.J.; Roberts, G.P.; Shanks, R.P.; Phillips, R.J.; Hooper, A. Determining histories of slip on normal faults with bedrock scarps using cosmogenic nuclide exposure data. *Tectonics* **2021**, *40*, e2020TC006457. [CrossRef]
39. Crone, A.J.; Machette, M.N.; Bonilla, M.G.; Lienkaemper, J.J.; Pierce, K.L.; Scott, W.E.; Bucknam, R.C. Surface Faulting Accompanying the Borah Peak Earthquake and Segmentation of the Lost River Fault, Central Idaho. *Bull. Seismol. Soc. Am.* **1987**, *77*, 739–770.
40. Stewart, I.S.; Hancock, P.L. Normal fault zone evolution and fault scarp degradation in the Aegean region. *Basin Res.* **1988**, *1*, 139–153. [CrossRef]
41. Hancock, P.L.; Barka, A.A. Kinematic indicators on active normal faults in western Turkey. *J. Struct. Geol.* **1987**, *9*, 573–584. [CrossRef]
42. Sümer, Ö.; Inci, U.; Sözbilir, H. Tectonic evolution of the Söke Basin: Extension-dominated transtensional basin formation in western part of the Büyük Menderes Graben, Western Anatolia, Turkey. *J. Geodyn.* **2013**, *65*, 148–175. [CrossRef]
43. Sözbilir, H.; Sarı, B.; Uzel, B.; Sümer, Ö.; Akkiraz, S. Tectonic implications of transtensional supradetachment basin development in an extension-parallel transfer zone: The Kocaçay Basin, western Anatolia, Turkey. *Basin Res.* **2011**, *23*, 423–448. [CrossRef]
44. Özkaymak, Ç.; Sözbilir, H. Stratigraphic and structural evidence for fault reactivation: The active Manisa fault zone, western Anatolia. *Turk. J. Earth Sci.* **2008**, *17*, 615–635.
45. Uzel, B.; Sözbilir, H.; Özkaymak, Ç.; Kaymakçı, N.; Langeris, C.G. Structural evidence for strike-slip deformation in the İzmir-Balıkesir Transfer Zone and consequences for late Cenozoic evolution of western Anatolia (Turkey). *J. Geodyn.* **2013**, *65*, 94–116. [CrossRef]
46. Özkaymak, Ç.; Sözbilir, H.; Uzel, B. Neogene–Quaternary evolution of the Manisa Basin: Evidence for variation in the stress pattern of the İzmir-Balıkesir Transfer Zone, western Anatolia. *J. Geodyn.* **2013**, *65*, 117–135. [CrossRef]

47. Duman, T.Y.; Çan, T.; Emre, Ö.; Kadirioğlu, F.T.; Başarır Baştürk, N.; Kılıç, T.; Arslan, S.; Özalp, S.; Kartal, R.F.; Kalafat, D.; et al. Seismotectonics database of Turkey. *Bull. Earthq. Eng.* **2018**, *16*, 3277–3316. [CrossRef]
48. Tan, O.; Tapirdamaz, M.C.; Yörük, A. The earthquakes catalogues for Turkey. *Turk. J. Earth Sci.* **2008**, *17*, 405–418.
49. Eyübagil, E.E.; Solak, H.İ.; Kavak, U.S.; Tiryakioglu, İ.; Sözbilir, H.; Aktuğ, B.; Özkaymak, Ç. Present day strike-slip deformation within the southern part of the İzmir-Balıkesir Transfer Zone based on GNSS data and implications for seismic hazard assessment in western Anatolia, Turk. *J. Earth Sci.* **2021**, *30*, 143–160.
50. Pavlides, S.; Caputo, R. Magnitude versus faults' surface parameters: Quantitative relationships from the Aegean Region. *Tectonophysics* **2004**, *380*, 159–188. [CrossRef]
51. Bozkurt, E.; Sozbilir, H. Evolution of the large-scale active manisa fault, Southwest Turkey: Implications on fault development and regional tectonics. *Geodin. Acta* **2006**, *19*, 427–453. [CrossRef]
52. Nicol, A.; Mouslopoulou, V.; Begg, J.; Oncken, O. Displacement accumulation and sampling of paleoearthquakes on active normal faults of Crete in the eastern Mediterranean. *Geochem. Geophys. Geosyst.* **2020**, *21*, e2020GC009265. [CrossRef]
53. United States Geological Survey National Earthquake Information Center. Available online: https://earthquake.usgs.gov/earthquakes/eventpage/us7000c7y0/executive (accessed on 26 October 2021).
54. McKenzie, D. Active tectonics of the Alpine-Himalayan belt: The Aegean Sea and surrounding regions. *Geophys. J. Int.* **1978**, *55*, 217–254. [CrossRef]
55. Ambraseys, N.N.; Jackson, J.A. Faulting associated with historical and recent earthquakes in the Eastern Mediterranean region. *Geophys. J. Int.* **1998**, *133*, 390–406. [CrossRef]
56. Bozkurt, E. Neotectonics of Turkey: A synthesis. *Geodin. Acta* **2001**, *14*, 3–30. [CrossRef]
57. Ambraseys, N.N. Engineering seismology. *Earthq. Eng. Struct. Dyn.* **1988**, *17*, 51–105. [CrossRef]
58. Guidoboni, E.; Comastri, A.; Triana, G. *Catalogue of Ancient Earthquakes in the Mediterranean Area up to the 10th Century*; Istituto Nazionale di Geofisica: Rome, Italy, 1994; ISBN 88-85213-06-5.
59. Özkaymak, Ç.; Sözbilir, H.; Uzel, B.; Akyüz, H.S. Geological and palaeoseismological evidence for Late Pleistocene-Holocene activity on the Manisa Fault Zone, western Anatolia. *Turk. J. Earth Sci.* **2011**, *20*, 449–474.
60. Özkaymak, Ç.; Sözbilir, H. Tectonic Geomorphology of the Spildağı High Ranges, Western Anatolia. *Geomorphology* **2012**, *173–174*, 128–140. [CrossRef]
61. Allen, C.R. Geological criteria for evaluating seismicity. *Geol. Soc. Am. Bull.* **1974**, *86*, 1041–1057. [CrossRef]
62. Paton, S. Active normal faulting, drainage patterns and sedimentation in southwestern Turkey. *J. Geol. Soc.* **1992**, *149*, 1031–1044. [CrossRef]
63. Hanson, K.L.; Kelson, K.I.; Angell, M.A.; Lettis, W.R. *Techniques for Identifying Faults and Determining Their Origins*; Contract Rep. NUREG/CR-5503; Nuclear Regulatory Commission: Washington, DC, USA, 1999; p. 504.
64. McCalpin, J.P. (Ed.) Paleoseismology. In *International Geophysics*, 2nd ed.; Academic Press: Cambridge, MA, USA, 2009; p. 629.
65. Yeats, R. Active Faults Related to Folding. In *Active Tectonics: Studies in Geophysics*; National Academy Press: Washington, DC, USA, 1986.
66. Ergin, K.; Güçlü, U.; Uz, Z. *A Catalogue of Earthquakes for Turkey and Surrounding Area (11 AD.-1964)*; Technical Report, no. 24; İstanbul Technical University: İstanbul, Turkey, 1967.
67. Ambraseys, N.N.; Finkel, C.F. *The Seismicity of Turkey and Adjacent Areas: A Historical Review, 1500–1800*; Eren Publishing & Book Trade: İstanbul, Turkey, 1995; p. 240.
68. Papazachos, B.; Papazachou, C. *The Earthquakes of Greece, Technical Books Edition*; Thessaloniki: Thessaloniki, Greece, 1997; p. 304.
69. Kandilli Observatory and Earthquake Research Institute. Boğaziçi University. Available online: http://www.koeri.boun.edu.tr/new/en (accessed on 26 October 2021).
70. Stone, J.O.; Allan, G.L.; Fifield, L.K.; Cresswell, R.G. Cosmogenic chlorine-36 from calcium spallation. *Geochim. Cosmochim. Acta* **1996**, *60*, 679–692. [CrossRef]
71. Ivy-Ochs, S.; Synal, H.A.; Roth, C.; Schaller, M. Initial results from isotope dilution for Cl and Cl-36 measurements at the PSI/ETH Zurich AMS facility. *Nucl. Instrum. Methods Phys. Res. Sect. B Beam Interact. Mater. At.* **2004**, *223–224*, 623–627. [CrossRef]
72. Ivy-Ochs, S.; Poschinger, A.V.; Synal, H.A.; Maisch, M. Surface exposure dating of the Flims landslide, Graubunden, Switzerland. *Geomorphology* **2009**, *103*, 104–112. [CrossRef]
73. Elmore, D.; Ma, X.; Miller, T.; Mueller, K.; Perry, M.; Rickey, F.; Sharma, P.; Simms, P.; Lipschutz, M.; Vogt, S. Status and plans for the PRIME Lab AMS facility. *Nucl. Instrum. Methods Phys. Res. Sect. B Beam Interact. Mater. At.* **1997**, *123*, 69–72. [CrossRef]
74. Evans, J.M.; Stone, J.O.H.; Fifield, L.K.; Cresswell, R.G. Cosmogenic chlorine-36 production in K-feldspar. *Nucl. Instrum. Methods Phys. Res. Sect. B Beam Interact. Mater. At.* **1997**, *123*, 334–340. [CrossRef]
75. Fink, D.; Vogt, S.; Hotchkis, M. Cross-sections for Cl-36 from Ti at E-p=35–150 MeV: Applications to in-situ exposure dating. *Nucl. Instrum. Methods Phys. Res. Sect. B Beam Interact. Mater. At.* **2000**, *172*, 861–866. [CrossRef]
76. Stone, J.O. Terrestrial chlorine-36 production from spallation of iron. In Proceedings of the 10th International Conference on Accelerator Mass Spectrometry, Berkeley, CA, USA, 5–10 September 2005.
77. Alfimov, V.; Ivy-Ochs, S. How well do we understand production of Cl-36 in limestone and dolomite? *Quat. Geochronol.* **2009**, *4*, 462–474. [CrossRef]
78. Stone, J.O. Air pressure and cosmogenic isotope production. *J. Geophys. Res.* **2000**, *105*, 23753–23759. [CrossRef]

79. Heisinger, B.; Lal, D.; Jull, A.J.T.; Kubik, P.; Ivy-Ochs, S.; Neumaier, S.; Knie, K.; Lazarev, V.; Nolte, E. Production of selected cosmogenic radionuclides by muons: 1. Fast muons. *Earth Planet. Sci. Lett.* **2002**, *200*, 345–355. [CrossRef]
80. Heisinger, B.; Lal, D.; Jull, A.J.T.; Kubik, P.; Ivy-Ochs, S.; Knie, K.; Nolte, E. Production of selected cosmogenic radionuclides by muons: 2. Capture of negative muons. *Earth Planet. Sci. Lett.* **2002**, *200*, 357–369. [CrossRef]
81. Wells, D.L.; Coppersmith, J.K. New empirical relationships among magnitude, rupture length, rupture width, rupture area, and surface displacement. *Bull. Seismol. Soc. Am.* **1994**, *84*, 974–1002.
82. Wesnousky, S.G. Displacement and Geometrical Characteristics of Earthquake Surface Ruptures: Issues and Implications for Seismic-Hazard Analysis and the Process of Earthquake Rupture. *Bull. Seismol. Soc. Am.* **2008**, *98*, 1609. [CrossRef]
83. Walsh, J.J.; Watterson, J. Analysis of the relationship between the displacements and dimensions of faults. *J. Struct. Geol.* **1988**, *10*, 239–247. [CrossRef]
84. Marrett, R.; Allmendinger, R.W. Estimates of strain due to brittle faulting: Sampling of fault populations. *J. Struct. Geol.* **1991**, *13*, 735–738. [CrossRef]
85. Kim, Y.S.; Sanderson, D.J. The relationship between displacement and length of faults: A review. *Earth-Sci. Rev.* **2005**, *68*, 317–334. [CrossRef]
86. Perrey, A. Note sur les tremblements de terre ressentis en 1847. *Mem. Acad. Sci. Arts Belles-Lett.* **1848**, *1847–1848*, 68–115.
87. Denton, G.H.; Anderson, R.F.; Toggweiler, J.R.; Edwards, R.L.; Schaefer, J.M.; Putnam, A.E. The Last Glacial Termination. *Science* **2010**, *328*, 1652–1656. [CrossRef] [PubMed]
88. Shakun, J.D.; Carlson, A.E. A global perspective on Last Glacial Maximum to Holocene climate change. *Quat. Sci. Rev.* **2010**, *29*, 1801–1816. [CrossRef]
89. Stern, J.V.; Lisiecki, L.E. Termination 1 timing in radiocarbon-dated regional benthic δ18O stacks. *Paleoceanography* **2014**, *29*, 1127–1142. [CrossRef]

geosciences

Article

Slope Failure in a Period of Increased Landslide Activity: Sennwald Rock Avalanche, Switzerland

Selçuk Aksay [1,2,3,*], Susan Ivy-Ochs [1,2], Kristina Hippe [2], Lorenz Grämiger [1] and Christof Vockenhuber [2]

1 Department of Earth Sciences, ETH Zürich, Sonneggstrasse 5, CH-8092 Zürich, Switzerland; ivy@phys.ethz.ch (S.I.-O.); graemiger.lorenz@gmail.com (L.G.)
2 Laboratory of Ion Beam Physics, ETH Zürich, Otto-Stern-Weg 5, CH-8093 Zürich, Switzerland; hippe@phys.ethz.ch (K.H.); cvockenh@ethz.ch (C.V.)
3 Soil Geography and Landscape Group, Wageningen University, P.O. Box 47, NL-6700 AA Wageningen, The Netherlands
* Correspondence: selcuk.aksay@wur.nl; Tel.: +31-317-485-465

Abstract: The Säntis nappe is a complex fold-and-thrust structure in eastern Switzerland, consisting of numerous tectonic discontinuities and a range of hillslopes prone to landsliding and large slope failures that modify the topography irreversibly. A slope failure, namely the Sennwald rock avalanche, occurred in the southeast wall of this fold-and-thrust structure due to the rock failure of Lower Cretaceous Helvetic limestones along the Rhine River valley. In this research, this palaeolandslide is examined in a multidisciplinary approach for the first time with detection and mapping of avalanche deposits, dynamic run-out modelling and cosmogenic nuclide dating. During the rock failure, the avalanche deposits were transported down the hillslope in a spreading-deck fashion, roughly preserving the original stratigraphic sequence. The distribution of landslide deposits and surface exposure age of the rock failure support the hypothesis that the landslide was a single catastrophic event. The ^{36}Cl surface exposure age of avalanche deposits indicates an age of 4.3 ± 0.5 ka. This time coincides with a notably wet climate period, noted as a conditioning factor for landslides across the Alps in the mid-Holocene. The contemporaneity of our event at its location in the Eastern Alps provide additional support for the contention of increased regional seismic activity in mid-Holocene.

Keywords: Sennwald rock avalanche; cosmogenic nuclide dating; run-out modelling

Citation: Aksay, S.; Ivy-Ochs, S.; Hippe, K.; Grämiger, L.; Vockenhuber, C. Slope Failure in a Period of Increased Landslide Activity: Sennwald Rock Avalanche, Switzerland. *Geosciences* **2021**, *11*, 331. https://doi.org/10.3390/geosciences11080331

Academic Editors: Ángel Rodés and Jesús Martínez-Frías

Received: 7 July 2021
Accepted: 2 August 2021
Published: 5 August 2021

1. Introduction

The Sennwald landslide is a rock avalanche associated with the Lower Cretaceous Helvetic limestones in the southeast wall of the Säntis nappe along the Rhine River valley. Similar catastrophic events constantly reshape the topography and permanently change landscapes, constituting a threat for human life in mountainous terrains such as the European Alps [1–4]. Understanding the patterns, triggering factors and failure mechanisms of landslides is a challenge, particularly for palaeolandslides as they exhibit insufficient topographic evidence to detect the fingerprints of landsliding [5]. It is well-known today that several factors can provide favourable conditions for slope failures and increase in landslide activity regionally, yet the final trigger is often strong ground shaking [2,6], or lengthy wet periods associated with heavy rainfall or storm events [7]. Although individual research on several landslides indicates both climate and seismic triggers, conditioning and triggering factors for landsliding at an Alpine scale are still under debate [7–13]. To address this problem and understand the landslide patterns, more palaeolandslide sites in the Alps such as the Sennwald landslide should be examined with field mapping, geochronological dating techniques and numerical modelling.

In the past decades, various types of landslides with different triggers have been examined with the recent development of age determination techniques [11]. The first dated catastrophic landslide using cosmogenic nuclides was the Koefels landslide in Austria [14,15], and it was followed by the Flims, Switzerland [16], and Fernpass, Austria [17], landslides. The investigation of landforms with dating techniques has contributed key aspects in our understanding of spatio-temporal landslide distribution patterns and their interaction with climate change in the Alps [9,11]. In particular, dating of mass movements with volumes larger than one million cubic meters using one or a combination of geochronological dating techniques brought a new dimension to the examination of past occurrences and allowed us to forecast future trigger events with respect to their frequency and size [5,11].

Earlier research showed that the inherited rock structure also plays an important role in rock slope stability and occurrence of landslides in various tectonic scales [18,19]. The position and geometry of the hillslope, rock type, orientation of bedding and sliding planes with respect to topographic surface and faults also influence the slope failure type and behaviour [20]. Regarding the rock type, high water supply due to heavy rainfall may result in fluid infiltration through faults and fractures [21] particularly in karstic rocks. In rocks affected by karstification, the maximum fluid pressure on the clay and mud permeability barrier on the sliding plane might generate gravitational catastrophic failures [7]. Understanding the structural complexity and geology of slope failure therefore is of great importance for modelling such occurrences with better accuracy. The modelling of such rock failures should be supported with meticulous reporting of field mapping data of both release and deposition areas. Mapping of landslide-related morphological features such as toma hills or the distribution pattern of landslide deposits in the deposition area is another key aspect in landslide research. A toma (i.e., cone-shaped hill) is a typical hill structure at various sizes made of landslide deposits and commonly seen in rock avalanche related events in the Alps [22]. Identifying such geomorphological features provides useful information in establishing landsliding patterns on a regional scale.

In landslides, both the release and deposition areas, and hillslope morphology are irreversibly affected in different ways depending on whether the landslide is triggered by storm and rainfall events, or earthquakes [23]. The permanent change in the topography can be identified for bedrock landslides triggered by different events at an orogen scale [2,13,23]. Earlier studies established that the topographic fingerprints of storm- and earthquake-related landslides can be distinguished with respect to the distribution of the landsliding in the hillslope at the time of the failure, whether sliding material clusters at ridge crests, toes of the hillslope or is uniformly distributed in the entire hillslope. Most Alpine landslides have not been investigated in terms of the topographic fingerprint identification for landslides with different triggers including the Sennwald landslide.

The Sennwald landslide area has been a site of curiosity for engineers and geologists over the last few decades. This site has been studied by geological companies previously, and they provided a good understanding of the subsurface of the landslide deposition area by borehole data. However, the timing and run-out behaviour of this palaeolandslide were not entirely investigated previously and there is a lack of a comprehensive field mapping. To address some of these open questions, we introduce a multidisciplinary examination of the Sennwald landslide with mapping of the landslide deposits, dynamic run-out modelling using DAN3D and surface exposure dating using cosmogenic ^{36}Cl as well as an analysis of potential triggers and causes of the landslide. We also present the estimated landslide volume and role of the stratigraphic position in slope failure. We believe that including this palaeolandslide in the inventory of Alpine landslides and rock avalanches with volumes larger than one million cubic meters is quite essential, especially with detailed field reporting and age control (i.e., cosmogenic nuclide dating). Additionally, the landscape change in the hillslope and Rhine River valley with respect to erosional processes and landsliding are explained in this work.

Tectonic Setting and Geology

The Sennwald landslide occurred in the Säntis nappe at the eastern end of the Helvetic nappes, which formed as a result of collision between the European and Adriatic plates [24] (Figure 1). The Helvetic nappe consists of allochthonous sediments on top that were overthrust in a northerly direction [25]. The entire nappe system was cut into two nappe systems by a major thrust, namely the Glarus thrust; the upper Glarus known as the Säntis nappe and the lower Glarus nappe [26,27]. The Säntis nappe is a complex fold-and-thrust (i.e., anticline-syncline folding) structure with axial planes dipping dominantly NW-SE in eastern Switzerland (Figure 1). It consists of numerous tectonic discontinuities [26,28,29] that form a range of hillslopes prone to large slope failures. The parallel-striking thrust and detachment faults of this complex structure caused propagation and folding mostly following the same parallel trend as the mountain chain [28]. The internal variety of the nappe structure is also due to the influence of numerous strike-slip faults and thrusts [28].

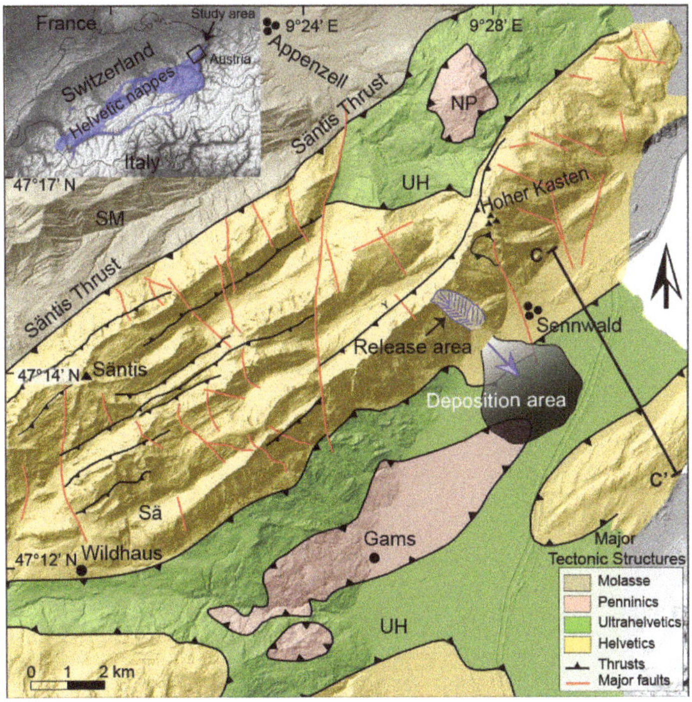

Figure 1. Tectonic map of the Säntis nappe including major tectonic structures and the landslide release and deposition area. Location bar is shown in the top left corner with the Helvetic nappes indicated with blue. Shaded relief image superimposed by a color-coded DEM (swissALTI3D) reproduced with the authorisation of swisstopo (JA100120). The tectonic map is modified following [28].

The release area of the landslide is within the Lower Cretaceous Helvetic limestones. The simplified stratigraphic unit from bottom to top is Vitznau Marl, Betliskalk, Helvetischer Kieselkalk, Drusberg Formation, Schrattenkalk, and Helvetischer Gault. The stratigraphic description of each rock unit in the landslide area is given below.

- Vitznau Marl (*Valanginian*) [29] consists of mainly marly limestones and is a fossiliferous and clay-rich unit at the base.
- Betliskalk (formerly *Valanginienkalk*) [29] is mostly bio-pelsparites and biomicrites with fine grain, brownish to greyish and weathered sand with chert nodules. The weathering colour is grey. Betliskalk is stratigraphically on top of Öhrlikalk and Vitznau-Mergel.
- Helvetischer Kieselkalk (*Valanginian-Hauterivian*) [30,31] consists of siliceous limestone from dark grey to bluish grey colour and is known to be rich in glauconite and pyrite.

It shows a mixture of calcareous particles, calcified sponge spicules and detrital quartz in a matrix of micritic to microsparitic calcite.

- Drusberg Formation (*Barremian*) consists of dark blue-grey calcareous shales regularly alternating with thin-bedded light-brownish limestones. Drusberg layers represent autochthonous fauna that consists of sponge spicules, echinoderm fragments and some radiolarians [32]. Oysters and other bivalves, ammonites and nautilids are also noted [33].
- Schrattenkalk (*Early Barremian-Early Aptian*) [34] is a grey to light grey marine limestone with bioclastic wackestone-packstone, and well-sorted grainstones. Coral, rudist, Nerinea, brachiopod, Porifera, Orbitolinid, echinoid, bivalve and brachiopod are noted in Schrattenkalk [33]. In some parts, detrital quartz and sometimes brecciated lithoclasts are seen in this formation.
- Helvetischer Gault (*Late Aptian-Middle Cenomanian*), Reference [29], also known as Garschella Formation, consists of greenish to greyish glauconitic sandstones, marls and nodular and sparitic carbonates. Ammonoids, bivalve, belemnite, nautilids, brachiopods and planktonic foraminifera are identified.

2. The Sennwald Landslide

2.1. The Release and Deposition Areas

The examination of the release and deposition areas in the landslide site was conducted with field mapping supported by the GIS analysis using high resolution (2 m ± 0.5 resolution) swissALTI3D digital elevation models, local geological map of the Säntis sheet of the Swiss Geological Atlas Nr. 78 with explanatory notes [35,36] and literature [26,28]. The analysis of the stratigraphy and rock structure was also made available to provide an insight to demonstrate the failure type of the landslide [37–40].

The cross section of the inherited rock structure shows that the dip angle of the bedding plane of limestone units in the release area varies between 50–65° SE, and each unit is parallel to each other (Figure 2). The dip angle of the sliding plane is also 50° SE, which is almost parallel to the bedding plane. The parallel position of the bedding and sliding planes to the surface topography is a major component that works for the favour of slope failure as shown in Figure 2.

The landslide material covers an area of about 6 km^2 on the alluvial plain along the Rhine River valley (Figure 3). Landslide boulders and blocks of different rock units from the Lower Cretaceous Helvetic limestone are identified at various dimensions at different locations in the landslide deposition site (Figure 4); $2 \times 2 \times 1$ m^3 being the smallest and $9 \times 8 \times 7$ m^3 the largest boulder. Most of them are covered with lichen and vegetation (Figure 4). As mentioned in the introduction, several hill-like structures, namely toma, are also observed in several locations in the alluvial terrace [41–43] (Figure 4). Tomas are thought to have formed due to the interaction of the spreading moving mass with the substrate in Alpine rock avalanche events. A wet alluvial substrate, as here in the Rhine River plain, is especially conducive to the formation of such extensional surface features [22].

Sediments of the post-failure ongoing erosional activity on the hillslope cover most of the deposition area today. Soft sediments and loose material were continuously transported from the mountain slope since the failure event forming alluvial fans and gullies at the toe of the hillslope. Forestation and vegetation dominate the general landscape in the entire landslide deposition area as well as swamps in places. Fourteen samples from limestone boulders in the deposition area were collected during the field mapping for lithological analysis and ^{36}Cl surface exposure dating (Figure 5). Further explanation on sampling and sample processing is given in "*surface exposure dating using cosmogenic nuclide ^{36}Cl*" section.

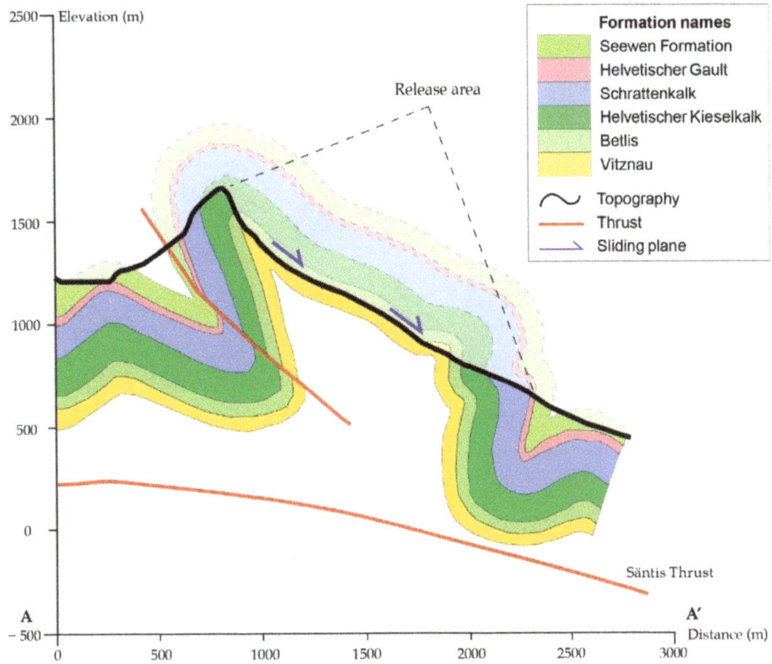

Figure 2. Cross section across the southern wall of the Säntis nappe showing the sliding plane of the Sennwald landslide (modified from [28]). Cross section line (*A-A′*) is shown in the Surface Map.

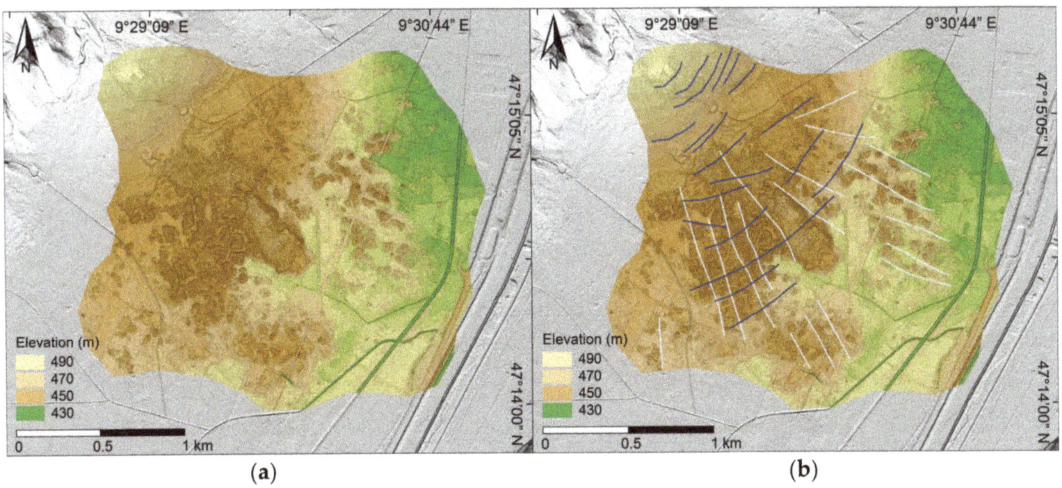

Figure 3. (**a**) Deposition area of the Sennwald landslide (green-brown) shown in detail on a shaded black and white relief image of Sennwald. (**b**) Transverse (blue lines) and linear (white lines) ridges indicate flow patterns. Shaded relief image superimposed by a color-coded DEM 15 (swissALTI3D) reproduced with the authorisation of swisstopo (JA100120).

(a) (b)

Figure 4. Field photos showing landslide boulders of various size and surface features in the landslide deposition area. (**a**) **Left**: 47°15′29 N–9°29′03 E, 559 m a.s.l., photo direction: 241°. **Left top**: Sample SW-13: 47°14′39.02 N–9°29′30.02 E, 442 m a.s.l., photo direction: 210°, boulder dimension: 2 × 3 × 1 m. **Left bottom**: Sample SW-1: 47°15′05.72 N–9°29′54.74 E, 443 m a.s.l., Boulder dimension: 2 × 7 × 3 m. (**b**) Tomas located on the southwestern side of the landslide deposition area. 47°14′63.38 N–30 9°29′10.33 E, 447 m a.s.l., photo direction: 210°.

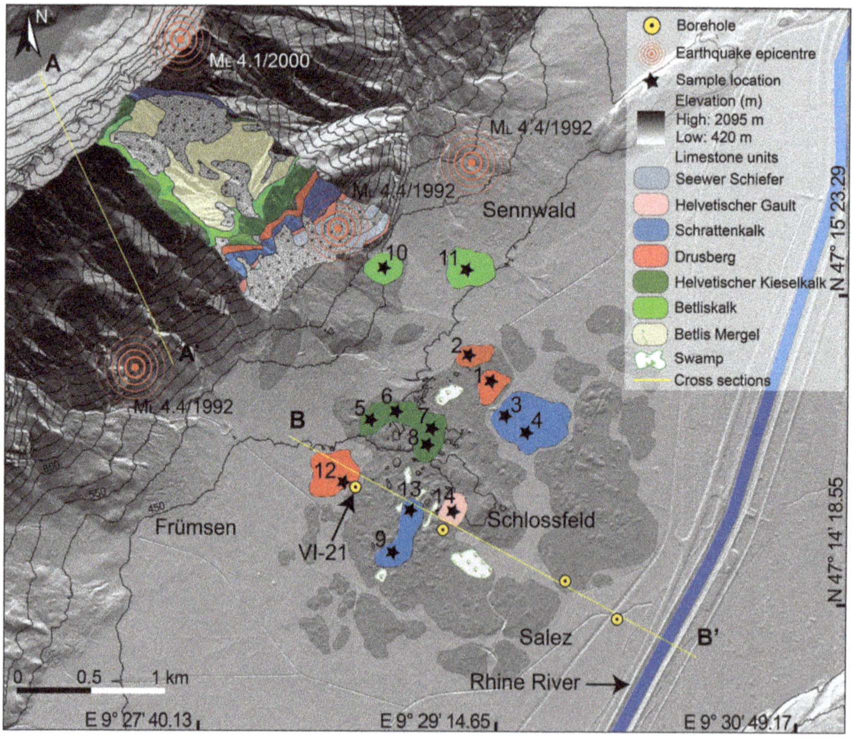

Figure 5. Surface map showing landslide deposits (grey), the dominant boulder lithologies in each sector (blue, yellow, red, green and pink) in the deposition area and the sampling locations (black asterisks indicated with sample numbers). Moderate earthquake epicentres (red asterisks) are also indicated. Altitude is given by colour code. Shaded relief image of the Sennwald region superimposed by a color-coded DEM (swissALTI3D) reproduced by the permission of Swisstopo.

2.2. Distribution of the Landslide Material in the Deposition Area

A comprehensive lithological identification of limestone units was necessary to understand the post-failure distribution pattern of these units in the landslide deposition area. The identification of different limestone units was made based on the textural properties of each limestone unit with hand specimens in the field and thin sections under the microscope (Figure 6). The textural and lithological properties are correlated with the published data in the literature for cross validation (also see *"tectonic setting and geology"* section). The examination of thin sections and hand specimens shows that five limestone units were involved in the rock failure, namely (stratigraphically from bottom to top), Betliskalk, Kieselkalk, Drusberg, Schrattenkalk and Helvetischer Gault (Figures 2 and 5).

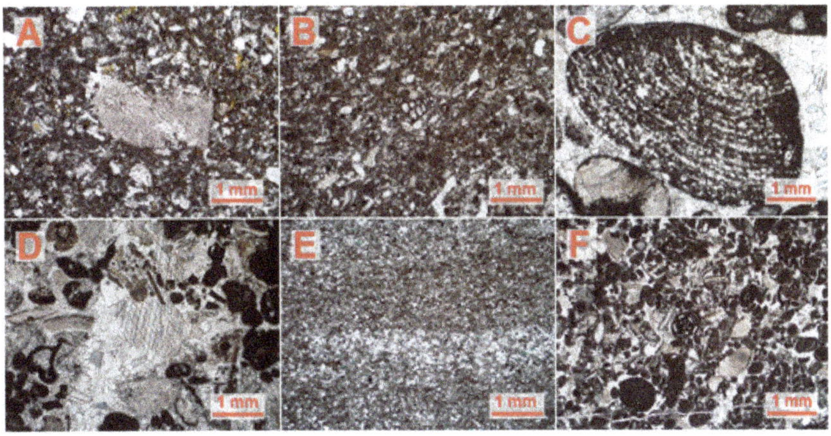

Figure 6. Selection of thin sections from landslide deposits (**A**) Drusberg, sample no: SW-1. (**B**) Drusberg, SW-2. (**C**) Schrattenkalk, SW-3. (**D**) Schrattenkalk, SW-4. (**E**) Helvetischer Kieselkalk, SW-7. (**F**) Schrattenkalk, SW-9. See also text.

Based on field observations, landslide deposits form radial and linear hills (i.e., longitudinal ridges) in plan view in the deposition area (Figure 3) (also see [12,44]). The linear hills are orientated radial and transverse to the direction of the flow. Hills range up to 5–10 m in height above the general ground surface. Further, the landslide material deposited in a spreading-deck fashion, preserving the original stratigraphic positions between strata [45]. Blocks from stratigraphically the top layer (i.e., Schrattenkalk) in the rock formation travelled the farthest, and blocks derived from stratigraphically the lowest layer (i.e., Betliskalk) travelled the least during the landslide. This release pattern results from shearing along boundaries between geological units (e.g., zones of weakness). This information provides a good understanding for the slope failure pattern and distribution of the landslide material.

The spreading-deck distribution of the landslide deposits might also indicate a single and rapid failure. If the failure would have occurred in multiple events, limestones of various lithological units would form an irregularly distributed mass of lithologies in the deposition area rather than forming a pattern preserving the original stratigraphic position after the failure. Some irregularly distributed smaller rock bodies of different limestone lithologies are found in the landslide deposition site, which might contradict the spreading-deck distribution style hypothesis of the landslide material. However, this might be due to the rapid movement of the rock mass which entangled and dragged relatively smaller limestone blocks of different lithologies.

2.3. Cross Profiles for the Release and Deposition Areas

Previously produced cross section of the landslide deposition area based on borehole data [46] shows earlier depositional environments in the landslide deposition area (Figure 7). These data provide useful information for our understanding of the past landscape and sediment transport dynamics from the hillslope to the valley as well as the impact of landsliding on reshaping the topography.

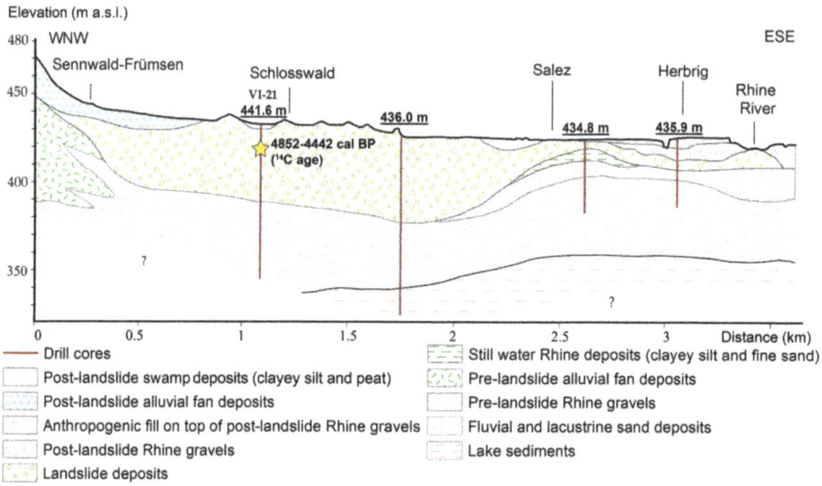

Figure 7. Cross section of the landslide deposition area. Cross section line (B-B′) and boreholes are shown in Figure 5. Modified after [46].

After slope failure, the rapid mass movement in Sennwald changed the morphology of the Rhine River valley as landslide deposits ploughed into the existing pre-landslide Rhine gravels and fluvial/lacustrine sediments. Similar events, where landslide deposits block river valleys and change fluvial regimes, are addressed with case studies in the literature [47,48]. Consequently, the river channel shifted towards E-SE, which enabled post-landslide Rhine gravels to be deposited along the present-day riverbank and the channel (Figure 7).

As shown in the cross section, the debris spreads over fluvial and lacustrine sediments during the landslide event. Afterwards, younger fluvial sediments continuously accumulated on landslide deposits around the present-day valley channel. These marked fluvial fluctuations below and above the landslide deposits indicate an ongoing sediment discharge and load levels of the Rhine River for that time. The slope material (i.e., alluvial fan and debris flow) both below and above the landslide deposits indicates pre- and post-failure ongoing erosional processes (i.e., annual sediment transfer via mountain creeks). The borehole data indicates some sediments with unknown thickness of lacustrine environment below the fluvial and lacustrine sediments, possibly indicating that the valley floor used to be lacustrine environment with transitions to delta and fluvial environments before the landslide event. This information is also consistent with suggested palaeo-lake and -river levels of Quaternary valleys in northern and eastern Switzerland [49].

Extensive work to reconstruct the shape of the overdeepened valleys in the Swiss Alps, formed during the course of glacial/interglacial cycles, provides a good overview of solid rock surface-Quaternary cover boundaries [50,51]. Despite the lack of seismic reflection data in the Sennwald area [52,53], the bedrock surface-Quaternary cover boundary (i.e., depth of the rock surface) for the Sennwald area along the Rhine valley is fairly well-known based on both the reconstructions mentioned above (Figure 8). The geological cross section, which shows the complexity of the rock structure, was made available based on the local geological map, the Säntis sheet of the Swiss Geological Atlas No. 78 [35,36]. As shown in the geological cross section and indicated in those overdeepening reconstructions, the bedrock in the region of the landslide is 10–50 m below the ground surface.

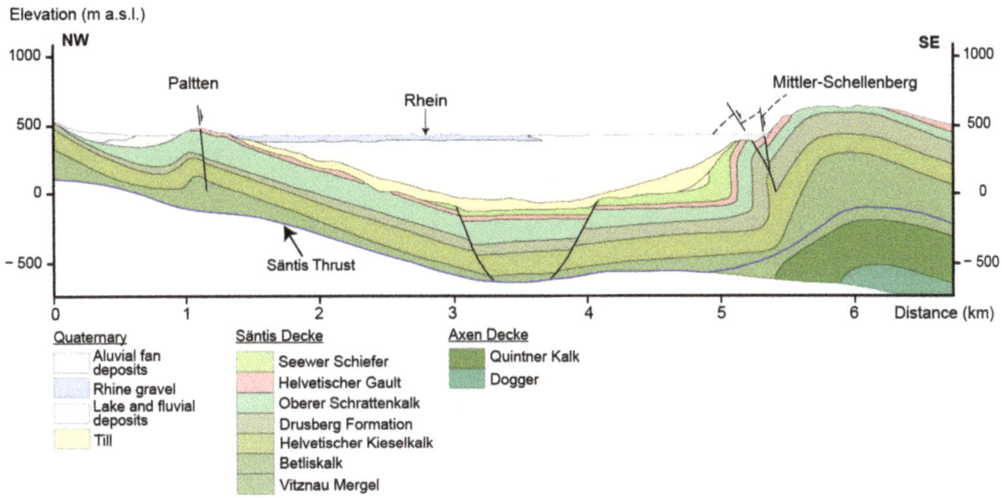

Figure 8. Säntis geological cross section—bedrock surface and Quaternary deposits boundary. Cross section line C-C′ is shown in Figure 1. Modified after [36].

3. Surface Exposure Dating Using Cosmogenic ^{36}Cl

Sampling locations for cosmogenic nuclide dating (Figure 5) were chosen uniformly in the deposition area to evaluate the surface exposure age distribution of landslide boulders. Following the sampling with criterion given by Ivy-Ochs and Kober (2008) [54], samples were prepared as suggested by Ivy-Ochs et al. (2004) [55]. Total Cl and ^{36}Cl were determined at ETH AMS facility with isotope dilution methods as suggested in the literature [56]. The ETH internal standard K381/4N with a value of 17.36×10^{-12} is used to normalise ^{36}Cl/^{35}Cl ratios, and the stable ^{37}Cl/^{35}Cl ratios are standardised to the natural ^{37}Cl/^{35}Cl ratio 31.98% [57,58]. Surface exposure ages are calculated with a numerical code [59]. The production rate of 54 ± 3.5 ^{36}Cl atoms $g_{(Ca)}^{-1}yr^{-1}$ defined for limestones, with a muon contribution of 9.6%, is used for exposure age calculations [60,61]. We used a surface erosion rate for karstified limestones of 0.5 cm/ka for all samples [62–64].

Cl concentrations vary between 4.45 ppm and 19.56 ppm as shown in Table 1. These are rather low total Cl concentration values, which means that the production of ^{36}Cl was dominantly from Ca. Thus, the low energy neutron capture ^{36}Cl production pathway was negligible. The elemental analysis (XRF) is performed to reveal the trace and major element concentrations, and thus the influence of major and trace elements to ^{36}Cl production. The results of XRF analysis also show that trace element concentrations are not significant in ^{36}Cl production in our samples (see Appendix A Table A1). Samples SW-5, 6, 7 and 8 (i.e., rocks of Helvetischer Kieselkalk) showed no reaction with the 2M nitric acid (HNO_3). Therefore, these four samples were not considered for further procedure.

Table 1. AMS-measured ^{36}Cl concentrations and exposure ages with error margins for the Sennwald landslide.

Sample Code	Elevation (m)	Latitude	Longitude	Shielding	^{36}Cl 10^6 Atoms/g Rock	Cl in Rock (ppm)	Exposure Ages (Years)
SW-1	443	47.2516	9.4985	0.992	0.112 ± 0.015	4.5 ± 0.1	3850 ± 520
SW-2	446	47.2530	9.4970	0.991	0.065 ± 0.005	19.6 ± 0.2	2340 ± 120
SW-4	439	47.2484	9.5017	0.985	0.151 ± 0.008	9.7 ± 0.2	5030 ± 310
SW-9	445	47.2415	9.4900	0.990	0.197 ± 0.012	13.6 ± 0.3	6480 ± 440
SW-10	503	47.2582	9.4910	0.980	0.140 ± 0.007	13.1 ± 0.2	4360 ± 260
SW-12	446	47.2458	9.4858	0.990	0.114 ± 0.006	11.5 ± 0.2	3770 ± 220
SW-13	442	47.2442	9.4917	0.990	0.130 ± 0.007	16.0 ± 0.2	4210 ± 260
SW-14	443	47.2439	9.4952	0.990	0.142 ± 0.007	15.5 ± 0.2	4440 ± 250

The mean ^{36}Cl surface exposure age is 4.3 ka with a mean error margin ±0.5 ka. Calculated ages show a uniform distribution in the deposition area with similar ages ranging from 3.7 ka to 5.0 ka except for two samples. Sample SW-2 reveals a relatively shorter exposure age, 2.3 ± 0.1 ka, whereas sample SW-9 reveals a longer exposure age, 6.5 ± 0.4 ka. SW-9, Helvetischer Gault stratigraphically the top layer, was possibly exposed to cosmic rays previously due to its position on the hillslope before the rock failure. The surface exposure age of sample SW-10, Betliskalk (stratigraphically bottom layer), is 4.4 ± 0.3 ka. This indicates that samples (except for SW-2 and 9) show similar surface exposure ages regardless of their (i) distribution in the deposition area and (ii) original stratigraphic position on the hillslope. The similarity of exposure ages provides support for the hypothesis that the Sennwald landslide was a single rock failure.

The mean surface exposure age is also in agreement with the ^{14}C age, 4150 ± 80 ^{14}C yr [65], which we calibrated using OxCal online v4.2.4 [66,67]. Our calibration of ^{14}C ages from previously dated wood pieces taken from the 21–21.5 m depth of borehole VI-21 (for borehole location see Figures 5–7) reveal 4852–4442 cal BP with 95.4% probability. The consistency of surface exposure ages and radiocarbon age also supports the single rock failure hypothesis.

4. Dynamic Run-Out Modelling

4.1. Pre-Failure Topography and Volume Estimation

Reconstruction of the pre-failure topography is a key step for volume estimation and dynamic run-out modelling. Our reconstruction of the pre-failure surface is based on the modern topography in the release and deposition areas with fieldwork observations, thickness estimates from the borehole data and GIS examination. The borehole data and our field measurements show that present-day maximum elevation in the deposition area is 460 m a.s.l., and the average thickness of the landslide deposits is ca. 35 m a.s.l., whereas the maximum thickness is ca. 85 m a.s.l. This shows that the maximum elevation of the pre-failure topography in the deposition area in the Rhine Valley was 425 m a.s.l. Furthermore, using the borehole data, reconstruction of pre-failure elevations and boundaries of the deposition area was made with transitional steps as we adjusted each chosen point based on each neighbouring point accordingly.

For the reconstruction of the pre-failure topography, we estimated the volume of the landslide material by subtracting the reconstructed pre-failure topography from the post-failure (present) topography using the borehole data. The subtraction between the pre- and post-failure topographies reveals the height difference (i.e., deposit thickness). The estimated volume of landslide material in the deposition area is 123 million m^3 and 92 million m^3 in the release area. About 25% of fragmentation bulking factor was calculated following the volume estimates for the deposition and release areas. The bulking factor describes the volume increase due to material dispersion or incorporated material during the run-out on a rock failure and is often suggested to be between 25–30% for similar rock movements [68–70].

4.2. Run-Out Simulation

The numerical run-out modelling is performed for the Sennwald landslide using a dynamic simulation modelling software, namely DAN3D, which is generated for mass movements such as landslides and debris flows [71]. DAN3D is a well-tested tool used for constructing landslide events by simulating the landslide material like a flowing rock avalanche [70,72–76]. The software works on the basis of certain inputs such as the elevation and rheology of rock [71]. The purpose of dynamic modelling is to understand the run-out behaviour, acquire and compare landslide deposit thickness with our estimates and confirm our single rock failure hypothesis.

Parameters used for the landslide simulation following criteria in the literature (References therein [70]) have been listed alongside the modelling results in Table 2. We assumed a unit weight of 24 kN m^{-3} for limestone to perform the modelling in DAN3D. Rheology parameters such as the internal friction angle (which affects the degree of the spreading material) and the friction angle (which represents the basal flow resistance) have been adjusted by trial-and-error to find the best fit for the total run-out distance and lateral spreading of the landslide material in the deposition area [77]. We set the basal friction angle to 16° and internal friction angle 30° after numerous trial runs. The basal friction angle that we applied is within the range of commonly used values for rock avalanches on the basis of earlier research [73,77] (also see [70]).

Table 2. Parameters used for run-out simulation and results.

Parameter	Value
Release volume	92×10^6 (m^3)
Deposition area	6 (km^2)
Unit weight for limestone	24.0 (kN/m^3)
Internal friction angle	35 (°)
Coefficient of friction	0.25
Mean velocity	50 (m/s)
Max-min thickness	60–20 (m)
Plan travel distance	1611 (m)
Total emplacement time	150 (s)

Figure 9 shows the visual run-out modelling results of selected time-steps with simulated deposit thicknesses and landslide deposition area. The landslide run-out direction is to the southeast. The duration of the simulated run-out is 150 s. The maximum velocity is 93 m/s, and mean velocity is approximately 50 m/s. From the top of the hillslope to the farthest point that landslide deposits reached is 4500 m.

The mean deposit thickness retrieved from the simulation, 24 m a.s.l., is within the expected range of present-day mean thickness in the deposition area measured from the borehole data. Although, the observed mean deposit thickness spatially varies in the deposition area and is slightly greater than the simulated deposit thickness. The thickness variation in observed deposits could be explained by inherited pre-existing morphological features of the pre-landslide topography and estimated bulking factor (i.e., 25%). Additionally, simulated deposits show a more evenly distributed pattern, since DAN3D simulates landslide deposits like a whole flowing material. To partially justify these differences, simulated and estimated velocities have been compared. The minimum velocity (v_{min}) required to reach the observed runup height (h) has been estimated following the equation below [78]:

$$v_{min} = \sqrt{(2gh)}$$

where g is gravitational acceleration, and h is the observed maximum deposits thickness (h = 85 m). Our estimated v_{min} is 41 m/s, which agrees well with the simulated model velocity of 50 m/s. This shows that the simulated velocity and deposit thickness are consistent with our predicted model and observed thickness.

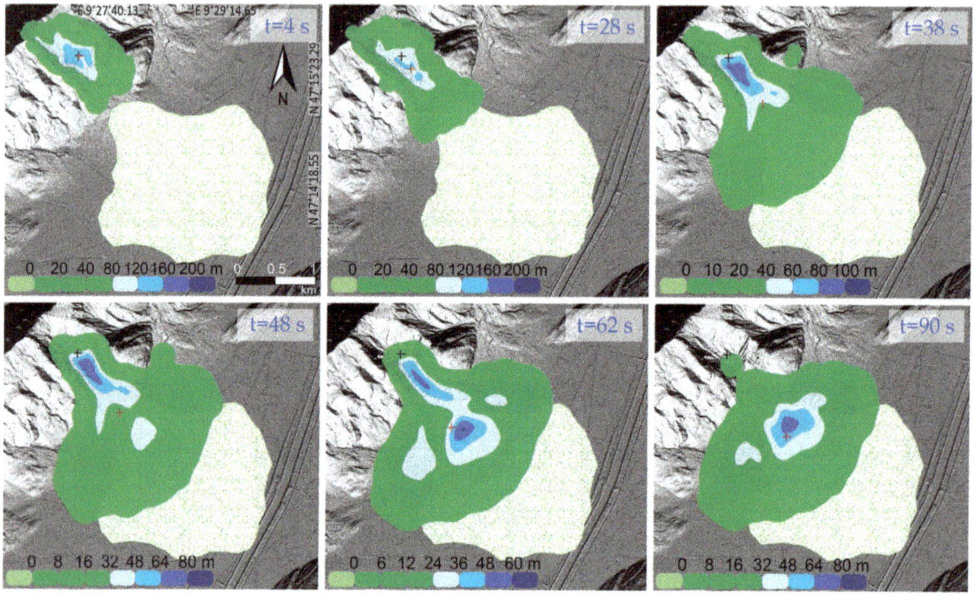

Figure 9. Dynamic run-out modelling showing the thickness of landslide deposits with selected time steps. The dotted area indicated the landslide deposition area determined using DEM in support of field observations. DAN3D numerical modelling code after [71,73].

The lateral spreading of simulated landslide deposits does not entirely resemble the field observations in given time-steps in Figure 9. This difference between the simulated and observed extents is partly due to the bulking factor. DAN3D does not take the bulking factor into account for simulation. Therefore, the spread in the present-day topography is greater than the simulated results. Additionally, there were artefact deposits that move independently from the defined run-out path in the simulation particularly after 90 s (also see [76]). Therefore, only representative visual results have been shown in Figure 9. These artefacts are due to the adjustable parameters (e.g., stiffness coefficient, margin cut-off thickness) within the code to provide flexibility to modify the simulation until the best fit is found, and they, in fact, do not indicate any change in the volume. Overall, the results of our run-out simulation support the hypothesis of a catastrophic single failure for the Sennwald rock avalanche.

5. Discussion: Potential Triggers

Examining the spatio-temporal clustering of landslides of similar size is a key aspect for a better understanding of potential enhancing causes of landslides in the context of Alpine (palaeo)landsliding, such as the Sennwald landslide. Most palaeolandslides in the Alpine setting had previously been associated with the after effects of the last deglaciation (i.e., after around 18,000 years ago) before geochronological dating techniques have been widely used [79]. However, examination of around 40 landslides with dating techniques in the Alps showed that they in fact occurred within two periods; around 11–9 ka year (i.e., within the Preboreal period) and around 5−3 ka year (i.e., within the Subboreal period) (Figure 10) [7,12]. In addition, two major climate fluctuations are identified at a European scale: (i) the mostly dry period at 8.2 ka and (ii) the period of long-lasting heavy rainfall at 4.2 ka [80]. Based on ^{36}Cl and ^{14}C data in our work, the Sennwald landslide occurred at 4.3 ± 0.5 ka, which shows a link with this suggested heavy rainfall period. Although there is a debate for the 4.2 ka period as a trigger for mass movements (see [7]), the remote spatial distribution of similar landslide events in various geographical locations in the Alps around this time [12] indicates a climate change influence possibly as a conditioning factor.

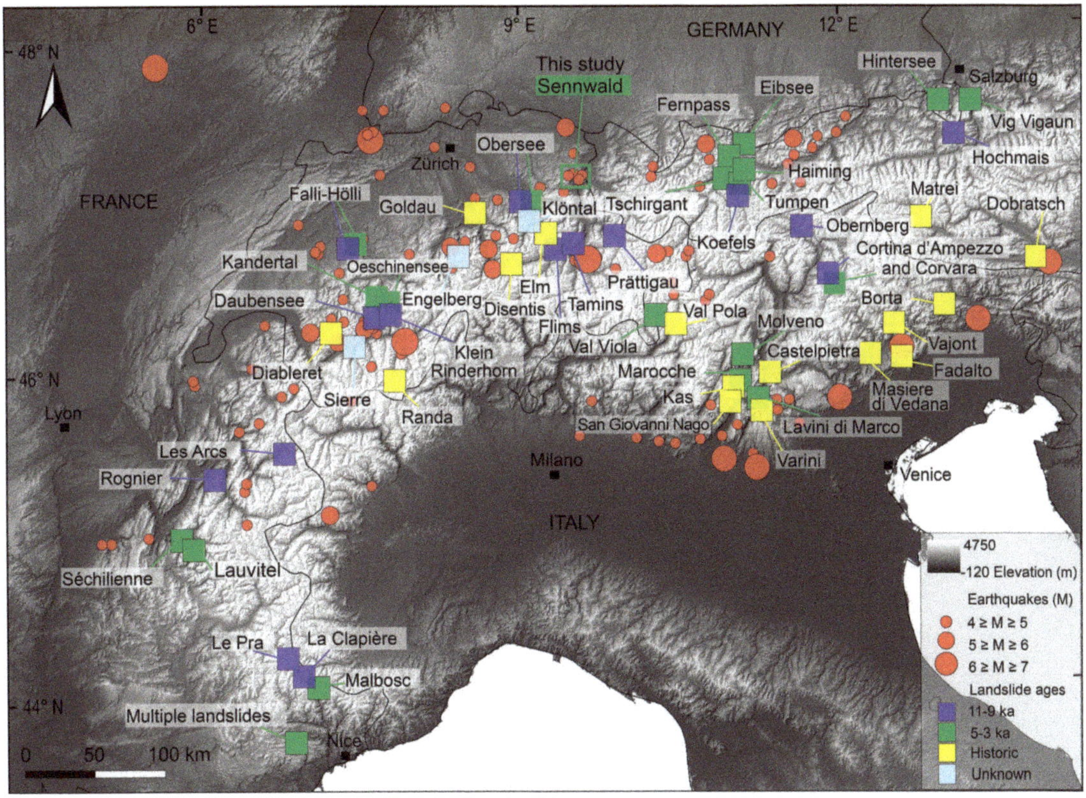

Figure 10. Spatio-temporal distribution of Alpine palaeolandslides shown with instrumental and historical earthquakes epicentres indicated with red circles [81,82]. Circle size indicates the magnitude of the earthquake. Landslides are listed from the oldest to the youngest as follows: around 11–9 ka (blue star): Prättigau 11.5–10.3 ka [83], La Clapiere 10.7–7.1 ka [84,85], Klein Rinderhorn and Daubensee 9.8 ka [70], Koefels 9.5 ka [86], Flims 9.4 ka [87], Tamins 9.4 ka [12], Hochmais 9.3–8.2 [88], Rognier 8.8 ka [89], Obernberg 8.6 ka [90]; around 5–3 ka (green rectangular), Haiming and Vig Vigaun (after [91]), Sechilienne 6.4–0 ka [92], Marocca Principale 5.3 ka [12], Multiple landslides 5.1–3.3 ka [7], Malbosc 5.0–3.0 ka (Braucher, unpublished; from [7]), Lauvitel 4.7 ka [93], Eibsee 4.1 ka [94], Fernpass 4.1 ka [17], Hintersee 3.8 ka [95], Tumpen 3.6 ka [88], Kandertal 3.2 ka [96], Tschirgant 3.0 ka [97], Molveno 4.8 ka [98], Lavini di Marco 3.0–0.8 ka [58], Oeschinensee 2.3 ka [99]; historic landslides (yellow triangle), Matrei, Disentis, Castelpietra, San Giovanni, Masiere di Vedana and Val Pola (after [91]), Fadalto 365 CE [100], Varini 565–970 CE [101], Kas 1100 CE [12], Dobratsch 1348 CE [102], Diableret 1714, 1749 CE, Borta 1692 [103], Goldau 1806 CE, Elm 1881 CE, Vajont 1963 CE, Randa 1991. Multiple events are also shown: Obersee 13.4–5.1, 7.7–3.6 ka [76], Le Pra 12–11, 9–7, 5.0–2.5 ka [104], Les Arcs 11–8.5, 5.5–3.2 ka [105], Cortina and Corvara 10.7–8.4, 8.2–6.9, 5.8–4.5, 4.0–2.1 ka [10], Falli-Hölli 10–9, 5.5–3.2 ka [93]; and landslides with unknown ages (light blue hexagon), Sierre, Engelberg, Klöntal. Hillshade map based on Swisstopo (JA100120).

In Sennwald, landslide deposits were stratigraphically on top of the Vitznau Marl (i.e., the sliding plane of the landslide), a Valanginian clay-rich limestone-marl alternation [29], on pre-failure state (as indicated in the "*tectonic setting and geology*" section). Clays are the weakest and most unstable material on slopes at >10° angle due to the parallel internal realignment of the clay particles which reduces the internal friction over time. The contact between this unstable barrier and weakened (i.e., karstified) limestone package might have prepared the conditions for a slope failure with the presence of water possibly with heavy rainfall influence. It should be noted that heavy rainfall periods should occur over a long period of time to be considered as a trigger or enhancing factor for such large landslides [10,106]. As pointed out, mid-Holocene climate changes have had significant effects on hydrological processes in Europe [107,108], influencing the European Alps as a trigger or enhancing factor for large landslides [7].

Although spatio-temporal clustering of landslides is well-identified at an Alpine scale, especially with the research on the past few decades, seismic triggers leading to large-scale slope failures remain not fully understood. Earthquakes may influence the slope stability reportedly in a number of ways based on field research in various landslide sites since the 18th century [6]. Earlier research suggested a link between landsliding and periods of increased neotectonic activity in Eastern Alps in the mid-Holocene [9,13]. Moreover, it is not impossible to associate palaeolandslides with earthquakes using historical earthquake data or seismic interpretation and core analysis on specific layers in lake sediments (see Figure 10) [109,110]. In a recent study, a spatial correlation between large rock avalanches (i.e., Eibsee and Fernpass) and the largest palaeoearthquake imprints recorded in the lakes was made in the Eastern Alps corresponding to 4.1 ± 0.1 ka BP [13]. This site is about 100 km away from Sennwald. In fact, moderate earthquakes occur around Sennwald even today (Figure 5) [111,112]. The earthquake catalogue [81,82] shows several moderate earthquakes (M_L = 4–5) with 5–10 km depth associated with the Säntis Thrust in Sennwald (Figures 5 and 10). The spatio-temporal contemporaneity of the Sennwald landslide with other palaeolandslides in the Eastern Alps support our hypothesis for an earthquake-triggered landslide.

The inherited rock structure plays an important role in rock slope stability and occurrence of rock slides as shown in earlier research [18,19]. The position of the hillslope, orientation of bedding and sliding planes and lithological differences may also contribute to the rock failure in various mountainous settings. The vulnerability to fragmentation has mostly been observed in large earthquake-related areas with thrusting, and it especially increases in slopes located on the hanging wall of fault systems [113]. The landslide material in Sennwald similarly originated from the hanging wall (Figure 2) of the Säntis thrust during the rock failure. The seismic shaking most likely caused the reduction of the frictional strength along the southeast dipping sliding plane (i.e., almost parallel to the southeast dipping hillslope) in the hanging wall. Therefore, it was crucial to examine the rock structure and geology with meticulous field mapping for a better understanding of our landslide event.

Furthermore, examining the topographic fingerprints of landslides is of importance to understand landsliding patterns and map palaeolandslides with far greater accuracy and in far greater numbers by using topographic as well as image data. Previous studies statistically examined topographic fingerprints of various landslides associated with climate and earthquake triggers on hillslopes, and they identified different patterns for storm and seismic shaking related events [2]. For example, as mentioned in the introduction, it is suggested that slope failures related to permafrost degradation would have high elevation release areas [114,115], whereas earthquake-triggered landslides may have their bedrock niche along ridge crests (seismic amplification) rather than down in the slope [23]. Heavy precipitation related landslides are expected to originate lower down in the slope [23], and rainfall-induced landslides may originate from the entire slope [2]. These landslide-trigger patterns provide useful insights to distinguish earthquake-triggered landslides from climate-triggered landslides. Our observation on present day topography and analysis on the position of rocks in the pre-failure state with respect to surface topography in the hillslope shows that the landsliding was originated in the ridge crest on the entire slope where the susceptibility to slope failure was greatest. This might support our hypothesis that the Sennwald landslide was possibly linked to heavy rainfall as an enhancing factor in a period of increased earthquake activity.

6. Conclusions

This research is a multidisciplinary examination of a palaeolandslide-related surface change in the southeast slope of the Säntis nappe in eastern Switzerland. The detailed investigation of the time, failure and run-out behaviour of the Sennwald palaeolandslide in this overdeepened Alpine valley are presented for the first time in this work.

Our analysis indicates that the orientation of the bedding and sliding planes worked in the favour of rock failure in Sennwald. Surface exposure ages, performed using cosmogenic nuclide ^{36}Cl from eight samples in the landslide deposition area, reveal that the rock failure occurred in the mid-Holocene with the calculated mean age of 4.3 ± 0.5 ka. The radiocarbon dating age, ca. 4.9−4.4 ka cal BP, supports the surface exposure age of the rock failure. The numerical 3D simulation of the landslide, made based on estimated volume, slope angle and friction angle, shows that 92×10^6 m^3 of material moved down the slope in 150 s. The landslide material travelled ca. 2000 m with a maximum velocity of 93 m/s and currently covers an area of about 6 km^2 in the deposition area.

The spreading-deck-like release mechanism of the landslide material with almost preserved stratigraphy, results of the DAN3D simulation and the uniform distribution of surface exposure ages in the deposition area support the single-failure hypothesis. The topographic fingerprint of the Sennwald landslide points out earthquake and heavy rainfall influence. Coincidentally, the time of the rock failure corresponds to the increased neotectonic activity in Eastern Alps, which increases our confidence for the hypothesis that the Sennwald landslide was associated with earthquake activity during a heavy rainfall period.

Author Contributions: Conceptualization, S.A. and S.I.-O.; methodology, S.A, S.I.-O., K.H., L.G. and C.V.; validation, S.A., S.I.-O., K.H., L.G. and C.V.; formal analysis, S.A. and S.I.-O.; investigation, S.A. and S.I.-O.; resources, S.I.-O.; data curation, S.A.; writing—original draft preparation, S.A.; writing—review and editing, S.A., S.I.-O., K.H. and L.G.; visualization, S.A.; supervision, S.I.-O.; project administration, S.I.-O.; funding acquisition, S.I.-O. All authors have read and agreed to the published version of the manuscript.

Funding: Fieldwork, laboratory work and AMS measurements were funded by the Laboratory of Ion Beam Physics, ETH Zürich.

Institutional Review Board Statement: Not applicable.

Informed Consent Statement: Not applicable.

Data Availability Statement: The authors declare that the data supporting the findings of this study are available within the article.

Acknowledgments: All members of the Ion Beam Physics group of ETHZ, led by Arno Synal, are thanked for their support of this work, especially those who did shifts during the measurements. We thank Peter Zwahlen for sharing his unpublished ^{14}C dating data taken from boreholes in the Sennwald area.

Conflicts of Interest: The authors declare no conflict of interest.

Appendix A

Table A1. XRF results for chemical compounds with LOI corrected values and trace elements with LOI corrected values used in the age calculations.

Sample	SW-1	SW-2	SW-4	SW-9	SW-10	SW-12	SW-13	SW-14
SiO_2	1.32419	11.614	0.69853	0.86071	0.8894	0.71356	1.1239	1.36607
TiO_2	0.00909	0.02581	0.01176	0.00905	0.00905	0.00846	0.0085	0.0161
Al_2O_3	0.09483	0.58497	0.09747	0.09896	0.09505	0.10379	0.08502	0.24805
Fe_2O_3	0.75522	0.35147	0.10195	0.1295	0.14201	0.16358	0.13262	0.1437
MnO	0.01249	0.00983	0.0028	0.00509	0.00679	0.00395	0.00397	0.00477
MgO	0.26518	0.46577	0.29073	0.55647	0.47695	0.38245	0.47098	0.4973
CaO	54.896	48.0414	55.3741	55.383	55.4178	55.5288	55.3443	57.7568
Na_2O	0.09767	0.02458	0.01008	0.02092	0.03621	0.01072	0.05214	0.01491
K_2O	0.02385	0.14624	0.01568	0.02036	0.02207	0.01861	0.02154	0.05188
P_2O_5	0.00738	0.02581	0.01008	0.01357	0.01075	0.01015	0.017	0.00716
Cr_2O_3	0	0	0	0	0	0	0	0.00012
NiO	0.00137	0.00127	0.00113	0.00096	0.0011	0.00133	0.00125	0.00122
LOI	43.7522	38.4559	44.4527	43.8691	43.8288	44.0069	43.77	40.6972
Total	**101.239**	**99.7471**	**101.067**	**100.968**	**100.936**	**100.952**	**101.031**	**100.805**
B	0.00000	0.00000	0.00000	0.00000	0.00030	0.00030	0.00030	0.00030
Sm	0.00000	0.00000	0.00000	0.00000	0.00006	0.00006	0.00004	0.00004
Gd	0.00000	0.00000	0.00000	0.00000	0.00006	0.00006	0.00005	0.00005
U	0.00000	0.00000	0.00000	0.00000	0.00040	0.00040	0.00050	0.00050
Th	0.003210	0.003310	0.003940	0.003490	0.000020	0.000020	0.000050	0.000050

References

1. Stark, C.P.; Hovius, N. The characterization of landslide size distributions. *Geophys. Res. Lett.* **2001**, *28*, 1091–1094. [CrossRef]
2. Meunier, P.; Hovius, N.; Haines, J.A. Topographic site effects and the location of earthquake induced landslides. *Earth Planet. Sci. Lett.* **2008**, *275*, 221–232. [CrossRef]
3. Korup, O. Earth's portfolio of extreme sediment transport events. *Earth-Sci. Rev.* **2012**, *112*, 115–125. [CrossRef]
4. Davies, T. *Landslide Hazards, Risks, and Disasters: Introduction*; Elsevier Inc.: Amsterdam, The Netherlands, 2015; ISBN 9780123964755.
5. Malamud, B.D.; Turcotte, D.L.; Guzzetti, F.; Reichenbach, P. Landslide inventories and their statistical properties. *Earth Surf. Process. Landf.* **2004**, *29*, 687–711. [CrossRef]
6. Keefer, D.K. Investigating landslides caused by earthquakes—A historical review. *Surv. Geophys.* **2002**, *23*, 473–510. [CrossRef]
7. Zerathe, S.; Lebourg, T.; Braucher, R.; Bourlès, D. Mid-Holocene cluster of large-scale landslides revealed in the Southwestern Alps by 36Cl dating. Insight on an Alpine-scale landslide activity. *Quat. Sci. Rev.* **2014**, *90*, 106–127. [CrossRef]
8. Corsini, A.; Marchetti, M.; Soldati, M. Dinamica olocenica dei versanti nell'area di Corvara in Badia (Dolomiti, Italia): Cronologia e significato paleoclimatico di alcuni fenomeni franosi. *Geogr. Fis. E Din. Quat.* **2001**, *24*, 127–139.
9. Prager, C.; Zangerl, C.; Brandner, R.; Patzelt, G. Increased rockslide activity in the middle Holocene? New evidence from the Tyrolean Alps (Austria). In *Landslides and Climate Change: Challenges and Solutions*; Taylor & Francis: Oxford, UK, 2007; pp. 25–34.
10. Borgatti, L.; Soldati, M. Landslides as a geomorphological proxy for climate change: A record from the Dolomites (northern Italy). *Geomorphology* **2010**, *120*, 56–64. [CrossRef]
11. Pánek, T. Recent progress in landslide dating: A global overview. *Prog. Phys. Geogr.* **2014**, *39*, 168–198. [CrossRef]
12. Ivy-Ochs, S.; Martin, S.; Campedel, P.; Hippe, K.; Alfimov, V.; Vockenhuber, C.; Andreotti, E.; Carugati, G.; Pasqual, D.; Rigo, M.; et al. Geomorphology and age of the Marocche di Dro rock avalanches (Trentino, Italy). *Quat. Sci. Rev.* **2017**, *169*, 188–205. [CrossRef]
13. Oswald, P.; Strasser, M.; Hammerl, C.; Moernaut, J. Seismic control of large prehistoric rockslides in the Eastern Alps. *Nat. Commun.* **2021**, *12*. [CrossRef]

14. Ivy-Ochs, S.; Heuberger, H.; Kubik, P.; Kerschner, H.; Bonani, G.; Frank, M.; Schlüchter, C. The age of the Köfels event. Relative, 14C and cosmogenic isotope dating of an early Holocene landslide in the Central Alps (Tyrol, Austria). *Z. Für Gletsch. Und Glazialgeol.* **1998**, *34*, 57–68.
15. Kubik, P.W.; Ivy-Ochs, S.; Masarik, J.; Frank, M.; Schlüchter, C. 10Be and 26Al production rates deduced from an instantaneous event within the dendro-calibration curve, the landslide of Köfels, Ötz Valley, Austria. *Earth Planet. Sci. Lett.* **1998**, *161*, 231–241. [CrossRef]
16. Ivy-Ochs, S.; Poschinger, A.V.; Synal, H.A.; Maisch, M. Surface exposure dating of the Flims landslide, Graubünden, Switzerland. *Geomorphology* **2009**, *103*, 104–112. [CrossRef]
17. Prager, C.; Ivy-Ochs, S.; Ostermann, M.; Synal, H.-A.; Patzelt, G. Geology and radiometric 14C-, 36Cl- and Th-/U-dating of the Fernpass rockslide (Tyrol, Austria). *Geomorphology* **2009**, *103*, 93–103. [CrossRef]
18. Moore, J.R.; Sanders, J.W.; Dietrich, W.E.; Glaser, S.D. Influence of rock mass strength on the erosion rate of alpine cliffs. *Earth Surf. Process. Landf.* **2009**, *34*, 1339–1352. [CrossRef]
19. Stead, D.; Wolter, A. A critical review of rock slope failure mechanisms: The importance of structural geology. *J. Struct. Geol.* **2015**, *74*, 1–23. [CrossRef]
20. Ambrosi, C.; Crosta, G.B. Valley shape influence on deformation mechanisms of rock slopes. *Geol. Soc. Spec. Publ.* **2011**, *351*, 215–233. [CrossRef]
21. Sweeting, M.M. *Karst Landforms*; Columbia University Press: New York, NY, USA, 1973.
22. Prager, C.; Zangerl, C.; Nagler, T. Geological controls on slope deformations in the Köfels rockslide area (Tyrol, Austria). *Austrian J. Earth Sci.* **2009**, *102*, 4–19.
23. Densmore, A.L.; Hovius, N. Topographic fingerprints of bedrock landslides. *Geology* **2000**, *28*, 371–374. [CrossRef]
24. Pfiffner, O.A. The Helvetic nappe system and landscape evolution of the Kander. *Bull. Fuer Angew. Geol.* **2010**, *15*, 53–61.
25. Pfiffner, A.O. *Geology of the Alps*, 2nd ed.; John Wiley & Sons: Sussex, UK, 2014; ISBN 978-1-118-70812-5.
26. Pfiffner, A. Fold-and-thrust tectonics in the Helvetic Nappes (E Switzerland). *Thrust Nappe Tectonics. Geol. Soc. Lond.* **1981**, 319–327. [CrossRef]
27. Pfiffner, O.A. The structure of the Helvetic nappes and its relation to the mechanical stratigraphy. *J. Struct. Geol.* **1993**, *15*, 511–521. [CrossRef]
28. Sala, P.; Pfiffner, O.A.; Frehner, M. The Alpstein in three dimensions: Fold-and-thrust belt visualization in the Helvetic zone, eastern Switzerland. *Swiss J. Geosci.* **2014**, *107*, 177–195. [CrossRef]
29. Burger, H.; Strasser, A. Lithostratigraphische Einheiten der untersten Helvetischen Kreide in der Zentral- und Ostschweiz. *Eclogae Geol. Helv.* **1981**, *74*, 529–560.
30. Funk, H. Typusprofile der helvetischen Kieselkalk-Formation und der Altmann-Schichten. *Eclogae Geol. Helv.* **1969**, *62*, 191–203.
31. Funk, H. The origin of authigenic quartz in the Helvetic Siliceous Limestone (Helvetischer Kieselkalk), Switzerland. *Sedimentology* **1975**, *22*, 299–306. [CrossRef]
32. Staeger, D. Geologie der Wilerhorngruppe zwischen Brienz und Lungern. *Eclogae Geol. Helv.* **1944**, *37/1*, 99–188.
33. Kursteiner, P. Field Guide to the excursions. In Proceedings of the 9th International Symposium, Cephalopods-Present and Past; Christian Klug, W.E., Ed.; Paläontologisches Institut und Museum, Universität Zürich: Zurich, Switzerland, 2014.
34. Ziegler, M.A. A Study of the Lower Cretaceous Facies Developements in the Helvetic Border Chain, North of the Lake of Thun (Switzerland). *Eclogae Geol. Helv.* **1967**, *60*, 509–527.
35. Eugster, H.; Forrer, M.; Frohlicher, H.; Kempf, T.; Schlatter, L.; Blaser, R.; Funk, H.; Langenegger, H.; Spoerri, M.; Habicht, K. *Säntis (map sheet 1115) Geological Atlas of Switzerland 1:25.000*; Federal Office of Topography (Swisstopo): Bern, Switzerland, 1982.
36. Funk, H.; Habicht, K.J.; Hantke, R.; Pfiffner, A.O. *Blatt 1115 Säntis—Geologischer Atlas der Schweiz 1:25'000, Erläuterungen 78*; Bundesamt für Wasser und Geologie: Bern, Switzerland, 2000.
37. Gutiérrez-Santolalla, F.; Acosta, E.; Ríos, S.; Guerrero, J.; Lucha, P. Geomorphology and geochronology of sackung features (uphill-facing scarps) in the Central Spanish Pyrenees. *Geomorphology* **2005**, *69*, 298–314. [CrossRef]
38. Cossart, E.; Braucher, R.; Fort, M.; Bourlès, D.L.; Carcaillet, J. Slope instability in relation to glacial debuttressing in alpine areas (Upper Durance catchment, southeastern France): Evidence from field data and 10Be cosmic ray exposure ages. *Geomorphology* **2008**, *95*, 3–26. [CrossRef]
39. Audemard, F.A.; Beck, C.; Carrillo, E. Deep-seated gravitational slope deformations along the active Boconó Fault in the central portion of the Mérida Andes, western Venezuela. *Geomorphology* **2010**, *124*, 164–177. [CrossRef]
40. Pánek, T.; Tábořík, P.; Klimeš, J.; Komárková, V.; Hradecký, J.; Št'astný, M. Deep-seated gravitational slope deformations in the highest parts of the Czech Flysch Carpathians: Evolutionary model based on kinematic analysis, electrical imaging and trenching. *Geomorphology* **2011**, *129*, 92–112. [CrossRef]
41. Furrer, G.; Fitze, P. Beitrag zum Permafrostproblem in den Alpen. *Vierteljahresschr. Der Nat. Ges. Zürich* **1970**, *115*, 353–368.
42. Reitner, J.M. Bericht 2000 über geologische Aufnahmen im Quartär auf Blatt 179 Lienz. *Jahrb. Der Geol. Bundesanst.* **2003**, *143*, 389–395.
43. Von Poschinger, A. Weitere Erkenntnisse und Weitere Fragen zum Flimser Bergsturz. *Bull. Fuer Angew. Geol.* **2006**, *11*, 35–43.
44. Dufresne, A.; Davies, T.R. Longitudinal ridges in mass movement deposits. *Geomorphology* **2009**, *105*, 171–181. [CrossRef]

45. Strom, A. Morphology and internal structure of rockslides and rock avalanches: Grounds and constraints for their modelling. In *Landslides from Massive Rock Slope Failure*; Springer Netherlands: Dordrecht, The Netherlands, 2006; Volume 49, pp. 305–326, ISBN 978-1-4020-4035-1.
46. Kobel, M.; Hantke, R. Zur Hydrogeologie des Rheintales von Sargans bis zum Bodensee (Exkursion E am 19. April 1979). *Jahresber. Und Mitt. Des Oberrh. Geol. Ver.* **1979**, 79–86. [CrossRef]
47. García-García, F.; Sánchez-Gómez, M.; Navarro, V.; Pla, S. Formation, infill, and dissection of a latest-Pleistocene landslide-dammed reservoir (Betic Cordillera, Southern Spain): Upstream and downstream geomorphological and sedimentological evidence. *Quat. Int.* **2011**, *233*, 61–71. [CrossRef]
48. Korup, O.; Wang, G. Multiple Landslide-Damming Episodes. In *Landslide Hazards, Risks and Disasters*; Shroder, J.F., Davies, T., Eds.; Elsevier: Amsterdam, The Netherlands, 2015; pp. 241–261, ISBN 9780123964755.
49. Preusser, F.; Reitner, J.M.; Schlüchter, C. Distribution, geometry, age and origin of overdeepened valleys and basins in the Alps and their foreland. *Swiss J. Geosci.* **2010**, *103*, 407–426. [CrossRef]
50. Dehnert, A.; Lowick, S.E.; Preusser, F.; Anselmetti, F.S.; Drescher-Schneider, R.; Graf, H.R.; Heller, F.; Horstmeyer, H.; Kemna, H.A.; Nowaczyk, N.R.; et al. Evolution of an overdeepened trough in the northern Alpine Foreland at Niederweningen, Switzerland. *Quat. Sci. Rev.* **2012**, *34*, 127–145. [CrossRef]
51. Dürst Stucki, M.; Schlunegger, F. Identification of erosional mechanisms during past glaciations based on a bedrock surface model of the central European Alps. *Earth Planet. Sci. Lett.* **2013**, *384*, 57–70. [CrossRef]
52. Pfiffner, A.; Frei, W.; Finckh, P.; Valasek, P. Deep seismic reflection profiling in the Swiss Alps: Explosion seismology results for line NFP 20-East. *Geology* **1988**, *16*, 987. [CrossRef]
53. Valasek, P.; Mueller, S.; Frei, W.; Holliger, K. Results of NFP 20 seismic reflection profiling along the Alpine section of the European Geotraverse (EGT). *Geophys. J. Int.* **1991**, *105*, 85–102. [CrossRef]
54. Ivy-Ochs, S.; Kober, F. Surface exposure dating with cosmogenic nuclides. *Quat. Sci. J. (Eiszeitalt. Und Ggw.)* **2008**, *57*, 179–209. [CrossRef]
55. Ivy-Ochs, S.; Synal, H.A.; Roth, C.; Schaller, M. Initial results from isotope dilution for Cl and 36Cl measurements at the PSI/ETH Zurich AMS facility. *Nucl. Instrum. Methods Phys. Res. Sect. B Beam Interact. Mater. At.* **2004**, *223–224*, 623–627. [CrossRef]
56. Christl, M.; Vockenhuber, C.; Kubik, P.W.; Wacker, L.; Lachner, J.; Alfimov, V.; Synal, H.-A. The ETH Zurich AMS facilities: Performance parameters and reference materials. *Nucl. Instrum. Methods Phys. Res. Sect. B Beam Interact. Mater. At.* **2013**, *294*, 29–38. [CrossRef]
57. Synal, H.-A.; Bonani, G.; Döbeli, M.; Ender, R.M.; Gartenmann, P.; Kubik, P.W.; Schnabel, C.; Suter, M. Status report of the PSI/ETH AMS facility. *Nucl. Instrum. Methods Phys. Res. Sect. B Beam Interact. Mater. At.* **1997**, *123*, 62–68. [CrossRef]
58. Martin, S.; Campedel, P.; Ivy-Ochs, S.; Viganò, A.; Alfimov, V.; Vockenhuber, C.; Andreotti, E.; Carugati, G.; Pasqual, D.; Rigo, M. Lavini di Marco (Trentino, Italy): 36Cl exposure dating of a polyphase rock avalanche. *Quat. Geochronol.* **2014**, *19*, 106–116. [CrossRef]
59. Alfimov, V.; Ivy-Ochs, S. How well do we understand production of 36Cl in limestone and dolomite? *Quat. Geochronol.* **2009**, *4*, 462–474. [CrossRef]
60. Stone, J.O.; Allan, G.L.; Fifield, L.K.; Cresswell, R.G. Cosmogenic chlorine-36 from calcium spallation. *Geochim. Et Cosmochim. Acta* **1996**, *60*, 679–692. [CrossRef]
61. Stone, J.O.; Evans, J.M.; Fifield, L.K.; Allan, G.L.; Cresswell, R.G. Cosmogenic Chlorine-36 Production in Calcite by Muons. *Geochim. Et Cosmochim. Acta* **1998**, *62*, 433–454. [CrossRef]
62. Ivy-Ochs, S.; Kerschner, H.; Reuther, A.; Maisch, M.; Sailer, R.; Schaefer, J.; Kubik, P.W.; Synal, H.-A.; Schlüchter, C. The timing of glacier advances in the northern European Alps based on surface exposure dating with cosmogenic 10Be, 26Al, 36Cl, and 21Ne. *Geol. Soc. Am. Spec. Pap.* **2006**, *415*, 43–60.
63. Häuselmann, P.; Lauritzen, S.-E.; Jeannin, P.-Y.; Monbaron, M. Glacier advances during the last 400 ka as evidenced in St. Beatus Caves (BE, Switzerland). *Quat. Int.* **2008**, *189*, 173–189. [CrossRef]
64. Ivy-Ochs, S.; Schaller, M. Chapter 6 Examining Processes and Rates of Landscape Change with Cosmogenic Radionuclides. *Radioact. Environ.* **2009**, *16*, 231–294. [CrossRef]
65. Zwahlen, P.; Radiocarbondatierung. Dr. M. Kobel + Partner AG, Büro für Technische Geologie AG, Sargans, Switzerland. Personal communication, 1980.
66. Reimer, P.J.; Austin, W.E.N.; Bard, E.; Bayliss, A.; Blackwell, P.G.; Bronk Ramsey, C.; Butzin, M.; Cheng, H.; Edwards, R.L.; Friedrich, M.; et al. The IntCal20 Northern Hemisphere Radiocarbon Age Calibration Curve (0-55 cal kBP). *Radiocarbon* **2020**, *62*, 725–757. [CrossRef]
67. Ramsey, C.B. Bayesian Analysis of Radiocarbon Dates. *Radiocarbon* **2009**, *51*, 337–360. [CrossRef]
68. Hungr, O.; Evans, S.G. Entrainment of debris in rock avalanches: An analysis of a long run-out mechanism. *Bull. Geol. Soc. Am.* **2004**, *116*, 1240–1252. [CrossRef]
69. Pirulli, M. The Thurwieser rock avalanche (Italian Alps): Description and dynamic analysis. *Eng. Geol.* **2009**, *109*, 80–92. [CrossRef]
70. Grämiger, L.M.; Moore, J.R.; Vockenhuber, C.; Aaron, J.; Hajdas, I.; Ivy-Ochs, S. Two early Holocene rock avalanches in the Bernese Alps (Rinderhorn, Switzerland). *Geomorphology* **2016**, *268*, 207–221. [CrossRef]

71. Hungr, O. A model for the runout analysis of rapid flow slides, debris flows, and avalanches. *Can. Geotech. J.* **1995**, *32*, 610–623. [CrossRef]
72. Evans, S.G.; Hungr, O.; Enegren, E.G. The Avalanche Lake rock avalanche, Mackenzie Mountains, Northwest Territories, Canada: Description, dating, and dynamics. *Can. Geotech. J.* **1994**, *31*, 749–768. [CrossRef]
73. Hungr, O.; Evans, S. Rock avalanche runout prediction using a dynamic model. *Landslides* **1996**, 21.
74. Hungr, O.; Evans, S. A dynamic model for landslides with changing mass. *Eng. Geol. Environ.* **1997**, *41*, 719–722.
75. Davies, T.R.; McSaveney, M.J. Runout of dry granular avalanches. *Can. Geotech. J.* **1999**, *36*, 313–320. [CrossRef]
76. Nagelisen, J.; Moore, J.R.; Vockenhuber, C.; Ivy-Ochs, S. Post-glacial rock avalanches in the Obersee Valley, Glarner Alps, Switzerland. *Geomorphology* **2015**, *238*, 94–111. [CrossRef]
77. Sosio, R.; Crosta, G.B.; Hungr, O. Numerical modeling of debris avalanche propagation from collapse of volcanic edifices. *Landslides* **2012**, *9*, 315–334. [CrossRef]
78. Evans, S.G.; Hungr, O.; Clague, J.J. Dynamics of the 1984 rock avalanche and associated distal debris flow on Mount Cayley, British Columbia, Canada; Implications for landslide hazard assessment on dissected volcanoes. *Eng. Geol.* **2001**, *61*, 29–51. [CrossRef]
79. Ivy-Ochs, S. Glacier variations in the European Alps at the end of the last glaciation. *Cuad. De Investig. Geográfica* **2015**, *41*, 295. [CrossRef]
80. Walker, M.J.C.; Berkelhammer, M.; Björck, S.; Cwynar, L.C.; Fisher, D.A.; Long, A.J.; Lowe, J.J.; Newnham, R.M.; Rasmussen, S.O.; Weiss, H. Formal subdivision of the Holocene Series/Epoch: A Discussion Paper by a Working Group of INTIMATE (Integration of ice-core, marine and terrestrial records) and the Subcommission on Quaternary Stratigraphy (International Commission on Stratigraphy). *J. Quat. Sci.* **2012**, *27*, 649–659. [CrossRef]
81. Fäh, D.; Giardini, D.; Kästli, P.; Deichmann, N.; Gisler, M.; Schwarz-Zanetti, G.; Alvarez-Rubio, S.; Sellami, S.; Edwards, B.; Allmann, B.; et al. ECOS-09 Earthquake Catalogue of Switzerland Release 2011. Report and Database -Events from AD 250. Public Catalogue. Swiss Seismological Service ETH Zürich. 2011. Available online: http://ecos09.seismo.ethz.ch/query_sum (accessed on 15 July 2015).
82. USGS ComCat Earthquake Hazards Program. The ANSS Comprehensive Earthquake Catalog (ComCat). Available online: http://earthquake.usgs.gov/earthquakes/search/ (accessed on 20 May 2021).
83. Dapples, F.; Oswald, D.; Raetzo, H.; Lardelli, T.; Zwahlen, P. New records of Holocene landslide activity in the Western and Eastern Swiss Alps: Implication of climate and vegetation changes. *Ecol. Geol. Helv.* **2003**, *96*, 1–9. [CrossRef]
84. Bigot-Cormier, F.; Braucher, R.; Bourlès, D.; Guglielmi, Y.; Dubar, M.; Stéphan, J.-F. Chronological constraints on processes leading to large active landslides. *Earth Planet. Sci. Lett.* **2005**, *235*, 141–150. [CrossRef]
85. El Bedoui, S.; Bois, T.; Jomard, H.; Sanchez, G.; Lebourg, T.; Trics, E.; Guglielmi, Y.; Bouissou, S.; Chemenda, A.; Rolland, Y.; et al. Paraglacial gravitational deformations in the SW Alps: A review of field investigations, ^{10} Be cosmogenic dating and physical modelling. *Geol. Soc. Lond. Spec. Publ.* **2011**, *351*, 11–25. [CrossRef]
86. Nicolussi, K.; Spötl, C.; Thurner, A.; Reimer, P.J. Precise radiocarbon dating of the giant Köfels landslide (Eastern Alps, Austria). *Geomorphology* **2015**, *243*, 87–91. [CrossRef]
87. von Poschinger, A.; Haas, U. Der Bergsturz von Flims, doch ein warmzeitliches Ereignis. *Bull. Für Angew. Geol.* **1997**, *2*, 35–46.
88. Prager, C.; Zangerl, C.; Patzelt, G.; Brandner, R. Age distribution of fossil landslides in the Tyrol (Austria) and its surrounding areas. *Nat. Hazards Earth Syst. Sci.* **2008**, *8*, 377–407. [CrossRef]
89. Hippolyte, J.-C.; Brocard, G.; Tardy, M.; Nicoud, G.; Bourlès, D.; Braucher, R.; Ménard, G.; Souffaché, B. The recent fault scarps of the Western Alps (France): Tectonic surface ruptures or gravitational sackung scarps? A combined mapping, geomorphic, levelling, and 10Be dating approach. *Tectonophysics* **2006**, *418*, 255–276. [CrossRef]
90. Ostermann, M.; Sanders, D.; Ivy-Ochs, S.; Alfimov, V.; Rockenschaub, M.; Römer, A. Early Holocene (8.6ka) rock avalanche deposits, Obernberg valley (Eastern Alps): Landform interpretation and kinematics of rapid mass movement. *Geomorphology* **2012**, *171–172*, 83–93. [CrossRef]
91. Martin, S.; Ivy-Ochs, S.; Viganò, A.; Campedel, P.; Rigo, M.; Vockenhuber, C.; Gabrieli, F.; Mair, V.; Rossato, S. LANDSLIDES OF THE WESTERN DOLOMITES: CASE STUDIES FROM THE ADIGE AND SARCA VALLEYS (NE ITALY). *Alp. Mediterr. Quat.* **2020**, *33*, 191–207. [CrossRef]
92. Le Roux, O.; Schwartz, S.; Gamond, J.F.; Jongmans, D.; Bourles, D.; Braucher, R.; Mahaney, W.; Carcaillet, J.; Leanni, L. CRE dating on the head scarp of a major landslide (Séchilienne, French Alps), age constraints on Holocene kinematics. *Earth Planet. Sci. Lett.* **2009**, *280*, 236–245. [CrossRef]
93. Delunel, R.; Hantz, D.; Braucher, R.; Bourlès, D.L.; Schoeneich, P.; Deparis, J. Surface exposure dating and geophysical prospecting of the Holocene Lauvitel rock slide (French Alps). *Landslides* **2010**, *7*, 393–400. [CrossRef]
94. Jerz, H.; von Poschinger, A. Neuere Ergebnisse zum Bergsturz Eibsee-Grainau. *Geol. Bavarica* **1995**, *99*, 383–398.
95. von Poschinger, A.; Thom, P. Bergsturz Hintersee/Ramsau (Berchtesgadener land): Neue Untersuchungsergebnisse. *Quartär Südbayern* **1995**, *99*, 399.
96. Singeisen, C.; Ivy-Ochs, S.; Wolter, A.; Steinemann, O.; Akçar, N.; Yesilyurt, S.; Vockenhuber, C. The Kandersteg rock avalanche (Switzerland): Integrated analysis of a late Holocene catastrophic event. *Landslides* **2020**, *17*, 1297–1317. [CrossRef]

97. Ostermann, M.; Ivy-Ochs, S.; Sanders, D.; Prager, C. Multi-method (^{14}C, ^{36}Cl, ^{234}U/ ^{230}Th) age bracketing of the Tschirgant rock avalanche (Eastern Alps): Implications for absolute dating of catastrophic mass-wasting. *Earth Surf. Process. Landf.* **2017**, *42*, 1110–1118. [CrossRef]
98. von Wartburg, J.; Ivy-Ochs, S.; Aaron, J.; Martin, S.; Leith, K.; Rigo, M.; Vockenhuber, C.; Campedel, P.; Viganò, A. Constraining the Age and Source Area of the Molveno landslide Deposits in the Brenta Group, Trentino Dolomites (Italy). *Front. Earth Sci.* **2020**, *8*, 164. [CrossRef]
99. Köpfli, P.; Grämiger, L.M.; Moore, J.R.; Vockenhuber, C.; Ivy-Ochs, S. The Oeschinensee rock avalanche, Bernese Alps, Switzerland: A co-seismic failure 2300 years ago? *Swiss J. Geosci.* **2018**, *111*, 205–219. [CrossRef]
100. Abele, G. Bergstürze in den Alpen: Ihre Verbreitung. In *Morphologie und Folgeerscheinungen, Wissenschaftliche Alpenvereinshefte 25*; Deutscher Alpenverein: München, Germany, 1974.
101. Orombelli, G.; Sauro, U. I Lavini di Marco: Un gruppo di frane oloceniche nel contesto morfoneotettonico dell'alta Val Lagarina (Trentino). *Suppl. Di Geogr. Fis. E Din. Quat.* **1988**, *1*, 107–116.
102. Brandt, A. Die Bergstürze an der Villacher Alpe (Dobratsch), Kärnten/ Östereich: Untersuchungen zur Ursache und Mechanik der Bergstürze. Ph.D. Thesis, University of Hamburg, Hamburg, Germany, 1981.
103. Bonnard, C. Technical and Human Aspects of Historic Rockslide-Dammed Lakes and Landslide Dam Breaches. Springer: Berlin/Heidelberg, Germany, 2011; pp. 101–122.
104. Sanchez, G.; Rolland, Y.; Corsini, M.; Braucher, R.; Bourlès, D.; Arnold, M.; Aumaître, G. Relationships between tectonics, slope instability and climate change: Cosmic ray exposure dating of active faults, landslides and glacial surfaces in the SW Alps. *Geomorphology* **2010**, *117*, 1–13. [CrossRef]
105. Hippolyte, J.-C.; Bourlès, D.; Braucher, R.; Carcaillet, J.; Léanni, L.; Arnold, M.; Aumaitre, G. Cosmogenic 10Be dating of a sackung and its faulted rock glaciers, in the Alps of Savoy (France). *Geomorphology* **2009**, *108*, 312–320. [CrossRef]
106. Soldati, M.; Corsini, A.; Pasuto, A. Landslides and climate change in the Italian Dolomites since the Late glacial. In Proceedings of the Catena; Elsevier: Amsterdam, The Netherlands, 2004; Volume 55, pp. 141–161.
107. Simonneau, A.; Chapron, E.; Vannière, B.; Wirth, S.B.; Gilli, A.; Di Giovanni, C.; Anselmetti, F.S.; Desmet, M.; Magny, M. Multidisciplinary distinction of mass-movement and flood-induced deposits in lacustrine environments: Implications for Holocene palaeohydrology and natural hazards (Lake Ledro, Southern Alps, Italy). *Clim. Past Discuss.* **2012**, *8*, 3205–3249. [CrossRef]
108. Magny, M.; Combourieu-Nebout, N.; De Beaulieu, J.L.; Bout-Roumazeilles, V.; Colombaroli, D.; Desprat, S.; Francke, A.; Joannin, S.; Ortu, E.; Peyron, O.; et al. North-south palaeohydrological contrasts in the central mediterranean during the holocene: Tentative synthesis and working hypotheses. *Clim. Past* **2013**, *9*, 2043–2071. [CrossRef]
109. Strasser, M.; Anselmetti, F.S.; Fäh, D.; Giardini, D.; Schnellmann, M. Magnitudes and source areas of large prehistoric norhern Alpine earthquakes revealed by slope failures in lakes. *Geology* **2006**, *34*, 1005–1008. [CrossRef]
110. Strasser, M.; Monecke, K.; Schnellmann, M.; Anselmetti, F.S. Lake sediments as natural seismographs: A compiled record of Late Quaternary earthquakes in Central Switzerland and its implication for Alpine deformation. *Sedimentology* **2013**, *60*, 319–341. [CrossRef]
111. Calais, E.; Nocquet, J.M.; Jouanne, F.; Tardy, M. Current strain regime in the Western Alps from continuous Global Positioning System measurements, 1996-2001. *Geology* **2002**, *30*, 651–654. [CrossRef]
112. Froitzheim, N.; Plasienka, D.; Schuster, R. Alpine tectonics of the Alps and Western Carpathians. *Geol. Cent. Eur.* **2008**, *2*, 1141–1232.
113. Brune, J.N. Shattered Rock and Precarious Rock Evidence for Strong Asymmetry in Ground Motions during Thrust Faulting. *Bull. Seismol. Soc. Am.* **2001**, *91*, 441–447. [CrossRef]
114. Stoffel, M.; Huggel, C. Effects of climate change on mass movements in mountain environments. *Prog. Phys. Geogr.* **2012**, *36*, 421–439. [CrossRef]
115. Deline, P.; Akçar, N.; Ivy-Ochs, S.; Kubik, P.W. Repeated Holocene rock avalanches onto the Brenva Glacier, Mont Blanc massif, Italy: A chronology. *Quat. Sci. Rev.* **2015**, *126*, 186–200. [CrossRef]

geosciences

MDPI

Article

Reconstructing the Gorte and Spiaz de Navesele Landslides, NE of Lake Garda, Trentino Dolomites (Italy)

Giacomo Ruggia [1], Susan Ivy-Ochs [1,2], Jordan Aaron [1], Olivia Steinemann [2], Silvana Martin [3], Manuel Rigo [3], Sandro Rossato [3,*], Christof Vockenhuber [2], Giovanni Monegato [4] and Alfio Viganò [5]

1 Department of Earth Sciences, ETH Zürich, Sonneggstrasse 5, 8092 Zürich, Switzerland; giacomo.ruggia@sunrise.ch (G.R.); ivy@phys.ethz.ch (S.I.-O.); jordan.aaron@erdw.ethz.ch (J.A.)
2 Laboratory of Ion Beam Physics, ETH Hönggerberg, Otto-Stern-Weg 5, 8093 Zürich, Switzerland; okronig@phys.ethz.ch (O.S.); vockenhuber@phys.ethz.ch (C.V.)
3 Department of Geosciences, University of Padova, Via Gradenigo 6, 35131 Padova, Italy; silvana.martin@unipd.it (S.M.); manuel.rigo@unipd.it (M.R.)
4 Institute of Geosciences and Earth Resources, National Research Council of Italy, Via Gradenigo 6, 35131 Padova, Italy; giovanni.monegato@igg.cnr.it
5 Servizio Geologico, Provincia Autonoma di Trento, Via Zambra 42, 38121 Trento, Italy; alfio.vigano@provincia.tn.it
* Correspondence: sandro.rossato@unipd.it

Citation: Ruggia, G.; Ivy-Ochs, S.; Aaron, J.; Steinemann, O.; Martin, S.; Rigo, M.; Rossato, S.; Vockenhuber, C.; Monegato, G.; Viganò, A. Reconstructing the Gorte and Spiaz de Navesele Landslides, NE of Lake Garda, Trentino Dolomites (Italy). *Geosciences* **2021**, *11*, 404. https://doi.org/10.3390/geosciences11100404

Academic Editors: Hans-Balder Havenith and Jesus Martinez-Frias

Received: 25 August 2021
Accepted: 21 September 2021
Published: 25 September 2021

Publisher's Note: MDPI stays neutral with regard to jurisdictional claims in published maps and institutional affiliations.

Abstract: We applied a multi-method approach to reconstruct the Gorte rock avalanche (85–95 Mm^3) located at the northeastern end of Lake Garda. The combination of field mapping, characterization of bedrock discontinuities, Dan3D-Flex runout modeling and dating of boulders with cosmogenic ^{36}Cl supports the conclusion that the deposits stem from a single rock avalanche at 6.1 ± 0.8 ka. The Gorte event may have triggered the Spiaz de Navesele–Salto della Capra landslide (3.2 Mm^3), whose deposits cover the southern end of the Gorte deposits. First-order controls on detachment were the NNE–SSW- and WNW–ESE-oriented fractures in the limestone bedrock, related to the Giudicarie and Schio-Vicenza fault systems, respectively. Dan3D-Flex runout modeling sufficiently reproduced the Gorte rock avalanche, which involved detachment and sliding of a quasi-intact block, likely along marly interbeds, followed by rapid disintegration. The frictional rheology in the source area and the turbulent frictional rheology (Voellmy) in the remaining part best replicate the observed deposit extent and thickness. Heavy precipitation that occurred at that time may have contributed to failure at Gorte. Nonetheless, its timing overlaps with the nearby (<15 km) Dosso Gardene (6630–6290 cal BP) and Marocca Principale (5.3 ± 0.9 ka) landslides, making a seismic trigger plausible.

Keywords: European Alps; rock avalanche; cosmogenic ^{36}Cl; Dan3D-Flex runout modeling

1. Introduction

Large-scale gravitational events are relevant processes for understanding landscape evolution in mountainous regions and can pose a serious risk to people and infrastructure [1–3]. In the Alps, landslides, including rock avalanches, are responsible for a tragic number of casualties and economic losses. In recent decades, a growing interest in these phenomena has developed, and they are increasingly studied in all of their aspects. In particular, if landslides can runout or fall into natural lakes [4,5], or into artificially dammed lakes as in the case of Vajont (NE Italy) [6] or hydroelectric basins (Tibet, Yarlung Tsangpo Tibet 22 March 2021, https://sandrp.in/2021/03/30/massive-landslide-on-yarlung-tsangpo-on-march-22-2021/), they can produce hazard cascades, such as landslide-generated tsunamis.

Back-analysis and modeling of past rock slope failures [7,8], combined with studies on landforms, sedimentology and internal structures of landslide deposits [9–11], allow insight into movement processes and emplacement sequences [12–14]. Reconstructing the processes that occur during release and emplacement provides fundamental data on the reasons for past and possible future slope failures, and on the extent of possibly impacted areas [7].

Rock avalanches have also been studied as phenomena occurring in specific areas or time periods, moving the focus from a single event to a regional analysis. Clustering of some major rock slope failures in space and time has been recognized in the Alps [15–20], suggesting factors acting over vast areas and over relatively long timespans. Detailed studies on volume, release area and runout characteristics of pre-historic and historic landslides contribute to our understanding of post-glacial slope dynamics and landscape evolution [4,13,16,21–25]. Inaccurate interpretations of these features can lead to misleading conclusions regarding rock avalanche hazards and mechanisms.

In our study, we focused on two closely related rock avalanche deposits located in the Nago-Torbole region in the far northeastern corner of Lake Garda (Trentino Province, Italy) (Figure 1). These deposits belong to a group of landslides that are located along the shores of Lake Garda, the plain of the town of Nago and the Loppio area and are genetically related to landslides within the Sarca Valley up to the Brenta Dolomite group (Figure 2).

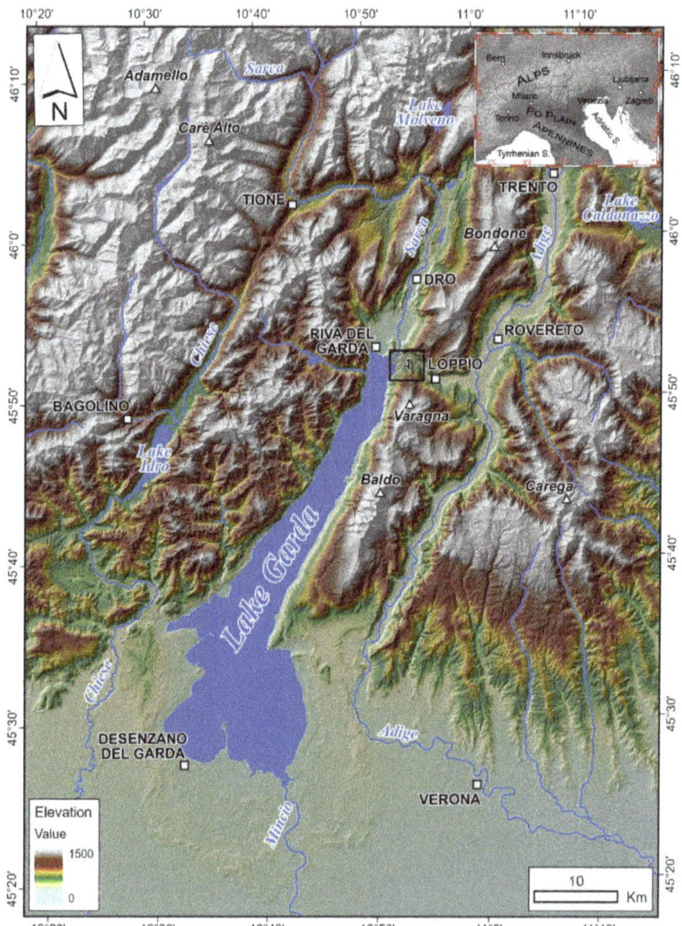

Figure 1. Geographical overview of the Alpine sector where the study area is. The major cities, peaks, rivers and lakes are shown, together with the location of Figure 4. The background was obtained from the Italian SRTM derived digital elevation model (25 m cells).

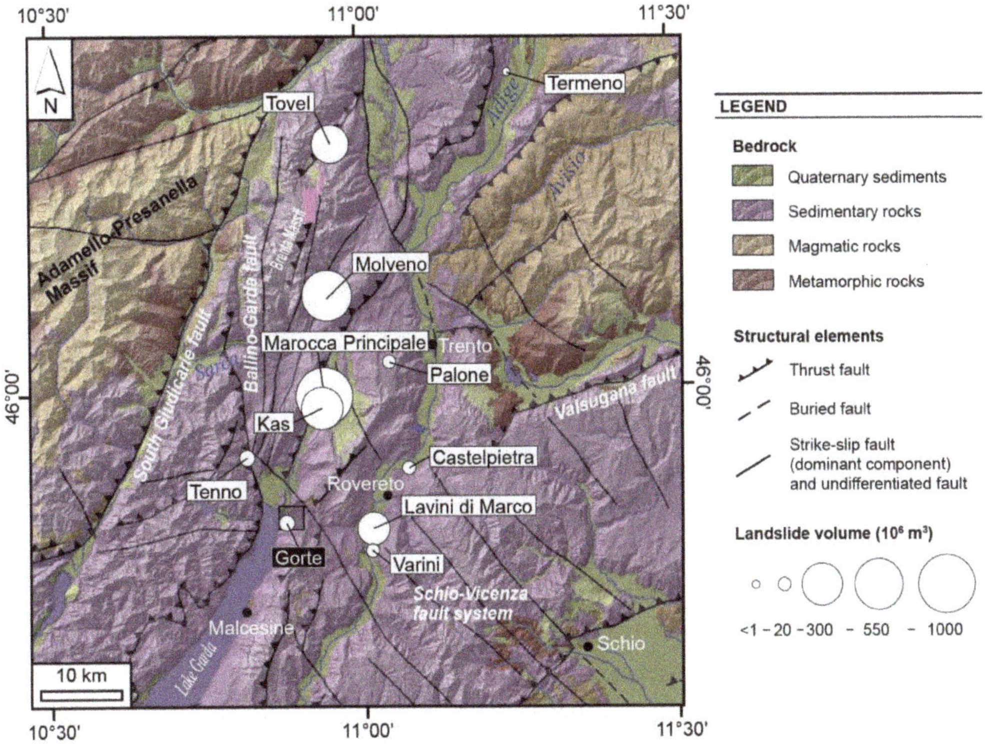

Figure 2. Simplified geological map of the Sarca and Adige Valleys (data from the Geological Survey of Trento, CARG project database) with the locations of the major landslides that have occurred in the area. Circle sizes are proportional to landslide volume (modified from [26]; Elsevier used with permission). The Tovel deposits are grouped in a single circle, despite the fact they are due to many events, with an estimated total volume of 300 Mm^3 [27]. The Gorte rock avalanche is marked in black. The gray square corresponds to the location of Figure 4.

Our study is focused on the largest deposit, the Gorte rock avalanche that detached from the Paternoster bedrock niche, and the deposit located at the foot of the Spiaz de Navesele–Salto della Capra release area (Figure 3). The aim of our study was to clarify whether the Gorte deposits were formed during a single event or during several events, as proposed by [28,29]. We used a multi-method approach combining field surveys supported by remote imagery, cosmogenic ^{36}Cl exposure dating and numerical runout modeling to reconstruct the large slope failures and the emplacement dynamics. Our new data can be compared with nearby isotopically dated deposits, for example, the Marocche di Dro [16], Lavini di Marco [30], Varini [31] and Molveno [12] rock avalanches (Figure 2), which allows an assessment of the most important preconditioning factors and possible triggers in a regional context. The numerical model provides insights into the dynamics of the rock avalanche, the characteristics of the path material and the runout behavior. Thus, this multi-method approach can provide a more complete picture about the landslide forcing factors in this part of the Alps and the possible resulting hazard.

Figure 3. Overview of the landslides of Gorte (on the left) and Marocche (on the right). The latter is due to the visible sliding planes of the Spiaz de Navesele and Salto della Capra. Mouth of the Sarca River seen on the far left of the image. Photo taken from the western side of Lake Garda (photo courtesy of Dr. Matteo Visintainer).

2. Setting and Previous Work

2.1. Geographical Setting

The study site is located within the NNE–SSW-oriented Lake Garda valley. This valley is linked to the tectonic evolution of the Giudicarie fault [32,33], while its depth is related to the Messinian sea level drop [34–36]. Lake Garda and the lower Sarca Valley are affected by present-day tectonic deformation [19,37,38], and earthquakes with an equivalent magnitude of $M_e = 5$ are common in the region [39]. These events are correlated with earthquake-triggered landslides on the lake shores and in the lake, as indicated by two seismoturbidite-homogenite beds (undated) that may be related to historic earthquakes [40].

Lake Garda is hosted within an asymmetric syncline. The eastern flanks are W-dipping parallel to the bedding and prone to sliding, while the western ones are steep and cut across the bedding orientation, giving origin to rockfalls [28]. The high topographic gradients of the eastern monoclinal shores and the western rocky cliff ranging in altitude from 100 to >2000 m a.s.l. favor slope failure [41] (Figure 1). The presence of numerous landslide scarps within the Mesozoic limestones around Lake Garda demonstrates that the surrounding slopes and lake shore must have suffered dangerous events in the past, including tsunamis, and could be sensitive to new events of the same type. The vulnerability of the slopes to failure is shown by the recent event (2 January 2021) at the locality of Tempesta on the eastern lake shore (http://www.protezionecivile.tn.it/territorio/primop_territorio/pagina211.html (accessed on 17 September 2021)).

North of the study area, the valley of Loppio connects the Adige Valley to the east and the Sarca Valley that opens to Lake Garda. The terminal tract of the Loppio Valley coincides with the Nago plain (~230 m a.s.l.) and is separated from the Sarca Valley by the rocky ridges of Santa Lucia and Doss Penede (Figures 3 and 4). The area of the present Nago alluvial plain is a basin receiving sediment from the adjacent slopes and from the Passo San Giovanni area drainages to the east. Minor gravitational events likely blocked the outlets to the north and west, allowing the accumulation of alluvial and lacustrine sediments. Geophysical surveys in the Passo San Giovanni area suggest a fill composed of 120 m of fine-grained sediments [42].

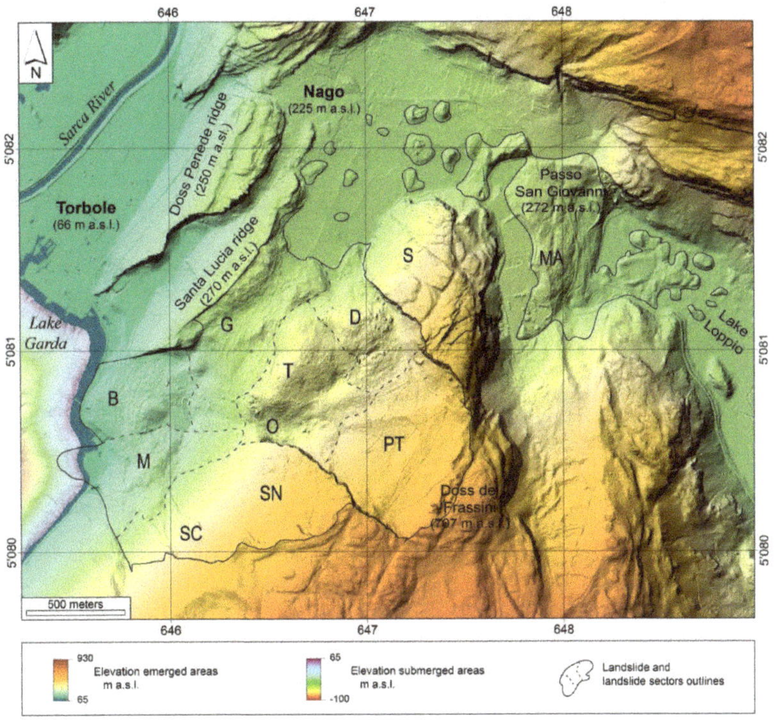

Figure 4. Shaded relief map of the study area with the most important toponyms, i.e., names, of the different areas mentioned in the text. **B**: Busatte, **D**: Daine, **G**: Gorte, **M**: Marocche, **MA**: Mala, **O**: Oltrezengol Valley, **PT**: Paternoster, **S**: Segron, **SC**: Salto della Capra, **SN**: Spiaz de Navesele, **T**: Tiro a Volo.

Several distinctive rock avalanche deposits are present at the junction between the paleovalley connecting the Loppio area with Lake Garda (Figure 4). According to [28], these include the following:

- The Gorte deposits, studied herein, with a release area along the western Doss dei Frassini slope (Paternoster);
- "Marocche" (this term is described more in detail below) deposits from the Spiaz de Navesele–Salto della Capra release areas which are also studied in the present work and will be referred to as Spiaz de Navesele;
- Tomas (conical hillocks) around the town of Nago interpreted to relate to the Segron release area;
- Deposits known as Mala located at Passo San Giovanni and interpreted to be sourced from the northern slope of Doss dei Frassini.

These deposits rest, at least partially, upon alluvial and/or lacustrine sediments accumulated inside the paleovalley of the Nago area and are, in turn, covered, to some extent, by the alluvial sediments forming the present-day Nago plain.

The prominent release areas in the surrounding bedrock slopes, for example, at Paternoster, and the blocky deposits in the Nago-Torbole region have been recognized and mapped for more than a century. In his Rovereto-Riva map (scale 1:75,000), Vacek [43] did not represent any landslide in the area. On the contrary, in his map, Habbe [44] drew three landslide deposits coming from Doss dei Frassini, by adding the deposit named here as Marocche located to the south of the Gorte deposits. In the Nago-Torbole area, Perna [28] mapped eight different landslide deposits, five of which correspond to the Daine, Tiro a Volo, Gorte,

Busatte and Marocche deposits as shown in Figure 4. Note that we use the term "Gorte" to refer to the whole of these deposits, excluding Marocche to the south.

Several names have been used to indicate the landslide deposits from the Doss dei Frassini western slope by previous authors. In Trentino, the word "Marocca" is used to indicate a chaotic mass of loose blocks. The word derives from the Paleo-European word mar (or kar), which means rock [28]. A "Marocca" is therefore a blocky landslide deposit. Based on observations of karst structures and the geomorphology of the deposits, [28,29] estimated that the main landslide event from the Doss dei Frassini western slope (Paternoster release area) occurred 10,000–5000 years ago, perhaps as three different events. In these estimations, the first event led to the formation of the Gorte and Busatte deposits (Figure 4), while the second led to the formation of the Daine and Tiro a Volo deposits. The third event involved only the walls delimiting the Paternoster release area, and the Daine and Oltrezengol Valley area. The volume of the total Gorte landslide release area (Paternoster) was estimated by [28] to be about 76 Mm^3, while for the deposits, he estimated a volume of 98 Mm^3. The landslide of Spiaz de Navesele–Salto della Capra is more recent as the blocky deposits overlie the deposit of Gorte. A volume of the bedrock released in the Spiaz de Navesele landslide was estimated at 2.5 Mm^3, while the deposit volume ("Marocche" in the topographic map, Figure 4) was estimated at around 3.2 Mm^3 [28].

2.2. Stratigraphic and Tectonic Setting

The Nago-Torbole landslide area is located 5–10 km to the east of the Ballino fault, which itself is connected with the Giudicarie fault (Figure 2). This fault divides the Lombardian basin, located to the west formed by deep-water Jurassic sediments, from the Trento carbonate platform, located to the east of the Ballino fault formed by shallow-water sediments. The area of Nago-Torbole is therefore located inside the Trento carbonate platform, which consists of limestones deposited during the Early and Middle Jurassic (Figure 5) [41,45].

At the base of the stratigraphy, there is the Rotzo Formation (RTZ), followed by the Massone Oolite Formation (OOM), which closes the Calcari Grigi Group ("Gray Limestones"). Above the Calcari Grigi, there are the Tofino Formation (TOF), particularly the member of Bocchetta Slavazi, and the San Vigilio Oolite Formation (OSV) on the southern boundary [41]. On the basis of field observations and thin sections, the RTZ Formation is a bioclastic peloidal gray micritic limestone (mudstone, in the samples identified as packstone) with mollusca fragments (e.g., bivalves, gastropods) and foraminifera in layers varying from some decimeters up to decameters, with dark marl interlayers rich in coal or black shales and oolitic calcarenites. The OOM Formation is a light gray to white oolitic limestone in thick layers (30–60 cm), locally cross-laminated, classified as grainstone including bioclasts such as calcispondes, bivalves, crinoids and echinoderm fragments. The TOF Formation is a micritic dark gray limestone (packstone to grainstone) with sponge spicules and radiolaria in medium-thin beds, brachiopods, crinoids and pelagic lamellibranchia. The OSV Formation is a yellowish crinoid oolitic limestone, in thick layers that are poorly stratified, sometimes cross-bedded. The OSV can be distinguished from the OOM because of the smaller oolites.

The tectonic setting of the Nago-Torbole area can be traced back to the post-collisional evolution of the Alps characterized by compression events of the Neogene [41]. However, some N–S-oriented morphological lineaments or fractures related to the Liassic faults, i.e., the Ballino line, are still well recognizable. The Valsugana event, which took place in the Serravalian-Tortonian (13–8 Ma), led to the formation of the NNE–SSW (NE–SW)-trending Giudicarie fault system [46]. The second fault system present in this area is the Schio-Vicenza, composed of subvertical NW–SE faults, linked to the Schio-Vicenza event which occurred during the Messinian-Pliocene (7–3 Ma) [46]. The Giudicarie and Schio-Vicenza have determined the tectonic setting of the area between Lake Loppio and the Sarca Valley, creating a system of lineaments, fractures and discontinuities that, in part, border and, in part, cross-cut the Nago-Torbole landslide area (Figure 2).

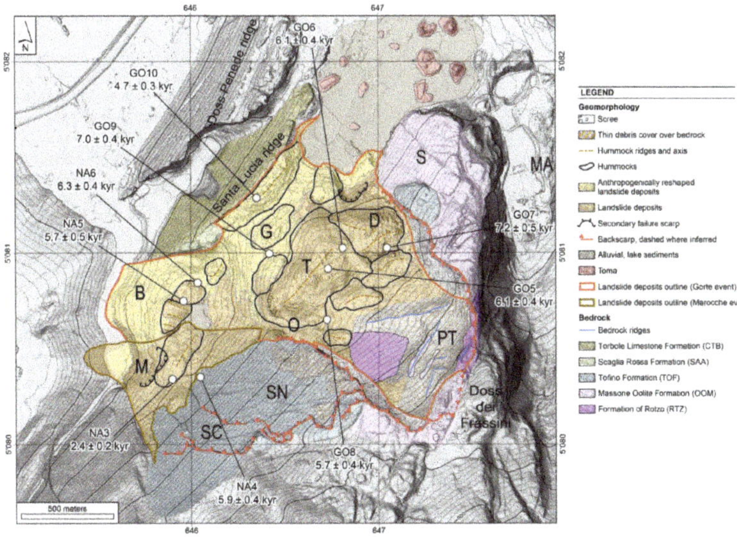

Figure 5. Geomorphological map of the study area. Boulders dated with ^{36}Cl and the calculated exposure ages are shown (Table 1). Samples NA4 (bedrock) and NA3 are from the Spiaz de Navesele event, and all others are boulders in the Gorte landslide deposits. **B**: Busatte, **D**: Daine, **G**: Gorte, **M**: Marocche, **MA**: Mala, **O**: Oltrezengol Valley, **PT**: Paternoster, **S**: Segron, **SC**: Salto della Capra, **SN**: Spiaz de Navesele, **T**: Tiro a Volo.

3. Methods

3.1. Field Survey and Remote Sensing

The field survey was completed using printed maps, supported by tablets with GIS Pro and Fieldmove apps. A project was created with ArcGIS Pro (ESRI) and integrated with LiDAR, a digital terrain model (DTM) and a digital surface model (DSM), with a resolution of 1 × 1 m [47], and high-resolution orthophotos acquired over different years. The data were analyzed in ArcGIS Pro to create maps such as slope, aspect and multidirectional hillshades (Raster Functions toolbox). In addition, topographic maps [47], the geological map [41] and the natural hazard map [48] were georeferenced and integrated. Furthermore, a bathymetric survey of Lake Garda (Torbole area) carried out in 2019 by the Italian Navy in collaboration with Cattolica University was integrated into the available maps. The hillshade map was created with the DTMs provided by Trento Province and the Italian Navy.

During fieldwork, we collected samples from the deposits and the release areas to investigate the stratigraphic characteristics and to confirm the provenance of the landslide boulders. We used the GIS Pro application to record points, lines and polygons, which we then exported in different formats (e.g., shapefile, or KML/GPX) and opened with software such as ArcGIS and Google Earth. Discontinuities (i.e., bedding, joint sets, faults, cleavage) were measured in the field and similarly located on the DEM. These data were analyzed using the cluster analysis tool of the Dips software using Fischer's density distribution. The position of the entered data was recorded through the GPS integrated in the device; the accuracy is on the order of a few meters.

We integrated all these data and further field observations to create a geomorphological map. This provided the basis for the analysis and interpretation of landforms and for the runout analysis. The DEM was further processed by drawing identified lineaments, which were further processed using the "Linear Directional Mean" tool in ArcGIS Pro to calculate the strike direction. These data were displayed as structural plots using the software Dips. Geomechanical measurements on Jurassic limestones conducted by the

Province of Trento laboratory were taken into consideration for analyzing the friction coefficient of the lithologies involved in the landslides to determine the predisposition to downward movement [42].

3.2. Cosmogenic ^{36}Cl Exposure Dating

Sixteen samples were taken for ^{36}Cl dating from sites all across the Gorte and Spiaz de Navesele landslide deposits. Ten boulders and six bedrock detachment surfaces were sampled with a battery-operated saw and hammer and chisel (locations shown in Figure 5). Boulders were sampled following the guidelines described in [49]: selected boulders were large in size (>2 m in height) to avoid displacement after the landslide, as massive as possible, and sampled from a surface not parallel to the bedding, in order to avoid pre-exposure (cf. [30]). The lithologies of all the samples are shown in Figure S1 of the Supplementary Material. Samples GO5-GO8 and GO10 were composed of OOM, while the GO9 and NA3-NA6 samples were composed of TOF.

Ten samples, nine from boulders in the deposits and one from the bedrock at the Spiaz de Navesele sliding plane (NA4), were prepared and measured with accelerator mass spectrometry (AMS) at the Laboratory of Ion Beam Physics, ETH Zurich (Table 1). Rock samples were crushed and sieved to <0.4 mm. After preliminary treatment with water and weak HNO_3, rock samples were dissolved with HNO_3 after addition of ^{35}Cl carrier following the procedures for the isotope dilution methodology presented in [50]. ^{36}Cl/Cl ratios were measured with the Ion Beam Physics ETH 6 MV TANDEM AMS system relative to the internal K382/4N ^{36}Cl/Cl standard (17.36×10^{-12}) and corrected for a procedural laboratory blank of $(1.94 \pm 0.34) \times 10^{-15}$ (Table 1) [51–53]. Major and trace elements were measured on aliquots of sample material with ICP-MS at Actlabs (Ontario, Canada) (Table 2).

Table 1. Sample name, AMS-measured ^{36}Cl concentrations and calculated apparent exposure ages (Figure 5).

Sample	Latitude North (°)	Longitude East (°)	Elevation (m a.s.l.)	Thickness (cm)	Topographic Shielding	^{36}Cl Concentration [1] (10^6 atoms/gram$_{rock}$)			Apparent Exposure Age (ka)		
GO5	45.8662	10.8903	370	2	0.956	0.168	±	0.009	6.05	±	0.37
GO6	45.8672	10.8914	368	2	0.992	0.173	±	0.009	6.10	±	0.37
GO7	45.8670	10.8942	387	1	0.989	0.208	±	0.011	7.17	±	0.46
GO8	45.8636	10.8903	355	1	0.967	0.157	±	0.010	5.66	±	0.40
GO9	45.8670	10.8863	277	2	0.933	0.171	±	0.008	6.96	±	0.41
GO10	45.8696	10.8856	274	2	0.958	0.122	±	0.007	4.73	±	0.32
NA3	45.8613	10.8796	196	2.5	0.969	0.057	±	0.005	2.42	±	0.24
NA4	45.8613	10.8812	205	2	0.968	0.144	±	0.008	5.93	±	0.37
NA5	45.8648	10.8805	172	2	0.978	0.131	±	0.011	5.65	±	0.49
NA6	45.8657	10.8815	165	2	0.968	0.146	±	0.008	6.29	±	0.40

[1] Measured against standard K382/4N $(17.36 \pm 0.35) \times 10^{-12}$ [51,52].

We used an in-house MATLAB code to calculate ^{36}Cl exposure ages. The code is based on the constants and equations presented in [54] (and references therein) and includes production through all pathways. Production rates were calculated individually for each sample based on the measured elemental concentrations (Table 2). We used the spallation production rate of 48.8 ± 3.4 ^{36}Cl atoms $(g_{Ca})^{-1}$ a^{-1} [55]. Treatment of muon production is described in detail in [54] and [56] and amounted to 9.6% at the rock surface. The contribution of neutron capture to ^{35}Cl-to-^{36}Cl production was calculated based on a value of 760 ± 150 neutrons $(g_{air})^{-1}$ a^{-1} [54]. These values are in excellent agreement with the recently published production rates of [57]. Topographic shielding correction was calculated with the MATLAB skyline function (http://stoneage.ice-d.org/math/skyline/skyline_in.html (accessed on 17 September 2021) (Balco 2018)). Production rates were scaled to the sample locations using the time-dependent scaling model (Lm) [58]. No correction was conducted for karst weathering of the boulder surfaces. Implementing a rate of 5 mm ka^{-1} ([59] and references therein) resulted in ~2% older ages. Final age uncertainties (Table 1 and Figure 5) included both analytical (one sigma) and production rate uncertainties.

Table 2. Elemental composition of leached samples. Cl values are from AMS measurements.

Sample	Al_2O_3 (%)	CaO (%)	Fe_2O_3 (%)	K_2O (%)	MgO (%)	MnO (%)	Na_2O (%)	P_2O_5 (%)	SiO_2 (%)	TiO_2 (%)	Sm (ppm)	Gd (ppm)	U (ppm)	Th (ppm)	Cl (ppm)		
GO5	0.1	54.94	0.05	0.03	0.39	0.006	0.03	0.02	0.28	0.003	0.1	0.1	0.6	<0.1	13.5	±	0.1
GO6	0.07	54.32	0.05	0.01	0.39	0.007	0.03	0.01	0.22	0.001	<0.1	0.1	0.7	<0.1	11.1	±	0.1
GO7	0.08	54.44	0.04	0.02	0.37	0.006	0.03	<0.01	0.21	0.001	0.1	0.1	0.6	<0.1	12.0	±	0.1
GO8	0.11	54.62	0.06	0.03	0.41	0.006	0.03	0.02	0.28	0.003	0.1	0.2	0.6	<0.1	13.4	±	0.1
GO9	0.18	53.78	0.17	0.05	0.65	0.018	0.03	0.01	0.66	0.008	1.3	1.6	1.5	0.3	14.2	±	0.1
GO10	0.11	54.74	0.05	0.03	0.48	0.005	0.03	0.02	0.37	0.003	0.1	0.2	0.5	<0.1	15.4	±	0.1
NA3	0.51	52.59	0.29	0.12	1.27	0.020	0.04	0.04	1.87	0.028	1.2	1.5	0.3	0.6	13.8	±	0.1
NA4	0.09	54.66	0.06	0.02	0.42	0.008	0.03	<0.01	0.32	0.002	0.2	0.2	0.6	<0.1	13.9	±	0.1
NA5	0.26	52.48	0.14	0.08	1.51	0.023	0.03	0.02	0.76	0.013	0.9	1.1	0.1	0.3	16.5	±	0.1
NA6	0.18	54.04	0.16	0.05	0.42	0.015	0.03	0.03	0.76	0.008	1	1.3	0.2	0.3	15.2	±	0.1

3.3. *Reconstruction of Pre-Failure Topography and Volume Estimation*

The topography of the valley prior to failure and the volume of the rock avalanche are key elements in analyzing the dynamic behavior of a landslide, and they are also an input for magnitude frequency analysis [60,61]. Both the initial volume of failed material (hereinafter referred to as "source" volume) and the deposit volume were estimated by reconstructing the pre-failure topography, and then differencing it from the present-day topography, accounting for any deposits that remain on the source zone rupture surface. It should be noted that the volume of deposits is expected to be approximately 25% higher than the source volume, as fragmentation and bulking lead to a volume increase [62]. The reconstruction was based on a newly created, present-day contour map obtained by combining a bathymetric survey of Lake Garda with the available LiDAR data [47]. We then integrated information from the geomorphological analysis and cosmogenic dating to interpret the pre-failure valley morphology, as further described in the Results section below.

3.4. *Runout Modeling*

We back-analyzed the dynamics of the Gorte rock avalanche using the semi-empirical runout model Dan3D-Flex [63]. This allowed us to test the plausibility of our reconstructed topography and our proposed failure scenario. Additionally, the back-analyzed basal shear strengths can be compared to values obtained at other rock avalanches, allowing us to contextualize the mobility of the Gorte rock avalanche.

Dan3D-Flex initially treats the failed mass as a "flexible block", which translates and rotates over the reconstructed topography. At a user-specified time, the mass fluidizes and is simulated as a frictional fluid whose behavior is governed by internal and basal rheologies [63,64]. In this study, the frictional rheology was used in the source area, where the rock avalanche moves over the rupture plane, and the Voellmy rheology was used for the path, consistent with the approach of [7]. These two rheologies are described in detail in [65] and will only be briefly descried here. The frictional rheology is a one-parameter rheology, and basal resistance is the product of bed-normal effective stress and the tangent of the friction angle, which is the calibrated parameter. The two-parameter Voellmy rheology combines a frictional term, proportional to a calibrated friction coefficient (f), with a velocity-dependent term, which is governed by a calibrated turbulence coefficient (ξ). The reconstructed pre-failure topographic surface is also a key input for runout modeling, and in the present work, we smoothed the DEM three times using a Gaussian low-pass filter, in order to aid the numerical stability.

To calibrate the model, we first performed a trial-and-error analysis to constrain the best-fit source zone friction angle and time spent as a flexible block. For this initial calibration step, the deposit volume in the source zone was the primary constraint used to assess the accuracy of simulations. Following this initial calibration, these values were held constant, and a posterior analysis, described in detail in [66], was used to constrain the two Voellmy parameters. Briefly, a posterior analysis computes the simulated results for a wide variety of parameter combinations and quantitatively compares the results to observations of the impact area and deposit volume. The algorithm then assigns the parameters which result in the closest match to field observations as more probable than those that result in a worse fit. In the present work, friction coefficients between 0.17 and 0.45, with steps of 0.1, and turbulence coefficients between 100 and 2000 m/s^2, with steps of 100, were used.

4. Results

4.1. Geomorphology and Age of the Gorte and the Spiaz de Navesele Landslides

The geomorphological analysis and deposit dating provide the basis for understanding the failure sequence of the two studied landslides and allow us to place these events in a regional context. Based on field relationships, we subdivided the deposits released from the western slope of Doss dei Frassini into the Gorte and the Spiaz de Navesele–Salto della Capra landslides, with the deposits of the latter overlying the former. Deposits of the Gorte landslide stem from the Paternoster release area, while the "Marocche" deposits relate to the Spiaz de Navesele–Salto della Capra source.

We mapped 730 morphological lineaments all over the study area (Figure 6). Discontinuities in the bedrock of the Paternoster and the Spiaz de Navesele–Salto della Capra release areas can be grouped into four different sets.

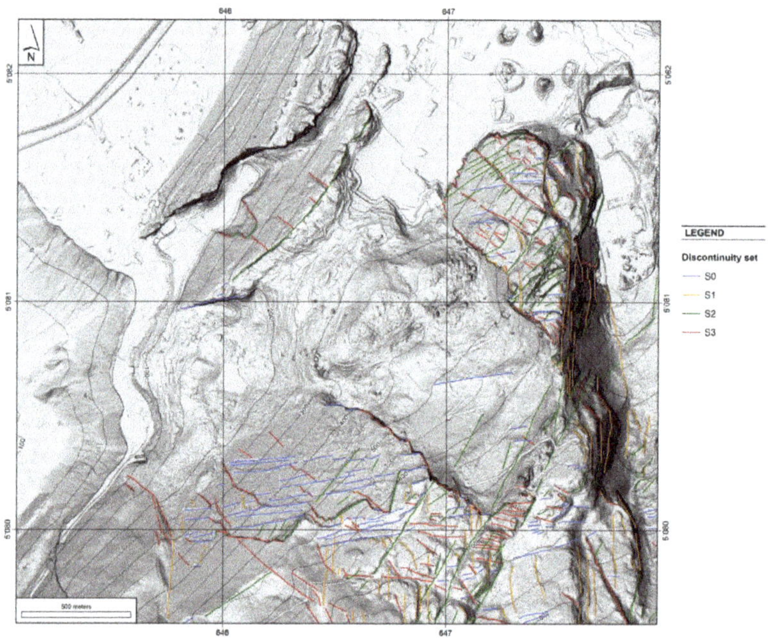

Figure 6. Structural lineaments identified in the study area. The measured discontinuities are grouped into four sets, identified by different colors. Colors of sets correspond to those of Figure 7.

The bedding (S_0) has an average orientation of 313°/27° (dip direction/dip angle), with dip angles varying from 20°–25° in the upper part to 25°–28° in the central sector. The other main discontinuities (S_1–S_3) are subvertical (dip angles > 70°). The S_1 set strikes NNW–SSE (072°/71°), with an average dip angle of 71°; S_2 strikes NNE–SSW (112°/74°), with an average dip angle of 74°; S_3 constitutes the fracture cleavage, is vertical and strikes WNW–ESE (019°/87°) (Figure 7).

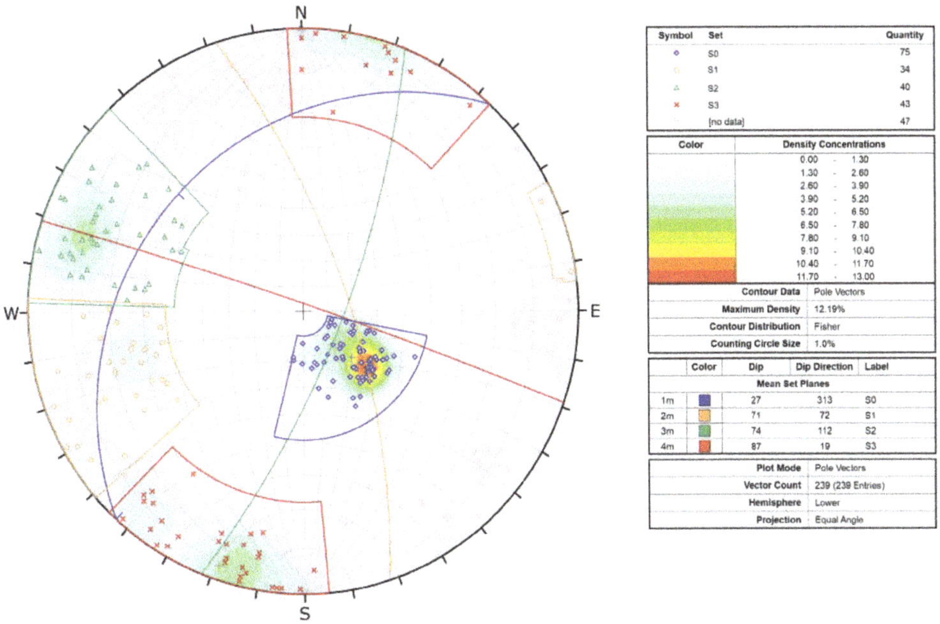

Figure 7. Stereonet of the structural measurements taken in the study area and shown in Figure 6. Colors of sets correspond to those of Figure 6.

4.1.1. The Gorte Rock Avalanche

The blocky deposits of the Gorte landslide detached from the Paternoster OOM bedrock (Figure 4). The Paternoster release zone extends from 380 to 700 m a.s.l. and covers an area of ~0.45 km^2. The release area niche is a semi-circular amphitheater that opens to the west. Impressive vertical scarps that are tens of meters high encircle the niche on the northern, eastern and southern sides (Figures 3 and 5). The northern scarp in the Paternoster release area is ~800 m long, 20 to 50 m high, WNW–ENE-aligned and mainly made of OOM, with a small patch of TOF cropping out in the northernmost part, at the top of the scarp (Figure 8A). The eastern scarp, which corresponds to the backscarp of the rock avalanche, is ~770 m long and 20–30 m high. The rock wall is made of RTZ in the northern side and of OOM in the southern side, the boundary between the two formations occurring at an elevation of ~650 m a.s.l. The southern scarp is ~1200 m long, 10 to 100 m high and WNW–ENE-aligned. Here, OOM and RTZ are both present, their boundary occurring at an elevation of ~520 m a.s.l. In the Paternoster release area, the majority of lineaments are oriented SE–NW or NE–SW, whereas a few E–W lineaments, but no N–S lineaments, are observed. The rocky walls show S_1 and S_2 joint sets and a pervasive fracture cleavage foliation. The spacing of the joints (about 50 cm) is greater than the spacing of the cleavage (about 5–10 cm).

The sliding plane of the rock avalanche is almost completely covered by a layer of debris up to 5 m thick, but some patches of bedrock are visible, especially on the southern side (Figure 5). The sliding plane is parallel to the bedding and is made of OOM in the upper part, whilst further west, some scarps cut the stratigraphy, and RTZ is exposed. There are no substantial differences in the orientation of the plane, being 290°/25° on average (S_0). Some undulations in the stratigraphy are present, and these form bedrock ridges that interrupt the flatness of the sliding plane. Two major bedrock ridges have been identified in the Paternoster sector: the easternmost (NNW–SSW-aligned) is located at ~500 m a.s.l. and is made of OOM, and the other is ENE–WSW-aligned and crosses the sliding plane from an elevation of ~500 to ~400 m a.s.l.

Figure 8. Photos from the study area. (**A**) The northern flank of the Paternoster release area, taken from the Tiro a Volo area. The wall in the upper left is 80 m high. Several rock spires are visible in the background. (**B**) View of the Navesele–Salto della Capra release area. The power line tower is about 30 m tall. The photos in the second and third lines are of samples from the Gorte rock avalanche deposit. (**C**) Boulder GO5 (6.1 ± 0.4 ka). The sample for ^{36}Cl dating was taken from the top surface. (**D**) Karren on the top sampled surface of boulder GO7 (7.2 ± 0.5 ka). (**E**) Boulder GO8 (5.7 ± 0.4 ka), where the top slightly sloping surface was sampled. (**F**) Boulder NA6 (6.3 ± 0.4 ka) is 6 m high. The bottom two photos are samples of the Navesele rock avalanche. (**G**) Sampled surface of boulder NA3 (2.4 ± 0.2 ka). (**H**) Sampled Navesele bedrock detachment surface of NA4 (5.9 ± 0.4 ka).

The Gorte rock avalanche deposit extends from 380 m a.s.l. in the eastern part to 65 m a.s.l. in the western sector and covers an area of ~0.53 km^2. It is bounded to the east by the Paternoster sliding plane, to the north by the Nago alluvial plain, to the west by the Santa Lucia bedrock ridge and Lake Garda and to the south by the Marocche deposit sourced at Spiaz de Navesele–Salto della Capra. The Santa Lucia ridge and the Doss Penede ridge (Figure 4) are ~150 m high and have an asymmetrical profile with gently sloping west-facing dip slopes and subvertical east-facing slopes. Based on location, debris lithology and morphological features, four sectors of the Gorte deposit can be distinguished: Daine, Tiro a Volo, Gorte and Busatte (Figure 4). The obtained ^{36}Cl exposure ages for Gorte rock avalanche boulders range from 4.7 ± 0.3 to 7.2 ± 0.5 ka (Table 1, Figures 5 and 8). These are discussed in more detail below within each specific site context.

The Daine blocky deposits comprise the northernmost sector of the rock avalanche, located next to the right scarp of the Paternoster release area. Here, the largest boulders (5–18 m in diameter) cluster next to the rocky wall, where the carapace is also particularly visible and pronounced. Moving away from the rock wall, the boulders are rarer and smaller (1–5 m diameter), but a few isolated boulders and pinnacles up to 15 m high are still present (Figure 8B). The internal structure of the deposit is clast-supported, mostly formed by angular pebbles and cobbles of about 5–10 cm, with a small amount of matrix that becomes even scarcer moving eastwards. All boulders show clear karst structures. Two hummocks are present (15–20 m high, 200–250 m long). They are NE–SW-aligned and thus perpendicular to the direction of the rock avalanche flow. The northernmost hummock shows a double ridge, but the overall orientation remains the same. An OOM boulder (GO7—Figure 8D) atop the easternmost hummock yielded an exposure age of 7.2 ± 0.5 ka. At the northernmost limit of the Daine area, near the Nago plain, a smooth scarp is present within the deposit, possibly marking a secondary (internal) collapse. However, the area is covered by dense vegetation, and the scarp may have been smoothed over the years, making it difficult to identify it unequivocally. The eventual deposit of this secondary failure is not visible, possibly covered by the Nago plain lacustrine sediments.

The Tiro a Volo sector is located south of Daine, from which it is separated by a ~40–50 m-wide and 15–30 m-deep depression (Figures 4 and 5). This depression is SE–NW-aligned and thus roughly parallel to the side walls of the niche. The Tiro a Volo area hosts the largest hummock of the whole deposit, being ~400 m long. It does not have a well-defined crest but has a rather flat surface, its west flank being the steepest, with an angle of ~30°–40°. The upper part of the hummock, as well as the part facing upstream, is characterized by the presence of many large boulders (1–2 m in diameter). There are fewer of the really huge blocks (>5 m) compared to the Daine area, even though there are some rare boulders up to 10 m in size. Two dates were obtained for boulders of OOM atop the described largest hummock. Two notably coherent ages of 6.1 ± 0.4 (GO5 and GO6—Figure 8C) were obtained. The northernmost part of the hummock and the side facing westwards are characterized by an almost total absence of clasts larger than 50 cm. Here, the deposit can be classified as sandy silty gravel with cobbles and isolated boulders. From this area, OOM boulder GO8 (Figure 8E) returned an age of 5.7 ± 0.4 ka. These characteristics are the same for the southeastern hummock, even if this one is much smaller (~200 m long). The southern limit of the Tiro a Volo sector coincides with the large (~200 m) and deep (~30–50 m) Oltrezengol Valley (Figures 4 and 5), which is reported to convey significant subsurface water flow [67]. At the head of this valley, a small (~100 m long) E–W-aligned hummock with a rather smooth surface is present, just below the Paternoster sliding plane. The head of the valley itself has an amphitheater shape, whilst the lower part of the Oltrezengol Valley is narrower (Figure 4). At the height of this narrowing, at the foot of the biggest Tiro a Volo hummock, there are several very large boulders with diameters of up to 20 m. The amphitheater shape, even in the absence of a marked backscarp that could have been smoothed over time, could have also formed during a secondary failure of the deposit in the immediate aftermath of the Gorte rock avalanche. However, a deposit of this collapse is not clearly visible.

The Gorte sector has a NE–SW elongated shape and extends to the north from the Nago alluvial plain to the south at the Busatte deposit sector. It is bounded to the west by the bedrock ridge of Santa Lucia. This sector is characterized by the extremely rare presence of large (5–10 m in diameter) boulders, also caused by the extensive reworking of the area by anthropogenic activities. No outcrops were located to evaluate the internal structure of the deposit. Next to the Santa Lucia ridge, a NE–SW-aligned hummock is present (~500 m long). Three other smaller hummocks are present in the Gorte area. The two biggest hummocks are NE–SW-aligned, ~350 m long, have an elliptical shape and a flat top and are less than 10 m high. The smallest hummock, ~250 m long, is located in the northern sector of Gorte and has a rough N–S alignment. The Gorte deposit is transversally cut by an incision (~30 m deep, ~200 m wide) that connects the Nago plain and the Busatte area. In the Gorte sector, two samples (GO9—TOF, GO10—OOM) were dated, returning ages of 7.0 $\pm$ 0.4 and 4.7 $\pm$ 0.3 ka, respectively (Figure 4). The GO10 age is the youngest of the rock avalanche deposit.

The Busatte area comprises the most distal sector of the Gorte rock avalanche deposit. It extends from ~160 to 65 m a.s.l., where it reaches Lake Garda. This deposit can be subdivided into two sectors: the upper one, rather flat, and the lower one, showing a mean slope of about 15°–20°. The latter has undergone marked anthropogenic modifications; various boulders up to 2 m in size are still visible, but no larger ones have been found, and no open sections are present to evaluate the internal structure of the deposit. The former is rather flat and shows small semi-circular hummocks, 5 m high. They are slightly ESE–WNW-aligned and almost parallel to the rock avalanche flow. They show a carapace, with boulders on top up to 5 m large. In the eastern side of the Busatte area, a plain sector ~0.1 km^2 wide is present. Here, fine sediments deposited over the rock avalanche thanks to the running waters conveyed by the Oltrezengol Valley that nowadays creates a swampy area. There, sands dominate, whilst boulders and rock fragments are almost absent. In the Busatte sector, exposure ages were obtained from two boulders, both of TOF. NA5 has been exposed since 5.7 $\pm$ 0.5 ka, and NA6 since 6.3 $\pm$ 0.4 ka.

4.1.2. Spiaz de Navesele Rock Avalanche

The deposit of the Spiaz de Navesele–Salto della Capra event is located between the base of these sliding planes and Lake Garda, in the southernmost sector of the study area. It is named "Marocche" and extends from 220 to 65 m a.s.l. The deposits cover an area of ~0.23 km^2.

The Spiaz de Navesele–Salto della Capra release area is located south of the Paternoster area (Figures 4 and 5). Despite having different names, these two sectors are adjacent and are separated only by a subvertical scarp (several meters high) that is roughly E–W-oriented. As a whole, the area is semi-circular in shape, with vertical scarps on the northern, eastern and southern sides (Figure 5). It extends over ~0.45 km^2, at an elevation ranging from 590 to 200 m a.s.l. The sliding plane, as well as the southern and eastern scarps, is made of TOF. The former is parallel to the bedding and has a dip direction of ~300–320, with a dip angle varying from 25° in the upper part to 35° in the lower sector. The sliding plane is crossed by several bedrock steps up to 10 m high that are roughly ENE–WSW-aligned (Figure 6). Lineaments striking E–W are common in this sector. An age of 5.9 $\pm$ 0.4 ka (NA4) was determined for a sample from the westernmost part of the TOF bedrock sliding plane (Figures 4 and 8H).

The deposits (Marocche) related to the Spiaz de Navesele–Salto della Capra dip-slope sliding plane are completely made of TOF debris and show distinct morphologies: a relatively flat (5°–10°) eastern part, and a steeper (30°–35°) western sector. The former is made of rock debris ranging from cobbles to pebbles, with a small amount of fine matrix. Several boulders ~1 m in diameter are present, with larger ones (up to 8 m high) clustered near the sliding plane. To the west, there is an NNE–SSW-oriented hummock that is several meters high. It is characterized by a clast-supported deposit, with boulders > 5 m. To the west of the hummock, a smooth scarp is present, possible due to a post-event (secondary) collapse of the Marocche deposit. The related deposit can be recognized under the lake in the bathymetric relief (Figures 4 and 5). The orientation and elevation of the hummock suggest that Marocche deposits overlie deposits of the Gorte landslide. This would be consistent with the exposure age of sample NA3 (TOF) collected from the flat top of a large boulder (Figure 8G), returning the age of 2.4 ± 0.2 ka. Nevertheless, the age from the sliding surface suggests a much older age for the Navesele event (see below).

4.2. Topographic Reconstruction and Volume Analysis

We used the procedure detailed in Section 3.3 to reconstruct the pre-failure topography and to estimate the source and deposit volumes. Starting with the source zone, we estimated that a volume of 70–75 Mm^3 was released from the Paternoster niche. Given the well-defined shape of the release area, the uncertainties mainly concern the location of the base of the landslide, as scree deposits, as well as rock avalanche deposits in the lower section, obscure the depth of the sliding surface. Between 680 m a.s.l. (the highest part of the Paternoster release area) and 380 m a.s.l., we removed a small volume of scree that was deposited after the landslide. Between 380 m a.s.l. and the bottom of the release area (labeled Tiro a Volo and Daine on Figure 4), thick deposits of rock avalanche material obscure the rupture surface. We therefore interpreted the sliding plane to have a lower slope (15°) with respect to the slope in the upper release area (25°).

For the deposit, we estimated a volume of 85–95 Mm^3, which leads to a bulking of ~26%, similar to the typical value of 25% estimated by [62]. We subdivided the deposit into three main areas (Figure 4): (1) a part remaining in the detachment niche that forms the hummocks of Daine and Tiro a Volo (Figure 5), (2) a part not visible today that is buried below the Nago plain and (3) a large part of the deposit that extends from the Nago plain to Lake Garda, in the locality of Gorte and Busatte (Figure 5).

The reconstruction of the surface below the Daine and Tiro a Volo is described above, and the other two deposit zones were reconstructed based on the following considerations. Core logs and geophysical data in the Passo San Giovanni area indicate that the transition from alluvial to lacustrine sediments is at an approximate elevation of 180 m a.s.l. [42]. Furthermore, the level of Lake Garda after the Last Glacial Maximum has not changed significantly in the last 10,000 years, reaching a maximum of 70 m a.s.l. and a minimum of 62 m a.s.l. (present level 65 m a.s.l.) [68]. Therefore, an alluvial plain extending from the Passo San Giovanni to Lake Garda was reconstructed. The slope of this alluvial plain was assumed constant between the two points whose elevations are known. The east side of the paleovalley was reconstructed with a slope similar to the mountainside, which increases to 35°–40° towards the lake. The west side of the paleovalley was reconstructed with near-vertical slopes, based on the bedrock outcrop visible just north of Busatte (Figure 4). This reconstructed morphology resembles the nearby ridges of Santa Lucia and Doss Penede.

4.3. Runout Analysis

Following the reconstruction of the pre-failure topography, we performed numerical runout modeling based on the methodology described in Section 3.4. The initial calibration step resulted in a best-fit bulk friction angle of 14° and a flexible block time of 10 s. As described above, the reconstructed rupture surface is compound, and thus movement over this surface requires internal deformation of the failed mass. We expect that fluidization of the initial failure occurred when the center of mass moved from the upper part, where the sliding plane is steeper, to the lower part with the less steep plane. A friction coefficient of 0.38 and turbulence coefficient of 700 m/s^2 were found to provide the best-fit Voellmy parameters between the simulated and observed runout. Interestingly, these values are comparable to back-analyses of case histories that overran the bedrock [7,24,69,70].

Figure 9 shows the simulation results obtained at different times using the best-fit parameter combination. At 20 s, the rigid sliding phase is already finished, and the landslide follows a frictional rheology inside the release area, while the landslide foreground part in the paleovalley of Gorte follows a Voellmy-type rheology. It can be seen that after

20 s, the toe of the landslide has already reached the bottom of the paleo valley, and at this moment, the landslide divides into two lobes deflected by the ridge of Santa Lucia, spreading in the SW direction towards Lake Garda and in the NE direction towards the current Nago plain. After 40 s, the landslide reaches the top of the ridge of Santa Lucia, and the Gorte paleovalley is already filled, with the depth of the deposit exceeding 100 m. A considerable part of the deposit is still inside the release area, with a significant thickness in the area of Daine. After 60 s, the landslide deposit begins to assume its final shape, and the two lobes achieve their maximum extensions, just reaching the shoreline of Lake Garda in the southern lobe. The Paternoster area is still partially covered by several tens of meters of debris. After 120 s, the movement of the rock avalanche can be considered complete, with the deposit taking its final outline.

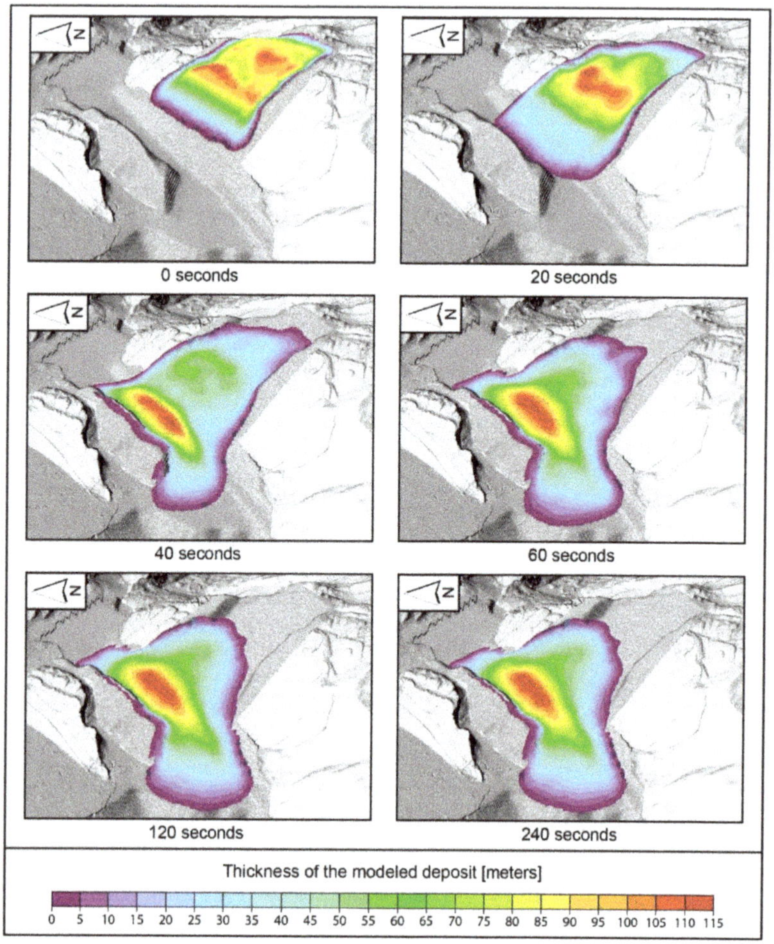

Figure 9. 3D overview of Dan3D-Flex modeling results for runout distance and thickness over time (for details, see text). The rock avalanche impacts and is deflected by the Santa Lucia ridge. For reference, the approximate length of this ridge is 1 km.

The modeled deposit extent is similar to the actual extent (Figure 10). The extent under the Nago plain is not known due to the lack of subsurface data. In our model, the western lobe in the direction of Lake Garda is larger than the eastern lobe. This is because the interpreted paleovalley descends steeply in the direction of the lake, while to the east towards Nago, the paleovalley rises gently. In the model results, the landslide deposit is thickest (114 m) in the paleovalley just to the east of the Santa Lucia ridge. The thickness of the modeled deposit decreases in the direction of the two lobes to the east and west.

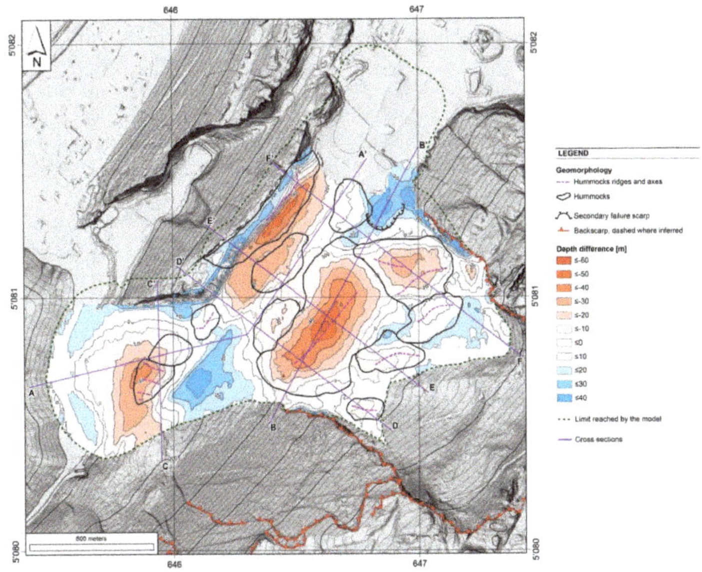

Figure 10. Map showing difference between the best-fit Voellmy parameter simulation of the landslide topography and the actual present-day topography. Negative values indicate that the simulated elevation is below the present topography. Extreme values around cliffs, as well as the part below the lake sediments, have been removed. In blue are the locations of the cross-sections shown in Figure 11.

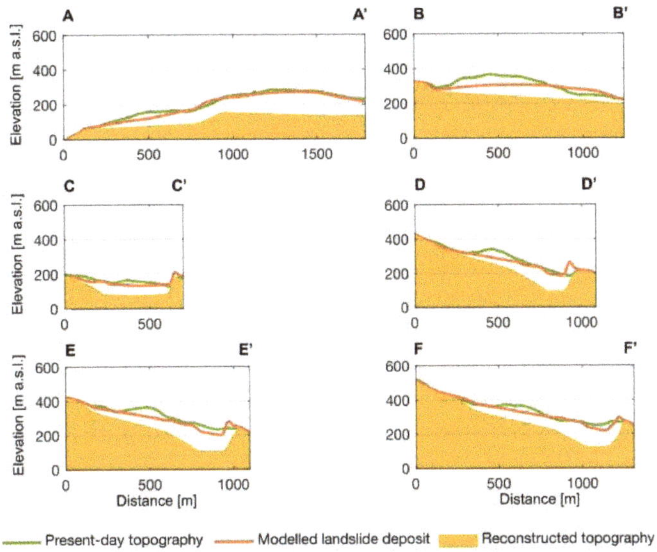

Figure 11. Cross-sections through the landslide deposit with the best-fit parameters of the Voellmy rheology. The topography based on modeling (red line), the present-day topography (green line) and the reconstructed pre-failure topography used for modeling are shown. The scale is the same for every segment. The location of the cross-sections is visible in Figure 10.

Figure 10 shows the elevation differences between the modeled landslide topography and the actual topography, and Figure 11 compares various cross-sections of the measured and estimated deposit depths. In Figure 10, the white colors show the area where the elevations of the modeled and actual landslide deposits are similar. In contrast, the red areas indicate where the modeled deposit is thinner than the actual one, while the blue areas indicate where it is thicker. Extreme values shown at the cliffs, particularly at the northern flank and in the southernmost corner of the Santa Lucia ridge, are not to be considered real differences but artifacts due to the presence of steep cliffs in these areas. The deposit below the lake sediments of the Nago plain has been removed, as, there, it is not possible to precisely define the real thickness of the lacustrine sediments. Overall, Figures 10 and 11 show that there is good agreement between the actual and modeled deposit thicknesses, which only differ where the large hummocks are present, especially in the Daine, Tiro a Volo and Gorte sectors. A similar observation has been noted in runout modeling results for the nearby Molveno rock avalanche [12].

5. Discussion

5.1. Release and Emplacement of the Two Studied Landslides

5.1.1. The Gorte Rock Avalanche

The eight determined ^{36}Cl boulder exposure ages for the Gorte deposit yield an average of 6.1 ± 0.8 ka (Figure 12), including all eight ages (GO5, GO6, GO7, GO8, GO9, GO10, NA5, NA6). Approximately 6100 years ago, an 800 × 1000 × 100 m block of OOM, TOF and RTZ limestones detached along the head scarp and side walls and began sliding down bedding planes. This was likely promoted and facilitated by interbedded thin marl layers in RTZ. Detachment was controlled by three recognized lineaments; the backscarp follows an S_2 surface with an NNE–SSW orientation and a 74° dip, and the two lateral scarps follow discontinuity set S_3 which is nearly vertical and is oriented WNW–ESE. Furthermore, our modeling supports the hypothesis that downslope movement was initially translational as a rock slide. After about 10 s, likely related to a kink in the bedrock plane, the mass began to break up but continued its movement in a northwest direction (Figure 9).

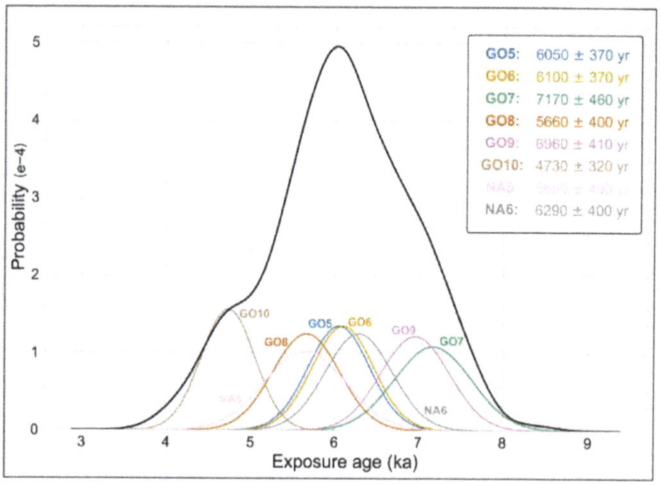

Figure 12. Camel plot for the eight ^{36}Cl exposure ages for boulders from the Gorte rock avalanche deposits. Colored lines show individual exposure ages and the Gaussian distribution of the uncertainties. The black line indicates the summed probability distribution for the dataset. The mean of the ages is 6.1 ± 0.8 ka.

Several hummocks dominate the topography of the Gorte deposits, providing additional information on the emplacement dynamics [11,13,71]. The largest are those of Daine and Tiro a Volo, which are aligned in an NE–SW direction and thus transverse to the flow direction of the landslide. In contrast, in the lower part of the deposit, hummocks are smaller in size, and are oriented longitudinally with respect to the flow direction (Figure 5). The former imply formation due to compressional forces connected to slowing during emplacement, perhaps related to steps in the underlying bedrock. The latter formed due to extensional processes related to spreading and differential velocities within the moving mass [71].

The morphology of the deposit (Figure 5) and the runout modeling results (Figure 9) indicate that the paleovalley located east of the Santa Lucia ridge (Figure 4) filled with blocky debris as the moving mass slowed due to impact with the steep eastern wall of that ridge. This impact deflected part of the moving debris to the west and part to the east. Neither field evidence nor modeling results indicate overtopping of the Santa Lucia ridge by the rock avalanche. The Gorte hummock, located right against the Santa Lucia ridge, is oriented parallel to both the bedrock ridge and the large transverse hummocks in the Daine and Tiro a Volo areas. We interpret that the elongated and narrow shape of this hummock is related to the blockage of the rock avalanche flow by the rocky ridge of Santa Lucia, which hindered its spreading to the west. Debris from the Gorte rock avalanche may have made it all the way into Lake Garda. Nevertheless, looking at the bathymetric data (Figure 5), no structures linked to deposits within the lake below the Busatte sector are evident. Some large (up to 18 m in diameter) isolated boulders are visible on a fan-delta-like morphology. However, given the absence of a direct connection with the other deposits described herein, their origin cannot be attributed with certainty.

The Gorte deposit blocked the westward-directed drainage of the paleovalley connecting the Passo San Giovanni area and Lake Garda (Figure 4). As a result, a lake formed. It filled up to the level of the outlet to the north, forming the Nago plain. Nevertheless, the age of infillings beneath the Nago plain and the possible occurrence of previous blockages related to earlier gravitational events cannot be discerned with the presently available data on the basin infill.

5.1.2. The Spiaz de Navesele Rock Avalanche

The blocky Marocche deposits of the Spiaz de Navesele event overlie and thus stratigraphically postdate the Gorte deposits. At present, chronological data do not allow us to fully decipher the timing of the Spiaz de Navesele event in comparison to the Gorte event. A single boulder was dated in the Marocche deposit at the foot of the Spiaz de Navesele sliding plane. It was located in an area that is strongly disturbed by anthropogenic activity, including numerous stone cairns. This single age (2.4 $\pm$ 0.2 ka) is difficult to interpret, and further dating would be necessary to determine if the Navesele event is several thousand years younger than the Gorte event or if the age of the bedrock sliding surface (5.9 $\pm$ 0.4 ka) provides a more realistic age. In that framework, the Gorte and Navesele events would have occurred almost simultaneously. Note that the right-hand boundary of the Spiaz de Navesele detachment surface stands as the left-hand steep side wall of the Gorte detachment niche. This is in line with the hypothesis that the Gorte event triggered the Spiaz de Navesele event.

The structural setting in Spiaz de Navesele–Salto della Capra is controlled by ENE–WSW lineaments (Figures 2 and 6). This allowed the landslide to flow and to spread in several directions, thus reducing the thickness of the deposit at Marocche (Figures 4 and 5) and explaining the lower slope in the Gorte southern area, which represents the transition zone between the lower paleovalley and the wide plain present at the time in the Busatte area. The large hummock in the Marocche deposit is oriented NE–SW and thus lies transverse to the direction of motion, implying limited mobility of the Spiaz de Navesele event.

5.1.3. Other Deposits in the Study Area

Our results from the Gorte and Spiaz de Navesele landslides provide an initial insight into the spatial and temporal relationships of the several landslide deposits in the Torbole-

Nago-Passo San Giovanni region. Yet, further study is required to understand the toma on the Nago plain and the blocky and hummocky deposits known as Mala located at Passo San Giovanni (Figure 4). Previous authors have suggested that the toma hills on the Nago plain (Figures 4 and 5) are related to a detachment from the Segron release area [28]. The toma hills appear to be buried along the base by the alluvial/lacustrine sediments of the Nago plain. This suggests that they are older than the timing of the lake formation but would also require that the lake was rather shallow. On the other hand, the tomas themselves may have formed during the Gorte event. Excessive runout of tomas (several kilometers) has been observed at the Flims [72] and Fernpass landslides [73]. Many of the tomas on the Nago plain are barely recognizable, having been strongly reshaped by human activity. No blocks suitable for exposure dating could be located.

To the east of these tomas are the Mala deposits, which are located near the hummocky terrain further to the east (Figure 4). The release area related to the Mala deposits is the north slope of Doss dei Frassini [28], and possibly the slopes to the north of Loppio Valley as well. The Mala deposits are not covered by Nago plain sediments; therefore, the Mala event is younger. The Mala deposits blocked the Loppio Valley, isolating the Nago plain from the eastern sector of the valley, where Lake Loppio formed and is still present as a very shallow lake. Finally, [74] reported that a minor detachment occurred in 1457 AD, whose deposit ended up on top of the larger Mala deposit at Passo San Giovanni.

5.2. Causes, Triggers and Relationship with Other Large Landslides in the Region

The reporting of a relatively large event near Passo San Giovanni only a few centuries ago and the recent (January 2021) event along the Lake Garda slope underline the importance of understanding the contributing causes and triggers for slope failures in this region. In the recent event, 800–900 m^3 of RTZ slid towards the lake on the dip slope and blocked the main road for several weeks. The bedrock slopes along both sides of Lake Garda have been undercut and oversteepened during repeated glaciations [75], which can predispose them to failure [24]. During the Last Glacial Maximum, the Nago-Torbole area was completely covered by the combined Sarca-Adige glacier lobe that flowed down the Sarca Valley. The frontal position, located just south of Lake Garda, was maintained between ~24.7 and ~17.5 ka cal BP [75,76]. In the region of our study site, the elevation of the glacier reached 750 m a.s.l. [41], completely submerging the Gorte release area beneath the glacier.

Faults and fractures in the rock related to the main regional fault systems, the Giudicarie and the Schio-Vicenza, have been shown to have played a decisive role in slope failures of the region [19]. In the Gorte release area, the identified S_2 and S_3 sets are linked to the regional tectonics of the Lake Garda and Sarca catchment. The S_2 set (NNE–SSW) is related the Giudicarie fault system. The Schio-Vicenza fault system, normally with a subvertical NW–SE orientation, is represented here by S_3 lineaments. The landslide backscarp is characterized by a cliff (about 60 m high) along an imposing NNE–SSW fault plane located northwest of Doss dei Frassini. This fault plane is likely connected to the Giudicarie fault system (Figures 2, 5 and 6). The northern and southern flanks, instead, are connected to the Schio-Vicenza NW–SE fault system. Two ridges in the Paternoster release area, which contributed to shaping the topography of the rupture plane, are oriented NNE–SSW and ENE–WSW. The S_0 set (bedding) shows inclinations between 25° and 35° and is one of the predisposing factors for the landslide (Figure 6).

The lithology played an important role because the rupture plane likely developed within the RTZ marly interlayers that are rich in coal or include black shales [41]. These interlayers were not observed in the Paternoster release area, having likely been eroded over the millennia after the collapse. Given the smaller friction angle of the marly RTZ interlayers, sliding was possible with bedding at an inclination of 25°, as it is seen in the Paternoster area. The marly interlayers have also been detected in the release areas of other landslides including the nearby Marocche of Dro and Molveno rock avalanches [12,16] and the well-known Vajont landslide [77]. Another predisposing factor of the landslide can be the presence of caves in the release area. As already discussed, several large caves (several decameters) are present in the Doss dei Frassini side walls [78]. These types of caves represent a further weakening factor of the rock mass [79–81], as in the case of the Masiere di Vedana rock avalanche, where several caves have been identified in the same lithological formation [17].

Connected to the caves, a karstic aquifer system is present along the relief, whose base level is represented by Lake Garda [45] with ephemeral springs inside the Busatte area ("Romani spring"), whereas the release areas and the sliding planes have neither springs nor surface runoff [67]. Underneath the Paternoster release area and the sliding planes of Spiaz de Navesele–Salto della Capra, the Adige-Garda tunnel was excavated to avoid the danger of flooding of the Adige Valley and to discharge part of the Adige River into Lake Garda [45].

The combination of the predisposition factors listed above requires a trigger to cause the rock slope failure. The first hypothesis is that the trigger was of a seismic origin. Seismic activity in the area has been extensively documented over the years, with earthquakes in the region of Lake Garda reaching, in historical times, a level on the Mercalli–Cancani–Sieberg intensity scale (MCS) of IX such as the earthquakes of Verona in 1117 AD and Brescia in 1222 AD [39]. More than one earthquake with an equivalent magnitude of around 5 with its epicenter in the area of Monte Baldo, about 15 km south of Nago-Torbole, has been recorded in the last two centuries [39]. The hypocenters of several recent earthquakes were located both along the Giudicarie belt and the Schio-Vicenza and other nearby minor faults [38,82]. Several historical landslides in the area such as Castelpietra, Kas, Prà da Lago and Varini (Lavini di Marco) [16,30], all within 15 km from the Nago-Torbole, were attributed to seismic crises, such as the "Middle Adige Earthquake" of 1046 AD, believed to be the cause of the Castelpietra landslide [26]. Recently, the hypo thesis of a common seismic trigger during the Middle Holocene has also been raised for the Marocche di Dro (Marocca Principale), 5.3 ± 0.9 ka, and the Marocca di Molveno, 4.8 ± 0.5 ka [12,83], located 13 and 25 km to the north, respectively.

The second trigger to be taken into consideration is the climate, particularly a humid climate with periods of heavy, persistent rainfall characterizing part of the Holocene. In the Holocene, three periods with a marked concentration of landslides were identified in the time intervals 10–9 ka, 5–3 ka and 2–1 ka, the last one especially for south of the Alps [15,16]. For the period 5–3 ka, an increased frequency of landslides can be traced back to a shift to a wetter climate, but also colder climatic conditions, whose extreme was reached around 4.2 ka, representing the transition from the Middle to the Late Holocene [15]. This period is characterized by both a general trend and the occurrence of individual extreme weather events, both locally and all over Europe [84,85]. The increase in landslides corresponding to extreme weather events and, in general, of humid periods is due to an increase in pore water pressure, which causes a reduction in the effective stress, thus inducing collapse [86,87]. The cyclicity of these events could then induce fatigue in the rock mass [86] that would lead to eventual failure. Our dating of the Gorte rock avalanche places it in the Middle Holocene period 7–5 ka, where Europe was characterized by warmer temperatures in summer and winter [88], and suggests a marked reduction in precipitation at the Alpine scale [89]. However, embedded in this period was an interval of frequent flooding on the northeastern Po Plain (Venetian plain) at 6900–6200 years ago [84], located just to the east of our study site.

Looking at the landslides already dated in the surroundings of the Gorte landslide (Figure 2), a very similar age has been calculated for the Dosso Gardene landslide, located adjacent to Lavini di Marco, about 11 km east of the study area. In this case, the buried soil below the landslide deposit was dated 6630–6290 cal BP [31], in agreement with the error of the Gorte rock avalanche (6.1 ± 0.8 ka), and at the limit of the landslide of Spiaz de Navesele (5.9 ± 0.4 ka). Additionally, the Marocca Principale of the Marocche di Dro (5.3 ± 0.9 ka) falls within the error range of the Gorte and the Spiaz de Navesele–Salto della Capra landslides. Nonetheless, the Marocca Principale has an age more similar to the

Molveno landslide (4.8 ± 0.5 ka; [12]), which, for the enormous volume of both, suggests a common trigger. The distance between the Molveno and Marocca Principale, visible in Figure 2, is about 13 km and is the same in the opposite direction to the Nago-Torbole area. Considering the increasingly evident correlation of several landslides in the area linked to seismic events, it is possible that the Gorte and Spiaz de Navesele–Salto della Capra landslides were triggered by an earthquake, which may have also been responsible for the landslides of Dosso Gardene or Marocca Principale.

6. Conclusions

The region of Nago-Torbole at the northeast end of Lake Garda is the site of numerous landslides. We applied geomorphological field mapping, dating with cosmogenic ^{36}Cl, topography reconstruction, volume estimation and runout modeling to study the landslides and deposits of Gorte and Spiaz de Navesele–Salto della Capra, for which we can present the following results:

- Despite the various uncertainties linked to the topographical reconstruction, the Gorte rock avalanche has a bedrock volume in the release area of about 70–75 Mm^3, with a deposit volume of about 85–95 Mm^3. The lithologies involved in the Gorte landslide are the Massone Oolite, the Rotzo and the Tofino Formations.
- The release area (Paternoster) is strongly affected by fractures related to regional-scale fault systems. The structural setting controls the bedding inclination to the west (25°–35°). The Schio-Vicenza fault system controls the WNW–ESE-oriented flanks of the landslide, whereas the backscarp is formed by an important NNE–SSW fault belonging to the Giudicarie fault system. The flow was initially a translational rock slide with associated toppling from the surrounding steep scarps.
- The Gorte rock avalanche deposit is characterized by large hummocks, both longitudinal and transversal to the rock mass flow. Within the deposit, secondary failures occurred. The blocky carapace is not homogeneously distributed, but there are zones with abundant large blocks, while other areas are completely boulder-free.
- Our ^{36}Cl boulder exposure dates underpin the hypothesis that the Gorte rock avalanche happened in a single event. The age obtained is 6.1 ± 0.8 ka.
- The Gorte rock avalanche dates to a relatively warm and dry period of the Middle Holocene. Nevertheless, a period of frequent flooding at 6900–6200 was recognized for the region and overlaps the timing of the Gorte rock avalanche. The age is comparable to the age of the Dosso Gardene landslide (6630–6290 cal BP), as well as the Marocca Principale (5.3 ± 0.9 ka), both within 15 km. The important seismic activity still ongoing nowadays makes it likely that the trigger of the landslide may have been seismic activity.
- The Gorte landslide was characterized by initial rock sliding followed by disintegration and spreading. To simulate the flow of the rock avalanche, using Dan3D-Flex, two different rheologies were tested. The model that reproduced the best results used a frictional rheology in the source area with a friction angle of 14°, and a turbulent frictional rheology (Voellmy) in the rest of the area, with a friction coefficient of 0.38 and a turbulence coefficient of 700 m/s^2.
- The sliding of Spiaz de Navesele–Salto della Capra took place south of Gorte, involving the Tofino Formation. The related deposit, named Marocche, formed a hummock over the Busatte area.
- The sliding of Spiaz de Navesele–Salto della Capra took place at the same time or immediately after the Gorte rock avalanche: it was dated to 5.9 ± 0.4 ka, but by means of a single bedrock sample. It is not to be excluded that the sliding took place in several phases, with more or less regular detachments over the millennia. A second sample, a boulder in the deposit, indicates an age of 2.4 ± 0.2 ka, but anthropogenic influence cannot be excluded.

Supplementary Materials: The descriptions of the thin sections obtained from sampled boulders are available online at https://www.mdpi.com/article/10.3390/geosciences11100404/s1, Figure S1: Thin sections from sampled boulders.

Author Contributions: Conceptualization, S.I.-O. and S.M.; data curation, G.R., S.I.-O., O.S. and C.V.; funding acquisition, S.I.-O. and S.M.; investigation, G.R., S.I.-O., J.A., O.S., S.M., M.R., S.R., C.V., G.M. and A.V.; methodology, G.R., S.I.-O., J.A., O.S., M.R., S.R. and C.V.; resources, S.I.-O. and S.M.; supervision, S.I.-O. and S.M.; visualization, G.R., O.S. and S.R.; writing—original draft, G.R., S.I.-O., J.A., S.M., S.R. and G.M.; writing—review and editing, G.R., S.I.-O., J.A., O.S., S.M., M.R., S.R., C.V., G.M. and A.V. All authors have read and agreed to the published version of the manuscript.

Funding: This research was funded by the Department of Geosciences of the University of Padova, grant number "SID 2020"; project title: "Historic and prehistoric landslides in the Alps. Implications for new hazard maps in mountainous areas. Focus on the Lake Garda–Brenta Dolomites sector (NE Italy)".

Data Availability Statement: All data mentioned in this paper are included within the manuscript and the Supplementary Material.

Acknowledgments: The authors are grateful to the Province of Trento for the measurements on rock parameter data and the availability of the LiDAR DEM, and to the Marina Militare Italiana for bathymetry data. Fieldwork, laboratory work and AMS measurements were partially funded by Ion Beam Physics, ETH Zurich. Matteo Visintainer is acknowledged for the photo used in Figure 3.

Conflicts of Interest: The authors declare no conflict of interest. The funders had no role in the design of the study; in the collection, analyses, or interpretation of data; in the writing of the manuscript, or in the decision to publish the results.

References

1. Rossato, S.; Martin, S.; Ivy-Ochs, S.; Viganò, A.; Vockenhuber, C.; Rigo, M.; Surian, N.; Mozzi, P. Post-LGM Catastrophic Landslides in the Dolomites: When, Where and Why. *Alp. Mediterr. Quat.* **2018**, *31*, 239–242.
2. Rossato, S.; Ghirotti, M.; Gabrieli, F.; Livio, F.; Bovo, F.; Brezzi, L.; Campedel, P.; Cola, S.; Ivy-Ochs, S.; Martin, S.; et al. Learning from the Past to Face the Future: Landslides in the Piave Valley (Eastern Alps, Italy). *Alp. Mediterr. Quat.* **2020**, *33*, 209–228. [CrossRef]
3. Strouth, A.; McDougall, S. Societal Risk Evaluation for Landslides: Historical Synthesis and Proposed Tools. *Landslides* **2021**, *18*, 1071–1085. [CrossRef]
4. Kremer, K.; Simpson, G.; Girardclos, S. Giant Lake Geneva Tsunami in AD 563. *Nat. Geosci.* **2012**, *5*, 756–757. [CrossRef]
5. Köpfli, P.; Grämiger, L.M.; Moore, J.R.; Vockenhuber, C.; Ivy-Ochs, S. The Oeschinensee Rock Avalanche, Bernese Alps, Switzerland: A Co-Seismic Failure 2300 Years Ago? *Swiss J. Geosci.* **2018**, *111*, 205–219. [CrossRef]
6. Semenza, E.; Ghirotti, M. History of the 1963 Vaiont Slide: The Importance of Geological Factors. *Bull. Eng. Geol. Environ.* **2000**, *59*, 87–97. [CrossRef]
7. Aaron, J.; McDougall, S. Rock Avalanche Mobility: The Role of Path Material. *Eng. Geol.* **2019**, *257*, 105126. [CrossRef]
8. McDougall, S.; Hungr, O. Dynamic Modelling of Entrainment in Rapid Landslides. *Can. Geotech. J.* **2005**, *42*, 1437–1448. [CrossRef]
9. Weidinger, J.T.; Korup, O.; Munack, H.; Altenberger, U.; Dunning, S.A.; Tippelt, G.; Lottermoser, W. Giant Rockslides from the Inside. *Earth Planet. Sci. Lett.* **2014**, *389*, 62–73. [CrossRef]
10. Dufresne, A.; Dunning, S.A. Process Dependence of Grain Size Distributions in Rock Avalanche Deposits. *Landslides* **2017**, *14*, 1555–1563. [CrossRef]
11. Wang, F.; Okeke, A.C.U.; Kogure, T.; Sakai, T.; Hayashi, H. Assessing the Internal Structure of Landslide Dams Subject to Possible Piping Erosion by Means of Microtremor Chain Array and Self-Potential Surveys. *Eng. Geol.* **2018**, *234*, 11–26. [CrossRef]
12. Von Wartburg, J.; Ivy-Ochs, S.; Aaron, J.; Martin, S.; Leith, K.; Rigo, M.; Vockenhuber, C.; Campedel, P.; Viganò, A. Constraining the Age and Source Area of the Molveno Landslide Deposits in the Brenta Group, Trentino Dolomites (Italy). *Front. Earth Sci.* **2020**, *8*, 164. [CrossRef]
13. Singeisen, C.; Ivy-Ochs, S.; Wolter, A.; Steinemann, O.; Akçar, N.; Yesilyurt, S.; Vockenhuber, C. The Kandersteg Rock Avalanche (Switzerland): Integrated Analysis of a Late Holocene Catastrophic Event. *Landslides* **2020**, *17*, 1297–1317. [CrossRef]
14. Mitchell, A.; McDougall, S.; Nolde, N.; Brideau, M.A.; Whittall, J.; Aaron, J.B. Rock Avalanche Runout Prediction Using Stochastic Analysis of a Regional Dataset. *Landslides* **2020**, *17*, 777–792. [CrossRef]
15. Zerathe, S.; Lebourg, T.; Braucher, R.; Bourlès, D. Mid-Holocene Cluster of Large-Scale Landslides Revealed in the Southwestern Alps by 36Cl Dating. Insight on an Alpine-Scale Landslide Activity. *Quat. Sci. Rev.* **2014**, *90*, 106–127. [CrossRef]
16. Ivy-Ochs, S.; Martin, S.; Campedel, P.; Hippe, K.; Alfimov, V.; Vockenhuber, C.; Andreotti, E.; Carugati, G.; Pasqual, D.; Rigo, M.; et al. Geomorphology and Age of the Marocche Di Dro Rock Avalanches (Trentino, Italy). *Quat. Sci. Rev.* **2017**, *169*, 188–205. [CrossRef]

17. Rossato, S.; Ivy-Ochs, S.; Martin, S.; Viganò, A.; Vockenhuber, C.; Rigo, M.; Monegato, G.; de Zorzi, M.; Surian, N.; Campedel, P.; et al. Timing, Drivers and Impacts of the Historic Masiere Di Vedana Rock Avalanche (Belluno Dolomites, NE Italy). *Nat. Hazards Earth Syst. Sci.* **2020**, *20*, 2157–2174. [CrossRef]
18. Oswald, P.; Strasser, M.; Hammerl, C.; Moernaut, J. Seismic Control of Large Prehistoric Rockslides in the Eastern Alps. *Nat. Commun.* **2021**, *12*, 1–8. [CrossRef]
19. Viganò, A.; Rossato, S.; Martin, S.; Ivy-Ochs, S.; Zampieri, D.; Rigo, M.; Monegato, G. Large Landslides in the Alpine Valleys of the Giudicarie and Schio-Vicenza Tectonic Domains (NE Italy). *J. Maps* **2021**, *17*, 197–208. [CrossRef]
20. Prager, C.; Zangerl, C.; Patzelt, G.; Brandner, R. Age Distribution of Fossil Landslides in the Tyrol (Austria) and Its Surrounding Areas. *Nat. Hazards Earth Syst. Sci.* **2008**, *8*, 377–407. [CrossRef]
21. Hovius, N.; Stark, C.P.; Allen, P.A. Sediment Flux from a Mountain Belt Derived by Landslide Mapping. *Geology* **1997**, *25*, 231–234. [CrossRef]
22. Korup, O.; Clague, J.J. Natural Hazards, Extreme Events, and Mountain Topography. *Quat. Sci. Rev.* **2009**, *28*, 977–990. [CrossRef]
23. Huggel, C.; Gruber, S.; Korup, O. Landslide Hazards and Climate Change in High Mountains. *Treatise Geomorphol.* **2013**, *13*, 288–301. [CrossRef]
24. Grämiger, L.M.; Moore, J.R.; Vockenhuber, C.; Aaron, J.; Hajdas, I.; Ivy-Ochs, S. Two Early Holocene Rock Avalanches in the Bernese Alps (Rinderhorn, Switzerland). *Geomorphology* **2016**, *268*, 207–221. [CrossRef]
25. Aksay, S.; Ivy-Ochs, S.; Hippe, K.; Grämiger, L.; Vockenhuber, C. Slope Failure in a Period of Increased Landslide Activity: Sennwald Rock Avalanche, Switzerland. *Geosciences* **2021**, *11*, 331. [CrossRef]
26. Ivy-Ochs, S.; Martin, S.; Campedel, P.; Hippe, K.; Vockenhuber, C.; Carugati, G.; Rigo, M.; Pasqual, D.; Viganò, A. Geomorphology and Age of Large Rock Avalanches in Trentino (Italy): Castelpietra. In Proceedings of the 4th World Landslide Forum: Landslide Research and Risk Reduction for Advancing the Culture of Living with Natural Hazards, Ljubljana, Slovenia, 29 May–2 June 2017; pp. 347–353. [CrossRef]
27. Oetheimer, C. Etude Géomorphologique Des Éboulements Remaniés de La Vallée de Tovel: Contribution à La Connaissance Des «marocche» Du Trentin (Italie). Studi trentini di scienze naturali. *Acta Geol.* **1989**, *65*, 3–47.
28. Perna, G. Itinerari geologici: La Marocca Di Gorte. *Econ. Trent.* **1991**, *4*, 65–88.
29. Perna, G. Inquadramento Cronologico Delle Grandi Frane Quaternarie Delle Valli Dei Laghi e Del Basso Sarca. *Econ. Trent.* **1996**, *3*, 55–89.
30. Martin, S.; Campedel, P.; Ivy-Ochs, S.; Viganò, A.; Alfimov, V.; Vockenhuber, C.; Andreotti, E.; Carugati, G.; Pasqual, D.; Rigo, M. Lavini Di Marco (Trentino, Italy): 36Cl Exposure Dating of a Polyphase Rock Avalanche. *Quat. Geochronol.* **2014**, *19*, 106–116. [CrossRef]
31. Orombelli, G.; Sauro, U. I Lavini Di Marco: Un Gruppo Di Frane Oloceniche Nel Contesto Morfoneotettonico Dell'alta Val Lagarina (Trentino). *Suppl. Geogr. Fis. Din. Quat.* **1988**, *1*, 107–116.
32. Castellarin, A.; Vai, G.B.; Cantelli, L. The Alpine Evolution of the Southern Alps around the Giudicarie Faults: A Late Cretaceous to Early Eocene Transfer Zone. *Tectonophysics* **2006**, *414*, 203–223. [CrossRef]
33. Picotti, V.; Prosser, G.; Castellarin, A. Structures and Kinematics of the Giudicarie-Val Trompia Fold and Thrust Belt (Central Southern Alps, Northern Italy). *Mem. Sci. Geol.* **1995**, *47*, 95–109.
34. Bini, A.; Cita, M.B.; Gaetani, M. Southern Alpine Lakes—Hypothesis of an Erosional Origin Related to the Messinian Entrenchment. *Mar. Geol.* **1978**, *27*, 271–288. [CrossRef]
35. Finckh, P.G. Are Southern Alpine Lakes Former Messinian Canyons?—Geophysical Evidence for Preglacial Erosion in the Southern Alpine Lakes. *Mar. Geol.* **1978**, *27*, 289–302. [CrossRef]
36. Winterberg, S.; Picotti, V.; Willett, S.D. Messinian or Pleistocene Valley Incision within the Southern Alps. *Swiss J. Geosci.* **2020**, *113*, 1–14. [CrossRef]
37. Viganò, A.; Bressan, G.; Ranalli, G.; Martin, S. Focal Mechanism Inversion in the Giudicarie-Lessini Seismotectonic Region (Southern Alps, Italy): Insights on Tectonic Stress and Strain. *Tectonophysics* **2008**, *460*, 106–115. [CrossRef]
38. Viganò, A.; Scafidi, D.; Ranalli, G.; Martin, S.; della Vedova, B.; Spallarossa, D. Earthquake Relocations, Crustal Rheology, and Active Deformation in the Central-Eastern Alps (N Italy). *Tectonophysics* **2015**, *661*, 81–98. [CrossRef]
39. Guidoboni, E.; Ferrari, G.; Tarabusi, G.; Sgattoni, G.; Comastri, A.; Mariotti, D.; Ciuccarelli, C.; Bianchi, M.G.; Valensise, G. CFTI5Med, the New Release of the Catalogue of Strong Earthquakes in Italy and in the Mediterranean Area. *Sci. Data* **2019**, *6*, 1–15. [CrossRef]
40. Gasperini, L.; Marzocchi, A.; Mazza, S.; Miele, R.; Meli, M.; Najjar, H.; Michetti, A.M.; Polonia, A. Morphotectonics and Late Quaternary Seismic Stratigraphy of Lake Garda (Northern Italy). *Geomorphology* **2020**, *371*, 107427. [CrossRef]
41. Castellarin, A.; Trombetta, G.L.; Picotti, V.; Cantelli, L.; Claps, M.; Luigi, S.; Carton, A.; Borsato, A.; Daminato, F.; Nardin, M.; et al. *Note Illustrative Della Carta Geologica d' Italia- Foglio 080 Riva Del Garda*; APAT—Servizio Geologico d'Italia: Firenze, Italy, 2005.
42. Campedel, P.; Zumiani, M. *Realizzazione Del Collegamento Stradale Passo Di San Giovanni—Località Cretaccio. Relazione Geologica*; Pronvincia Autonoma di Trento: Trento, Italy, 2017.
43. Vacek, M. *Geologische Spezialkarte Der Im Reichsrate Vertretenen Königreiche Und Länder Der Österreichisch-Ungarischen Monarchie—Zone 22, Col. IV, Rovereto Und Riva*; K.K. Geologische Reichsanstalt: Wien, Austria, 1903.
44. Habbe, K.A. Die Würmzeitliche Vergletscherung Des Gardasee-Gebietes. *Freibg. Geogr. Arb.* **1969**, *3*, 254.

45. Valle, C.; Santuliana, E. *Opera Nr. 436. Collegamento "Loppio-Busa" Sulla S.S.240 e Circonvallazione Di Torbole. Progetto Preliminare e Studio V.I.A., Tavola Geologia, Relazione Geologica-Geotecnica*; Provincia Autonoma di Trento: Trento, Italy, 2008.
46. Castellarin, A.; Cantelli, L. Neo-Alpine Evolution of the Southern Eastern Alps. *J. Geodyn.* **2000**, *30*, 251–274. [CrossRef]
47. Provincia Autonoma di Trento LIDA RRilievo 2006/2007/2008. Available online: http://www.territorio.provincia.tn.it/portal/server.pt/community/lidar/847/lidar/23954 (accessed on 21 August 2021).
48. *Provincia Autonoma di Trento Carta Della Pericolosità e Carta Di Sintesi Della Pericolosità-Section N. 80110-Riva Del Garda*; Provincia Autonoma di Trento: Trento, Italy, 2020.
49. Ivy-Ochs, S.; Kober, F. Surface Exposure Dating with Cosmogenic Nuclides. *EG Quat. Sci. J.* **2008**, *57*, 179–209. [CrossRef]
50. Ivy-Ochs, S.; Synal, H.A.; Roth, C.; Schaller, M. Initial Results from Isotope Dilution for Cl and 36Cl Measurements at the PSI/ETH Zurich AMS Facility. *Nucl. Instrum. Methods Phys. Res. Sect. B Beam Interact. Mater. At.* **2004**, *223*, 623–627. [CrossRef]
51. Christl, M.; Vockenhuber, C.; Kubik, P.W.; Wacker, L.; Lachner, J.; Alfimov, V.; Synal, H.A. The ETH Zurich AMS Facilities: Performance Parameters and Reference Materials. *Nucl. Instrum. Methods Phys. Res. Sect. B Beam Interact. Mater. At.* **2013**, *294*, 29–38. [CrossRef]
52. Vockenhuber, C.; Miltenberger, K.U.; Synal, H.A. 36Cl Measurements with a Gas-Filled Magnet at 6 MV. *Nucl. Instrum. Methods Phys. Res. Sect. B Beam Interact. Mater. At.* **2019**, *455*, 190–194. [CrossRef]
53. Synal, H.A.; Bonani, G.; Döbeli, M.; Ender, R.M.; Gartenmann, P.; Kubik, P.W.; Schnabel, C.; Suter, M. Status Report of the PSI/ETH AMS Facility. *Nucl. Instrum. Methods Phys. Res. Sect. B Beam Interact. Mater. At.* **1997**, *123*, 62–68. [CrossRef]
54. Alfimov, V.; Ivy-Ochs, S. How Well Do We Understand Production of 36Cl in Limestone and Dolomite? *Quat. Geochronol.* **2009**, *4*, 462–474. [CrossRef]
55. Stone, J.O.; Allan, G.L.; Fifield, L.K.; Cresswell, R.G. Cosmogenic Chlorine-36 from Calcium Spallation. *Geochim. Cosmochim. Acta* **1996**, *60*, 679–692. [CrossRef]
56. Stone, J.O.H.; Evans, J.M.; Fifield, L.K.; Allan, G.L.; Cresswell, R.G. Cosmogenic Chlorine-36 Production in Calcite by Muons. *Geochim. Cosmochim. Acta* **1998**, *62*, 433–454. [CrossRef]
57. Marrero, S.M.; Phillips, F.M.; Caffee, M.W.; Gosse, J.C. CRONUS-Earth Cosmogenic 36Cl Calibration. *Quat. Geochronol.* **2016**, *31*, 199–219. [CrossRef]
58. Balco, G.; Stone, J.O.; Lifton, N.A.; Dunai, T.J. A Complete and Easily Accessible Means of Calculating Surface Exposure Ages or Erosion Rates from 10Be and 26Al Measurements. *Quat. Geochronol.* **2008**, *3*, 174–195. [CrossRef]
59. Steinemann, O.; Ivy-Ochs, S.; Grazioli, S.; Luetscher, M.; Fischer, U.H.; Vockenhuber, C.; Synal, H.-A. Quantifying Glacial Erosion on a Limestone Bed and the Relevance for Landscape Development in the Alps. *Earth Surf. Process. Landf.* **2020**, *45*, 1401–1417. [CrossRef]
60. Scheidegger, A.E. On the Prediction of the Reach and Velocity of Catastrophic Landslides. *Rock Mech.* **1973**, *5*, 231–236. [CrossRef]
61. Hungr, O.; Evans, S.G.; Hazzard, J. Magnitude and Frequency of Rock Falls and Rock Slides along the Main Transportation Corridors of Southwestern British Columbia. *Can. Geotech. J.* **1999**, *36*, 224–238. [CrossRef]
62. Hungr, O.; Evans, S.G. Entrainment of Debris in Rock Avalanches: An Analysis of a Long Run-out Mechanism. *Bull. Geol. Soc. Am.* **2004**, *116*, 1240–1252. [CrossRef]
63. Aaron, J.; Hungr, O. Dynamic Simulation of the Motion of Partially-Coherent Landslides. *Eng. Geol.* **2016**, *205*, 1–11. [CrossRef]
64. McDougall, S.; Hungr, O. A Model for the Analysis of Rapid Landslide Motion across Three-Dimensional Terrain. *Can. Geotech. J.* **2004**, *41*, 1084–1097. [CrossRef]
65. Hungr, O.; McDougall, S. Two Numerical Models for Landslide Dynamic Analysis. *Comput. Geosci.* **2009**, *35*, 978–992. [CrossRef]
66. Aaron, J.; McDougall, S.; Nolde, N. Two Methodologies to Calibrate Landslide Runout Models. *Landslides* **2019**, *16*, 907–920. [CrossRef]
67. Provincia Autonoma di Trento Reticolo Idrografico. Available online: https://webgis.provincia.tn.it/ (accessed on 21 August 2021).
68. Baroni, C. Lake Garda: An Outstanding Archive of Quaternary Geomorphological Evolution. In *Landscapes and Landforms of Italy*; Soldati, M., Marchetti, M., Eds.; Springer: Berlin/Heidelberg, Germany, 2017; pp. 169–179.
69. Sosio, R.; Crosta, G.B.; Hungr, O. Complete Dynamic Modeling Calibration for the Thurwieser Rock Avalanche (Italian Central Alps). *Eng. Geol.* **2008**, *100*, 11–26. [CrossRef]
70. Si, P.; Aaron, J.; McDougall, S.; Lu, J.; Yu, X.; Roberts, N.J.; Clague, J.J. A Non-Hydrostatic Model for the Numerical Study of Landslide-Generated Waves. *Landslides* **2017**, *15*, 711–726. [CrossRef]
71. Dufresne, A.; Davies, T.R. Longitudinal Ridges in Mass Movement Deposits. *Geomorphology* **2009**, *105*, 171–181. [CrossRef]
72. Von Poschinger, A.; Ruegg, T. Die Churer Tomahügel, Ein Besonderes Zeugnis Der Landschaftsgenese. *Jahresbericht der Naturforschenden Gesellschaft Graubünden* **2012**, *117*, 93–100.
73. Prager, C.; Ivy-Ochs, S.; Ostermann, M.; Synal, H.A.; Patzelt, G. Geology and Radiometric 14C-, 36Cl- and Th-/U-Dating of the Fernpass Rockslide (Tyrol, Austria). *Geomorphology* **2009**, *103*, 93–103. [CrossRef]
74. Montandon, F. Chronologie Des Grands Éboulements Alpins Du Début de l'ère Chrétienne à Nos Jours. *Matériaux Pour l'étude des Calamités* **1933**, *32*, 271–340.
75. Monegato, G.; Scardia, G.; Hajdas, I.; Rizzini, F.; Piccin, A. The Alpine LGM in the Boreal Ice-Sheets Game. *Sci. Rep.* **2017**, *7*, 2078. [CrossRef]

76. Ravazzi, C.; Pini, R.; Badino, F.; de Amicis, M.; Londeix, L.; Reimer, P.J. The Latest LGM Culmination of the Garda Glacier (Italian Alps) and the Onset of Glacial Termination. Age of Glacial Collapse and Vegetation Chronosequence. *Quat. Sci. Rev.* **2014**, *105*, 26–47. [CrossRef]
77. Wolter, A.; Stead, D.; Ward, B.C.; Clague, J.J.; Ghirotti, M. Engineering Geomorphological Characterisation of the Vajont Slide, Italy, and a New Interpretation of the Chronology and Evolution of the Landslide. *Landslides* **2016**, *13*, 1067–1081. [CrossRef]
78. Provincia Autonoma di Trento Catasto Grotte. Available online: https://webgis.provincia.tn.it/ (accessed on 21 August 2021).
79. Pánek, T.; Hradecký, J.; Šilhán, K.; Smolková, V.; Altová, V. Time Constraints for the Evolution of a Large Slope Collapse in Karstified Mountainous Terrain of the Southwestern Crimean Mountains, Ukraine. *Geomorphology* **2009**, *108*, 171–181. [CrossRef]
80. Sauro, F.; Zampieri, D.; Filipponi, M. Development of a Deep Karst System within a Transpressional Structure of the Dolomites in North-East Italy. *Geomorphology* **2013**, *184*, 51–63. [CrossRef]
81. Gutiérrez, F.; Parise, M.; de Waele, J.; Jourde, H. A Review on Natural and Human-Induced Geohazards and Impacts in Karst. *Earth Sci. Rev.* **2014**, *138*, 61–88. [CrossRef]
82. Viganò, A.; Scafidi, D.; Martin, S.; Spallarossa, D. Structure and Properties of the Adriatic Crust in the Central-Eastern Southern Alps (Italy) from Local Earthquake Tomography. *Terra Nova* **2013**, *25*, 504–512. [CrossRef]
83. Martin, S.; Ivy-Ochs, S.; Viganò, A.; Campedel, P.; Rigo, M.; Vockenhuber, C.; Gabrieli, F.; Mair, V.; Rossato, S. Landslides of the Western Dolomites: Case Studies from the Adige and Sarca Valleys (NE Italy). *Alp. Mediterr. Quat.* **2020**, *33*, 191–207. [CrossRef]
84. Rossato, S.; Fontana, A.; Mozzi, P. Meta-Analysis of a Holocene 14C Database for the Detection of Paleohydrological Crisis in the Venetian-Friulian Plain (NE Italy). *Catena* **2015**, *130*, 34–45. [CrossRef]
85. Benito, G.; Macklin, M.G.; Panin, A.; Rossato, S.; Fontana, A.; Jones, A.F.; Machado, M.J.; Matlakhova, E.; Mozzi, P.; Zielhofer, C. Recurring Flood Distribution Patterns Related to Short-Term Holocene Climatic Variability. *Sci. Rep.* **2015**, *5*, 1–8. [CrossRef]
86. Gischig, V.; Preisig, G.; Eberhardt, E. Numerical Investigation of Seismically Induced Rock Mass Fatigue as a Mechanism Contributing to the Progressive Failure of Deep-Seated Landslides. *Rock Mech. Rock Eng.* **2016**, *49*, 2457–2478. [CrossRef]
87. Loew, S.; Gschwind, S.; Gischig, V.; Keller-Signer, A.; Valenti, G. Monitoring and Early Warning of the 2012 Preonzo Catastrophic Rockslope Failure. *Landslides* **2017**, *14*, 141–154. [CrossRef]
88. Bartlein, P.J.; Harrison, S.P.; Brewer, S.; Connor, S.; Davis, B.A.S.; Gajewski, K.; Guiot, J.; Harrison-Prentice, T.I.; Henderson, A.; Peyron, O.; et al. Pollen-Based Continental Climate Reconstructions at 6 and 21 Ka: A Global Synthesis. *Clim. Dyn.* **2011**, *37*, 775–802. [CrossRef]
89. Routson, C.C.; McKay, N.P.; Kaufman, D.S.; Erb, M.P.; Goosse, H.; Shuman, B.N.; Rodysill, J.R.; Ault, T. Mid-Latitude Net Precipitation Decreased with Arctic Warming during the Holocene. *Nature* **2019**, *568*, 83–87. [CrossRef]

geosciences

Article

Empirical Evidence for Latitude and Altitude Variation of the In Situ Cosmogenic $^{26}Al/^{10}Be$ Production Ratio

Christopher T. Halsted [1,*], Paul R. Bierman [1] and Greg Balco [2]

[1] Rubenstein School of Environment and Natural Resources, Gund Institute, University of Vermont, Burlington, VT 05405, USA; Paul.Bierman@uvm.edu
[2] Berkeley Geochronology Center, Berkeley, CA 94709, USA; balcs@bgc.org
* Correspondence: Christopher.Halsted@uvm.edu; Tel.: +1-802-656-2911

Abstract: We assess if variations in the in situ cosmogenic $^{26}Al/^{10}Be$ production ratio expected from nuclear physics are consistent with empirical data, knowledge critical for two-isotope studies. We do this using 313 samples from glacially transported boulders or scoured bedrock with presumed simple exposure histories in the Informal Cosmogenic-nuclide Exposure-age Database (ICE-D) from latitudes between 53°S to 70°N and altitudes up to 5000 m above sea level. Although there were small systematic differences in Al/Be ratios measured in different laboratories, these were not significant and are in part explained by differences in elevation distribution of samples analyzed by each laboratory. We observe a negative correlation between the $^{26}Al/^{10}Be$ production ratio and elevation (p = 0.0005), consistent with predictions based on the measured energy dependence of nuclear reaction cross-sections and the spatial variability in cosmic-ray energy spectra. We detect an increase in the production ratio with increasing latitude, but this correlation is significant only in a single variate model, and we attribute at least some of the correlation to sample elevation bias because lower latitude samples are typically from higher elevations (and vice versa). Using 6.75 as the $^{26}Al/^{10}Be$ production ratio globally will bias two-isotope results at higher elevations and perhaps higher latitudes. Data reported here support using production rate scaling that incorporates such ratio changes, such as the LSDn scheme, to minimize such biases.

Keywords: cosmogenic nuclides; nuclide production; burial dating

Citation: Halsted, C.T.; Bierman, P.R.; Balco, G. Empirical Evidence for Latitude and Altitude Variation of the In Situ Cosmogenic $^{26}Al/^{10}Be$ Production Ratio. *Geosciences* **2021**, *11*, 402. https://doi.org/10.3390/geosciences11100402

Academic Editors: Naki Akçar, Susan Ivy-Ochs, Fritz Schlunegger and Jesus Martinez-Frias

Received: 30 July 2021
Accepted: 14 September 2021
Published: 23 September 2021

1. Introduction

Paired-nuclide, in situ cosmogenic nuclide analyses are valuable tools for investigating complex landscape histories, including burial after and/or during exposure. In situ cosmogenic nuclides are formed in minerals at the Earth's surface when exposed to the high-energy particle cascade produced during interactions between cosmic radiation and atmospheric gasses [1]. Differences in production and decay ratios between multiple in situ cosmogenic radionuclides are used to estimate burial/exposure durations and erosion histories—with applications ranging from non-erosive glacier histories [2–4], to long-term fluvial incision [5,6] and archaeological investigations [7–9].

An essential component of this methodology is knowing with certainty the production ratio between measured in situ cosmogenic radionuclides. While the decay rates of cosmogenic radionuclides have been empirically constrained [10–12], the production rates of cosmogenic nuclides, and thus their production ratios, are estimated using models of the relevant physics [13] and validated with calibration studies that rely on independent age constraints of landscape features [14,15].

Two of the most-used in situ cosmogenic radionuclides in dual-nuclide studies are ^{26}Al and ^{10}Be. Both are produced in quartz, and their ratio has been measured since the 1980s [16]. The near-ubiquity of quartz across the world and the improved analytical precision for ^{10}Be—and more recently ^{26}Al [17]—measurements make these nuclides the "go to" for dual-nuclide studies. The improvement in measurement precision will allow for more useful interpretation of the data, but only if the production ratio is well constrained [18].

Physics-based nuclide production models suggest that the $^{26}Al/^{10}Be$ surface production ratio should decrease with elevation and increase with latitude [19,20]; however, most analyses of $^{26}Al/^{10}Be$ data assume a globally constant surface production ratio. Empirical evidence from high-latitude sites [21] suggest that this assumption may not be valid. A surface production ratio that changes with latitude and/or elevation would mean that many of the studies using this dual-nuclide methodology contain systematic biases in their results because the assumed surface production ratio may differ from the actual ratio at the sampling site. As cosmogenic nuclide measurements become more precise, better constraining of the production ratio becomes more important.

Here, we assess if changes in the $^{26}Al/^{10}Be$ production ratio with latitude and/or elevation are detectable in empirical data using a compilation of 313 previously published in situ $^{26}Al/^{10}Be$ ratios from samples spanning a wide range of latitudes and elevations. We applied selection criteria to increase the chances that these glacially related samples have experienced simple exposure histories—that is, only one short (<25 kyr) period of exposure and no burial, so that the measured ratio ($^{26}Al/^{10}Be$) represents the surface production ratio. We apply deductive statistical analyses to this compilation to determine if $^{26}Al/^{10}Be$ ratio variations with altitude, latitude, and sample processing laboratory are present at a statistically significant level. This analysis allows us to test whether there are detectable production ratio variations, thus necessitating application of physics-based nuclide production models in dual-nuclide studies, or if the nominal ratio of 6.75 is suitable at all latitudes and elevations.

2. Background

2.1. In Situ Cosmogenic ^{26}Al and ^{10}Be

In this study, we focus on the production of in situ cosmogenic ^{26}Al and ^{10}Be in quartz. ^{26}Al and ^{10}Be are produced primarily through spallation reactions in quartz (>95% at sea level and high latitude) with minor production from muon interactions [1,11,12]. The ratio of muonic to spallation production increases with depth below Earth's surface [1] and decreases at higher elevations. ^{26}Al has a half-life of 0.705 million years [10] and ^{10}Be has a half-life of 1.39 Ma [11,12].

The accuracy of measured $^{26}Al/^{10}Be$ ratios as a proxy for the $^{26}Al/^{10}Be$ ratio at production is controlled by both laboratory procedures and the geologic history of sample sites. Nuclides inherited from prior periods of exposure both at the surface and at depth influence the concentration of ^{26}Al and ^{10}Be in surface samples and thus the measured $^{26}Al/^{10}Be$ ratio. Muon-induced production produces relatively few nuclides but does so at a ratio higher than surface production, which is dominated by neutrons [22]. Storage of previously exposed material at depths below the penetration depth of most neutrons can lower measured ratios as ^{26}Al decays more quickly than ^{10}Be [10].

Laboratory concerns include the measurement of cosmogenically produced isotopes (^{26}Al and ^{10}Be) by Accelerator Mass Spectrometry (AMS), which often sets the limit on precision, and quantification of stable isotopes (^{27}Al and ^{9}Be), which is critical for accuracy. Low-energy AMS machines may be unable to completely reject isobaric interferences encountered in ^{26}Al analyses, thus artificially increasing calculated ^{26}Al concentrations and the $^{26}Al/^{10}Be$ ratio [23]. When complexed with fluoride during HF digestion, Al can be difficult to get back into solution, thus leading to underestimation of ^{26}Al and consequently low $^{26}Al/^{10}Be$ ratios [24]. Stable beryllium is added as a carrier (isotope dilution) but stable aluminum is native to the quartz being digested, meaning that full retention and recovery of that aluminum is critical to accurately quantifying the concentration of ^{26}Al. Stable aluminum quantification errors can arise from chemical processing steps, including aliquot measurements after, rather than before, drying dissolved samples, adding sulfuric acid to digestion solutions, and systematic offsets in calibration of inductively coupled plasma optical emission spectrometers for measuring ^{27}Al concentrations [25]. Low recovery of ^{27}Al will result in lower than actual ^{26}Al/^{10}Be ratios [24]. Significant variation in measured ^{26}Al between chemistry labs has been observed during inter-lab comparisons, and much of this variation is attributed to differences in methodology for quantifying ^{27}Al [26–28]. A small number of quartz mineral separates contain significant amounts of stable ^{9}Be, which, if unaccounted for, would result in spuriously low ^{10}Be concentrations and high ^{26}Al/^{10}Be ratios, and few laboratories routinely measure beryllium in quartz [29].

2.2. Applications

Paired in situ cosmogenic ^{26}Al and ^{10}Be are used in a wide variety of studies seeking to understand burial and erosion histories. Early applications included the history of enigmatic Libyan desert glass [16], the age of ancient glacial deposits in the Sierra Nevada mountains [30], and the glacial history of Antarctica [31]. Bierman et al. [2] demonstrated the utility of the paired-nuclide approach for calculating minimum total durations of exposure and burial for complex glacial histories in temperate regions—those involving multiple periods of advance and retreat. Investigations of complex glacial histories continue to be a common application of ^{26}Al/^{10}Be [32,33], including cases with minimal glacial erosion due to cold-based ice cover [4,34] and glacial histories inferred from marine sediment records [35,36]. ^{26}Al/^{10}Be has also been used to evaluate long-term erosion rates in arid regions [37], histories of tectonic uplift [38], river incision [5,6], and paleosol burial [39,40]. ^{26}Al/^{10}Be is also used for age control in archaeological investigations of hominin evolution [7–9,41], providing burial ages of bones and artifacts.

2.3. Previous Constraints on ^{26}Al/^{10}Be Production Ratio

The production ratio of ^{26}Al/^{10}Be at Earth's surface has been constrained in two ways: experimental measurements—sampling surfaces with 'known' exposure histories—and physics-based models that simulate interactions between cosmic radiation, atmospheric atoms, and terrestrial atoms in target minerals. See Corbett et al. ([21] and Table S1 therein) for a summary of these studies and the production ratios they calculated and measured.

Studies published prior to 1991 estimated an ^{26}Al/^{10}Be production ratio of ~6.1 [30,42], which was updated to 6.75 following updates to accelerator standards and refinements of the ^{10}Be half-life [11,12,43]. Early models of ^{26}Al and ^{10}Be production indicated a spallation production ratio of 6.05 that did not change with elevation (latitude changes in in situ ^{26}Al production were not modelled [44]). Without robust empirical evidence to support the use of more complex numerical models for nuclide production, an ^{26}Al/^{10}Be production ratio of 6.75 is typically assumed to be constant over all latitudes and elevations in many scaling schemes used for cosmogenic nuclide data interpretation [45].

2.4. Indication of Spatial Variability in the ^{26}Al/^{10}Be Production Ratio

Cross-sections for nuclide production from spallation reactions suggest changes in the ^{26}Al/^{10}Be production ratio with latitude and elevation [19,20,46–48]. Although ^{26}Al and ^{10}Be are both produced in quartz at Earth's surface primarily through spallation reactions, the cross section—or likelihood of reaction—for spallation production is different for each nuclide. ^{26}Al has a lower energy threshold for production than ^{10}Be, so neutron fluxes with different energy spectra produce ^{26}Al and ^{10}Be at different ratios [47,48]. Earth's geomagnetic field deflects lower-energy components of the primary cosmic ray flux more readily at lower latitudes [46]. The energy spectrum of the secondary neutron flux that reaches Earth's surface therefore differs with latitude, implying a lower ^{26}Al/^{10}Be production ratio at low latitudes, where the more energetic neutron flux favors ^{10}Be production, and a higher production ratio at high latitudes, where the less energetic neutron flux favors ^{26}Al production [19].

A similar dynamic is expected with changes in elevation due to changes in the neutron flux energy spectrum with depth in the atmosphere. The cosmogenically derived neutron flux loses energy with increasing atmospheric depth due to interactions with atmospheric gas atoms [1,42]. Thus, the $^{26}Al/^{10}Be$ production ratio should be lower at high elevations, where the neutron flux has higher energy and favors ^{10}Be production, and highest at low elevations, where the lower energy neutron flux favors ^{26}Al production [46].

3. Materials and Methods

3.1. Data Sources and Sample Selection

To determine if the assumption of a constant $^{26}Al/^{10}Be$ production ratio is an oversimplification, we test for spatial heterogeneity in the $^{26}Al/^{10}Be$ production ratio using previously published samples in the Informal Cosmogenic-nuclide Exposure-age Database (ICE-D; ice-d.org, n = 313; ref. [49]). Within the ICE-D database, we extracted data from ICE-D: Alpine, data from alpine glacial landforms (n = 243), and ICE-D: Calib, samples used to calibrate cosmogenic-nuclide production rates by assuming exposure ages based on other geologic constraints (n = 70). ^{26}Al and ^{10}Be concentration measurements for all samples were normalized to the KNSTD and 07KNSTD standards, respectively [10,43]. In both sub-databases, we targeted samples that likely experienced simple exposure histories, such that all ^{26}Al and ^{10}Be are from a single exposure extending to the present day with no nuclides remaining from periods of prior exposures. In other words, we presume that the measured $^{26}Al/^{10}Be$ concentration ratios equal the $^{26}Al/^{10}Be$ production ratios for the samples we selected.

We applied the following criteria: 1. We selected samples in ICE-D: Alpine with reported ^{26}Al and ^{10}Be concentration measurements and with exposure ages under 25 ka. Querying for ages under 25 ka ensures that measured concentrations are from a single, short period of near-surface exposure. We do not include samples from Antarctica, where prolonged burial that alters measured $^{26}Al/^{10}Be$ ratios is evident in many samples (e.g., [50]). 2. We extracted all sample data from ICE-D: Calib and calculated exposure ages using the reported ^{10}Be concentrations, LSDn scaling, and the default exposure age calculator settings in version three of the online exposure age calculator described by Balco et al. [45] (i.e., without the reference production rate from the calibration site). We kept samples in our analysis if their calculated ages using these exposure age calculator settings matched the expected ages from nearby geologic calibration sites, indicating little inherited ^{10}Be (and by association ^{26}Al) was present in these samples.

To avoid samples affected by geologic and/or laboratory processes that can skew ratios, we discarded samples from our initial query with physically unreasonable $^{26}Al/^{10}Be$ ratios. To account for the inevitable scatter in ratios due to analytical uncertainty of ^{26}Al and ^{10}Be measurements, we first fit a normal distribution to the $^{26}Al/^{10}Be$ ratio uncertainties in our compilation (Figures S2 and S3) and calculate the ratio uncertainty mean and standard deviation. We use the uncertainty mean (10.1%) plus one standard deviation (9.0%) as a threshold, beyond which we deemed the ratios physically unreasonable. Applying this 19% analytical uncertainty threshold to the canonical $^{26}Al/^{10}Be$ production ratio value of 6.75 gives a range of 5.47 to 8.03 for accepted $^{26}Al/^{10}Be$ ratios (details in Supplement). We assume that outlier samples were affected by geologic and/or laboratory processes that skew measured $^{26}Al/^{10}Be$ concentration ratios such that they do not reflect the surface production ratio. Our final tally of samples used in statistical analyses (n = 313) does not include outliers (n = 48) removed from the original ICE:D query (Figure S1).

3.2. Statistical Analyses

We use single and bi-variate linear models, Monte Carlo simulations, and analysis of variance (ANOVA) tests to determine if variations in the measured $^{26}Al/^{10}Be$ concentration ratios in our compilation are correlated with elevation and/or latitude. We first divide the sample population into three latitude transects and five elevation transects to isolate latitude and elevation as variables and create sample groups for ANOVA testing. Each elevation transect is a bin of samples from similar latitudes but spanning a range of elevations, while each latitude transect is a bin of samples with similar elevations but varying latitudes (Figure 1).

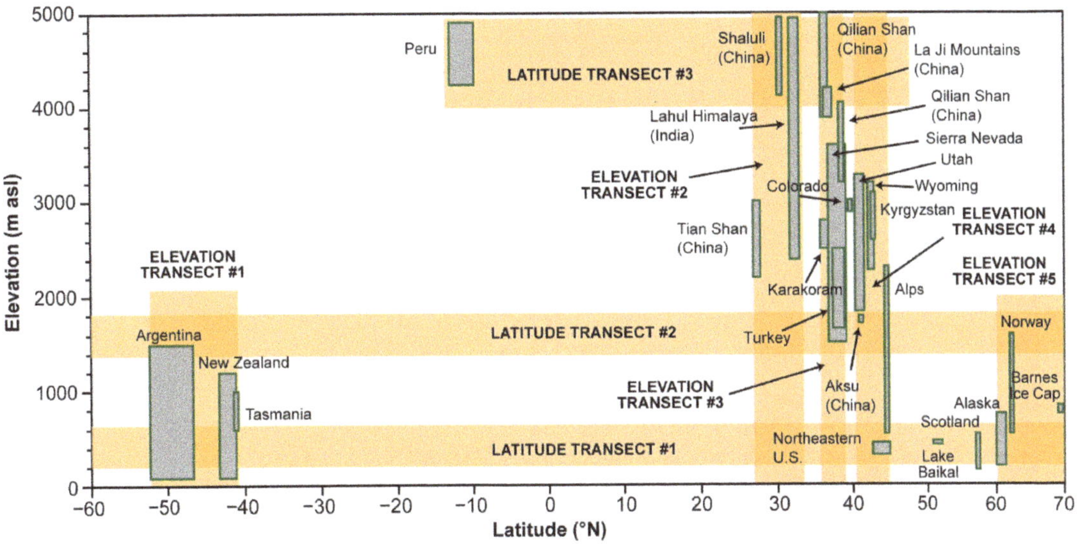

Figure 1. Sample locations and study design.

We use single and bi-variate linear models to determine if there are statistically significant correlations between the $^{26}Al/^{10}Be$ production ratio, latitude, and elevation. We run single regression linear models for elevation and latitude (as absolute latitude) vs. measured $^{26}Al/^{10}Be$ ratios first with the entire compilation and then with samples separated into transects, calculating 95% confidence intervals, correlation coefficients (r), and *p*-values against a null model (no variation in production ratio) for each iteration. We assume no uncertainty on latitude or elevation measurements. The bi-variate regression model includes elevation and latitude variables and is run for the entire compilation of samples.

We run Monte Carlo analyses to assess the influence of $^{26}Al/^{10}Be$ concentration ratio uncertainties on linear regressions. In each Monte Carlo analysis, we run 1000 iterations of linear regression with samples randomly adopting an $^{26}Al/^{10}Be$ concentration ratio value from within their uncertainty bounds (assuming a Gaussian uncertainty distribution) in each iteration. Monte Carlo analyses allow us to constrain a population of regressions using the uncertainties on each data point, producing another type of confidence interval that incorporates data uncertainty.

To assess if $^{26}Al/^{10}Be$ concentration ratios differ between transects, we perform ANOVA testing with transects as groups. If ANOVA testing indicates that one or more transect $^{26}Al/^{10}Be$ concentration ratio means are different at a 5% significance level, we perform multiple pairwise comparison of the concentration ratio means [51] to determine which transects differ and the statistical significance of differences. To assess if different cosmogenic nuclide sample preparation labs have an influence on measured $^{26}Al/^{10}Be$ concentration ratios, we also perform ANOVA testing with the five labs that processed the greatest number of samples in this compilation. The five labs are located at Lawrence Livermore National Laboratory (LLNL), the University of Washington (UW), the Swiss Federal Institute of Technology in Zürich (ETH), the Australian Nuclear Science and Technology Organization (ANSTO), and the Purdue Rare Isotopes Measurement Laboratory (PRIME). Together, these labs are responsible for the ^{10}Be and ^{26}Al extraction of 70% of the samples in our compilation.

4. Results

4.1. Compilation Statistics

$^{26}Al/^{10}Be$ concentration ratios in our sample compilation approximate a normal distribution with $\mu = 6.57$ and $\sigma = 0.52$ (Figure 2). However, our compilation has spatial bias. While samples are present at most elevations between sea level and 5000 m asl (Figure 3B), they are biased towards the mid-latitudes in both hemispheres, with particular density around the northern mid-latitudes (Figure 3B). There are no samples from low latitude/low elevation or high latitude/high elevation locations (Figure 1). Sample processing year does not have an observable impact on the measured $^{26}Al/^{10}Be$ ratio (Figure S10).

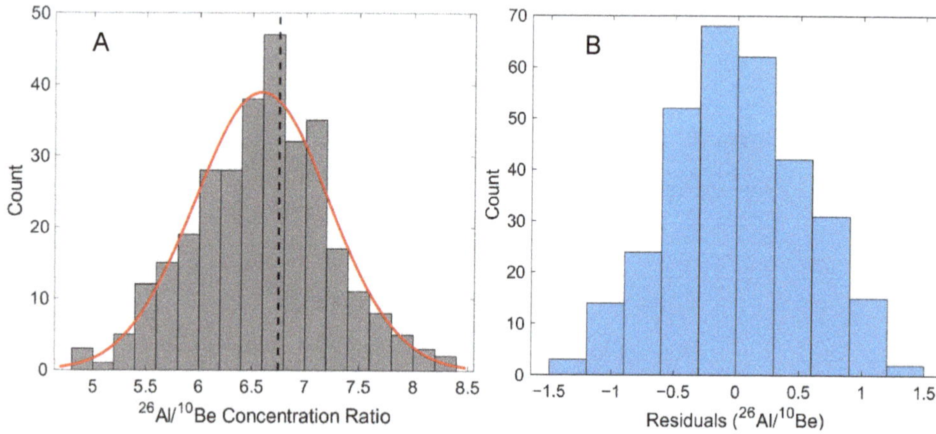

Figure 2. (**A**) Histogram and normal distribution approximation of $^{26}Al/^{10}Be$ concentration ratios in the compilation analyzed here. Dashed vertical line shows the currently accepted production ratio value of 6.75. (**B**) Residuals of bi-variate regression model predicting $^{26}Al/^{10}Be$ variations due to elevation and latitude.

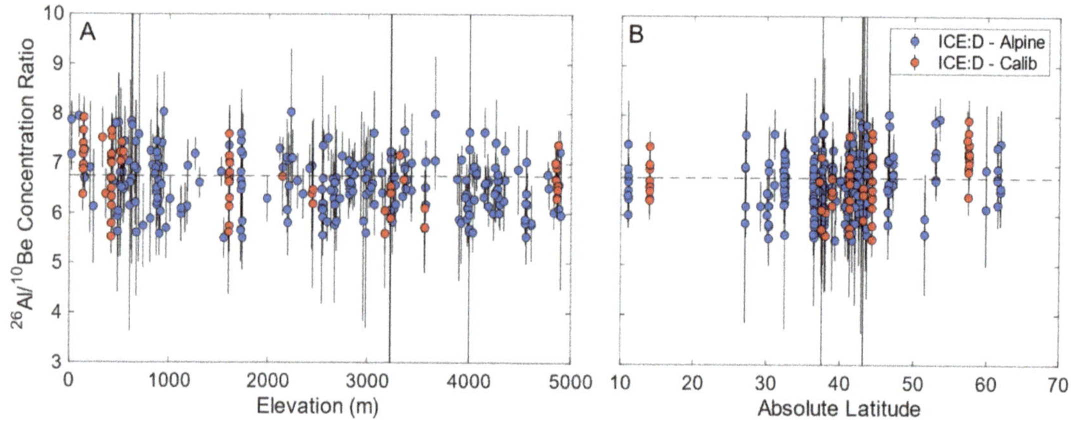

Figure 3. Measured $^{26}Al/^{10}Be$ concentration ratios plotted against elevation (**A**) and latitude (**B**). Dashed horizontal line shows the currently accepted production ratio value of 6.75. Samples from the ICE:D—Alpine database are colored blue, samples from ICE:D—Calib are red.

4.2. Regression Statistics

A simple linear regression of measured $^{26}Al/^{10}Be$ ratios vs elevation is consistent with lowering of the $^{26}Al/^{10}Be$ production ratio with elevation. The elevation regression shows a statistically significant (p = 0.000028) negative correlation (r = −0.23) between elevation and measured $^{26}Al/^{10}Be$ ratios (Figure 4A). Although large, the residuals from the elevation/ratio regression are normally distributed; there is no evidence of heteroscedasticity. The regression exhibits good fit to the data, as indicated by a reduced chi-squared test, which accounts for scatter caused by uncertainty in the data (χ^2_ν = 1.25). The Monte Carlo regressions support a negative correlation, with every regression exhibiting a negative slope and with the 95% confidence interval (95% of regressions) overlapping the ratio change expected with elevation in nuclide production models (change in ratio = −0.083 per km elevation, calculated from [20]).

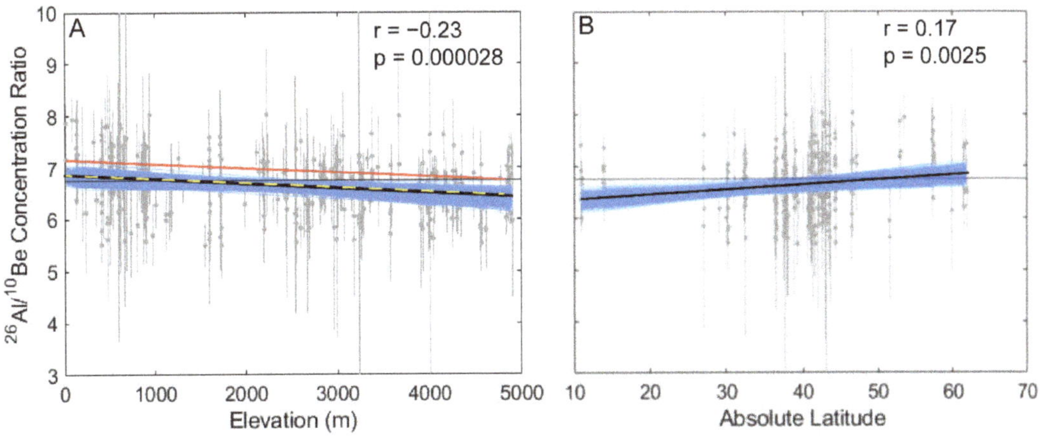

Figure 4. Linear regressions correlating measured $^{26}Al/^{10}Be$ concentration ratios to elevation (**A**) and latitude (**B**). Gray data points are the same as Figure 3. Central, solid black line in each figure is the most likely regression. Thin, blue lines are individual Monte Carlo regressions. Gray horizontal line shows a ratio of 6.75. Dashed yellow line in (**A**) shows the expected change in ratio with elevation from LSDn scaling at 5° latitude. Solid red line in (**A**) is the same but at 60° latitude.

A linear model with measured $^{26}Al/^{10}Be$ ratios and latitude (as absolute latitude) supports an increase in the production ratio with increasing latitude. The latitude/ratio regression shows a statistically significant (p = 0.0025) positive correlation (r = 0.17) between latitude and $^{26}Al/^{10}Be$ ratios (Figure 4B). This latitude/ratio regression has normally distributed residuals with no evidence of heteroscedasticity and exhibits good fit to the data (χ^2_ν = 1.27). The 95% confidence interval of the change in ratio with latitude as provided by Monte Carlo regressions overlaps the change expected from nuclide production models (change in ratio = 0.0053 per degree latitude, calculated from [20]).

Our bi-variate linear model with elevation and latitude as variables (Table 1) is a statistically significant improvement over a null model (p = 0.00016) and supports a negative correlation between elevation and $^{26}Al/^{10}Be$ production ratio but does not support a positive correlation between latitude and $^{26}Al/^{10}Be$ production ratio (Table 1). The bi-variate model has normally distributed residuals (Figures 2B and S5), no evidence of heteroscedasticity (Figure S7), and fits the data well (χ^2_ν = 1.24), offering a marginal improvement over the elevation-only regression. The change in $^{26}Al/^{10}Be$ ratio with elevation in the bi-variate model is within the 95% Monte Carlo confidence range of the elevation-only model, close to the change predicted by a nuclide production model [20], and is statistically significant (p = 0.004). The change in $^{26}Al/^{10}Be$ ratio with latitude in the bi-variate model is not statistically significant (p = 0.947). This model indicates that there is variation in measured $^{26}Al/^{10}Be$ concentration ratios over space, and that elevation differences appear to have the strongest correlation to these changes.

Table 1. Bivariate regression statistics table with elevation and latitude as variables ($y = x_1 + x_2 \times elev + x_3 \times lat$).

	Estimate	SE	tStat	*p* Value
Intercept	6.86	0.24	28.26	8.74×10^{-88}
Elevation	-8.97×10^{-5}	3.09×10^{-5}	−2.91	0.004
Latitude	-3.02×10^{-4}	0.005	−0.07	0.947

F-statistic vs. constant model: 9, *p*-value = 0.00016.

Linear regressions in elevation and latitude transects support $^{26}Al/^{10}Be$ ratio variations in only one transect. Measured $^{26}Al/^{10}Be$ ratios in elevation transect 1 (spanning latitudes 40°–50°S; Figure 1) exhibit a statistically significant ($p = 0.011$) negative correlation ($r = -0.34$) with elevation. Correlations in every other transect are not significant at the 5% level and exhibit wide 95% confidence intervals (Figure S4).

4.3. ANOVA

ANOVA tests indicate that there is a statistically significant difference in measured $^{26}Al/^{10}Be$ ratios between samples from the highest and lowest elevations. Both the mean and median $^{26}Al/^{10}Be$ concentration ratio from latitude transects 1 (200 to 600 m asl) and 3 (4000 to 5000 m asl) are significantly different ($p = 4.96 \times 10^{-5}$; Figure 5). Moreover, the ratio differences are as predicted by nuclear physics models, with the lower elevations having a higher $^{26}Al/^{10}Be$ ratio (mean ± SE = 6.80 ± 0.12, median = 6.95) than higher elevations (mean = 6.49 ± 0.11, median = 6.48, Figure 5). The mean and median $^{26}Al/^{10}Be$ concentration ratio from latitude transect 2, covering the 1400 to 1800 m asl elevation band, is different than latitude transect 1 at the 5%, but not 1%, significance level ($p = 0.025$) and is not significantly different than latitude transect 3 (Figure 5).

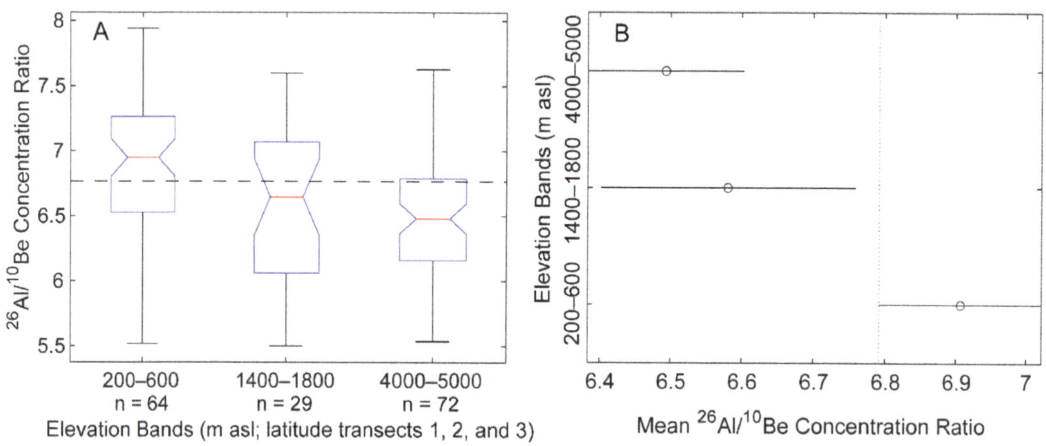

Figure 5. (**A**) Boxplots from ANOVA testing of latitude transects. Each boxplot shows the median (central red line), 25th and 75th percentile values (bottom and top edges of box), and max/min values that are not considered outliers (whiskers). The notches in each box represent median comparison intervals; two boxes with notches that do not overlap have medians that are different at the 5% significance level. For more information see [52]. (**B**) Multiple comparison of means from ANOVA testing. Each circle is the mean $^{26}Al/^{10}Be$ concentration ratio from the latitude transects. The line extending horizontally out from this point is the standard error of the mean. Vertical line is superimposed to illustrate the difference between groups.

Mean $^{26}Al/^{10}Be$ concentration ratios from elevation transects 1 (6.76 ± 0.14) and 3 (6.45 ± 0.12), which cover 40° to 50°S and 36° to 39°N, respectively, are significantly different ($p = 0.008$; Figure 6). Every other elevation transect is statistically similar. Aside from transect 3, all elevation transects also have mean $^{26}Al/^{10}Be$ concentration ratios that overlap the canonical value of 6.75 within the envelope of mean standard errors (Figure 6B).

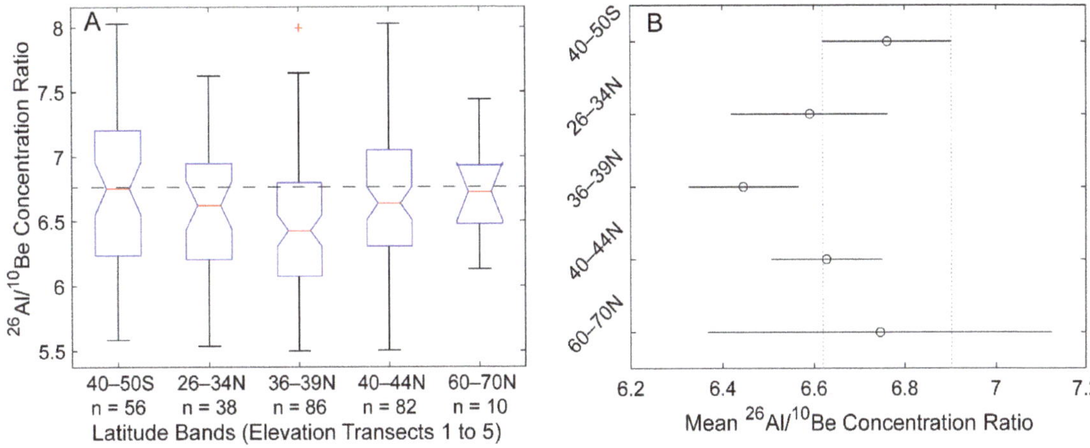

Figure 6. ANOVA boxplots (**A**) and multiple comparison of means (**B**) for elevation transects. See Figure 5 for explanation of each figure. Red cross in (**A**) is an outlier.

Sample preparation lab ANOVA testing revealed that samples from the University of Washington Cosmogenic Nuclide Laboratory (UW) have higher mean and median $^{26}Al/^{10}Be$ concentration ratios than other labs (Figure 7A). Sample ratios from UW were different than samples processed at Lawrence Livermore National Laboratories (LLNL), but not other labs, at a statistically significant level ($p = 0.006$).

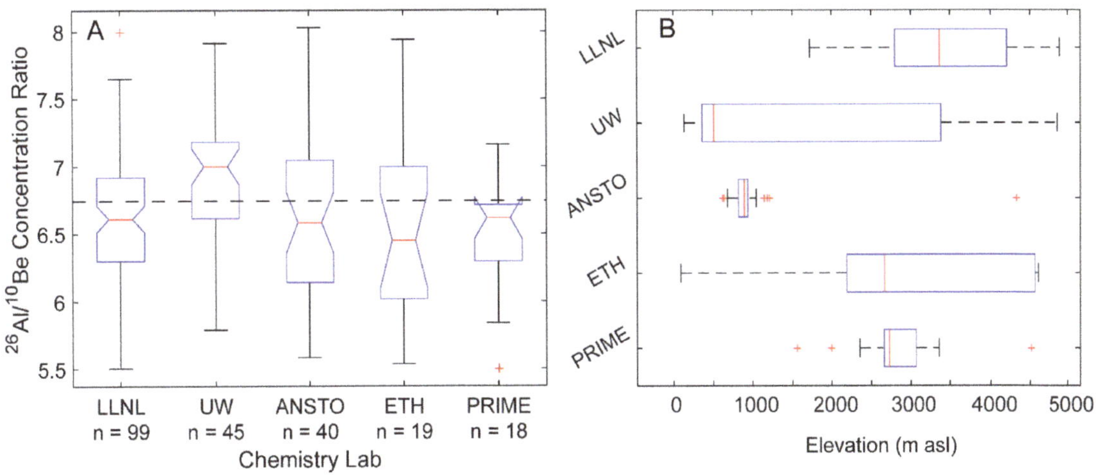

Figure 7. (**A**) ANOVA boxplots for sample preparation labs, see Figure 5 for explanation. Note that the PRIME box appears distorted because the upper limit of the median comparison interval exceeds the 75th percentile value. (**B**) Elevations of samples processed by five different sample preparation laboratories.

5. Discussion

The $^{26}Al/^{10}Be$ production ratio changes with elevation as numerical models of the underlying nuclear physics predict. Our analysis of 313 glacially eroded and exposed samples from around the world, for which we assume simple exposure histories, supports the calculations and conclusions of Lifton et al. [20] and Argento et al. [19]. These numerical models predict a decrease in the $^{26}Al/^{10}Be$ production ratio from sea level to 5000 m asl of ~4% and 2.5%, respectively (see Figure 8 in [20]), and our ANOVA results agree well with the model predictions (4.6 ± 0.7% difference between latitude transects 1 and 3). Our bi-variate regression produces a good fit to measured $^{26}Al/^{10}Be$ ratios (χ^2_ν = 1.24), and the negative correlation between elevation and ratio is highly significant (Table 1).

The positive correlation between the $^{26}Al/^{10}Be$ ratio and latitude is less robust in our data than the negative correlation with elevation. The statistically significant correlation observed in the latitude/ratio regression (Figure 4B) is not replicated in the bi-variate model (Table 1) and ANOVA tests are inconclusive, with only two latitude bands of the five elevation transects exhibiting a statistically significant difference in $^{26}Al/^{10}Be$ (Figure 6). These two elevation transects are not substantially different in terms of absolute latitude, with transect 1 covering 40°–50°S and transect 3 covering 36°–39°N (Figure 1), and we attribute at least some of the difference in ratios to sample elevation differences between the transects. Despite the intention for transects to isolate latitude and elevation as variables, these two transects contain samples from different elevations. Samples in elevation transect 1 range from sea level to ~1500 m asl; samples in elevation transect 3 range from ~1500 to 5000 m asl. Numerical model predictions and the statistically significant elevation/ratio correlation in our analyses suggest that $^{26}Al/^{10}Be$ ratios in elevation transect 1 should be several percent higher than ratios in elevation transect 3 just due to elevation differences. Thus, the 1 to 5% difference in mean $^{26}Al/^{10}Be$ ratios observed between these transects in ANOVA results is at least partially due to elevation-related differences in production.

The higher ratios observed in samples processed at UW do not skew our interpretation of elevation and latitude influences on the $^{26}Al/^{10}Be$ production ratio, and we attribute the higher ratios partially to differences in the elevation of samples processed in these two labs. To assess the leverage of the higher-ratio UW samples, we created a bi-variate regression model with these samples removed and found no significant difference in our results. Both the elevation/ratio correlation and the model itself were still statistically significant, although the elevation/ratio correlation was not as robust as when the UW samples are included (p = 0.02 vs. p = 0.004; Table S1). Samples processed at UW are from low elevations, with more than 50% from below 500 m asl, while the only lab with significantly different ratios, LLNL, has samples from predominantly high elevation locations (Figure 7B). The negative correlation between elevation and $^{26}Al/^{10}Be$ ratios demonstrated here and predicted by Lifton et al. [20] could thus be partially responsible for the observed difference between UW and LLNL results.

Differences in sample processing techniques may also explain some of the difference in measured ratios between labs. Data from our initial query (i.e., before setting cutoff values to constrain "reasonable" ratios) show that UW has less variance in measured $^{26}Al/^{10}Be$ ratios than any other chemical processing lab (relative standard deviation, RSD = 9.6% compared to RSD > 11% at other labs and 17.7% at LLNL), and all but two UW samples were within the cutoff ratio bounds. LLNL ratios from the initial query are skewed low, indicating perhaps an underestimation of native ^{27}Al in samples and thus the calculated ^{26}Al concentrations. We fit our bi-variate model to the measured ratios from each of the major chemical processing labs to assess this hypothesis and indeed found a more left-skewed residuals distribution from LLNL and a tighter fit (smaller residuals) from UW (Figure 8). Thus, we attribute the difference in measured $^{26}Al/^{10}Be$ ratios between LLNL and UW to both elevation differences between samples and more variable and low-skewed $^{26}Al/^{10}Be$ measurements from LLNL.

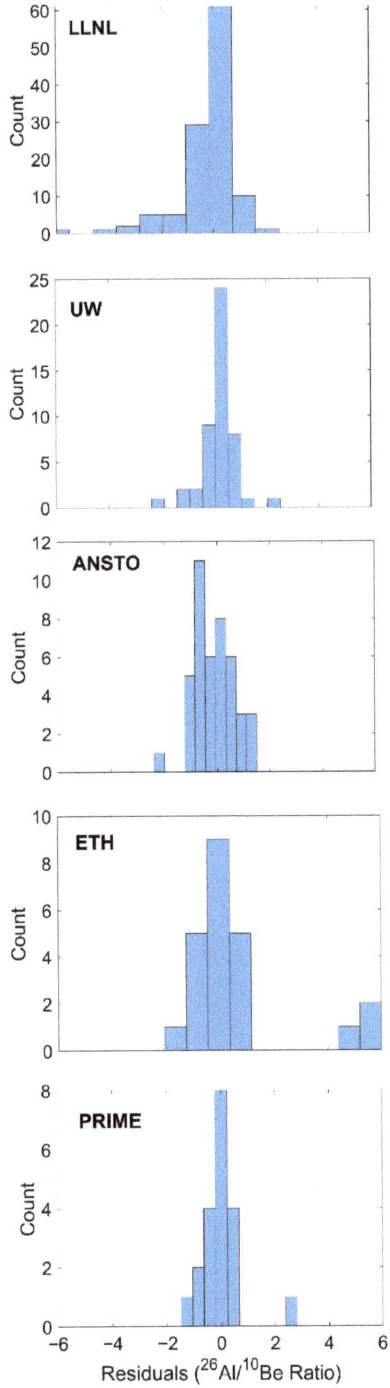

Figure 8. Residual histograms demonstrating the fit of our bi-variate model to samples from each of the major chemical processing labs.

6. Implications

Our analysis suggests that dual-nuclide studies that assume a spatially invariant ^{26}Al/^{10}Be production ratio of 6.75 contain small but systemic biases in their data interpretations. The change in the ^{26}Al/^{10}Be production ratio (5–6% between the equator and the poles and between sea level and mountain landscapes) is similar to the current analytical uncertainty of well-measured ^{26}Al/^{10}Be ratios. Using the nominal ratio (6.75) at high latitudes and low elevations will underestimate burial times at higher latitudes and lower elevations.

Using a nuclide-specific production rate spatial scaling model (such as the LSDn scaling scheme from Lifton et al. [20]) will improve the accuracy of dual-nuclide studies. The LSDn scaling scheme fits the ^{26}Al/^{10}Be ratio data in this compilation nearly as well as the bi-variate regression (χ^2_ν = 1.46; Figures 9, S6, S8 and S9) and predicts ratio variations with elevation that are consistent with the empirical data (Figure 4A). The LSDn scaling scheme also predicts an increase in the ^{26}Al/^{10}Be production ratio with latitude, which agrees with our single variate latitude/ratio regression but is not observed in our bi-variate regression.

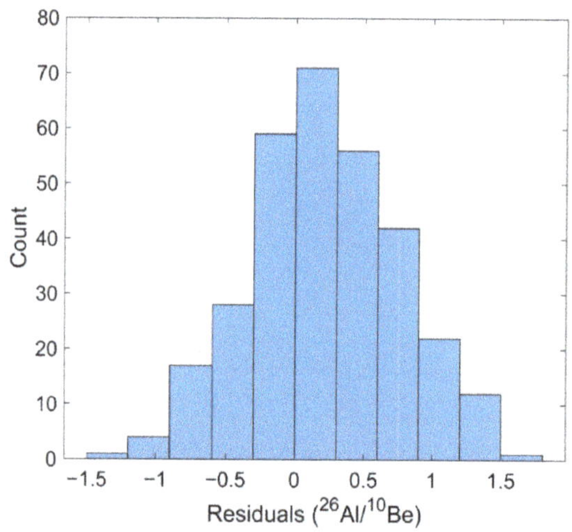

Figure 9. Residuals histogram showing differences between production ratios predicted by the LSDn scaling scheme for samples in the ICE:D compilation and measured ^{26}Al/^{10}Be concentration ratios. The similarity in residuals distribution seen here compared to Figure 2B, the residuals from our bi-variate model based solely on measured ratios, demonstrates the good fit of LSDn scaling to empirical data.

Our analysis is limited by spatial gaps in the data, particularly at low latitude/low altitude and high latitude/high altitude locations and thus may be biased by elevation-dependence of sites from different latitudes. Filling these gaps is essential to improving our understanding of ^{26}Al/^{10}Be production ratio variations, but will be challenging. Glaciers did not occupy low latitude/low altitude sites, but other episodically exposed surfaces, such as those from rock falls, could be useful. Sampling high latitude/high altitude sites is logistically difficult, and many of these sites which have been sampled show evidence for significant concentrations of nuclides inherited from prior periods of exposure.

Analysis of these compiled data indicate the need for improving the precision and accuracy of ^{26}Al/^{10}Be measurements and thus their application to geochronology and understanding landscape dynamics. Of particular concern are measurements of stable ^{27}Al. Use of internal laboratory standards for quality control can help assure the quality of both ^{26}Al and ^{10}Be data. These are available as liquid standards [53], homogenized glass sand powder [25], and as purified quartz [26,28]. Improved precision of ^{26}Al concentration measurements will also be critical to constraining ^{26}Al/^{10}Be ratios [17].

Supplementary Materials: The following are available online at https://www.mdpi.com/article/10.3390/geosciences11100402/s1, Extended methods and results, Figures S1–S10 and Table S1. Figure S1: Distribution of ^{26}Al/^{10}Be concentration ratios from the initial ICE:D query. Vertical dashed line marks a ratio of 6.75. Figure S2: Distribution of concentration ratio uncertainties (%) from the initial ICE:D query. Figure S3: Standard boxplot of concentration ratio uncertainties from the initial ICE:D query. Box limits are the 25th and 75th percentile values, center red line is the median, whiskers are the high and low values not considered outliers, red crosses show outliers. Figure S4: Linear regressions (central lines) and 95% confidence intervals (upper and lower lines) for elevation and latitude transects. Note: 95% confidence interval lines for latitude band 2 are outside the y-axis bounds. Figure S5: Model check for normality in bi-variate linear regression correlating the measured ^{26}Al/^{10}Be concentration ratios to elevation and latitude. X-axis shows fitted values (ratios) from this model, while y-axis shows the measured ratios. Blue line is the 1:1 reference line. Figure S6: Model check for homoscedasticity in the bi-variate regression model. X-axis is fitted ratio values from the model, y-axis is residuals of the model compared to the data. No clear pattern in residuals is observed, indicating homoscedasticity. Figure S7: Model check for normality in LSDn scaling model against the measured ^{26}Al/^{10}Be concentration ratios. Axes are same as in Figure S4, but x-axis now shows fitted ratio values from LSDn scaling. Figure S8: Model check for homoscedasticity in the LSDn scaling model compared to concentration ratio data. Axes are same as in Figure S5, but x-axis shows fitted values from the LSDn scaling model. Figure S9: Comparison of the bi-variate linear model from this study (bottom, gray) and the LSDn scaling model (top, multi-colored) against the data in this compilation (blue dots). The decrease in concentration ratio with increasing elevation is nearly identical between the two models, but the LSDn model shows a more pronounced increase in ratio with latitude. Figure S10: Exploration of the influence of sample collection year on measured ratios and ratio uncertainties. Table S1: the statistics table for the bivariate regression run without UW samples.

Author Contributions: Conceptualization, C.T.H., P.R.B. and G.B.; methodology, C.T.H., P.R.B. and G.B.; software, C.T.H.; validation, C.T.H., P.R.B. and G.B.; formal analysis, C.T.H.; data curation, C.T.H.; writing—original draft preparation, C.T.H.; writing—review and editing, P.R.B. and G.B.; visualization, C.T.H.; supervision, P.R.B. All authors have read and agreed to the published version of the manuscript.

Funding: This research was supported by funding provided to P.R.B. through the National Science Foundation (Award #1733887).

Data Availability Statement: ICE:D-Alpine and ICE:D-Calib data can be found at ice-d.org. The compilation used here is available upon request to the first author.

Conflicts of Interest: The authors declare no conflict of interest. The funders had no role in the design of the study; in the collection, analyses, or interpretation of data; in the writing of the manuscript, or in the decision to publish the results.

References

1. Gosse, J.C.; Phillips, F.M. Terrestrial in situ cosmogenic nuclides: Theory and application. *Quat. Sci. Rev.* **2001**, *20*, 1475–1560. [CrossRef]
2. Bierman, P.R.; Marsella, K.A.; Patterson, C.; Davis, P.T.; Caffee, M. Mid-Pleistocene cosmogenic minimum-age limits for pre-Wisconsinan glacial surfaces in southwestern Minnesota and southern Baffin Island: A multiple nuclide approach. *Geomorphology* **1999**, *27*, 25–39. [CrossRef]
3. Briner, J.P.; Miller, G.H.; Davis, P.T.; Bierman, P.R.; Caffee, M. Last glacial maximum ice sheet dynamics in Arctic Canada inferred from young erratics perched on ancient tors. *Quat. Sci. Rev.* **2003**, *22*, 437–444. [CrossRef]
4. Corbett, L.B.; Bierman, P.R.; Rood, D.H. Constraining multi-stage exposure-burial scenarios for boulders preserved beneath cold-based glacial ice in Thule, northwest Greenland. *Earth Planet. Sci. Lett.* **2016**, *440*, 147–157. [CrossRef]
5. Granger, D.E.; Kirchner, J.W.; Finkel, R.C. Quaternary downcutting rate of the New River, Virginia, measured from differential decay of cosmogenic ^{26}Al and ^{10}Be in cave-deposited alluvium. *Geology* **1997**, *25*, 107–110. [CrossRef]
6. McPhillips, D.; Hoke, G.D.; Liu-Zeng, J.; Bierman, P.R.; Rood, D.H.; Niedermann, S. Dating the incision of the Yangtze River gorge at the First Bend using three-nuclide burial ages. *Geophys. Res. Lett.* **2016**, *43*, 101–110. [CrossRef]

7. Carbonell, E.; Bermúdez De Castro, J.M.; Parés, J.M.; Pérez-González, A.; Cuenca-Bescós, G.; Ollé, A.; Mosquera, M.; Huguet, R.; Van Der Made, J.; Rosas, A.; et al. The first hominin of Europe. *Nature* **2008**, *452*, 465–469. [CrossRef]
8. Gibbon, R.J.; Pickering, T.R.; Sutton, M.B.; Heaton, J.L.; Kuman, K.; Clarke, R.J.; Brain, C.K.; Granger, D.E. Cosmogenic nuclide burial dating of hominin-bearing Pleistocene cave deposits at Swartkrans, South Africa. *Quat. Geochronol.* **2014**, *24*, 10–15. [CrossRef]
9. Granger, D.E.; Gibbon, R.J.; Kuman, K.; Clarke, R.J.; Bruxelles, L.; Caffee, M.W. New cosmogenic burial ages for Sterkfontein Member 2 Australopithecus and Member 5 Oldowan. *Nature* **2015**, *522*, 85–88. [CrossRef]
10. Nishiizumi, K. Preparation of ^{26}Al AMS standards. In *Proceedings of the Nuclear Instruments and Methods in Physics Research, Section B: Beam Interactions with Materials and Atoms*; Elsevier: Amsterdam, The Netherlands, 2004; Volume 223–224, pp. 388–392. [CrossRef]
11. Chmeleff, J.; von Blanckenburg, F.; Kossert, K.; Jakob, D. Determination of the ^{10}Be half-life by multicollector ICP-MS and liquid scintillation counting. *Nucl. Instrum. Methods Phys. Res. Sect. B Beam Interact. Mater. Atoms* **2010**, *268*, 192–199. [CrossRef]
12. Korschinek, G.; Bergmaier, A.; Faestermann, T.; Gerstmann, U.C.; Knie, K.; Rugel, G.; Wallner, A.; Dillmann, I.; Dollinger, G.; von Gostomski, C.L.; et al. A new value for the half-life of ^{10}Be by Heavy-Ion Elastic Recoil Detection and liquid scintillation counting. *Nucl. Instrum. Methods Phys. Res. Sect. B Beam Interact. Mater. Atoms* **2010**, *268*, 187–191. [CrossRef]
13. Lifton, N.A.; Bieber, J.W.; Clem, J.M.; Duldig, M.L.; Evenson, P.; Humble, J.E.; Pyle, R. Addressing solar modulation and long-term uncertainties in scaling secondary cosmic rays for in situ cosmogenic nuclide applications. *Earth Planet. Sci. Lett.* **2005**, *239*, 140–161. [CrossRef]
14. Fenton, C.R.; Niedermann, S.; Dunai, T.; Binnie, S.A. The SPICE project: Production rates of cosmogenic 21Ne, ^{10}Be, and 14C in quartz from the 72 ka SP basalt flow, Arizona, USAu. *Quat. Geochronol.* **2019**, *54*, 101019. [CrossRef]
15. Balco, G.; Briner, J.; Finkel, R.C.; Rayburn, J.A.; Ridge, J.C.; Schaefer, J.M. Regional beryllium-10 production rate calibration for late-glacial northeastern North America. *Quat. Geochronol.* **2009**, *4*, 93–107. [CrossRef]
16. Klein, J.; Giegengack, R.; Middleton, R.; Sharma, P.; Underwood, J.R.; Weeks, R.A. Revealing Histories of Exposure Using In Situ Produced ^{26}Al and ^{10}Be in Libyan Desert Glass. *Radiocarbon* **1986**, *28*, 547–555. [CrossRef]
17. Caffee, M.W.; Granger, D.E.; Woodruff, T.E. The Gas-Filled-Magnet at PRIME Lab: Increased Sensitivity of Cosmogenic Nuclide Measurements. In *Proceedings of the American Geophysical Union, Fall Meeting*; American Geophysical Union: Washington, DC, USA, 2015; Volume 2015.
18. Gillespie, A.R.; Bierman, P.R. Precision of terrestrial exposure ages and erosion rates estimated from analysis of cosmogenic isotopes produced in situ. *J. Geophys. Res. Solid Earth* **1995**, *100*, 24637–24649. [CrossRef]
19. Argento, D.C.; Stone, J.O.; Reedy, R.C.; O'Brien, K. Physics-based modeling of cosmogenic nuclides part II—Key aspects of in-situ cosmogenic nuclide production. *Quat. Geochronol.* **2015**, *26*, 44–55. [CrossRef]
20. Lifton, N.; Sato, T.; Dunai, T.J. Scaling in situ cosmogenic nuclide production rates using analytical approximations to atmospheric cosmic-ray fluxes. *Earth Planet. Sci. Lett.* **2014**, *386*, 149–160. [CrossRef]
21. Corbett, L.B.; Bierman, P.R.; Rood, D.H.; Caffee, M.W.; Lifton, N.A.; Woodruff, T.E. Cosmogenic ^{26}Al/^{10}Be surface production ratio in Greenland. *Geophys. Res. Lett.* **2017**, *44*, 1350–1359. [CrossRef]
22. Heisinger, B.; Lal, D.; Jull, A.J.T.; Kubik, P.; Ivy-Ochs, S.; Neumaier, S.; Knie, K.; Lazarev, V.; Nolte, E. Production of selected cosmogenic radionuclides by muons 1. Fast muons. *Earth Planet. Sci. Lett.* **2002**, *200*, 345–355. [CrossRef]
23. Müller, A.M.; Christl, M.; Lachner, J.; Synal, H.A.; Vockenhuber, C.; Zanella, C. ^{26}Al measurements below 500 kV in charge state 2+. *Nucl. Instrum. Methods Phys. Res. Sect. B Beam Interact. Mater. Atoms* **2015**, *361*, 257–262. [CrossRef]
24. Bierman, P.R.; Caffee, M. Cosmogenic exposure and erosion history of Australian bedrock landforms. *GSA Bull.* **2002**, *114*, 787–803. [CrossRef]
25. Fujioka, T.; Fink, D.; Mifsud, C. Towards improvement of aluminium assay in quartz for in situ cosmogenic ^{26}Al analysis at ANSTO. *Nucl. Instrum. Methods Phys. Res. Sect. B Beam Interact. Mater. Atoms* **2015**, *361*, 346–353. [CrossRef]
26. Binnie, S.A.; Dewald, A.; Heinze, S.; Voronina, E.; Hein, A.; Wittmann, H.; von Blanckenburg, F.; Hetzel, R.; Christl, M.; Schaller, M.; et al. Preliminary results of CoQtz-N: A quartz reference material for terrestrial in-situ cosmogenic ^{10}Be and ^{26}Al measurements. *Nucl. Instrum. Methods Phys. Res. Sect. B Beam Interact. Mater. Atoms* **2019**, *456*, 203–212. [CrossRef]
27. Phillips, F.M.; Argento, D.C.; Balco, G.; Caffee, M.W.; Clem, J.; Dunai, T.J.; Finkel, R.; Goehring, B.; Gosse, J.C.; Hudson, A.M.; et al. The CRONUS-Earth Project: A synthesis. *Quat. Geochronol.* **2016**, *31*, 119–154. [CrossRef]
28. Jull, A.J.T.; Scott, E.M.; Bierman, P. The CRONUS-Earth inter-comparison for cosmogenic isotope analysis. *Quat. Geochronol.* **2015**, *26*, 3–10. [CrossRef]
29. Portenga, E.W.; Bierman, P.R.; Duncan, C.; Corbett, L.B.; Kehrwald, N.M.; Rood, D.H. Erosion rates of the Bhutanese Himalaya determined using in situ-produced ^{10}Be. *Geomorphology* **2015**, *233*, 112–126. [CrossRef]
30. Nishiizumi, K.; Winterer, E.L.; Kohl, C.P.; Klein, J.; Middleton, R.; Lal, D.; Arnold, J.R. Cosmic ray production rates of ^{10}Be and ^{26}Al in quartz from glacially polished rocks. *J. Geophys. Res. Solid Earth* **1989**, *94*, 17907–17915. [CrossRef]
31. Nishiizumi, K.; Kohl, C.P.; Arnold, J.R.; Klein, J.; Fink, D.; Middleton, R. Cosmic ray produced ^{10}Be and ^{26}Al in Antarctic rocks: Exposure and erosion history. *Earth Planet. Sci. Lett.* **1991**, *104*, 440–454. [CrossRef]
32. Christ, A.J.; Bierman, P.R.; Schaefer, J.M.; Dahl-Jensen, D.; Steffensen, J.P.; Corbett, L.B.; Peteet, D.M.; Thomas, E.K.; Steig, E.J.; Rittenour, T.M.; et al. A multimillion-year-old record of Greenland vegetation and glacial history preserved in sediment beneath 1.4 km of ice at Camp Century. *Proc. Natl. Acad. Sci. USA* **2021**, *118*, e2021442118. [CrossRef] [PubMed]

33. Schaefer, J.M.; Finkel, R.C.; Balco, G.; Alley, R.B.; Caffee, M.W.; Briner, J.P.; Young, N.E.; Gow, A.J.; Schwartz, R. Greenland was nearly ice-free for extended periods during the Pleistocene. *Nature* **2016**, *540*, 252–255. [CrossRef] [PubMed]
34. Stroeven, A.P.; Fabel, D.; Hättestrand, C.; Harbor, J. A relict landscape in the centre of Fennoscandian glaciation: Cosmogenic radionuclide evidence of tors preserved through multiple glacial cycles. *Geomorphology* **2002**, *44*, 145–154. [CrossRef]
35. Bierman, P.R.; Shakun, J.D.; Corbett, L.B.; Zimmerman, S.R.; Rood, D.H. A persistent and dynamic East Greenland Ice Sheet over the past 7.5 million years. *Nature* **2016**, *540*, 256–260. [CrossRef] [PubMed]
36. Shakun, J.D.; Corbett, L.B.; Bierman, P.R.; Underwood, K.; Rizzo, D.M.; Zimmerman, S.R.; Caffee, M.W.; Naish, T.; Golledge, N.R.; Hay, C.C. Minimal East Antarctic Ice Sheet retreat onto land during the past eight million years. *Nature* **2018**, *558*, 284–287. [CrossRef]
37. Clapp, E.M.; Bierman, P.R.; Schick, A.P.; Lekach, J.; Enzel, Y.; Caffee, M. Sediment yield exceeds sediment production in arid region drainage basins. *Geology* **2000**, *28*, 995–998. [CrossRef]
38. Erlanger, E.D.; Granger, D.E.; Gibbon, R.J. Rock uplift rates in South Africa from isochron burial dating of fl uvial and marine terraces. *Geology* **2012**, *40*, 1019–1022. [CrossRef]
39. Granger, D.E.; Muzikar, P.F. Dating sediment burial with in situ-produced cosmogenic nuclides: Theory, techniques, and limitations. *Earth Planet. Sci. Lett.* **2001**, *188*, 269–281. [CrossRef]
40. Balco, G.; Rovey, C.W. An isochron method for cosmogenic-nuclide dating of buried soils and sediments. *Am. J. Sci.* **2008**, *308*, 1083–1114. [CrossRef]
41. Shen, G.; Gao, X.; Gao, B.; Granger, D.E. Age of Zhoukoudian Homo erectus determined with $^{26}Al/^{10}Be$ burial dating. *Nature* **2009**, *458*, 198–200. [CrossRef] [PubMed]
42. Lal, D. Cosmic ray labeling of erosion surfaces: In situ nuclide production rates and erosion models. *Earth Planet. Sci. Lett.* **1991**, *104*, 424–439. [CrossRef]
43. Nishiizumi, K.; Imamura, M.; Caffee, M.W.; Southon, J.R.; Finkel, R.C.; McAninch, J. Absolute calibration of ^{10}Be AMS standards. *Nucl. Instrum. Methods Phys. Res. Sect. B Beam Interact. Mater. Atoms* **2007**, *258*, 403–413. [CrossRef]
44. Masarik, J.; Reedy, R.C. Terrestrial cosmogenic-nuclide production systematics calculated from numerical simulations. *Earth Planet. Sci. Lett.* **1995**, *136*, 381–395. [CrossRef]
45. Balco, G.; Stone, J.O.; Lifton, N.A.; Dunai, T.J. A complete and easily accessible means of calculating surface exposure ages or erosion rates from ^{10}Be and ^{26}Al measurements. *Quat. Geochronol.* **2008**, *3*, 174–195. [CrossRef]
46. Argento, D.C.; Reedy, R.C.; Stone, J.O. Modeling the earth's cosmic radiation. *Nucl. Instrum. Methods Phys. Res. Sect. B Beam Interact. Mater. Atoms* **2013**, *294*, 464–469. [CrossRef]
47. Caffee, M.W.; Nishiizumi, K.; Sisterson, J.M.; Ullmann, J.; Welten, K.C. Cross section measurements at neutron energies 71 and 112 MeV and energy integrated cross section measurements (0.1 < En < 750 MeV) for the neutron induced reactions $O(n,x)^{10}Be$, $Si(n,x)^{10}Be$, and $Si(n,x)^{26}Al$. *Nucl. Instrum. Methods Phys. Res. Sect. B Beam Interact. Mater. Atoms* **2013**, *294*, 479–483. [CrossRef]
48. Reedy, R.C. Cosmogenic-nuclide production rates: Reaction cross section update. In *Proceedings of the Nuclear Instruments and Methods in Physics Research, Section B: Beam Interactions with Materials and Atoms*; Elsevier: Amsterdam, The Netherlands, 2013; Volume 294, pp. 470–474.
49. Balco, G. Technical note: A prototype transparent-middle-layer data management and analysis infrastructure for cosmogenic-nuclide exposure dating. *Geochronology* **2020**, *2*, 169–175. [CrossRef]
50. Bentley, M.J.; Fogwill, C.J.; Kubik, P.W.; Sugden, D.E. Geomorphological evidence and cosmogenic $^{10}Be/^{26}Al$ exposure ages for the Last Glacial Maximum and deglaciation of the Antarctic Peninsula Ice Sheet. *GSA Bull.* **2006**, *118*, 1149–1159. [CrossRef]
51. Hochberg, Y.; Tamhane, A.C. *Multiple Comparison Procedures*; Wiley: New York, NY, USA, 1987.
52. McGill, R.; Tukey, J.W.; Larsen, W.A. Variations of box plots. *Am. Stat.* **1978**, *32*, 12–16. [CrossRef]
53. Corbett, L.B.; Bierman, P.R.; Woodruff, T.E.; Caffee, M.W. A homogeneous liquid reference material for monitoring the quality and reproducibility of in situ cosmogenic ^{10}Be and ^{26}Al analyses. *Nucl. Instrum. Methods Phys. Res. Sect. B Beam Interact. Mater. Atoms* **2019**, *456*, 180–185. [CrossRef]

geosciences

Article

The NUNAtak Ice Thinning (NUNAIT) Calculator for Cosmonuclide Elevation Profiles

Ángel Rodés

Scottish Universities Environmental Research Centre, University of Glasgow, East Kilbride G75 0QF, UK; angel.rodes@glasgow.ac.uk

Abstract: Cosmogenic nuclides are widely used to constrain the landscape history of glaciated areas. At nunataks in continental polar regions with extremely arid conditions, cosmogenic nuclides are often the only method available to date the ice thinning history of the glacier. However, the amount of cosmogenic isotopes accumulated at the surface of nunataks depends not only on the length of time that rock has been exposed since the last deglaciation but also on the full history of the surface, including muon production under ice, exposure during previous interglacials, subaerial weathering rate, glacial erosion rate, and uplift rate of the nunatak. The NUNAtak Ice Thinning model (NUNAIT) simulates the cosmonuclide accumulation on vertical profiles, fitting the aforementioned parameters to a set of multi-isotope apparent ages from samples taken at different elevations over the ice-sheet surface. The NUNAIT calculator is an easy-to-use tool that constrains parameters that describe the geological history of a nunatak from a set of surface exposure ages.

Keywords: nunatak; cosmonuclides; ^{10}Be; ^{26}Al; ^{21}Ne; ^{3}He; ^{36}Cl; ^{14}C; MATLAB; Octave

Citation: Rodés, Á. The NUNAtak Ice Thinning (NUNAIT) Calculator for Cosmonuclide Elevation Profiles. *Geosciences* **2021**, *11*, 362. https://doi.org/10.3390/geosciences11090362

Academic Editors: Naki Akçar, Susan Ivy-Ochs, Fritz Schlunegger and Jesus Martinez-Frias

Received: 15 July 2021
Accepted: 23 August 2021
Published: 26 August 2021

1. Introduction

Quantifying the changes in the thickness of the Greenland and Antarctic ice sheets is key to understanding future sea-level rise [1]. Cosmogenic nuclides are widely used for the quantification of glacial chronologies. However, the climatic interpretation of the existing cosmonuclide data sets requires accounting for geologic processes that cause apparent exposure ages on glacial landforms to differ from the age of deglaciation [2].

Nunataks, the mountains emerging from polar ice sheets, have been used as vertical dipsticks that record past changes in the thickness of the polar ice sheets (e.g., [3]). Cosmogenic signatures at the surface of nunataks are the result of the intermittent exposure of the surfaces to cosmic radiation through the glacial cycles (e.g., [4]), glacial erosion (e.g., [5,6]), and the subaerial weathering of these surfaces (e.g., [7]). Therefore, the abundance of one or more cosmonuclides in one of these surfaces can be explained by the combination of multiple possible scenarios [8].

Stroeven et al. [4] modelled the accumulation of ^{10}Be and ^{26}Al in tors. The model they used is based on complex exposure-burial histories forced along the ice-free/ice-covered conditions provided by a marine oxygen isotope δ^{18}O proxy glacial record. In this model, a given δ^{18}O cutoff value defines when the surface of the tor is exposed or shielded from cosmic radiation. Li et al. [9] developed a method to solve the cutoff value in the marine oxygen isotope record that satisfies a set of ^{10}Be and ^{26}Al concentrations considering fixed values of glacial erosion and subaerial weathering. Knudsen et al. [10] described a method to solve not only the δ^{18}O cutoff value but also the glacial and interglacial erosion rates from a set of multiple cosmonuclide concentrations that can include ^{10}Be, ^{26}Al, ^{14}C, and/or ^{21}Ne data.

The models described by Stroeven et al. [4], Li et al. [9], and Knudsen et al. [10] are designed to be applied on a single site, and therefore one cutoff δ^{18}O value can be solved at a time. To solve the elevation of the ice surface during the glaciations, several samples should be used to obtain an elevation profile of δ^{18}O cutoff values, which would allow the reconstruction of the ice sheet thickness with time. If the elevation of the ice surface is known, a second iteration of modelling would allow calculating how cosmonuclides accumulate during glacial times by reconstructing the muonic production cross-section under the ice sheet for any time.

In summary, the interpretation of cosmonuclide concentrations from nunataks often requires accounting for the effects of surface and subglacial erosion, glacial dynamics, and tectonic activity. The models described in the literature are focused on solving one or two of the parameters that emulate these processes, usually using data from a single sample. Here, I describe an easy-to-use method to solve up to five parameters that emulate the glacial history, surface erosion, and tectonic uplift using a set of surface exposure ages.

The NUNAtak Ice Thinning (NUNAIT) calculator presented here solves (1) the elevation history of the ice surface, (2) the glacial erosion rate, (3) the subaerial weathering rate, and (4) the nunatak uplift rate from a multi-sample (elevation profile) and multi-isotope (^{10}Be, ^{26}Al, ^{21}Ne, ^{3}He, ^{36}Cl, and/or ^{14}C) data set. The calculator does not require the input of production rates, as the default inputs are not cosmogenic concentrations but apparent surface exposure ages, and approximate muon cross sections are calculated using the latitude and elevation of the sampling sites.

2. Method Details

Here, I present a set of MATLAB®/GNU Octave© scripts that form the NUNAIT calculator and their mathematical descriptions. All scripts needed to run the NUNAIT calculator (Supplementary Materials) are freely accessible at https://github.com/angelrodes/NUNAIT (accessed on 19 August 2021).

When running the script `START.m`, the user is asked to run the calculator or select previous data to display the text and output.

If the first option is selected (Run simulation), two types of files can be selected:

- A `.csv` file containing basic input data;
- A `.mat` file containing full input data, including apparent concentrations and apparent production rates. A `*_sampledata.mat` is generated every time a `.csv` is processed. This allows, for example, changing the distribution of the production rates before running the simulations by editing the `*_sampledata.mat` file.

If the second option is selected (Display results), a `.mat` file containing previously calculated data is required. This type of file is generated at the end of each fitting session with the same name as the input file and `_model.mat`.

2.1. Input Data

Site data have to be inputted in individual comma separated files (`.csv`) for each measurement. Some examples of input files are included in the folder "Examples". The input file contains the following headers (first line) that we recommend are not changed:

1. name: Sample name without spaces or symbols.
2. lat: Latitude used to calculate the muon contributions (decimal degrees).
3. site_elv: Elevation of the sample above sea level (m).
4. isotope: Mass of the cosmogenic isotope. Currently accepting 3, 10, 14, 21, 26, and 36 for ^{3}He, ^{10}Be, ^{14}C, ^{21}Ne, ^{26}Al, and ^{36}Cl, respectively.
5. base_level: Current elevation of the glacier surface above sea level at the sampling site (m). This is used to calculate the ice position through time.
6. apparent_years: Apparent surface exposure age calculated with any cosmogenic calculator, any scaling scheme, and any production rate reference.
7. dapparent_years: External uncertainty of the previous age.

Apparent concentrations (C) are calculated from apparent surface exposure (T) ages following Lal [11]:

$$C = \frac{1}{\lambda} \cdot \left(1 - e^{-\lambda T}\right) \quad (1)$$

where λ is the decay constant of the isotope considered. The values of λ are stored in `constants.m`.

Note that the concentrations described in Equation (1) are scaled to site production rates. Therefore, they should be expressed in time units (years).

To reduce computing time, conditional statements are avoided in the code by considering all cosmonuclides radioactive. To do this, stable isotopes are assigned values of λ corresponding to 100 times the age of the Earth. As $T << 1/\lambda$ for stable isotopes, Equation (1) results in $C \simeq T$.

The calculated concentrations, together with the muon relative contributions described in Sections 2.3 and 2.4, are stored in a `.mat` file with the same name as the original `.csv` file and the suffix `_sampledata`. If the user needs to change the calculated concentrations or relative production rates, this file can be modified and used as an input file.

2.2. Climate Curves

The scripts `make_climatecurves.m` and `make_climatecurves_ant.m` generate a time series of δ^{18}O values that will be used to calculate the vertical position of the glacial surface over the samples.

The curves from Lisiecki and Raymo [12] and Zachos [13] are combined and scaled with NGRIP data [14] or Five-core data [15] in Antarctica.

All records are arbitrarily scaled to the LR04 stack data [12]. As the δ^{18}O values generated will finally be transformed into elevations by the model described in Section 2.6, the choice of one data set as reference is irrelevant.

To reduce the number of calculations and the computing time while representing the ice changes relevant to the cosmogenic accumulation, the data are interpolated for ages every 10 years for the last century, every 100 years until 20 ka, every 200 years until 50 ka, every 500 years until 100 ka, and every 1% increase for ages older than 100 ka. The resulting simplified curve is shown in Figure 1.

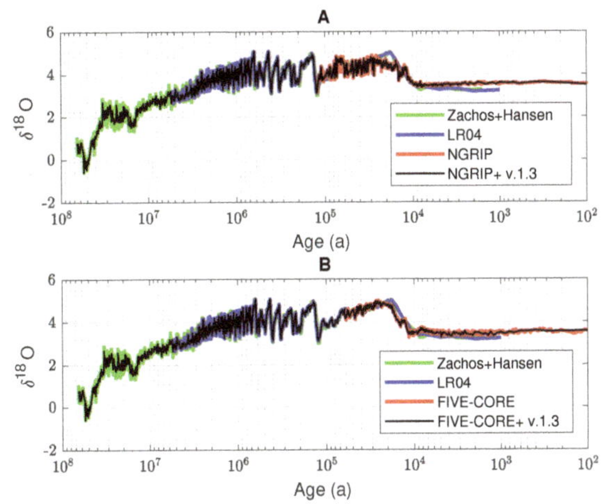

Figure 1. δ^{18}O glacial proxies. Combination of scaled δ^{18}O curves from Lisiecki and Raymo [12], Zachos [13], NGRIP [14], and Buizert et al. [15], depicted with colours. Black lines show the simplified curves used by NUNAIT for latitudes north (**A**) and south (**B**) of latitude 55° S.

2.3. Muon Contributions

The function `muon_contribution.m` generates the muon contribution and its uncertainty based on latitude (*lat*) and elevation (*elv*) for a given nuclide. If either latitude or elevation is not a number, a global average is given. A single value of latitude and elevation is used to calculate the contribution of muons to the total surface production of ^{10}Be. All other productions are scaled accordingly.

The contribution of muons to the total surface ^{10}Be production ($R_\mu(^{10}Be)$) is calculated as

$$\frac{P_\mu(^{10}Be)}{P_{total}(^{10}Be)} = \frac{1}{100}\cdot\left(1.29+\frac{lat}{900}+1.056\cdot e^{-\left(\frac{lat+1}{30.31}\right)^2}\right)\cdot\left(0.1+0.9\cdot e^{\frac{-elv}{2000}}\right) \quad (2)$$

This approximation is based on the ^{10}Be production at 1678 sites equally distributed on land areas according to `ETOPO1_Bed_g_geotiff.tif` [16] and calculated using `P_mu_total_alpha1.m` and `stone2000.m` from Balco [17] and Balco et al. [18], respectively. The fitting of this approximation is shown in Figure 2. This formula fits the original data within a 5% standard deviation.

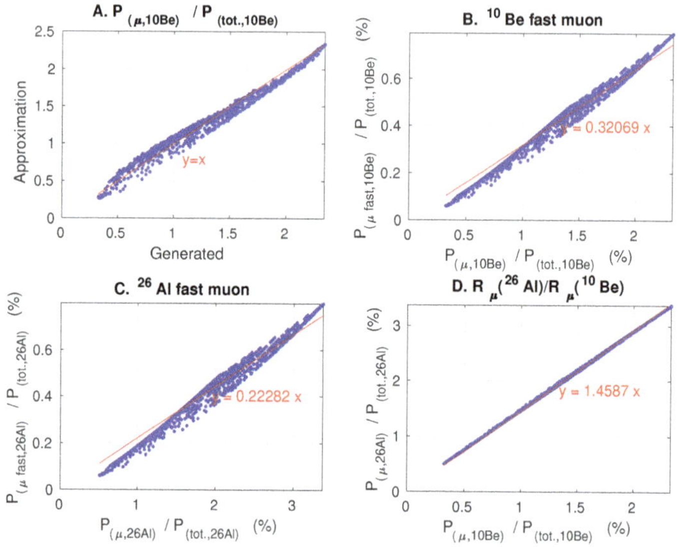

Figure 2. ^{10}Be and ^{26}Al surface muon contributions. (**A**) Percentage of ^{10}Be muon production rates with respect to the total muon production rate generated using `P_mu_total_alpha1.m` [17] for 1678 land sites, and the approximation calculated using Equation (2) for the same sites. (**B**,**C**) Share of ^{10}Be and ^{26}Al fast muon production with respect to the total muon production at the surface. (**D**) Ratio between the ^{26}Al and the ^{10}Be muon shares.

Considering that `P_mu_total_alpha1.m` fits the empirical data available within a ~5% and a ~13% for the ^{10}Be and ^{26}Al muon production rates, respectively [17], the uncertainty of the calculated muon contributions based on Equation (2) should be at least a 7% for ^{10}Be and 14% for ^{26}Al.

The calculation of the muon contributions for other nuclides are based on the following ratios:

- $R_\mu(^{26}Al)/R_\mu(^{10}Be) = 1.4587$. See Figure 2.
- $R_\mu(^{36}Cl)/R_\mu(^{10}Be) = 3.2720$, according to Heisinger and Nolte [19].
- $R_\mu(^{21}Ne)/R_\mu(^{10}Be) = 4.086$, according to Balco and Shuster [20].
- $R_\mu(^{3}He)/R_\mu(^{10}Be) = 1$, consistent with Blard et al. [21].
- $R_\mu(^{14}C)/R_\mu(^{10}Be) = 8.2767$, according to Heisinger and Nolte [19].

As the uncertainties of these ratios are unknown, this script assigns a conservative 20% uncertainty for muon contributions calculated using Equation (2) to cover both the uncertainties at the surface and the subsurface extrapolations described in Section 2.4.

All these data can be changed in the files `constants.m` and `muon_contribution.m`.

2.4. Muon Cross Sections

To simulate production under ice and rock surfaces, muon production was approximated as three exponential functions of depth [22]. A total of 1678 ^{10}Be and ^{26}Al muon production rates generated using `P_mu_total_alpha1.m` [17] were analysed to fit three exponential decays with attenuation lengths of 850, 5000, and 500 g cm^{-2} (Figure 3). These attenuation lengths correspond to 75% of the fast muon, 25% of the fast muon, and the negative muon productions at the surface, respectively.

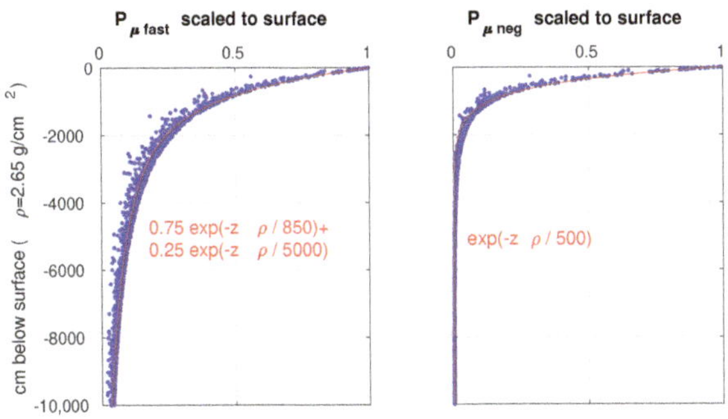

Figure 3. ^{10}Be muon cross sections. Fast and negative muon production rates scaled to surface values calculated using `P_mu_total_alpha1.m` [17], for 1678 land sites from `ETOPO1_Bed_g_geotiff.tif` [16], and random depths between 0 and 100 m below the surface (blue dots). Red lines represent the exponential decay approximations used in this work.

The share of surface fast muon production with respect to the total muon production ($P_{\mu fast}/P_{\mu total}$) considered for each isotope is:

- ^{10}Be: $P_{\mu fast}/P_{\mu total} = 0.32069$. See Figure 2.
- ^{26}Al: $P_{\mu fast}/P_{\mu total} = 0.22282$. See Figure 2.
- ^{36}Cl: $P_{\mu fast}/P_{\mu total} = 0.0620$, according to Heisinger and Nolte [19].
- ^{21}Ne: $P_{\mu fast}/P_{\mu total} = 1$, according to Balco and Shuster [20].
- ^{3}He: $P_{\mu fast}/P_{\mu total} = 0.32069$, consistent with Blard et al. [21].
- ^{14}C: $P_{\mu fast}/P_{\mu total} = 0.0672$, according to Heisinger and Nolte [19].

The uncertainties of these approximations are within the uncertainties described in Section 2.3 for ^{10}Be and ^{26}Al.

All these data can be changed in the file `constants.m`.

2.5. Densities

A density of $\rho_{ice} = 0.917$ g cm^{-3} is considered for ice [23], and a density of $\rho = 2.65$ cm^{-3} for bedrock.

2.6. Nunatak Accumulation Model

The nunatak accumulation model (`nuna_model.m`) considers the depth of the sample under the bedrock surface (z) and the thickness of the ice on top of the surface (z_{ice}) based on the input conditions (weathering w, glacial erosion rate, and maximum and current ice levels) for each time range (Δt) defined by the climate curve and for each sample. The model concentration is calculated as

$$C_i = \frac{P}{\lambda + w \cdot \rho/\Lambda} \cdot e^{-(z \cdot \rho + z_{ice} \cdot \rho_{ice})/\Lambda} \cdot \left(1 - e^{-\Delta t \cdot (\lambda + w \cdot \rho/\Lambda)}\right) \cdot e^{-\lambda \cdot t} \quad (3)$$

where P is the production rate considered (spallation and each of the muon types), Λ is the attenuation length for the production rate considered, λ is the decay constant of the nuclide, and t is the age corresponding to the end of the time range defined by the climate curves.

The final concentration for each sample C_{model} is calculated by adding all the C_i for all production types and time ranges.

The effect of the glacial erosion rate is ignored inside each time range (Δt), as usually C_i is much more sensitive to z_{ice} than to the change in position of the sample under the bedrock surface due to glacial erosion during ice-covered periods.

2.7. Model Fitting

The fitting of the model described in Section 2.6 is performed by the script `fit_nuna_model.m`.

The script asks the user to set maximum and minimum values for the parameters to be fitted: ice-free weathering rate, glacial erosion rate, ice-thinning since maximum glacier extension, deviation of the current ice surface, and uplift rate. As weathering and erosion rates are simulated in logarithmic space, minimum values of 0.1 mm/Ma are assumed.

It also allows changing the fit type. With a value of `0`, the script will try to fit the model to the data normally. With a value of `1`, models with concentrations below the sample concentrations will be ignored. With a value of `2`, models with concentrations above the sample concentrations will be ignored. A value of `3` is used to represent the models within the stated parameter limits ignoring the sample concentrations. If fit type `3` is used, the script assumes that all generated models fit the data.

The script selects the climate reference based on the average latitude of the samples. The Antarctic curves described in Section 2.2 are used for latitudes south of 55° S.

The degrees of freedom (ν) are calculated by subtracting the number of parameters with an initial range greater than 0 from the number of data in the input file (section 2.1). A minimum ν of 1 is always considered.

The script calculates concentrations corresponding to the sample positions for random parameter values between the parameter limits. Randomisation of the weathering and erosion rate values is performed logarithmically. A combination of random parameter values is computed in each iteration. The goodness of fit is defined by the chi-squared function:

$$\chi^2 = \sum_{i=1}^{n} \left(\frac{C_{model} - C_i}{\sqrt{\sigma^2_{C_{model}} + \sigma^2_{C_i}}} \right)^2 \tag{4}$$

where C_i and σ_{C_i} are the sample concentrations and their uncertainties derived from the apparent surface exposure ages (Section 2.1), C_{model} are the model concentrations corresponding to sample i (Section 2.6), and $\sigma_{C_{model}}$ is the model uncertainty corresponding to the uncertainty of the muon produced concentration (Section 2.3) plus the minimum age spacing of the climate data (10 years, as described in Section 2.2).

Models fitting the data within a 1σ confidence level are defined by the ones with $\chi^2 \leq \chi^2_{min.} + \nu$, and models fitting the data within a 2σ confidence level are defined by the ones with $\chi^2 \leq \chi^2_{min.} + 2 \cdot \nu$. Note that these formulas do not fully represent the chi-squared distribution described in Rodés et al. [24] (section 2.2.1). The method described in Rodés et al. [24] often yields infinite values when computing maximum fitting values ($\chi^2_{max.}$) for poor fittings and high ν. The formula $\chi^2_{max.} = \chi^2_{min.} + \mathrm{n} \cdot \nu$ is an approximation to the method described by Avni [25] for high degrees of freedom.

After a learning cycle of 3000 iterations (`consts.minmodelstoconverge` in `constants.m`), the limits of the randomised parameters start converging. Initially, the new limits converge to the models that fit the data within a 2σ confidence level, and within a 1σ confidence level for the last 1/3 of the total iterations. If the number of models fitting the data within this confidence level is lower than n_{conv}, this confidence range is increased to 4σ, 8σ, 16σ, etc. n_{conv} is initially equal to the desired fitting models and decreases exponentially with time. The new parameter limits are calculated every 100 iterations (`consts.convergencestep` in `constants.m`) from the models fitting the desired confidence level and expanded by 10% of the range to avoid missing fitting values at the limits of the 1σ range.

The script runs iterations until one of the following conditions are met:

- The simulations reach the maximum number of models to calculate: `consts.maxnmodels = 50,000` in `constants.m`.
- There are more than the desired fitting models that fit the data within the 1σ confidence level: `consts.targetnmodelsonesigma = 300` in `constants.m`.

To represent the results as probability density distributions of the parameters, the relative probability corresponding to each model is calculated as

$$P(\chi^2) \propto \sqrt{\frac{\nu}{\chi^2}} \cdot e^{\chi^2/(2\cdot\nu)} \tag{5}$$

which has a similar shape as the cumulative distribution function of the chi-squared distribution but can be computed avoiding zeros for high values of χ^2 and ν.

Finally, a set of fake samples covering a wide range of altitudes and all the fitted nuclides is generated. The parameter values of the models fitting the original data within a 1σ confidence level are used to generate altitudinal concentration profiles within the fake sample's data. Maximum and minimum concentration profiles are generated and used to plot the scatter of the fitting models.

2.8. Data Representation

A summary of the results is outputted in the command window by the script `display_results.m`.

Three figures are generated by `plot_results.m` as graphical output:

- The probability distribution of the models for each of the parameters.
- A representation of the ice surface evolution and the altitudinal trajectories of the samples (if no uplift rate is considered, these will be horizontal lines). Uncertainties corresponding to all 1σ models are also represented.
- Altitudinal profiles of the apparent exposure ages for all 1σ models and all nuclides, and the actual apparent exposure ages of the samples (model vs. data).

An example of the full graphical output generated by the NUNAIT calculator is shown in Section 3.2.

3. Examples

Two natural examples of input files are included in the folder "Examples". Inputs can be generated from new or published data. Data from published data can be easily generated from the ICE-D: ANTARCTICA database [26] (http://antarctica.ice-d.org, accessed on 19 August 2021), that compiles a large number of cosmogenic data sets, including updated exposure ages, organized by sites. The only datum required by the NUNAIT calculator that is missing in the ICE-D database is the current elevation of the ice surface. This value could be easily guessed by checking the lowest sampling site in the set, which often coincides with the ice surface.

3.1. Marble Hills

Marrero et al. [27] reported the first cosmogenic nuclide-derived erosion rates for carbonate rocks in Antarctica. Erosion rates were derived from carbonate bedrock samples at the Marble Hills field site (Ellsworth Mountains). I generated the exposure ages required by the NUNAIT calculator using the CRONUS online calculators [28] with the data included in Marrero et al. [27] (Appendix A).

In the original paper, Marrero et al. [27] calculated an apparent ^{36}Cl erosion rate of 0.22 ± 0.02 mm/ka from the samples above the elevation of $\sim$550 m above the present ice-surface, and apparent ^{36}Cl erosion rates >4 mm/ka at lower elevations. Marrero et al. [27]

interpreted that samples below 550 m over the present ice surface require complex exposure-burial histories to explain their composition.

As shown in Figure 4, the NUNAIT calculator predicts a subaerial weathering rate between 0.52 and 0.84, 2 to 4 times higher than the one calculated by Marrero et al. [27]; a glacial erosion rate below 37 mm/ka, concordant with the cold-based glacial processes expected in Marble Hills; and a maximum ice extension of 176–232 m above the present ice surface. These results suggest that the data from samples above 232 m could be compatible with simple exposure conditions, and the scatter of these data could be explained by inhomogeneous weathering ratios of the continuously exposed surfaces.

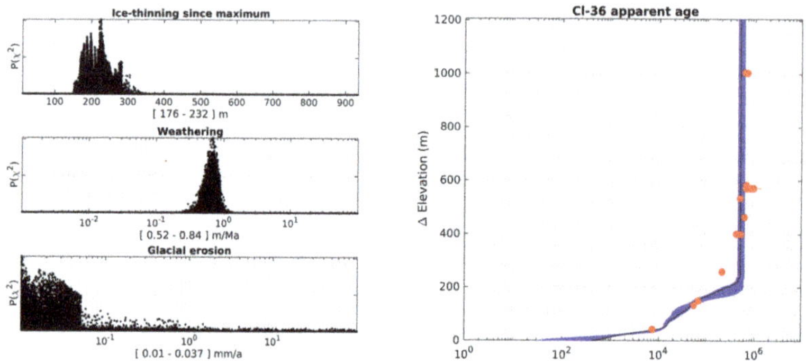

Figure 4. Marble Hills NUNAIT results. Results of fitting the NUNAIT model to Marble Hills' data. Right graph: data are depicted in red, and the models fitting the data within 1σ confidence level are plotted in blue. The best-fitting model, with a reduced chi-squared value of 7.2, is shown as a black line. Left graphs: probability distribution of the tested parameters.

The fit type 1 was used to constrain the minimum weathering rate that is compatible with these data (Figure 5), implicitly assuming that faster apparent weathering rates due to different lithologies, slopes, etc., can produce shorter apparent exposure ages in some surfaces. A (minimum) subaerial weathering rate between 0.19 and 0.21 m/Ma was obtained using this setup. These values agree with the value of 0.22 ± 0.02 mm/ka obtained by Marrero et al. [27].

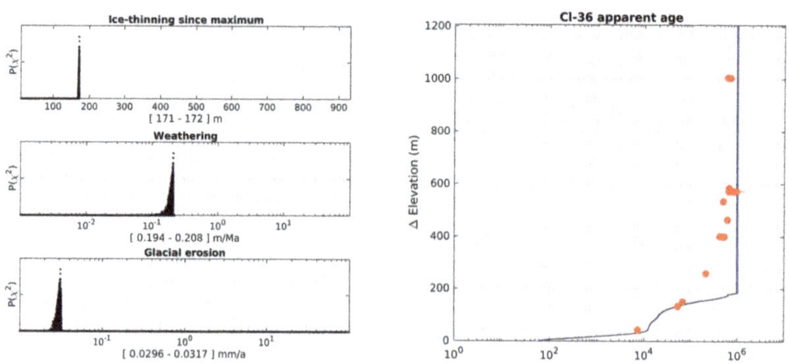

Figure 5. Marble Hills NUNAIT results using fit type 1. A reduced chi-squared value of 23 was obtained for the best fitting model. Note that when using fit type 1, models producing apparent exposure ages shorter than sample data are discarded. This results in no data being shown above a certain threshold in the first two graphs on the left column (ice-thinning and weathering plots).

3.2. Mount Hope

The longest nunatak elevation profile showing a wide range of cosmogenic isotope exposure ages in the ICE-D ANTARCTICA database is probably the site HOPE (Mt. Hope, Beardmore Glacier, Southern Ross Sea). Samples from the HOPE site appear in Spector et al. [29] and Spector [30]. The data set contains ^{10}Be, ^{26}Al, ^{21}Ne, ^{3}He, and ^{14}C exposure ages.

According to Spector et al. [29], Mt. Hope (836 m) remained ice-covered until 14.4 $\pm$ 0.5 ka. Several kilometres upstream from this position, two lateral moraines at 1050 and 1200 m mark the maximum elevation of ice during the Last Glacial Maximum.

Using only the cosmogenic exposure ages from the bedrock at Mt. Hope, the NUNAIT calculator yields a maximum elevation of ice during the Last Glacial Maximum of $\sim$1065 m above sea level (Figure 6), which seems to be in good agreement with the position of the lateral moraines described by Spector et al. [29].

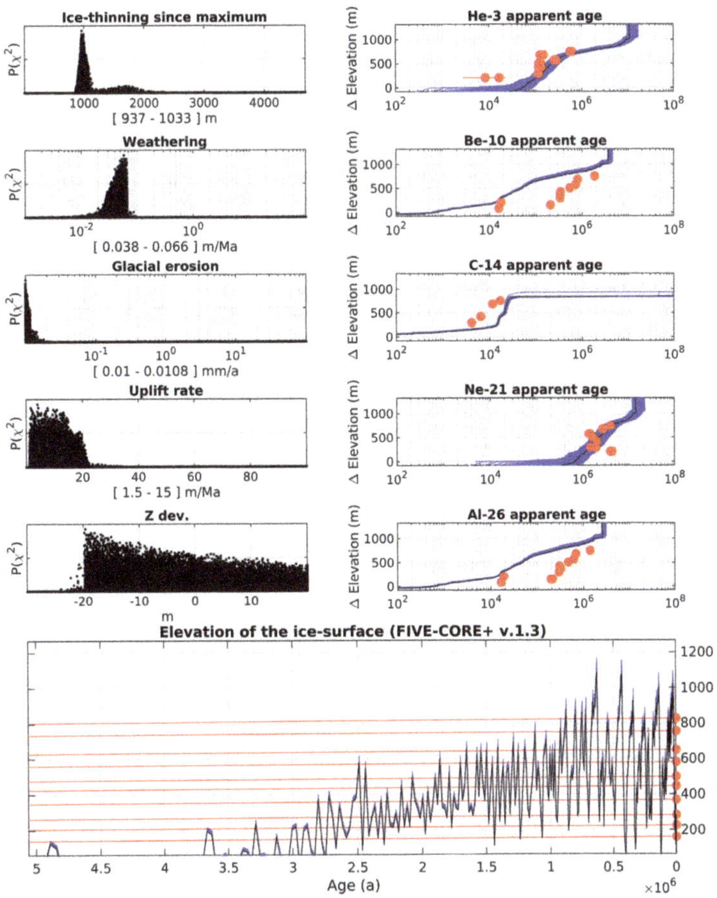

Figure 6. Full graphical output of the NUNAIT calculator for Mt. Hope data set. Left graphs show the probability distribution of the fitting parameters. Right graphs show the best model, the models fitting the data within 1σ, and the sample exposure ages in black, blue, and red, respectively. The best fit yielded a reduced chi-squared value of 78.7. The bottom graph shows the evolution of the ice surface and the position of the samples according to the model best fit and one sigma results.

Figure 6 shows that the optimum fit of the NUNAIT model mimics the distribution of the data for each isotope but does not fit the ratios between isotope data well.

As for the Marble Hills' data, the NUNAIT model fits this data set for very low weathering and glacial erosion rates and predicts a maximum uplift rate of 15 m/Ma.

4. Discussion and Conclusions

The NUNAIT and previous models described by Stroeven et al. [4], Li et al. [9], and Knudsen et al. [10] are based on the same principle: using a climate proxy (δ^{18}O record) to solve complex exposure-burial histories that fit the surface cosmogenic nuclide data. Previous models focused on solving the problem for sets of multiple isotope data from single samples. The method presented here focuses on solving the same problem but considering data from all the sampled sites on the nunatak, and yielding results that are consistent with the whole data set. Thus, the results obtained using the NUNAIT calculator are expected to be less precise than the results based on single samples but more robust, as the model can consider more possible scenarios by randomizing erosion rates, the position of the current ice surface, or uplift.

The NUNAIT calculator requires an input of cosmogenic data as apparent exposure ages with no surface erosion rate considered. The models fitting the data are also expressed as apparent exposure ages in the output. The use of input/output in this intuitive format has some advantages and disadvantages.

The user needs to calculate apparent exposure ages using a local or online exposure age calculator (e.g., [18,28,31]), allowing the user to consider any calculator, production rate reference, and scaling factor. This simplifies the use of the NUNAIT calculator, as the information about the production of cosmogenic isotopes (production rate, shielding, self-shielding factors, radiogenic produced concentrations, etc.) is implicitly included in the input data.

Although the user does not need to deal with production rates, the NUNAIT model works internally with scaled concentrations and constant production rates. Equation (1) assumes that the average production rate for the apparent (minimum) age equates to the constant production rate. This introduces differences with the time-dependent production models typically considered for the calculation of surface production rates. According to Balco et al. [18] (Figures 3 and 4), these differences should not exceed 10% of apparent exposure ages for most altitudes in polar regions. However, this uncertainty should be represented in the input data by the external uncertainty of the apparent exposure ages.

As the model is fitted using external uncertainty of the apparent exposure ages (Equation (4)), the fittings provided by the NUNAIT calculator are more sensitive to the spatial distribution of cosmogenic concentrations than to the ratios between different isotopes (e.g., Section 3.2), in contrast with the methods described by Stroeven et al. [4], Li et al. [9], and Knudsen et al. [10]. This effect is intentional and seeks to reflect the uncertainties of the cosmogenic surface production rate ratios realistically (e.g., [32–34]).

As the default input does not include any information on the muon contributions, these values need to be estimated as shown in Section 2.3. This approximation introduces an uncertainty of a similar magnitude as the one derived from the scaling scheme. The simplification of the muon cross-sections described in Section 2.4 introduces an additional uncertainty of 5% in the muon production rates under the ice sheet. The uncertainty of the muon produced cosmonuclides should also include the scatter of the global data available for the calibration of muon production under the surface, which is ~5% and ~14% for ^{10}Be and ^{26}Al, respectively, according to Balco [17] (Table 1). The NUNAIT calculator incorporates these uncertainties by considering a 20% uncertainty for all muon-produced concentrations.

According to the data summarized in Balco [17], the best predictions of the muon-produced ^{26}Al/^{10}Be ratios fit the empirical data within ~20% uncertainty (Figure 7). For other isotope pairs, the existing empirical data about their production rate ratios at depth are more scarce (e.g., [35]). Therefore, we should assign an uncertainty greater than 20% to our modelled concentration ratios at great depths. As the NUNAIT model considers a 20% uncertainty for all muon-produced concentrations, it assumes a 28% uncertainty

($\sqrt{20^2 + 20^2}$) for any synthetic concentration ratio under the ice-sheet, which is probably an overestimation of the $^{26}Al/^{10}Be$ uncertainties. However, similar to the surface predictions, this overestimation of the uncertainty makes the model less sensitive to the ratios between isotopes and therefore relatively more sensitive to the spatial distribution of the data.

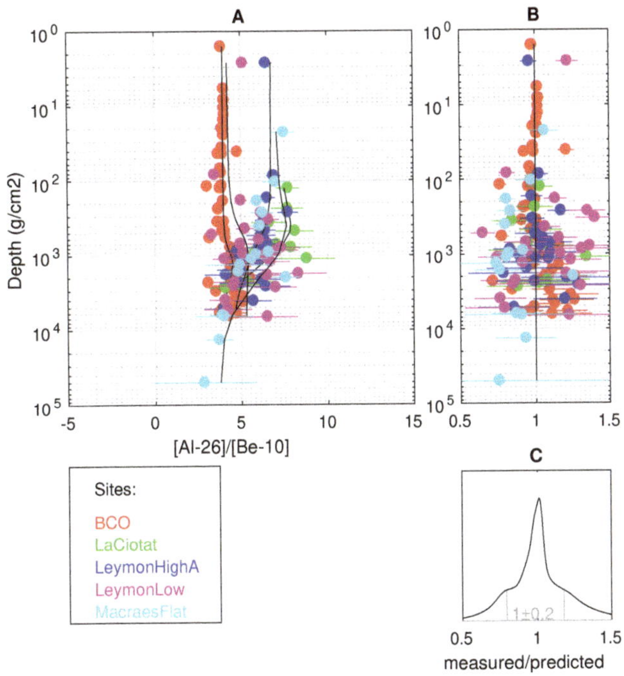

Figure 7. Subsurface $^{26}Al/^{10}Be$ ratios. (**A**) Plot of $^{26}Al/^{10}Be$ concentration ratios shown in Balco [17] (coloured circles with error bars) and cross sections predicted by `P_mu_total_alpha1.m` (black lines), also from Balco [17]. All concentration ratios (measured) scaled to their predicted ratios (**B**) are plotted as a camel-plot with 68% of its area at ~100 ± 20% (**C**).

The NUNAIT model considers a constant ice density for the ice column covering the samples during glaciations. This value can be adjusted by the user, and the effect of its uncertainty is not expected to exceed the 20% uncertainty considered for muon-produced concentrations.

When uplift is considered in the model, it is assumed to be a constant rate. Isostatic rebound is not emulated by the NUNAIT model. This should not greatly affect the distribution of glaciated and ice-free elevations through time, as the isostatic rebounds are expected to be coupled with the changes in the elevations of the ice surface. However, a constant fast uplift could result in surfaces accumulating cosmogenic nuclides at slower rates in the past due to the reduced production at lower elevations. This effect is not yet considered by the NUNAIT model. Therefore, this model could overestimate the concentration of stable isotopes in highly uplifted areas that have not been glaciated in the past.

During ice-free periods, the NUNAIT model considers a homogeneous weathering rate along with the elevation profile. This might not be very realistic for areas with intense periglacial processes that produce increased erosion rates in local areas (e.g., rock falls). When fitting the model to data from areas with evident periglacial processes, the minimum fitting type should be selected (fit type 1 described in Section 2.7).

Marrero et al. [27] described a systematic difference between bedrock and boulder samples, with boulder samples yielding systematically lower erosion rates. By default, the NUNAIT model considers a homogeneous erosion rate under the ice. Therefore, when fitting data from erratic boulders that could have been preserved during glacial periods, and hence maintaining a higher surface cosmonuclide concentration than the bedrock, the maximum fitting type should be selected (fit type 2 described in Section 2.7).

The examples in Section 3 show that the NUNAIT calculator yields results that go beyond the typical observations deduced from surface exposure dating, such as glacial erosion rates and uplift rates. Therefore, the NUNAIT calculator is presented as an easy-to-use tool that will help glaciologists to interpret cosmogenic data from nunataks, where exposure histories are usually complex. Moreover, the methods described in Section 2 can be used to develop new cosmogenic-based tools with intuitive and simplified inputs and outputs.

Supplementary Materials: All scripts discussed in Section 2 and the data discussed in Section 3 are freely accessible at https://github.com/angelrodes/NUNAIT (accessed on 19 August 2021, subject to the terms of the GNU General Public License, version 3, as published by the FreeSoftware Foundation).

Funding: This research received no external funding.

Data Availability Statement: All scripts and example data are freely accessible at https://github.com/angelrodes/NUNAIT (accessed on 19 August 2021).

Acknowledgments: I would like to thank Laura Medina-Ruiz for proofreading the manuscript and the three anonymous reviewers for their constructive insights during the review of the article.

Conflicts of Interest: The author declares no conflict of interest.

References

1. Alley, R.B.; Clark, P.U.; Huybrechts, P.; Joughin, I. Ice-Sheet and Sea-Level Changes. *Science* **2005**, *310*, 456–460. [CrossRef]
2. Balco, G. Contributions and unrealized potential contributions of cosmogenic-nuclide exposure dating to glacier chronology, 1990–2010. *Quat. Sci. Rev.* **2011**, *30*, 3–27. [CrossRef]
3. Stone, J.O.; Balco, G.A.; Sugden, D.E.; Caffee, M.W.; Sass, L.C.; Cowdery, S.G.; Siddoway, C. Holocene Deglaciation of Marie Byrd Land, West Antarctica. *Science* **2003**, *299*, 99–102. [CrossRef]
4. Stroeven, A.P.; Fabel, D.; Hättestrand, C.; Harbor, J. A relict landscape in the centre of Fennoscandian glaciation: cosmogenic radionuclide evidence of tors preserved through multiple glacial cycles. *Geomorphology* **2002**, *44*, 145–154. [CrossRef]
5. Davis, P.T.; Bierman, P.R.; Marsella, K.A.; Caffee, M.W.; Southon, J.R. Cosmogenic analysis of glacial terrains in the eastern Canadian Arctic: A test for inherited nuclides and the effectiveness of glacial erosion. *Ann. Glaciol.* **1999**, *28*, 181–188. [CrossRef]
6. Stroeven, A.P.; Fabel, D.; Harbor, J.; Hättestrand, C.; Kleman, J. Quantifying the erosional impact of the fennoscandian ice sheet in the torneträsk–narvik corridor, northern sweden, based on cosmogenic radionuclide data. *Geogr. Ann. Ser. A Phys. Geogr.* **2002**, *84*, 275–287. [CrossRef]
7. Nishiizumi, K.; Kohl, C.; Arnold, J.; Klein, J.; Fink, D.; Middleton, R. Cosmic ray produced ^{10}Be and ^{26}Al in Antarctic rocks: Exposure and erosion history. *Earth Planet. Sci. Lett.* **1991**, *104*, 440–454. [CrossRef]
8. Bierman, P.R.; Marsella, K.A.; Patterson, C.; Davis, P.; Caffee, M. Mid-Pleistocene cosmogenic minimum-age limits for pre-Wisconsinan glacial surfaces in southwestern Minnesota and southern Baffin Island: A multiple nuclide approach. *Geomorphology* **1999**, *27*, 25–39. [CrossRef]
9. Li, Y.; Fabel, D.; Stroeven, A.P.; Harbor, J. Unraveling complex exposure-burial histories of bedrock surfaces under ice sheets by integrating cosmogenic nuclide concentrations with climate proxy records. *Geomorphology* **2008**, *99*, 139–149. [CrossRef]
10. Knudsen, M.F.; Egholm, D.L.; Jacobsen, B.H.; Larsen, N.K.; Jansen, J.D.; Andersen, J.L.; Linge, H.C. A multi-nuclide approach to constrain landscape evolution and past erosion rates in previously glaciated terrains. *Quat. Geochronol.* **2015**, *30*, 100–113. [CrossRef]
11. Lal, D. Cosmic ray labeling of erosion surfaces: In situ nuclide production rates and erosion models. *Earth Planet. Sci. Lett.* **1991**, *104*, 424–439. [CrossRef]
12. Lisiecki, L.E.; Raymo, M.E. A Pliocene-Pleistocene stack of 57 globally distributed benthic δ18O records. *Paleoceanography* **2005**, *20*, PA1003. [CrossRef]
13. Zachos, J. Trends, Rhythms, and Aberrations in Global Climate 65 Ma to Present. *Science* **2001**, *292*, 686–693. [CrossRef]
14. North Greenland Ice Core Project Members. High-resolution record of Northern Hemisphere climate extending into the last interglacial period. *Nature* **2004**, *431*, 147–151. [CrossRef]
15. Buizert, C.; Sigl, M.; Severi, M.; Markle, B.R.; Wettstein, J.J.; McConnell, J.R.; Pedro, J.B.; Sodemann, H.; Goto-Azuma, K.; Kawamura, K. Abrupt ice-age shifts in southern westerly winds and Antarctic climate forced from the north. *Nature* **2018**, *563*, 681–685. [CrossRef]
16. Amante, C. *ETOPO1 1 Arc-Minute Global Relief Model: Procedures, Data Sources and Analysis*; National Geophysical Data Center: Washington, DC, USA, 2009; doi:10.7289/V5C8276M. [CrossRef]

17. Balco, G. Production rate calculations for cosmic-ray-muon-produced 10Be and 26Al benchmarked against geological calibration data. *Quat. Geochronol.* **2017**, *39*, 150–173. [CrossRef]
18. Balco, G.; Stone, J.O.; Lifton, N.A.; Dunai, T.J. A complete and easily accessible means of calculating surface exposure ages or erosion rates from ^{10}Be and ^{26}Al measurements. *Quat. Geochronol.* **2008**, *3*, 174–195. [CrossRef]
19. Heisinger, B.; Nolte, E. Cosmogenic in situ production of radionuclides: Exposure ages and erosion rates. *Nucl. Instrum. Methods Phys. Res. Sect. B Beam Interact. Mater. At.* **2000**, *172*, 790–795. [CrossRef]
20. Balco, G.; Shuster, D.L. Production rate of cosmogenic 21Ne in quartz estimated from 10Be, 26Al, and 21Ne concentrations in slowly eroding Antarctic bedrock surfaces. *Earth Planet. Sci. Lett.* **2009**, *281*, 48–58. [CrossRef]
21. Blard, P.H.; Braucher, R.; Lavé, J.; Bourlès, D. Cosmogenic 10Be production rate calibrated against 3He in the high Tropical Andes (3800–4900 m, 20–22° S). *Earth Planet. Sci. Lett.* **2013**, *382*, 140–149. [CrossRef]
22. Granger, D.; Smith, A. Dating buried sediments using radioactive decay and muogenic production of ^{26}Al and ^{10}Be. *Nucl. Instrum. Methods Phys. Res. Sect. B Beam Interact. Mater. At.* **2000**, *172*, 822–826. [CrossRef]
23. Harvey, A. Properties of Ice and Supercooled Water. In *CRC Handbook of Chemistry and Physics*; CRC Press: Boca Raton, FL, USA, 2019.
24. Rodés, A.; Pallàs, R.; Braucher, R.; Moreno, X.; Masana, E.; Bourlés, D.L. Effect of density uncertainties in cosmogenic 10Be depth-profiles: Dating a cemented Pleistocene alluvial fan (Carboneras Fault, SE Iberia). *Quat. Geochronol.* **2011**, *6*, 186–194. [CrossRef]
25. Avni, Y. Energy spectra of X-ray clusters of galaxies. *Astrophys. J.* **1976**, *210*, 642–646. [CrossRef]
26. Balco, G. Technical note: A prototype transparent-middle-layer data management and analysis infrastructure for cosmogenic-nuclide exposure dating. *Geochronology* **2020**, *2*, 169–175. [CrossRef]
27. Marrero, S.M.; Hein, A.S.; Naylor, M.; Attal, M.; Shanks, R.; Winter, K.; Woodward, J.; Dunning, S.; Westoby, M.; Sugden, D. Controls on subaerial erosion rates in Antarctica. *Earth Planet. Sci. Lett.* **2018**, *501*, 56–66. [CrossRef]
28. Marrero, S.M.; Phillips, F.M.; Borchers, B.; Lifton, N.; Aumer, R.; Balco, G. Cosmogenic nuclide systematics and the CRONUScalc program. *Quat. Geochronol.* **2016**, *31*, 160–187. [CrossRef]
29. Spector, P.; Stone, J.; Cowdery, S.G.; Hall, B.; Conway, H.; Bromley, G. Rapid early-Holocene deglaciation in the Ross Sea, Antarctica. *Geophys. Res. Lett.* **2017**, *44*, 7817–7825. [CrossRef]
30. Spector, P. Antarctic Glacial History Inferred from Cosmogenic-Nuclide Measurements in Rocks. Unpublished Ph.D. Dissertation, University of Washington, Washington, DC, USA, 2018.
31. Martin, L.; Blard, P.H.; Balco, G.; Lavé, J.; Delunel, R.; Lifton, N.; Laurent, V. The CREp program and the ICE-D production rate calibration database: A fully parameterizable and updated online tool to compute cosmic-ray exposure ages. *Quat. Geochronol.* **2017**, *38*, 25–49. [CrossRef]
32. Goethals, M.; Hetzel, R.; Niedermann, S.; Wittmann, H.; Fenton, C.; Kubik, P.; Christl, M.; von Blanckenburg, F. An improved experimental determination of cosmogenic ^{10}Be/^{21}Ne and ^{26}Al/^{21}Ne production ratios in quartz. *Earth Planet. Sci. Lett.* **2009**, *284*, 187–198. [CrossRef]
33. Balco, G.; Shuster, D.L. ^{26}Al-^{10}Be-^{21}Ne burial dating. *Earth Planet. Sci. Lett.* **2009**, *286*, 570–575. [CrossRef]
34. Phillips, F.M.; Argento, D.C.; Balco, G.; Caffee, M.W.; Clem, J.; Dunai, T.J.; Finkel, R.; Goehring, B.; Gosse, J.C.; Hudson, A.M.; et al. The CRONUS-Earth Project: A synthesis. *Quat. Geochronol.* **2016**, *31*, 119–154. [CrossRef]
35. Fernandez-Mosquera, D.; Hahm, D.; Marti, K. Calculated rates of cosmic ray muon-produced Ne in subsurface quartz. *Geophys. Res. Lett.* **2010**, *37*, L15403. [CrossRef]

geosciences

Article

New System for Measuring Cosmogenic Ne in Terrestrial and Extra-Terrestrial Rocks

Domokos Györe *, Luigia Di Nicola, David Currie and Finlay M. Stuart

Scottish Universities Environmental Research Centre (SUERC), East Kilbride G75 0QF, UK;
Luigia.DiNicola@glasgow.ac.uk (L.D.N.); d.currie.1@research.gla.ac.uk (D.C.); fin.stuart@glasgow.ac.uk (F.M.S.)
* Correspondence: Domokos.Gyore@glasgow.ac.uk

Abstract: Cosmogenic Ne isotopes are used for constraining the timing and rate of cosmological and Earth surface processes. We combined an automated gas extraction (laser) and purification system with a Thermo Fisher ARGUS VI mass spectrometer for high through-put, high precision Ne isotope analysis. For extra-terrestrial material with high cosmogenic Ne concentrations, we used multi-collection on Faraday detectors. Multiple measurements (n = 26) of 1.67×10^{-8} cm^3 air-derived ^{20}Ne yielded an uncertainty of 0.32%, and $^{21}Ne/^{20}Ne$ = 0.17% and $^{22}Ne/^{20}Ne$ = 0.09%. We reproduced the isotope composition of cosmogenic Ne in the Bruderheim chondrite and Imilac pallasite in a sub-ten mg sample. For lower Ne amounts that are typical of terrestrial samples, an electron multiplier detector was used in peak jumping mode. Repeated analysis of 3.2×10^{-11} cm^3 STP ^{20}Ne from air reproduced $^{21}Ne/^{20}Ne$ and $^{22}Ne/^{20}Ne$ with 1.1% and 0.58%, respectively, and ^{20}Ne intensity with 1.7% (n = 103) over a 4-month period. Multiple (n = 8) analysis of cosmogenic Ne in CREU-1 quartz yielded $3.25 \pm 0.24 \times 10^8$ atoms/g (2 s), which overlaps with the global mean value. The repeatability is comparable to the best data reported in the international experiments performed so far on samples that are 2–5× smaller. The ability to make precise Ne isotope determinations in terrestrial and extra-terrestrial samples that are significantly smaller than previously analysed suggests that the new system holds great promise for studies with limited material.

Keywords: cosmogenic ^{21}Ne; noble gas isotope; mass spectrometry; neon isotope; ^{21}Ne; Ne dating; Bruderheim chondrite; Imilac pallasite; CREU quartz; cosmogenic nuclides

Citation: Györe, D.; Di Nicola, L.; Currie, D.; Stuart, F.M. New System for Measuring Cosmogenic Ne in Terrestrial and Extra-Terrestrial Rocks. *Geosciences* **2021**, *11*, 353. https://doi.org/10.3390/geosciences11080353

Academic Editors: Jesus Martinez-Frias and Naki Akçar

Received: 15 July 2021
Accepted: 19 August 2021
Published: 23 August 2021

1. Introduction

Cosmogenic Ne was proven to be an adept recorder of the timing and rates of surface processes on the Earth and Moon [1–3] and the time of meteorite release from parent bodies [4]. Precise determination of neon isotopes in rocks and minerals are conventionally made using static gas magnetic sector mass spectrometers [5–8]. The low production rate at the Earth's surface and the presence of isobaric interferences at all Ne isotopes means that precise determination of terrestrial cosmogenic Ne is routinely measured in only a handful of laboratories worldwide [5]. Improvements in mass spectrometry in the last ten years, in particular the ability to resolve some of main isobaric interferences [8–10], will lead to better and faster cosmogenic Ne determinations.

Here, we report the use of a Thermo Fisher ARGUS VI mass spectrometer with an automated gas extraction and purification system for the determination of cosmogenic Ne in both extra-terrestrial and terrestrial material. It is a low resolution instrument [11], which requires low background levels and a good understanding of isobaric interferences [6]. Where these can be obtained, the high sensitivity and good instrument stability combine to allow high throughput cosmogenic Ne determinations on samples that are analysed by conventional instruments. Two protocols can be applied depending on Ne concentrations: multi-collection using Faraday detectors and peak-jumping mode using a compact discrete dynode (CDD) detector. We demonstrate the multi-collection Faraday technique with new analysis of cosmogenic Ne in sub-ten mg samples of Bruderheim chondrite and Imilac pallasite and the peak-jumping CDD analysis of 19.9 mg aliquots of CREU-1 quartz. We use this study to demonstrate how significant reduction in sample size affects uncertainty and the implication of that in cosmogenic Ne dating.

2. Analytical System

The Thermo Fisher ARGUS VI is a low resolution (R < 200), small volume (0.7 litre) 6-detector static vacuum mass spectrometer designed primarily for Ar isotope analysis [12]. It was recently shown to be capable of making high precision determinations of Ne, Kr and Xe isotopes [6,11,12]. The need to minimise isobaric interferences for Ne isotope determinations demanded several modifications to the mass spectrometer vacuum envelope. Gas equilibrated with the mass spectrometer is forced to enter the ion source via a SAES GP50 ZrAl alloy getter held at room temperature in order to minimise the H_2^+ level, which controls the rate of formation of Ar^{2+} and CO_2^{2+} [13], as well as $^{20}NeH^+$ [6]. A charcoal-filled finger on the source block is cooled with liquid nitrogen during analysis to reduce the level of residual Ar and CO_2 in the mass spectrometer.

The ion source operates at 4.5 kV acceleration potential, 250 μA trap current and 80 eV electron energy to minimise NeH^+ production in the source [6]. The source parameters (ion repeller voltage, extraction, focus and symmetry) were tuned for maximum sensitivity using ^{20}Ne. The magnet position was set to achieve the coincidence of the flat-topped peaks of all Ne isotopes (^{20}Ne, ^{21}Ne and ^{22}Ne) on the Faraday detectors (10^{12} Ohm resistance amplifiers), and of ^{22}Ne and ^{21}Ne on the combination of Faraday and a compact discrete dynode (CDD) detector, respectively (Figure 1). The release of CO_2 from the CDD requires that it is conditioned. In the initial phase of this work, the CDD was bombarded by a CO_2 beam and pumped by the mass spectrometer ion pump for several weeks. This resulted in a reduction in the dynamic CO_2 beam intensity from >100 fA to a normal operating level of 2.5 fA.

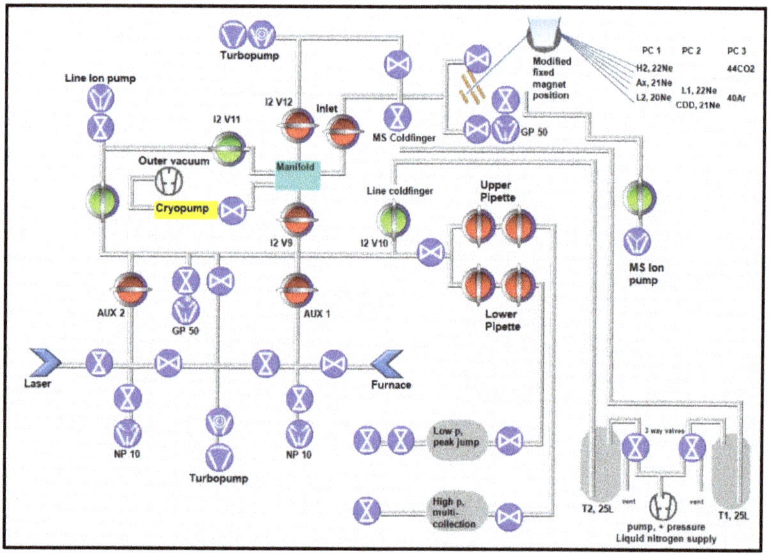

Figure 1. Schematic picture of the fully automatized Thermo Fisher ARGUS VI mass spectrometer and preparation system in SUERC. Two 2L air reservoirs, each equipped with a 0.1 cm^3 gas pipette, provided the means for calibration. The magnet position allowed multi collecting of Ne isotopes. PC1: (H2: ^{22}Ne—Ax: ^{21}Ne—L2: ^{20}Ne). PC2: (L1: ^{22}Ne—CDD: ^{21}Ne). PC3: (H2: $^{44}CO_2$—L2: ^{40}Ar). Blue valve: manual. Green: pneumatically actuated, open. Red: pneumatically actuated, closed. PC: peak coincidence.

High precision Ne isotope analysis is complicated by isobaric interferences at all peaks. Although new instruments can either fully or partially resolve some of the common interferences [8,9], in many cases, a correction is required [6,7,10,14,15]. This requires the measurement of several other peaks, which has an implication on measurement time and data precision. The optimisation of analysis procedures in mass spectrometer control software (Qtegra) required the use of four lab books, starting with hydrogen, followed by mass 18, 19, 40 and 44, prior to the analysis of Ne isotopes (Figure 2). The number of cycles required for Ne isotope analysis was optimised to overcome uncertainty introduced by the time delay. We developed both a multi-collection mode using the array of Faraday detectors where the smallest Ne beam was greater than ~7 fA and a peak jumping mode that used the CDD for smaller Ne beam intensities.

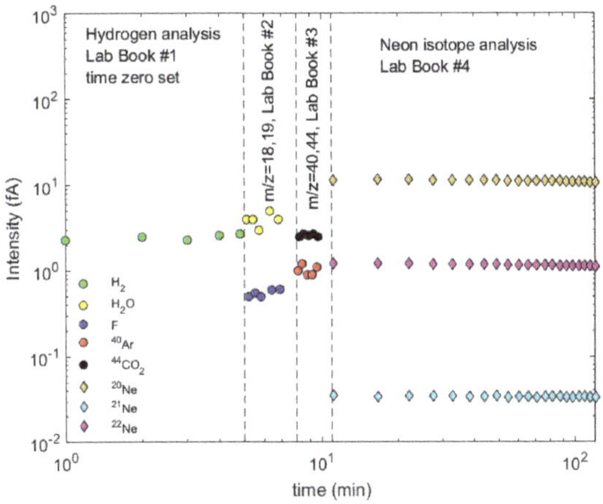

Figure 2. Schematic of the measurement protocol for high precision Ne analysis using the Thermo Fisher ARGUS VI mass spectrometer. The analysis was performed in four lab books in Qtegra to minimise magnet current movement when Ne isotopes were measured. Interfering compounds and hydrogen were measured first for 10 min. Ne isotope measurements were long enough to overcome the uncertainty introduced by the time gap.

3. Cosmogenic Ne in Extra-Terrestrial Material: Multi-Collection Faraday Technique

The multi-collection Faraday mode was developed for the determination of the $^{21}Ne/^{20}Ne$ of air [6] and precise analysis of the Ne isotope composition of natural gases [14]. The high sensitivity of the ARGUS system allowed this procedure to be used to determine cosmogenic Ne in small samples of extra-terrestrial material. Here, we report Ne isotopes in small samples (4.1 to 10.7 mg) of Bruderheim L-chondrite and the Imilac pallasite in order to establish the precision of Ne isotope determinations in meteorites.

Repeated analysis of air standard (n = 26) over the two-week analysis period yielded repeatability values of $^{21}Ne/^{20}Ne$ and $^{22}Ne/^{20}Ne$ of 0.17% and 0.09%, respectively (1σ), calculated by fitting a best Gaussian curve to the probability density distribution [15]. All isotope data are consistent with the combination of mass fractionation of air and the presence of NeH^+, as described earlier [6]. The repeatability of the intensity of ^{20}Ne (~3450 fA) was 0.32%. Isobaric interferences from $^{40}Ar^{2+}$ and $^{44}CO_2{}^{2+}$ were trivial (less than 0.3‰) and all other interferences were significantly smaller than 0.1‰ [6].

The meteorite samples were heated to 1500 °C for 10 min in a double-walled furnace. The gas purification and Ne isotope analysis were conducted using methods described earlier [16]. Furnace hot blanks were measured and the Ne isotopic composition was found to be isotopically indistinguishable from air, albeit less than 0.3% of sample beam intensities. All samples were reheated, which confirmed that all Ne was extracted in the heating step. For these analyses, we did not routinely monitor H_2O^+ and HF^+ as beam intensities were so low that it had a negligible (<0.1‰) effect on $m/z = 20$. While $^{20}NeH^{+}$ generation in the mass spectrometer ion source was significant for analyses of terrestrial Ne at partial pressures that are similar to those determined here, in this study, the contribution at $m/z = 21$ was insignificant as $^{21}Ne^{+}$ abundances in meteorites were approximately 30 times higher. We estimate that the $^{20}NeH^{+}$ contribution for $^{21}Ne^{+}$ was less than 0.1‰ in the meteorite analyses reported here.

The Ne isotope data for both meteorite samples are presented in Table 1 and Figure 3. The two Bruderheim samples yielded $^{20}Ne/^{22}Ne$ of 0.867 ± 0.002 and 0.852 ± 0.002 and $^{21}Ne/^{22}Ne$ values of 0.916 ± 0.002 and 0.925 ± 0.002. These plotted within the range determined by earlier studies ([15,16], (T. Graf, pers. comm.)), thus confirming our ability to replicate existing data. The difference in $^{21}Ne/^{20}Ne$ ratios of the two samples was small, but beyond the analytical uncertainty of each measurement. This may reflect variation in the contribution of the primordial Ne component within the samples. The Bruderheim sample had a well-established ^{22}Ne concentration of $12.14 \pm 0.11 \times 10^{-8}$ cm^3 STP/g (personal communication with T. Graf). On this basis, using our measurements, we calculated an instrument sensitivity of $5.01 \pm 0.06 \times 10^{-12}$ cm^3 STP Ne/fA, which was equivalent to $1.25 \pm 0.02 \times 10^{15}$ cps/cm^3 STP (where 1 cps = 1.6×10^{-19} A).

The isotopic compositions of Ne for the two samples of Imilac pallasite were indistinguishable from each other. The $^{20}Ne/^{22}Ne$ values were 0.896 ± 0.002 and 0.893 ± 0.002 and the $^{21}Ne/^{22}Ne$ values were 1.028 ± 0.003 and 1.030 ± 0.003. These values were extremely close to those determined earlier for the Imilac pallasite [17] (Figure 3). The measured ^{22}Ne concentrations of 66.52 ± 0.08 and $69.42 \pm 0.07 \times 10^{-8}$ cm^3 STP/g overlapped that of earlier determinations ($66.0 \pm 3.6 \times 10^{-8}$ cm^3 STP ^{22}Ne/g) [17].

The precision of the Ne isotope ratios of both meteorites was ±0.2–0.3%. This was ~50–100 and 5–10 times greater than the precision of $^{20}Ne/^{22}Ne$ and $^{21}Ne/^{22}Ne$ measurements of air, and was slightly higher than the repeatability of the air measurements (0.17% and 0.09% for $^{20}Ne/^{22}Ne$ and $^{21}Ne/^{22}Ne$, respectively). We explain this by the large amount of matrix present in the gas phase after melting of the mineral. The intensity determinations had an uncertainty of ± 0.02% at signal sizes of 200 fA, and this was identical to the precision of ^{22}Ne beams of a similar size (~335 fA) derived from air calibrations.

Table 1. Neon isotope data of newly measured meteorites.

Sample	Weight (mg)	$^{20}Ne/^{22}Ne$	$^{21}Ne/^{22}Ne$	^{22}Ne
		This study		
Bruderheim-1	9.13	0.867 (2)	0.916 (2)	N/A
Bruderheim-2	10.67	0.852 (2)	0.925 (2)	N/A
Imilac-1	10.10	0.896 (2)	0.893 (2)	66.52 (8)
Imilac-2	4.10	1.028 (3)	1.030 (3)	69.42 (7)
		Pers. comm. with T. Graf		
Bruderheim-1	49.2	0.914	0.851	11.98
Bruderheim-2	42.2	0.914	0.850	12.30
Bruderheim-3	25.5	0.911	0.844	12.18
Bruderheim-4	33.0	0.912	0.842	12.20
Bruderheim-5	24.3	0.912	0.849	12.15
Bruderheim-6	53.7	0.918	0.839	11.99
Bruderheim-7	32.8	0.922	0.828	12.12
Bruderheim-8	33.4	0.925	0.839	12.21

1σ are shown in parenthesis as last significant figures. Ne isotope concentrations are in cm^3 STP/g $\times 10^8$ where $p = 0.101$ MPa and T = 0 °C in accordance with [20]. N/A: not applicable as the Bruderheim, in this study, was used to determine mass spectrometer sensitivity using data from T. Graf (pers. comm.). Errors of data obtained from pers. comm. with T. Graf were not provided.

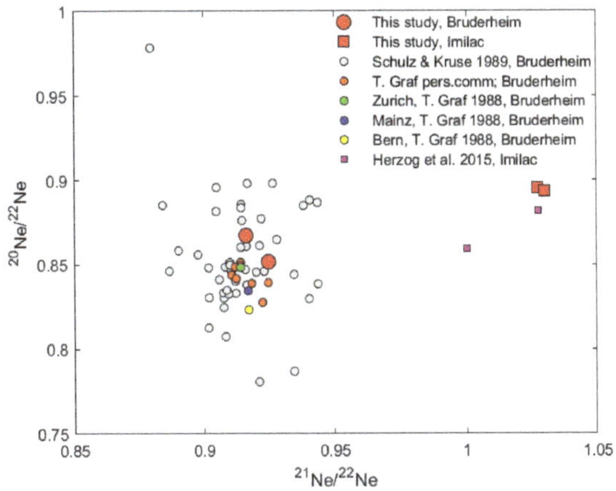

Figure 3. Neon isotopic composition of Bruderheim chondrite and Imilac pallasite analysed using ARGUS VI mass spectrometer. Bruderheim data overlapped with earlier determinations. Slight deviations in between the measurements may indicate variation in the contribution of primordial components. Imilac data overlapped with each other within the measurement error for both isotope ratios, and overlapped with earlier determinations. Measurement uncertainties are smaller than symbols. Other data are from Refs. [17–19] and personal communication with T. Graf.

Our data were obtained from a significantly smaller sample than earlier studies. The direct comparison of the quality of our data to earlier determinations is difficult because of the low number of analyses and the absence of precision in previous works. Our Bruderheim chondrite samples (9.1 and 10.7 mg) were between 2 and 6 times smaller than most recent measurements (T. Graf pers. comm.). Our Imilac pallasite samples (4.1 and 10.1 mg) were at least half the size of earlier determinations [17]. Incorporating the repeatability of air Ne isotope ratios into the overall isotope ratio uncertainty (~0.5%), we found a four-fold improvement on those measured in samples that were of at least 2–5 times greater mass [17].

4. Cosmogenic Ne in Terrestrial Material: Analysis by Peak Jumping Using Electron Multiplier

4.1. Analysis Procedure and Repeated Measurement of Low Quantities of Atmospheric Ne

The concentration of Ne in terrestrial rocks and minerals is significantly less than in meteorites, and typically requires the use of electron multiplier detectors [8–10]. We developed a peak jumping protocol using the CDD detector that was located in the L3 position. The peak measurement sequence followed that described above (Figure 2). Hydrogen ($m/z = 2$) and background ($m/z = 2.2$) were measured on the L2 Faraday detector with all other isotopes determined on the CDD. No NeH^+ was recorded during analysis, in line with findings using a Noblesse-HR at a similarly low Ne partial pressure [7].

Over the course of 4 months, the blanks were composed of fractionated air (n = 36). Isobaric interference from HF^+ always stayed below 1‰. Contributions at $m/z = 20$ from $H_2{}^{18}O^+$ (~1%) and $^{40}Ar^{++}$ (~5%) were small. In contrast, $^{44}CO_2{}^{++}$ dominated $m/z = 22$. The air-like blank in the mass spectrometers rendered blank correction for cosmogenic Ne redundant.

Calibration data (n = 103) were collected over the course of 4 months from aliquots of air at the level of $9.36 \pm 0.21 \times 10^{-14}$ cm^3 STP ^{21}Ne, which was equivalent to ^{21}Ne from 7.7 mg CREU quartz. The repeatability of the $^{21}Ne/^{20}Ne$ and $^{22}Ne/^{20}Ne$ of air was 1.10% and 0.62%, respectively (Figure 4) (1σ, 2 outliers), which was calculated by fitting a best-fit Gaussian curve to the probability density distribution. The isobaric interferences from H_2O^+, HF^+ and $^{40}Ar^{2+}$ at $m/z = 20$, $^{63}Cu^{3+}$ at $m/z = 21$ and $^{66}Zn^{3+}$ at $m/z = 22$ were trivial.

Interference from organic compounds was assumed to be negligible [6]. The contribution of CO_2^{2+} at $m/z = 22$ averaged around 7% over the 4-month period. Singly/doubly charged component determinations followed earlier practice [6]. The isotope ratios plotted slightly to the right of the mass fractionation line in Figure 4, indicating either a constant excess at $m/z = 21$ (termed excess Ne) (horizontal movement in Figure 4), which does not appear in the blank measurements, or a different CO_2 contribution to that of the dynamic blanks (vertical movement in Figure 4). However, doubly charged CO_2 would only increase with increasing partial pressure in the mass spectrometer [13,21], which would move our data to the lower $^{22}Ne/^{20}Ne$ regimes and would not explain our data. Therefore, we suggest the presence of excess Ne.

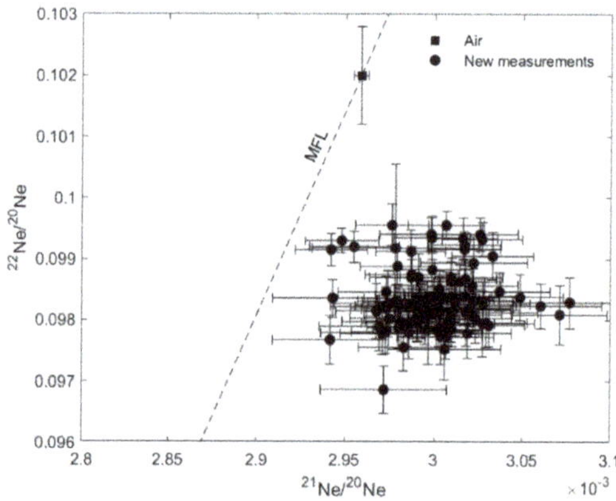

Figure 4. Ne isotope ratio from the repeated analysis of air using the ARGUS VI mass spectrometer. The delivery amount was $9.36 \pm 0.21 \times 10^{-14}$ cm^3 STP ^{21}Ne/aliquot, equivalent of ^{21}Ne from 7.7 mg CREU. The repeatability of $^{21}Ne/^{20}Ne$ was 1.1% (1σ) (n = 103), which was a significant achievement at this partial pressure of Ne. The repeatability of $^{22}Ne/^{20}Ne$ was 0.62% and was likely governed by the 7% CO_2 correction at mass 22. All data were located slightly right of the mass fractionation line (MFL), indicating a constant excess at $m/z = 21$, which was unchanged over an order of magnitude of Ne partial pressure and was corrected out (see text). Air is from Refs. [6,22]. Plotted uncertainties are 1σ.

Excess Ne, or, more precisely, excess mass 21 is the result of the liberation of material inside the mass spectrometer by the calibration (or sample) beam. This was not detected on the Faraday detectors at higher beam intensities, and the fact that the CDD is known to produce CO_2 implies that the electron multiplier is the source [6]. Test measurements revealed that the excess remains unchanged over an order of magnitude of Ne partial pressure in the mass spectrometer. This suggests that if the sample stays within this partial pressure range, the excess can simply be corrected out. In the worst-case scenario when the sample is outside of this range, which was not the case for this study, our accuracy could vary by the extent of the excess—thus, by at least $\pm 1\%$—and would leave the precision unchanged. Intensities of ^{20}Ne (i.e., sensitivity) were found to be variable in the short-term (days/weeks), but long-term (4 months) observation suggests that the variability is natural. ^{20}Ne reproduced by 1.13% (9 outliers out of 103, 1 σ) over this period (Figure 5). We suggest this this should be used as the best representation of the repeatability of a sample. The best two-week period exhibited repeatability that was as low as 0.16% (n = 10), obtained from calibration data acquisition with blank measurements before and after each, with no interruption of the pumping and baking laser pan volume and subsequent sample analysis.

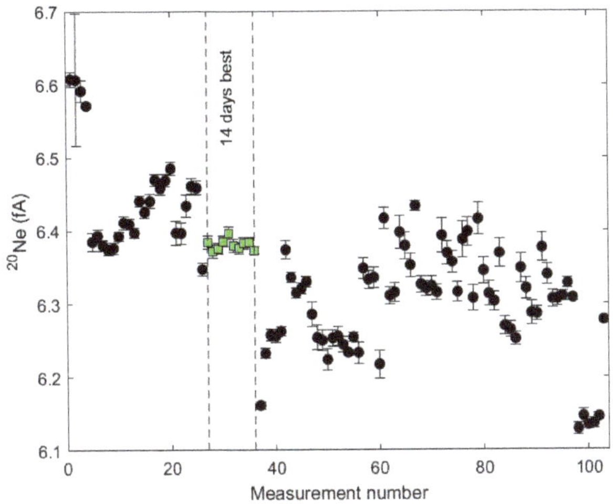

Figure 5. ^{20}Ne intensities from the repeated analysis of 3.2×10^{-11} cm^3 STP ^{20}Ne of air using the ARGUS VI mass spectrometer. Repeatability over the 4 months analytical period was 1.13% (ignoring 9 outliers out of 103). The best 14-day period (green squares) (n = 10) exhibited a repeatability of 0.16% when no interruption of sample analysis occurred. The ^{21}Ne concentration of air/calibration aliquot was equivalent to that of 7.7 mg CREU-1 quartz. Long-term variability seems to be the natural variability of sensitivity, which we think is the best representation of the repeatability of unknowns.

4.2. CREU-1 Quartz

Samples of ~19.9 mg of 250–500 µm CREU-1 quartz [5] were weighed in >99% pure, 20×2 mm Pt foil tubes. The tubes were crimped at both ends then placed into 1 cm^2 recesses in a fully degassed Cu pan and pumped to $<10^{-8}$ mbar prior to degassing at 80 °C for 12 h. A sapphire cover glass was used to avoid volatilized metal from adsorbing onto the sapphire viewport. Neon was extracted from the samples by heating to ~1350 °C for 10 min using a 75 W Fusions 970 (Photon Machines) diode laser (970 nm) [23]. Remote operation of the laser was developed during the COVID-related lockdown in late 2020. This, combined with automated gas purification and separation [6], allowed full remote analysis of 10 samples (one laser pan) without the need for laboratory attendance. This significantly increased sample throughput to around 20 samples (2 laser pans) per week and had the advantage that samples could be analysed more closely in time, thereby eliminating slight fluctuations in sensitivity (Figure 5). The cold blanks of the laser pan and the heat of the empty Pt tube (hot blank) were indistinguishable from the system blanks. Isobaric interferences were corrected as above and were similar to air calibrations.

All CREU quartz yielded Ne isotope data plotted on or close to the established spallation line (Figure 6A,B). The ^{21}Ne/^{20}Ne and ^{22}Ne/^{20}Ne ratios varied from 0.00836 ± 0.00007 to 0.01379 ± 0.00022 and from 0.1092 ± 0.00036 to 0.1145 ± 0.00063, respectively. The samples contained higher proportions of air than most samples reported in the international calibration exercise [5,24,25]. This may reflect the low temperature and short duration of the pre-extraction bake out.

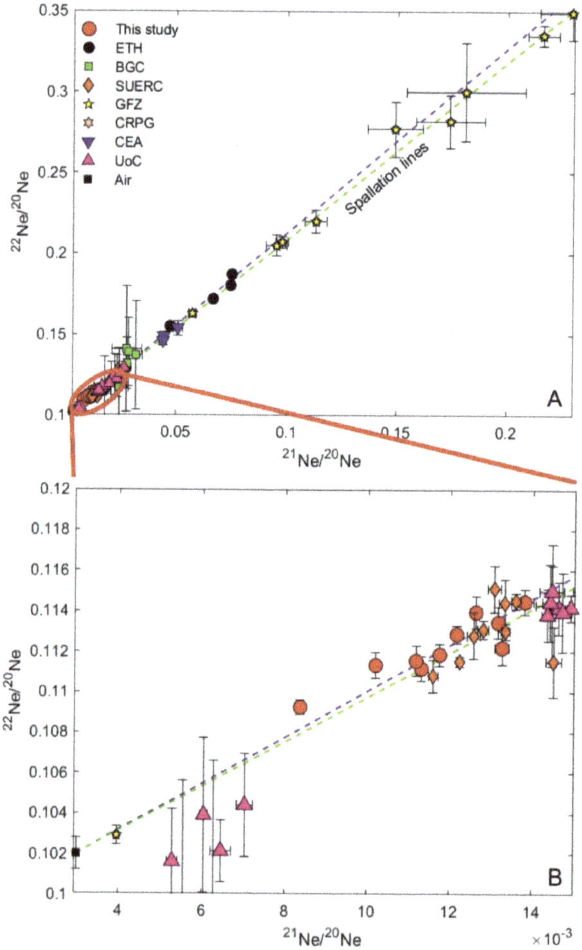

Figure 6. Neon isotope composition of 19.9 mg CREU-1 quartz using the ARGUS VI mass spectrometer. Data were plotted on or close to established cosmogenic spallation lines (**A**) ([26] blue and [27] green), with small deviations that were similar to earlier determinations (**B**). Data were produced over 4 months. Air value (black square) from [6,22]. Other data are from [5,24,25]. ETH (Switzerland), BGC (USA), SUERC (UK), GFZ (Germany), CRPG (France), CEA: China Earthquake Administration (PR China), UoC: University of Cologne (Germany).

The quality of the $^{21}Ne/^{20}Ne$ measurements of CREU quartz, which is the basis for calculating the cosmogenic ^{21}Ne content, is now being assessed against the values obtained in other laboratories. The measurement error is governed by the signal of ^{21}Ne, which is governed by two factors: (1) the degree of cosmogenic compound in the mineral versus atmospheric compound (e.g., how far the sample was located from air on the spallation line) and (2) the mass of the mineral analysed. Consequently, theory would suggest that a higher ^{21}Ne signal (via one of the two factors given above) would lead to a smaller relative error of $^{21}Ne/^{20}Ne$. Unfortunately, when data were plotted on a 3D plot of the relative error of $^{21}Ne/^{20}Ne$ and $^{21}Ne/^{20}Ne$, and the mass of CREU, we were unable to conclude this (Figure 7). We noticed no trend in the error with either increasing mass or $^{21}Ne/^{20}Ne$ and, at this stage, we were unable to compare our performance to that of other laboratories.

We suggest that there is at least one more factor that we have to take into account in order to assess our performance against that of other laboratories.

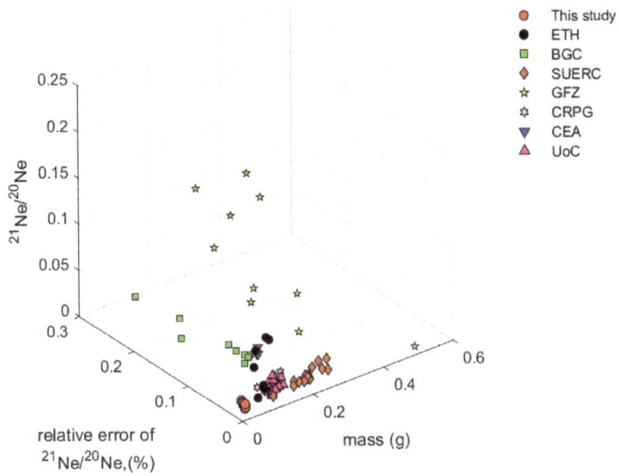

Figure 7. The error of $^{21}Ne/^{20}Ne$ from CREU-1 quartz with respect to the absolute isotope ratio value and amount of material analysed (mass) from a number of different laboratories. In theory, the relative error (1σ) should decrease with either increasing mass or increasing $^{21}Ne/^{20}Ne$ or the combination of both. No trend of this nature was observed, suggesting that other factors govern the relative error in the Ne isotope ratio of CREU. ETH (Switzerland), BGC (USA), SUERC (Scotland), GFZ (Germany), CRPG (France): CEA: China Earthquake Administration (PR China), UoC: University of Cologne (Germany). Data are from Refs. [5,24,25].

Using the established procedure [13], the cosmogenic ^{21}Ne content of the CREU-1 quartz samples ranged from 2.91 to 3.73 × 10^8 atoms/g, with the majority having a much narrower range from 3.08 to 3.50 × 10^8 atoms/g (Table 2). The mean cosmogenic ^{21}Ne concentration, calculated by fitting the best Gaussian curve (see above), was 3.25 ± 0.24 × 10^8 atoms/g (2σ, 2 outliers). While this was 6.6% lower than the accepted value of 3.48 ± 10 × 10^8 [5], it overlapped with 2σ. Further, it overlapped with the data from SUERC and GFZ, it was indistinguishable with the data reported by CRPG [5], and the mean value differed from the arithmetic mean to the same extent as the average BGC data (Figure 8). Regarding the outliers, we explain the highest value (3.73 × 10^8 atoms/g) by a memory effect in the system caused by inadequate pumping due to a failure of the pump for an unknown period of time overnight. The error on $^{21}Ne^*$ was estimated using the 4-month calibration data. As discussed above, we think this is the best representation of instrument performance and avoids underestimation of our error.

These data were obtained from 19.9 ± 0.5 mg CREU-1 quartz, and constituted, by a significant margin, the smallest amount yet reported (Table 2). Previously published measurements of <250 mg aliquots of CREU-1 quartz [5,24,25] binned in 10 mg groups for each laboratory are plotted in Figure 9. It can be seen that twenty one data groups were produced if we ignore data from the fine fraction from ref. [5] (Table S1). We applied the same statistics as above and ignored groups that would be made up by one data point only. The exception from this is CRPG data, which showed a significant underdispersion.

The cosmogenic ^{21}Ne generated by measurement in early generation static vacuum magnetic-sector mass spectrometers (VG5400 (CRPG, GFZ, CEA) and MAP 215-50 (SUERC)) showed repeatability of between 4% (80 mg) and 2.1% (>200 mg). Data obtained from state-of-the-art analytical systems, e.g., Thermo Fisher ARGUS VI (this study), MAP 215-50 with modern electronics (BGC) and Helix MC Plus (UoC), exhibited a significant improvement. The data (39 measurements incorporated into 9 groups) appeared to plot along an exponential curve. They implied that modern instrumentation generates a 5-fold improvement in repeatability compared to data produced by VG5400 and MAP 215-50 instruments. This improvement was the result of the combination of several factors. The development of new source electronics led to more stable ion sources. New control software (e.g., Qtegra in the case of the Thermo Fisher mass spectrometers) with built-in regression functions replaced in-house built data manipulation software for old generation mass spectrometers, which resulted in two key outcomes: (1) the generalization of a regression method with improved mathematics and error propagation, and (2) the opportunity for computer-controlled, automatized gas preparation. The latter means more precise reproduction of gas preparation (i.e., opening and closing valves with highly precise time intervals in between steps) and more precise temperature control of cold fingers. We were unable to resolve the question as to which of these factors played the most significant role in improving the repeatability of Ne isotope determinations. The custom-built mass spectrometer at ETH [28] showed the best performance, although one data point was able to fit the exponential model of repeatability and mass of the previous group, but this is not commercially available.

Table 2. Ne data of CREU quartz analysed using SUERC's ARGUS VI mass spectrometer.

Sample	Weight (mg)	$^{21}Ne/^{20}Ne$	$^{22}Ne/^{20}Ne$	$^{21}Ne^*$
CREU-A	19.67	0.0084 (1)	0.1092 (4)	332 (6)
CREU-B	19.45	0.0102 (1)	0.1113 (6)	329 (6)
CREU-C	19.36	0.0132 (1)	0.1134 (8)	373 (7)
CREU-D	20.08	0.0122 (1)	0.1129 (4)	350 (7)
CREU-E	20.88	0.0117 (1)	0.1119 (5)	323 (6)
CREU-F	20.29	0.0138 (2)	0.1145 (6)	327 (6)
CREU-G	19.64	0.0133 (2)	0.1122 (8)	308 (6)
CREU-H	19.98	0.0126 (1)	0.1139 (8)	318 (6)
CREU-I	19.18	0.0113 (1)	0.1112 (6)	324 (6)
CREU-J	19.48	0.0112 (1)	0.1116 (7)	291 (5)

1σ errors are shown in parenthesis as last significant figures. $^{21}Ne^*$ is the cosmogenic (non-atmospheric) ^{21}Ne. Concentration is given in cm^3 STP/g × 10^6, where p = 0.101 MPa and T = 0 °C in accordance with [20].

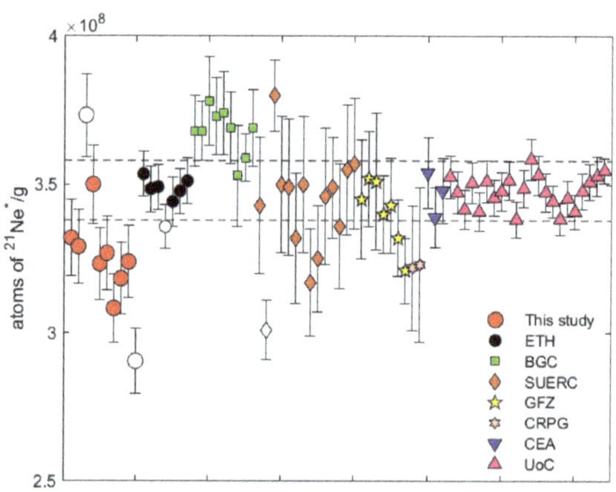

Figure 8. Cosmogenic ^{21}Ne in CREU-1 quartz measured using the ARGUS VI mass spectrometer. Mean value (3.25 ± 0.12 × 10^8 $^{21}Ne^*$ atoms/g, n = 8, 2 outliers) was lower by 6.6% than the accepted standard value (dashed line, 348 ± 10 × 10^8 $^{21}Ne^*$ atoms/g) but it overlapped it within 2σ. Data are after Refs. [5,24,25]. ETH (Switzerland), BGC (USA), SUERC (Scotland), GFZ (Germany), CRPG (France): CEA: China Earthquake Administration (PR China), UoC: University of Cologne (Germany). 2 sigma uncertainties are plotted. Empty: outlier.

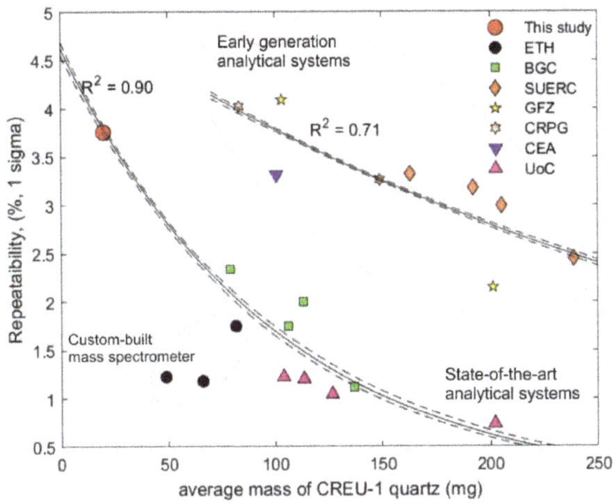

Figure 9. Repeatability of CREU-1 quartz measurements with respect to analysed amounts from a number of different laboratories worldwide. Data from Refs. [5,24,25] were split into groups by masses within 10 mg of material. The only exception was CPRG, which were significantly under dispersed. Data show that beside analysed mass (e.g., signal size of ^{21}Ne) the analytical system is a key factor in data quality. Data from early generation analytical systems (SUERC, CRPG, GFZ, CEA) grouped together and showed the lowest performance at a given mass. Data emerged from state-of-the-art analytical systems including modern electronics, and a high degree of automation (see text) (ARGUS VI in SUERC of this study, BGC and UoC) characterises another group, showing an excellent correlation (n = 9, 39 individual measurements) between the analysed amount of CREU-1 quartz and of ^{21}Ne*. Only the custom-built mass spectrometer (ETH) showed a better performance. We conclude that SUERC's ARGUS VI mass spectrometer analysed 19.9 mg of CREU-1 quartz performs as expected. Apart from three laboratories globally (ETH, BGC and UoC), all laboratories would require the analysis of 5 times or more material to reach the repeatability of this study. ETH (Switzerland), BGC (USA), SUERC (Scotland), GFZ (Germany), CRPG (France): CEA: China Earthquake Administration (PR China), UoC: University of Cologne (Germany). 1 sigma errors on masses are smaller than symbols. Curve errors are 1 sigma.

5. Conclusions and Future Research Directions

The Thermo Fisher ARGUS VI mass spectrometer, tuned for high precision Ne analysis, is now well characterised for the determination of cosmogenic Ne in terrestrial and extra-terrestrial rocks. We characterised the performance in both multi-collection Faraday and peak jumping mode using an electron multiplier. In multi-collection Faraday mode, we replicated the Ne isotope composition of Bruderheim L chondrite and Imilac pallasite. For extra-terrestrial material, we found a 4-fold improvement in the overall uncertainty of the Ne isotope ratio (0.5%) compared to that obtained using 2–6 times more material in earlier works. We suggest that a 5–10-fold reduction in the repeatability may be obtained, albeit more measurement would be needed to confirm this. Peak jumping on the CDD detector allowed the analysis of cosmogenic Ne in ~19.9 mg CREU-1 quartz, which was 3–5 times less than ever reported in previous studies and the acquisition of calibration data at the level of $9.36 \pm 0.21 \times 10^{-14}$ cm^3 STP ^{21}Ne (equivalent of 7.7 mg CREU). We reproduced ^{21}Ne/^{20}Ne with 1.1% (1σ) (n = 103), which was a significant achievement at this Ne partial pressure. Our CREU measurement yielded to ^{21}Ne* content of $325 \pm 12 \times 10^6$ ^{21}Ne* atoms/g (n = 8), which overlaps with the internationally accepted value within 2σ. The reproducibility of ^{21}Ne* (3.7%, 1 sigma) was exactly what the relationship of repeatability and sample mass would suggest, established by data obtained from similar, state-of-the-art analytical systems. This was a ~4 times improvement in comparison to early generation analytical systems. The remote operation of laser heating and the automation of gas purification, separation and analysis procedures increased sample throughput and exact repetition of the procedure and, by allowing more calibration and blank measurements, would likely produce better quality data. The ability to determine Ne isotopes in small amounts of material will be invaluable in studies where sample material is extremely limited, e.g., planetary and comet return missions, and when the appropriate mineral phase is in low abundances in terrestrial or extra-terrestrial rocks.

The incorporation of high gain Faraday amplifiers (10^{13} Ohm and beyond) may allow further improvements in data quality and reductions in sample size for the analysis of cosmogenic Ne-rich material. However, ^{21}Ne analysis from terrestrial samples will likely remain beyond the reach of Faraday detectors with existing amplifier technology and mass spectrometer sensitivity. For small sample analysis, effort will focus on reducing the background level of CO_2, which will prove difficult to resolve without the application of high-resolution instrumentation, e.g., that of [29], and the exploration of the advantages of multi-collection using combined CDD–Faraday.

Supplementary Materials: The following are available online at https://www.mdpi.com/article/10.3390/geosciences11080353/s1, Table S1. Air calibration measurements from aliquots of 9.4×10^{-14} cm^3 STP 21Ne.

Author Contributions: Conceptualization: F.M.S., D.G. and L.D.N. Analytical development: D.G. and L.D.N. Data acquisition: D.G., L.D.N. and D.C. Data interpretation: F.M.S., D.G., L.D.N. and D.C. First draft: D.G. and F.M.S. Writing and editing: all authors. All authors have read and agreed to the published version of the manuscript.

Funding: This research received no external funding.

Data Availability Statement: All data are available in Tables and Supplementary Materials.

Acknowledgments: The research has been funded by SUERC. Rainer Wieler is thanked for providing Bruderheim samples and Thomas Graf is thanked for providing unpublished Ne isotope data of Bruderheim. John Faithfull provided a sample of Imilac from the Hunterian Museum collection. Ángel Rodés is thanked for help in statistics.

Conflicts of Interest: The authors declare no conflict of interest.

References

1. Dunai, T.J.; González López, G.A.; Juez-Larré, J. Oligocene–Miocene age of aridity in the Atacama Desert revealed by exposure dating of erosion-sensitive landforms. *Geology* **2005**, *33*, 321–324. [CrossRef]
2. Füri, E.; Zimmermann, L.; Deloule, E.; Trappitsch, R. Cosmic ray effects on the isotope composition of hydrogen and noble gases in lunar samples: Insights from Apollo 12018. *Earth Planet. Sci. Lett.* **2020**, *550*, 116550. [CrossRef]
3. Sinclair, H.D.; Stuart, F.M.; Mudd, S.M.; McCann, L.; Tao, Z. Detrital cosmogenic 21Ne records decoupling of source-to-sink signals by sediment storage and recycling in Miocene to present rivers of the Great Plains, Nebraska, USA. *Geology* **2018**, *47*, 3–6. [CrossRef]
4. Wieler, R. Cosmic-Ray-Produced Noble Gases in Meteorites. *Rev. Mineral. Geochem.* **2002**, *47*, 125–170. [CrossRef]
5. Vermeesch, P.; Balco, G.; Blard, P.-H.; Dunai, T.J.; Kober, F.; Niedermann, S.; Shuster, D.L.; Strasky, S.; Stuart, F.M.; Wieler, R.; et al. Interlaboratory comparison of cosmogenic 21Ne in quartz. *Quat. Geochronol.* **2015**, *26*, 20–28. [CrossRef]
6. Györe, D.; Tait, A.; Hamilton, D.; Stuart, F.M. The formation of NeH+ in static vacuum mass spectrometers and re-determination of 21Ne/20Ne of air. *Geochim. Cosmochim. Acta* **2019**, *263*, 1–12. [CrossRef]
7. Saxton, J.M. The 21Ne/20Ne ratio of atmospheric neon. *J. Anal. At. Spectrom.* **2020**, *35*, 943–952. [CrossRef]
8. Farley, K.A.; Treffkorn, J.; Hamilton, D. Isobar-free neon isotope measurements of flux-fused potential reference minerals on a Helix-MC-Plus mass spectrometer. *Chem. Geol.* **2020**, *537*, 119487. [CrossRef]
9. Marrocchi, Y.; Burnard, P.G.; Hamilton, D.; Colin, A.; Pujol, M.; Zimmermann, L.; Marty, B. Neon isotopic measurements using high-resolution, multicollector noble gas mass spectrometer: HELIX-MC. *Geochem. Geophys. Geosyst.* **2009**, *10*. [CrossRef]
10. Honda, M.; Zhang, X.; Phillips, D.; Hamilton, D.; Deerberg, M.; Schwieters, J.B. Redetermination of the 21Ne relative abundance of the atmosphere, using a high resolution, multi-collector noble gas mass spectrometer (HELIX-MC Plus). *Int. J. Mass Spectrom.* **2015**, *387*, 1–7. [CrossRef]
11. Ruzié-Hamilton, L.; Clay, P.L.; Burgess, R.; Joachim, B.; Ballentine, C.J.; Turner, G. Determination of halogen abundances in terrestrial and extraterrestrial samples by the analysis of noble gases produced by neutron irradiation. *Chem. Geol.* **2016**, *437*, 77–87. [CrossRef]

12. Bai, X.; Qiu, H.; Liu, W.; Mei, L. Automatic 40Ar/39Ar dating techniques using multicollector ARGUS VI noble gas mass spectrometer with self-made peripheral apparatus. *J. Earth Sci.* **2018**, *29*, 408–415. [CrossRef]
13. Niedermann, S.; Graf, T.; Marti, K. Mass spectrometric identification of cosmic-ray-produced neon in terrestrial rocks with multiple neon components. *Geochim. Cosmochim. Acta* **1993**, *118*, 65–73. [CrossRef]
14. Gilfillan, S.M.V.; Györe, D.; Flude, S.; Johnson, G.; Bond, C.E.; Hicks, N.; Lister, R.; Jones, D.G.; Kremer, Y.; Haszeldine, R.S.; et al. Noble gases confirm plume-related mantle degassing beneath Southern Africa. *Nat. Commun.* **2019**, *10*, 1–7. [CrossRef]
15. Kirkup, L. *Data Analysis for Physical Scientists, Featuring Excel*, 2nd ed.; Cambridge University Press: Cambridge, UK, 2012.
16. Ritter, B.; Stuart, F.M.; Binnie, S.A.; Gerdes, A.; Wennrich, V.; Dunai, T.J. Neogene fluvial landscape evolution in the hyperarid core of the Atacama Desert. *Sci. Rep.* **2018**, *8*, 13952. [CrossRef] [PubMed]
17. Herzog, G.F.; Cook, D.L.; Cosarinsky, M.; Huber, L.; Leya, I.; Park, J. Cosmic-ray exposure ages of pallasites. *Meteorit. Planet. Sci.* **2015**, *50*, 86–111. [CrossRef]
18. Graf, T. Produktion Kosmogener Nuklide in Meteoriten. Ph.D. Thesis, ETH Zürich, Zürich, Switzerland, 1988; 136p. (In German).
19. Schultz, L.; Kruse, H. Helium, neon and argon in meteorites—A data compilation. *Meteoritics* **1989**, *24*, 155–172. [CrossRef]
20. Ozima, M.; Podosek, F.A. *Noble Gas Geochemistry*, 2nd ed.; Cambridge University Press: Cambridge, UK, 2002; p. 367.
21. Balco, G.; Shuster, D.L. Production rate of cosmogenic 21Ne in quartz estimated from 10Be, 26Al, and 21Ne concentrations in slowly eroding Antarctic bedrock surfaces. *Earth Planet. Sci. Lett.* **2009**, *281*, 48–58. [CrossRef]
22. Eberhardt, P.; Eugster, O.; Marti, K. A redetermination of the isotopic composition of atmospheric neon. *Z. Nat. A* **1965**, *20*, 623–624. [CrossRef]
23. Carracedo, A.; Rodés, Á.; Smellie, J.L.; Stuart, F.M. Episodic erosion in West Antarctica inferred from cosmogenic 3He and 10Be in olivine from Mount Hampton. *Geomorphology* **2019**, *327*, 438–445. [CrossRef]
24. Ritter, B.; Vogt, A.; Dunai, T.J. Noble gas extraction procedure and performance of the Cologne Helix MC Plus multi-collector noble gas mass spectrometer for cosmogenic neon isotope analysis. *Geochronology* **2021**, *3*, 421–431. [CrossRef]
25. Ma, Y.; Wu, Y.; Li, D.; Zheng, D. Analytical procedure of neon measurements on GV 5400 noble gas mass spectrometer and its evaluation by quartz standard CREU-1. *Int. J. Mass Spectrom.* **2015**, *380*, 26–33. [CrossRef]
26. Niedermann, S.; Graf, T.; Kim, J.S.; Kohl, C.P.; Marti, K.; Nishiizumi, K. Cosmic-ray-produced 21Ne in terrestrial quartz: The neon inventory of Sierra Nevada quartz separates. *Earth Planet. Sci. Lett.* **1994**, *125*, 341–355. [CrossRef]
27. Schäfer, J.M.; Ivy-Ochs, S.; Wieler, R.; Leya, I.; Baur, H.; Denton, G.H.; Schlüchter, C. Cosmogenic noble gas studies in the oldest landscape on earth: Surface exposure ages of the Dry Valleys, Antarctica. *Earth Planet. Sci. Lett.* **1999**, *167*, 215–226. [CrossRef]
28. Baur, H. Numerische Simulation und Praktische Erprobung einer Rotationssymmetrischen Ionenquelle für Gasmassenspektrometer. Ph.D. Thesis, ETH Zürich, Zürich, Switzerland, 1980. (In German).
29. Eiler, J.M.; Clog, M.; Magyar, P.; Piasecki, A.; Sessions, A.; Stolper, D.; Deerberg, M.; Schlueter, H.-J.; Schwieters, J. A high-resolution gas-source isotope ratio mass spectrometer. *Int. J. Mass Spectrom.* **2013**, *335*, 45–56. [CrossRef]

MDPI
St. Alban-Anlage 66
4052 Basel
Switzerland
Tel. +41 61 683 77 34
Fax +41 61 302 89 18
www.mdpi.com

Geosciences Editorial Office
E-mail: geosciences@mdpi.com
www.mdpi.com/journal/geosciences

www.ingramcontent.com/pod-product-compliance
Lightning Source LLC
Chambersburg PA
CBHW061826210926
51850CB00006B/824
* 9 7 8 3 0 3 6 5 7 1 4 8 5 *